Springer Series in Materials Science

Volume 186

For further volumes:
http://www.springer.com/series/856

The Springer Series in Materials Science covers the complete spectrum of materials physics, including fundamental principles, physical properties, materials theory and design. Recognizing the increasing importance of materials science in future device technologies, the book titles in this series reflect the state-of-the-art in understanding and controlling the structure and properties of all important classes of materials.

Handong Li • Zhiming M. Wang
Editors

Bismuth-Containing Compounds

Springer

Editors
Handong Li
State Key Laboratory of Electronic Thin Film and Integrated Devices
University of Electronic Science and Technology of China
Chengdu
China, People's Republic

Zhiming M. Wang
State Key Laboratory of Electronic Thin Film and Integrated Devices
University of Electronic Science and Technology of China
Chengdu
China, People's Republic

ISSN 0933-033X ISSN 2196-2812 (electronic)
ISBN 978-1-4939-4744-7 ISBN 978-1-4614-8121-8 (eBook)
DOI 10.1007/978-1-4614-8121-8
Springer New York Heidelberg Dordrecht London

Printed on acid-free paper

Springer is part of Springer Science+Business Media (www.springer.com)

Preface

As one typical heavy metal, bismuth has been known as the most diamagnetic of all metals, and the thermal conductivity is lower than any metal except mercury. It also has a high electrical resistance, and has the highest Hall effect of any metal. However, scientific interests in bismuth-containing compounds have gone beyond bismuth metal since the material families of bismides offer richer crystal structures and electronic properties than the element itself. Research on bismides with diverse crystalline structures varying from covalent zinc-blende to van der Waals type layered structures has not only enriched the content of traditional disciplines such as optoelectronics, optics, and photocatalysis but also bred many new areas such as multiferroics and topological insulators. Actually, research and development of bismuth-containing compounds has been one of the most promising fields in materials science. By providing a comprehensive understanding of these important materials and related applications, this book aims to convey the current status of research on most representative bismuth-containing compounds.

Among bismuth-contained compounds, dilute bismide semiconductors have received extensive attention recently and practical device applications have been demonstrated. Therefore, a significant portion of the book (Chaps. 1–10) deals with dilute bismuth-containing semiconductors. Chapter 1 reviews progress in material growth and characterization of dilute bismide semiconductors and their quantum structures whose potential applications in near- and mid-infrared photonic devices are summarized in Chap. 2. Since the novel optoelectronic properties of dilute bismide semiconductors are mainly determined by the band structure, the latter is of fundamental importance and theoretical investigations are described in Chap. 3. Chapters 4–6 cover the epitaxy and characterizations of dilute bismides by various thin film growth methods. Chapter 4 focuses on molecular beam epitaxy of dilute bismides and some unique physical properties such as thermoelectric behaviors are discussed. Chapter 5 presents detailed study on growth behaviors of dilute bismides by metalorganic vapor phase epitaxy. Chapter 6 highlights the growth procedures and characterizations of InSbBi and GaSbBi materials by economical liquid phase epitaxy.

Chapters 7 and 8 concentrate on optical and transport properties of dilute bismides. Spectroscopic ellipsometry studies on dilute bismides are described in Chap. 7, followed by the effect of Bi incorporation on transport and photoluminescence behavior of dilute bismide alloys (Chap. 8). Chapter 9 reports on the localized states in GaAsBi and GaAs/GaAsBi heterostructures. Surface science and engineering of bismuth-contained III–V semiconductor films are reviewed in Chap. 10.

Chapters 11–15 are dedicated to other novel bismuth-contained compound systems, including bismuth chalcogenides, bismuth ferrite, sodium bismuthate, and other complex bismides. Bismuth chalcogenides, which were recently revealed as topological insulators, are discussed in Chaps. 11 and 12. Epitaxial growth, angle-resolved photoemission spectroscopy, and magnetoresistance measurements of bismuth chalcogenide thin films are the subject of Chap. 11, while Chap. 12 focuses on the synthesis and electrostatic properties of nanostructured bismuth-based topological insulators by vapor phase deposition method. Chapter 13 investigates physical phenomena involving domain walls in bismuth ferrite which have many implications in multiferroics. In Chap. 14 bismuth-contained compounds with noncentrosymmetric structure, which are characterized as second harmonic generation materials, are reviewed. Finally, the authors of Chap. 15 discuss the science and applications of sodium bismuthate in heterogeneous photocatalysis.

We are grateful to all of the authors who have contributed chapters to this book. It took nearly 1 year to put this work together, but we are confident that the effort was worthwhile and that the book will be a useful reference not only for scientists and engineers exploring novel bismides but also for graduate and postgraduate students specializing in solid-state physics and materials science. Last but not least, we would like to thank Mr. Haiyuan Chen who provided the necessary editorial assistance and support throughout the process. The Editors also acknowledge the financial support of the National Natural Science Foundation of China through Grant NSFC-51272038.

Chengdu, People's Republic of China

Handong Li
Zhiming M. Wang

Contents

Contributors

Z. Batool Advanced Technology Institute and Department of Physics, University of Surrey, Guildford, Surrey, UK

Christopher A. Broderick Tyndall National Institute, Lee Maltings, Dyke Parade, Cork, Ireland

Department of Physics, University College Cork, Cork, Ireland

H. Cao Department of Physics, Purdue University, West Lafayette, IN, USA

Xiaofeng Chang Department of Applied Chemistry, College of Materials Science and Technology, Nanjing University of Aeronautics and Astronautics, Nanjing, China

Haiyuan Chen State Key Laboratory of Electronic Thin Film and Integrated Devices, University of Electronic Science and Technology of China, Chengdu, P.R. China

Y.P. Chen Department of Physics, Purdue University, West Lafayette, IN, USA

Wen-Dan Cheng State Key Laboratory of Structural Chemistry, Fujian Institute of Research on the Structure of Matter, Chinese Academy of Sciences, Fuzhou, Fujian, China

Sunanda Dhar Department of Electronic Science, University of Calcutta, Kolkata, India

M. Dobrowolska Department of Physics, University of Notre Dame, Notre Dame, IN, USA

Pernell Dongmo Department of Materials Science and Engineering, University of Delaware, Newark, DE, USA

J. Fan Department of Physics, Arizona State University, Tempe, AZ, USA

Hédi Fitouri Faculté des Sciences de Monastir, Université de Monastir-Unité de Recherche sur les Hétéro-Epitaxies et Applications, Monastir, Tunisia

J.K. Furdyna Department of Physics, University of Notre Dame, Notre Dame, IN, USA

Takuma Fuyuki Department of Electronics, Kyoto Institute of Technology, Kyoto, Japan

Lei Geng School of Physics and Electronic Information, Huaibei Normal University, Huaibei, Anhui, China

Mohammed Ashraf Gondal Laser Research Group, Physics Department and Center of Excellence in Nanotechnology, King Fahd University of Petroleum and Minerals, Dhahran, Saudi Arabia

Yi Gu State Key Laboratory of Functional Materials for Informatics, Shanghai Institute of Microsystem and Information Technology, Chinese Academy of Sciences, Shanghai, China

Guolin Hao Hunan Key Laboratory for Micro-Nano Energy Materials and Devices, Xiangtan University, Hunan, P.R. China

Laboratory for Quantum Engineering and Micro-Nano Energy Technology, Faculty of Materials and Optoelectronic Physics, Xiangtan University, Hunan, P.R. China

M.Z. Hasan Department of Physics, Princeton University, Princeton, NJ, USA

K. Hild Advanced Technology Institute and Department of Physics, University of Surrey, Guildford, Surrey, UK

T.J.C. Hosea Advanced Technology Institute and Department of Physics, University of Surrey, Guildford, Surrey, UK

N. Hossain Advanced Technology Institute and Department of Physics, University of Surrey, Guildford, Surrey, UK

Belgacem El Jani Faculté des Sciences de Monastir, Université de Monastir-Unité de Recherche sur les Hétéro-Epitaxies et Applications, Monastir, Tunisia

Guangbin Ji Department of Applied Chemistry, College of Materials Science and Technology, Nanjing University of Aeronautics and Astronautics, Nanjing, China

S.R. Jin Advanced Technology Institute and Department of Physics, University of Surrey, Guildford, Surrey, UK

Rajeev N. Kini Indian Institute of Science Education and Research Thiruvananthapuram (IISER-TVM), CET Campus, Engineering College, Kerala, India

Pekka Laukkanen Materials Research Laboratory, Department of Physics and Astronomy, University of Turku, Turku, Finland

Chen-Sheng Lin State Key Laboratory of Structural Chemistry, Fujian Institute of Research on the Structure of Matter, Chinese Academy of Sciences, Fuzhou, Fujian, China

C. Liu Department of Physics, Princeton University, Princeton, NJ, USA

Xinyu Liu Department of Physics, University of Notre Dame, Notre Dame, IN, USA

Zhong-Zhen Luo State Key Laboratory of Structural Chemistry, Fujian Institute of Research on the Structure of Matter, Chinese Academy of Sciences, Fuzhou, Fujian, China

I.P. Marko Advanced Technology Institute and Department of Physics, University of Surrey, Guildford, Surrey, UK

Angelo Mascarenhas National Renewable Energy Laboratory, Golden, CO, USA

Eoin P. O'Reilly Tyndall National Institute, Lee Maltings, Dyke Parade, Cork, Ireland

Department of Physics, University College Cork, Cork, Ireland

R.E. Pimpinella Department of Physics, University of Notre Dame, Notre Dame, IN, USA

Marko Punkkinen Materials Research Laboratory, Department of Physics and Astronomy, University of Turku, Turku, Finland

Xiang Qi Hunan Key Laboratory for Micro-Nano Energy Materials and Devices, Xiangtan University, Xiangtan, Hunan, P.R. China

Laboratory for Quantum Engineering and Micro-Nano Energy Technology, Faculty of Materials and Optoelectronic Physics, Xiangtan University, Xiangtan, Hunan, P.R. China

Ahmed Rebey Faculté des Sciences de Monastir, Université de Monastir-Unité de Recherche sur les Hétéro-Epitaxies et Applications, Monastir, Tunisia

Nebiha Ben Sedrine Campus Tecnológico e Nuclear, Instituto Superior Técnico, Universidade Técnica de Lisboa, Sacavèm, Portugal

Jan Seidel School of Materials Science and Engineering, University of New South Wales, Sydney, Australia

Jun Shao National Laboratory for Infrared Physics, Shanghai Institute of Technical Physics, Chinese Academy of Sciences, Shanghai, China

D.J. Smith Department of Physics, Arizona State University, Tempe, AZ, USA

Yuxin Song Photonics Laboratory, Department of Microtechnology and Nanoscience (MC2), Chalmers University of Technology, Göteborg, Sweden

S.J. Sweeney Advanced Technology Institute and Department of Physics, University of Surrey, Guildford, Surrey, UK

Muhammad Usman Tyndall National Institute, Lee Maltings, Dyke Parade, Cork, Ireland

Shumin Wang Photonics Laboratory, Department of Microtechnology and Nanoscience (MC2), Chalmers University of Technology, Göteborg, Sweden

State Key Laboratory of Functional Materials for Informatics, Shanghai Institute of Microsystem and Information Technology, Chinese Academy of Sciences, Shanghai, China

T.-L. Wu Department of Physics, Purdue University, West Lafayette, IN, USA

Zain Hassan Abdallah Yamani Center of Excellence in Nanotechnology, King Fahd University of Petroleum and Minerals, Dhahran, Saudi Arabia

Masahiro Yoshimoto Department of Electronics, Kyoto Institute of Technology, Kyoto, Japan

Hao Zhang State Key Laboratory of Structural Chemistry, Fujian Institute of Research on the Structure of Matter, Chinese Academy of Sciences, Fuzhou, Fujian, China

Wei-Long Zhang State Key Laboratory of Structural Chemistry, Fujian Institute of Research on the Structure of Matter, Chinese Academy of Sciences, Fuzhou, Fujian, China

Y.-H. Zhang Center for Photonics Innovation, Arizona State University, Tempe, AZ, USA

School of Electrical, Computer and Energy Engineering, Arizona State University, Tempe, AZ, USA

Jianxin Zhong Hunan Key Laboratory for Micro-Nano Energy Materials and Devices, Xiangtan University, Hunan, P.R. China

Laboratory for Quantum Engineering and Micro-Nano Energy Technology, Faculty of Materials and Optoelectronic Physics, Xiangtan University, Hunan, P.R. China

Yujun Zhong Department of Materials Science and Engineering, University of Delaware, Newark, DE, USA

Joshua Zide Department of Materials Science and Engineering, University of Delaware, Newark, DE, USA

Chapter 1
Dilute Bismides for Mid-IR Applications

Yuxin Song, Yi Gu, Jun Shao, and Shumin Wang

Abstract Dilute bismides are a group of emerging materials with unique properties. Incorporation of a small amount of Bi in common III–V host materials results in large band-gap reduction and strong spin-orbit splitting, leading to potential applications in mid-infrared (Mid-IR) optoelectronics. In this chapter, we review recent progresses on epitaxy and characterizations of novel bismides, i.e., $GaSb_{1-x}Bi_x$, $InSb_{1-x}Bi_x$, $InAs_{1-x}Bi_x$, and InAsSbBi. Although these dilute bismides have been successfully grown, to obtain high Bi incorporations and retain high crystal quality is still very challenging.

Y. Song
Photonics Laboratory, Department of Microtechnology and Nanoscience (MC2),
Chalmers University of Technology, Gothenburg 412 96, Sweden
e-mail: yuxin.song@chalmers.se

Y. Gu
State Key Laboratory of Functional Materials for Informatics,
Shanghai Institute of Microsystem and Information Technology,
Chinese Academy of Sciences, Shanghai 200050, P. R. China
e-mail: ygu@mail.sim.ac.cn

J. Shao
National Laboratory for Infrared Physics, Shanghai Institute of Technical Physics,
Chinese Academy of Sciences, Shanghai 200083, P. R. China
e-mail: jshao@mail.sitp.ac.cn

S. Wang (✉)
Photonics Laboratory, Department of Microtechnology and Nanoscience (MC2),
Chalmers University of Technology, Gothenburg 412 96, Sweden

State Key Laboratory of Functional Materials for Informatics,
Shanghai Institute of Microsystem and Information Technology,
Chinese Academy of Sciences, Shanghai 200050, P. R. China
e-mail: Shumin@chalmers.se

H. Li and Z.M. Wang (eds.), *Bismuth-Containing Compounds*, Springer Series in Materials Science 186, DOI 10.1007/978-1-4614-8121-8_1,

1.1 Introduction

Bismuth (Bi) is the heaviest and the least studied group-V element. When a small amount of Bi atoms are incorporated in common III–V compounds, for example, arsenides and antimonides, Bi impurity states will be formed close to the valence band edge of the host semiconductor. The interaction between the valence band states and the Bi states is expected to lead to a large energy band-gap reduction [1, 2] and strong spin-orbit splitting [3]. To the contrary of dilute nitride, which is well-known for its large band-gap bowing effect by lowering of the conduction band edge, the influence of Bi on band structure occurs only in the valence band if strain is not accounted. This provides a new degree of freedom to engineer band structure of semiconductors for potential electronic and optoelectronic applications. Furthermore, electron transport in dilute bismide materials is much less influenced than those in dilute nitride materials since little perturbation occurs in the conduction band [4]. Table 1.1 summarizes the band-gap reduction data for dilute bismide materials and other common III–V compounds for comparison. It can be found that incorporation of Bi is more efficient in reducing the band-gap of hosting binary III–V compounds than all the listed ternaries except dilute nitrides. By combining incorporation of N with Bi, it is possible to modify the conduction and valence bands independently, thus opening a new dimension of freedom of band engineering.

Intensive research has been implemented on $GaAs_{1-x}Bi_x$ and related materials after its first realization of epitaxial growth by metal organic vapor phase epitaxy (MOVPE) in 1998 [5] and by molecular beam epitaxy (MBE) in 2003 [6]. The band-gap bowing effect [2] and the spin-orbit splitting effect [7] were experimentally observed afterwards. The valence band edge reduction can lead to a reduced temperature dependence of the band-gap, which is attractive for fabricating

Table 1.1 Summary of band-gap reduction in dilute bismides and some common III–V ternary compounds at each percent of Bi or other incorporated element (x%)

Dilute bismides	Band-gap reduction (meV/%)
$GaAs_{1-x}Bi_x$	83 (300 K) [2]
$GaSb_{1-x}Bi_x$	32 (77 K) [9]
	100 (4 K) [10]
$InAs_{1-x}Bi_x$	20 (77 K) [1]
$InSb_{1-x}Bi_x$	23 (77 K) [1]
Common III–V ternary compounds	(300 K) [11]
$In_xGa_{1-x}As$	15
$In_xGa_{1-x}Sb$	10
GaN_xAs_{1-x}	156 [12]
$GaAs_xSb_{1-x}$	19
$InAs_xSb_{1-x}$	9

The data is valid only for small x-values

temperature-insensitive lasers, optical amplifiers, etc. for telecommunications. The large spin-orbit splitting is also proposed to suppress inter-valence band Auger recombination processes and thus increasing the characteristic temperature of 1.55 μm telecom lasers on GaAs [8]. Up to date, applications for dilute bismide-arsenides are proposed for less than 2 μm wavelength range since the host GaAs has a relative large band-gap. Applications of other III–V-bismides for longer wavelengths are much less studied.

Antimonide-based III–V semiconductor compounds are very attractive for opto-electronic devices working in mid-infrared (MIR), such as lasers and detectors, as well as high speed electronic devices, for example, bipolar transistors [13]. Incorporation of Bi in III-Sb will have several potential benefits. First, Bi incorporation in GaSb can reduce the band-gap and thus extend light emission wavelength more effectively than utilizing InGaSb. Similarly, Bi in InSb or InAsSb can extend the wavelength to 8–12 μm which is one of the most important atmospheric transmission windows. The mature III–V processing technologies will enable these materials competitive to HgCdTe which is dominating in this wavelength range for photodetection nowadays. Second, one major problem for >3 μm GaSb-based type-I quantum well (QW) lasers is insufficient valence band offset [14] due to the use of heavily compressively strained InGaAsSb QWs, resulting in a significant hole leakage current. By employing Bi, the large valence band offset can suppress the hole leakage. Third, the spin-orbit splitting energy of GaSb at 300 K is about 0.735 eV, very close to its direct band-gap of 0.75 eV. It requires a small amount of Bi in $GaSb_{1-x}Bi_x$ to reverse the two energy values to suppress inter-valence band Auger recombination processes in optoelectronic devices.

Several attempts to reach the narrowest possible band-gap of III–V materials for long wavelength infrared detectors have been carried out since the late 1970s by synthesizing $InSb_{1-x}Bi_x$ [15–17] and $InAs_{1-x}Bi_x$ bulk materials [18], while $GaSb_{1-x}Bi_x$ is rarely explored [9, 19]. Very recently, two groups have reported growth of $GaSb_{1-x}Bi_x$ thin films by MBE [20] and by liquid phase epitaxy (LPE) [10].

In this chapter, we will summarize the status of synthesis and characterizations of dilute $GaSb_{1-x}Bi_x$, $InSb_{1-x}Bi_x$, $InAs_{1-x}Bi_x$, and their quaternary compound InAsSbBi aiming for MIR applications. Finally, optical characterizations of Bi containing QW structures are presented.

1.2 GaSbBi

1.2.1 Early Works

To the best of our knowledge, the earliest works on synthesis of $GaSb_{1-x}Bi_x$ bulk crystals, published in Russian, were carried out by Zinov'ev et al. and Germogenov et al. in USSR utilizing LPE in the mid 1980s [9]. The band-gap of $GaSb_{1-x}Bi_x$ as a function of Bi content was for the first time quantitatively determined (shown in Fig. 1.1) for Bi concentration up to 0.6 % [9] by photoluminescence (PL) method.

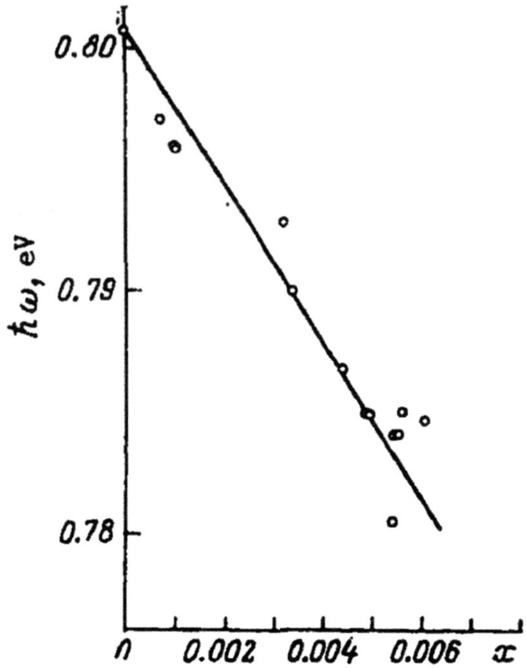

Fig. 1.1 Dependence of the maximum energy position of the band edge PL at 77 K on the composition of Bi from $GaSb_{1-x}Bi_x$ solid solutions, reproduced from [9]

Bismuth was attempted as dopants in GaSb using LPE method to reduce background *p*-doping by German groups in the 1990s [19, 21]. A band-gap reduction of 0.8 meV was found by PL, and the Bi concentration was estimated to be 0.015 % [19].

In the beginning of 2012, two papers on $GaSb_{1-x}Bi_x$ growth were published almost simultaneously. One is by Das et al. from an Indian/British group using LPE method [10] and the other is from our group in Sweden utilizing MBE method [20]. Both papers will be summarized below. Epitaxy of $GaSb_{1-x}Bi_x$ is still in its infancy stage. Significant progress on material quality is essential for future devices.

1.2.2 $GaSb_{1-x}Bi_x$ Thin Films Grown by MBE

A group of 200 nm thick $GaSb_{1-x}Bi_x$ samples are grown on undoped (100) GaSb substrates by MBE. All these samples are grown at 0.1 μm/h with a Ga beam equivalent pressure (BEP) of $2.0e^{-8}$ Torr and an Sb BEP of $8.8e^{-8}$ Torr. The growth temperature is varied from 330 to 390 °C and the Bi BEP is set to be close to the vapor pressure at a particular growth temperature, i.e., $1e^{-8}$, $4e^{-8}$, $7e^{-8}$, $1e^{-7}$, and $2e^{-7}$ Torr at 330, 360, 370, 380, and 390 °C, respectively.

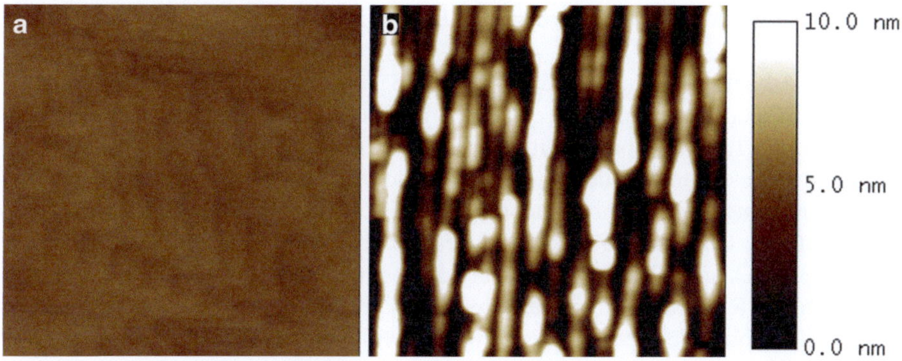

Fig. 1.2 1 × 1 μm² AFM height images showing the typical surface morphologies of smooth surface with triangle-shaped features (**a**) and rough surface (**b**)

1.2.2.1 Strategy to Enhance Bi Incorporation in MBE Growth

V/III ratio and growth temperature, T_g, are two crucial parameters to enhance Bi incorporation. Due to the higher bonding energy of Ga–Sb than that of Ga–Bi, the Sb/Ga flux ratio should be kept low to be close to the Ga-rich growth condition while the Bi flux should be as high as possible to facilitate Bi incorporation. However, a low Sb/Ga flux ratio may cause formation of Ga droplets. Bismuth is also metallic and is different from other group-V elements. Excess Bi atoms on the growth surface will form Bi droplets if they are neither incorporated nor evaporated. All these processes, like surface migration, incorporation, desorption, and droplet formation, highly depend on growth temperature. A low growth temperature facilitates Bi staying on surface but can have a risk for formation of Bi droplets if not incorporated. A high growth temperature reduces such a risk but leads to less effective Bi incorporation and potential Ga droplets. Based on the above discussions, our strategy is to use a low growth rate and keep a low constant Sb flux (BEP of $8.8e^{-8}$ Torr), but vary the Bi flux which is set to be equal to the vapor pressure at a particular growth temperature in the range of 330–390 °C. The Bi vapor pressure follows exponentially with temperature and is about $1e^{-8}$ and $2e^{-7}$ Torr at 330 °C and 390 °C, respectively.

1.2.2.2 Surface Morphology

The reference sample without Bi shows smooth and uniform surface with atomic steps. The GaSbBi samples grown at 360 °C or lower show similar high quality surface with atomic steps forming triangle-shaped features as shown in Fig. 1.2a and have slightly larger RMS roughness than that of the reference sample. For the samples grown at high temperatures, the surface becomes rough with elongated dots as shown in Fig. 1.2b. The shape of elongated dots indicates anisotropy of surface diffusion of adatoms along the two orthogonal crystal directions.

1.2.2.3 Confirmation of Bi Incorporation

Rutherford backscattering spectroscopy (RBS) measurements are carried out to examine Bi incorporation in these samples and the results of four samples are presented in Fig. 1.3a. Compared with the reference sample, a peak/step-like signal is observed at the channel value of around 480 and its intensity increases with T_g. This signal is a strong evidence of Bi incorporation. From the simulations, the Bi concentration of 0.2, 0.7, and 0.7 % is deduced for the $GaSb_{1-x}Bi_x$ samples grown at 360 °C, 370 °C, and 390 °C, respectively, indicating a trend of Bi saturation at high temperatures. RBS channeling measurement is performed for the sample grown at 390 °C. The channeled spectrum is compared with two random spectra as shown in Fig. 1.3b. By evaluating the yield difference, about 65 % of the incorporated Bi atoms are estimated to be at substitutional sites. The rest Bi atoms are either interstitials or accumulated on surface.

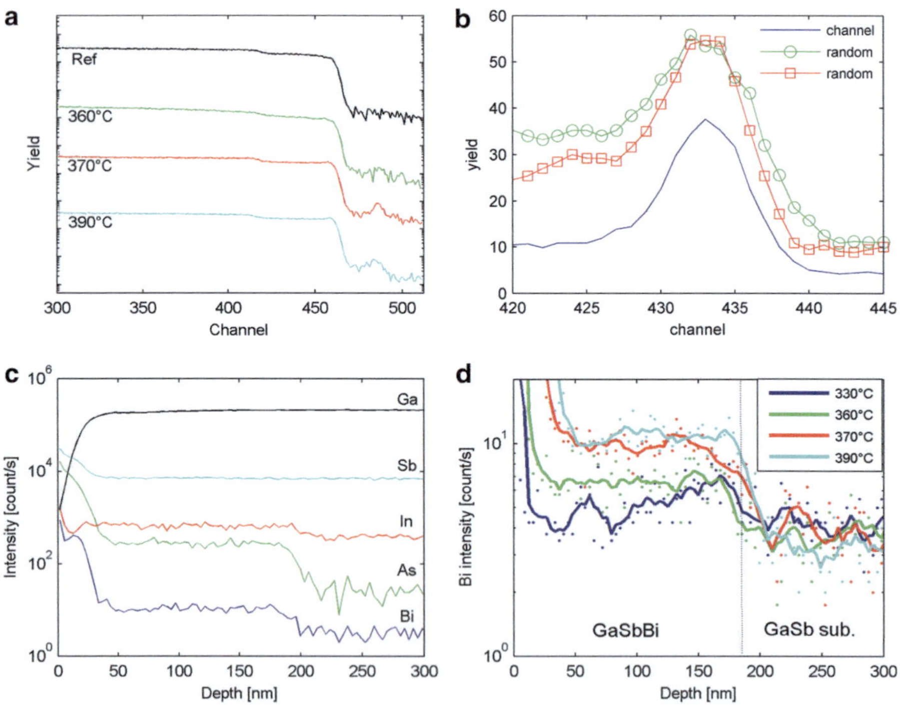

Fig. 1.3 (**a**) RBS spectra of the reference sample and the samples grown at 360, 370, and 390 °C. (**b**) RBS channeling results (focusing on the Bi signal) for the sample grown at 390 °C. The *green-circled and red-squared curves* are from two random measurements and the *blue solid curve* is from the channeled measurement. (**c**) SIMS profiles of different elements in the sample grown at 390 °C. (**d**) Bi profiles measured from SIMS in the $GaSb_{1-x}Bi_x$ samples. *Dots* are measurement data and the *solid curves* are obtained from five-point-smoothening of the measurement data

Secondary ion mass spectrometry (SIMS) measurements are carried out in parallel with the RBS to show the relative counts of elements along the depth in a sample. Figure 1.3c shows depth profiles of the three key elements Ga (69), Sb (121), and Bi (209) as well as In (115) and As (75) which may remain in the MBE growth chamber as background species. The initial intensity changes within 20–25 nm are artifacts. A clear step of Bi intensity at the interface between the epitaxial $GaSb_{1-x}Bi_x$ layer and the GaSb substrate, corroborated by the background As and In steps at the same depth, ensures that the Bi signal is reliable although its intensity is relatively low. It is possible for the Bi signal to be interfered by Ga isotopes due to atomic mass superimposition (Ga 69 + Ga 70 + Ga 70 = 209). However, no correlation can be found between the Ga isotopes and the Bi curves, proving again the reliability of the Bi results. Background As is detected from the growth chamber. It is about one order of magnitude higher than the value found in GaSb substrates and is similar for both the Ref sample and the GaSbBi samples grown in the range of 330–390 °C.

Figure 1.3d summarizes the Bi distribution profile in four samples showing a step for the Bi intensity at the same position. Strong Bi surface accumulation is clearly observed for the samples grown at 370 and 390 °C. The Bi level increases following T_g (also Bi BEP) up to the maximum growth temperature of 390 °C. However, the difference between 370 and 390 °C is rather small, indicating saturation of Bi incorporation although the Bi flux is higher at 390 °C than at 370 °C. This saturation results from the balance between the increased Bi BEP, the weak Ga–Bi bonding, and the enhanced Bi segregation at high growth temperatures. Due to the lack of pure GaBi crystals for calibration, it is very difficult to obtain the exact Bi concentration in these samples. Comparisons are made relatively. For example, the average incorporated Bi level in the $GaSb_{1-x}Bi_x$ layer grown at 370 °C is 1.45 times compared with that grown at 360 °C. This difference is smaller than that deduced from the RBS measurement confirming the relatively high percentage of Bi surface accumulation in the sample grown at 370 °C. We can thus conclude that, by combining the results of RBS and SIMS measurements, Bi incorporation is unambiguously demonstrated. Bismuth concentration in the $GaSb_{1-x}Bi_x$ layer increases with the Bi BEP and tends to saturate at the growth temperature above 370 °C.

1.2.2.4 Lattice Contraction

As GaBi is predicted to have a larger lattice constant (6.324 Å [22]) than that of GaAs (5.653 Å) and GaSb (6.096 Å), the lattice constant of $GaAs_{1-x}Bi_x$ and $GaSb_{1-x}Bi_x$ is expected to be between the values of the constituent binaries. Fitting of $GaAs_{1-x}Bi_x$ layer peaks in an X-ray diffraction (XRD) rocking curve has been utilized as a convenient and accurate tool to measure the Bi composition in $GaAs_{1-x}Bi_x$ materials [6, 23]. However, the $GaSb_{1-x}Bi_x$ thin film signals in XRD (004) rocking curves are found at the right side of the GaSb substrate peak as shown in Fig. 1.4, indicating smaller lattice constants of $GaSb_{1-x}Bi_x$ thin films than that of GaSb along the growth direction. We attribute this to vacancies at group-V sites

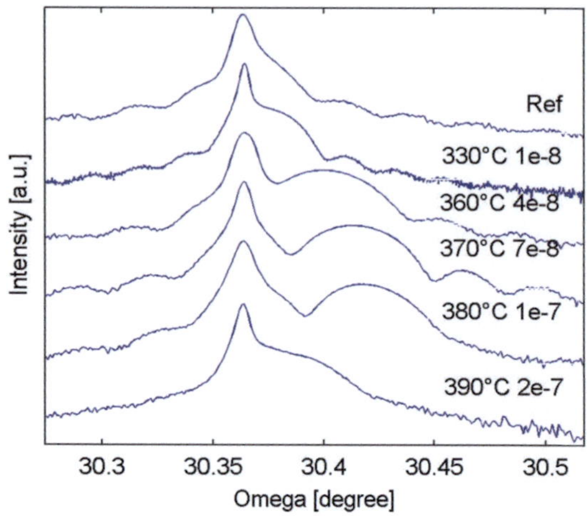

Fig. 1.4 XRD (004) rocking curves of the $GaSb_{1-x}Bi_x$ samples (logarithmic scale). Labels to the *low-right* of each *curve* indicate the growth temperature and the Bi BEP

caused by Bi segregation. During growth of $GaSb_{1-x}Bi_x$, Bi most likely takes group-V sites like Sb. Surface segregation is a well-known phenomenon for ternary III–V compounds [24]. Bi is semimetallic, has a larger atomic radius, larger atomic mass, and weaker bonding energy with group-III elements. Therefore, Bi should be very easy to segregate to the surface. In order to realize Bi incorporation, a very low Sb flux was used, leading to a quasi-Ga-rich growth condition for $GaSb_{1-x}Bi_x$. When a proportion of the incorporated Bi atoms are segregated to the surface, there are not enough excess Sb atoms to fill in the vacant sites left by the segregated Bi. Then, there are two possibilities; either Ga atoms will take the sites forming Ga–Ga anti-site defects or the vacancies will be buried in the $GaSb_{1-x}Bi_x$ layer. Both cases will lead to a smaller lattice constant as it would be for $GaSb_{1-x}Bi_x$. Vacancies are more effective in reducing lattice constant than Ga–Ga anti-sites. If the amount of vacancies is considerably high, they can eventually compensate the incorporated Bi causing lattice contraction. A thick sample with $GaSb_{1-x}Bi_x$ layer thickness of 500 nm was grown under similar growth conditions as the sample grown at 380 °C. To the contrary, the $GaSb_{1-x}Bi_x$ layer peak appears at the left side of the GaSb substrate, which is shown as the top curve in Fig. 1.5, indicating a larger lattice constant than that of GaSb. We then etched this sample with four different depths. It can be found in Fig. 1.5 that, when the $GaSb_{1-x}Bi_x$ layer thickness is reduced after etching, the $GaSb_{1-x}Bi_x$ peak moves towards the GaSb substrate peak and eventually to the right side, becoming lower and broader at the same time. This phenomenon supports the vacancy hypothesis discussed above. In the beginning of the growth of the thick $GaSb_{1-x}Bi_x$ sample, similar processes happen as for the thin-layer samples, leading to lattice contraction. The Bi segregation will stop when the growth surface is covered by a complete Bi layer and further growth leads to a constant Bi incorporation causing lattice dilation. This lattice contraction phenomenon may cause uncertainties when extracting Bi concentration from XRD measurements based on the Vegard's law.

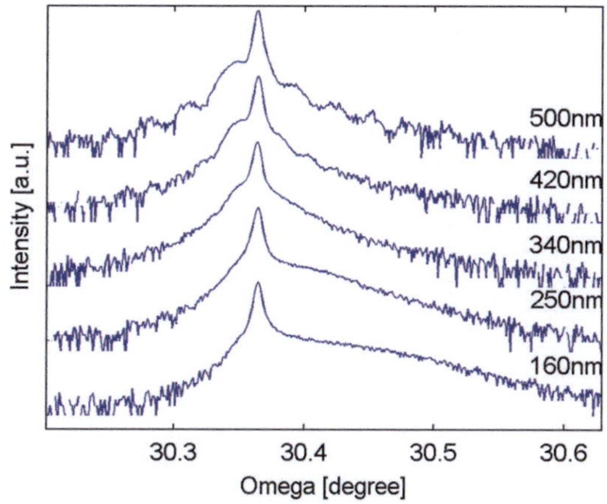

Fig. 1.5 XRD (004) rocking curves of a thick $GaSb_{1-x}Bi_x$ sample (500 nm) with different etch depths. The labels above each *curve* indicate the remaining layer thickness from the interface

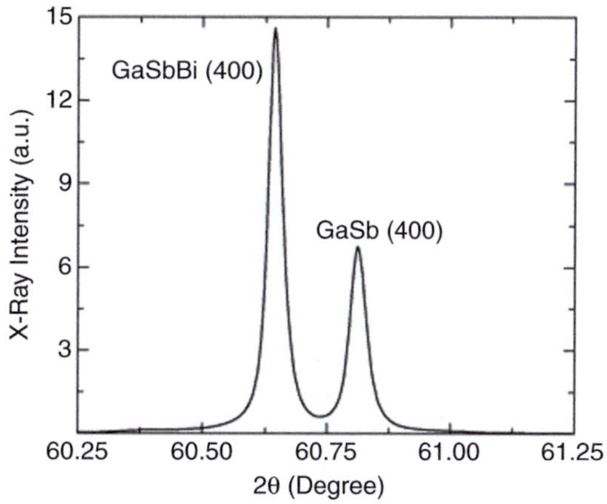

Fig. 1.6 XRD (004) rocking curves for an LPE-grown $GaSb_{1-x}Bi_x$ thick film on a GaSb substrate, reprinted with permission from Elsevier [10]

In summary, MBE growth of $GaSb_{1-x}Bi_x$ thin films is demonstrated for the first time. Bi incorporation is confirmed by RBS and SIMS measurements with the highest incorporated Bi composition of 0.7 %. Lattice contraction is observed from XRD measurements as a result of Bi segregation-induced vacancies. This poses a challenge for growing GaSbBi QWs using MBE.

1.2.3 $GaSb_{1-x}Bi_x$ Grown by LPE

Das et al. reported LPE growth of $GaSb_{1-x}Bi_x$ thick films of 4–5 μm from Ga solutions on GaSb (100) substrates [10]. Bismuth incorporation is evidenced by energy dispersive X-ray (EDX), XRD, and Raman spectroscopy measurements. Figure 1.6 shows the XRD rocking curve from one of the samples. A sharp peak corresponding

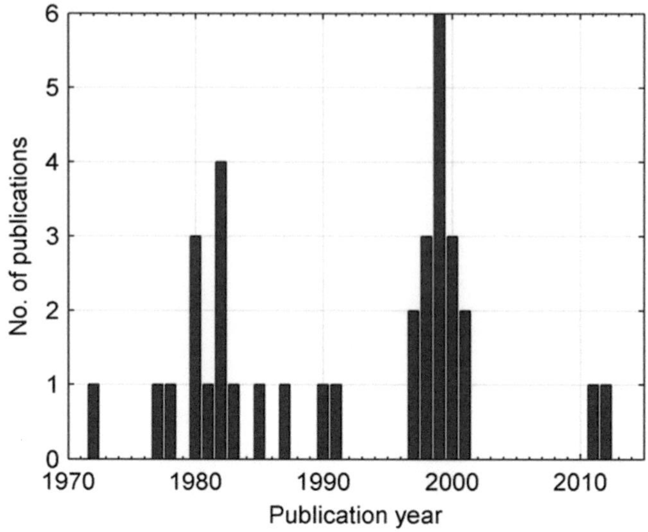

Fig. 1.7 Statistics of publications on InSbBi

to $GaSb_{1-x}Bi_x$ can be found at a lower diffraction angle to that of GaSb substrate, indicating lattice expansion for $GaSb_{1-x}Bi_x$. The maximum Bi composition is measured to be 0.4 % by assuming the validity of the Vegard's law.

PL measurements at 4 K were successfully carried out for these samples. All the $GaSb_{1-x}Bi_x$ layers showed a band-to-band PL together with a second peak at lower energy attributed to the Bi-related acceptor defects. A band-gap reduction of $GaSb_{1-x}Bi_x$ compared to GaSb was observed. The largest band-gap reduction was 40 meV measured at 0.4 % Bi incorporation, corresponding to as large as 100 meV/%Bi. Room temperature Raman spectroscopy measurements showed weak features from the GaBi phonon mode and from free Bi atoms.

1.3 InSbBi

The first $InSb_{1-x}Bi_x$ growth was demonstrated by Joukoff and Jean-Louis in 1972 using the Czochralski technique [25]. Figure 1.7 shows statistics of publications on $InSb_{1-x}Bi_x$. By counting the number of publications, $InSb_{1-x}Bi_x$ is still an under-developing field. Two peak times can be found. The first one is around the 1980s. Several groups have tried different epitaxy methods to grow $InSb_{1-x}Bi_x$. Zilko and Greene et al. implemented systemic studies of phase stability of this new material [26, 27]. The next peak arose about 20 years later, dominating by Razeghi et al. who tried to demonstrate $InSb_{1-x}Bi_x$ photodetection in the 8–12 μm range [28–31]. For a review of $InSb_{1-x}Bi_x$ growth up to 1998, readers are referred to the review article in ref. [28].

After this period, there has been no publication until 2011 when we presented growth of $InSb_{1-x}Bi_x$ thin films using MBE [32]. Another paper by LPE is published by Das et al. [33].

A solid solubility limit of 2.6 % was found due to the fact that InSb has a zinc-blende crystal structure while InBi is predicted to be tetragonal [28]. In order to break the solubility limit of Bi in InSb, non-equilibrium growth techniques have been employed [34]. MBE [16, 17, 35], MOVPE [29, 36–38], multi-target sputtering [15, 26], and LPE [33, 39] are the most commonly used growth techniques. Although up to 14 % Bi concentration has been claimed [36], most Bi atoms are incorporated not at substitutional sites and don't contribute to the band-gap reduction. All the studies employing modern MBE and MOVPE which are indispensible for growing nanometer scale heterostructures were sporadically tried in the time span of 20 years from 1981 to 2000. The growth mechanisms and Bi incorporation are still not well investigated. In the following section we review the recent results using MBE and LPE.

1.3.1 *$InSb_{1-x}Bi_x$ Thin Films Grown by MBE*

Samples were grown on undoped (100) GaAs substrates in a Riber Compact21 MBE system. A 50 nm GaAs buffer layer was grown to smoothen the surface judged by a streaky RHEED pattern. Then the growth temperature was reduced to 510 °C and a 200 nm GaSb layer was grown utilizing the "interfacial misfit array" method to obtain a low threading dislocation density and smooth surface. This GaSb layer is almost fully relaxed. After that, the growth temperature was further reduced to 280 °C and a 30 nm InSb buffer layer was grown above the GaSb layer to release the 6.3 % lattice mismatch. From RHEED, InSb quantum dots were formed at the initial growth and the surface becomes smooth again after deposition of 30 nm InSb. Then, the growth temperature was increased to a desired value ranging between 355 and 395 °C and a 150 nm $InSb_{1-x}Bi_x$ layer was grown. The strategy to enhance Bi incorporation is the same as discussed above for growth of $GaSb_{1-x}Bi_x$.

1.3.1.1 Surface Morphology

Figure 1.8 shows 10×10 μm^2 AFM scans of both reference and $InSb_{1-x}Bi_x$ samples. Smooth surface with visible atomic steps in 1×1 μm^2 scans (not shown) is observed for the $InSb_{1-x}Bi_x$ sample grown at 355 °C and the reference sample grown at 360 °C with a RMS roughness value of 0.26 and 0.34 nm, respectively. A few holes are observed from the InSb reference sample but not from the $InSb_{1-x}Bi_x$ sample, indicating a surfactant effect when using Bi. The surface is getting rough with 3D features at high growth temperatures in both cases. Discrete platform-like features together with holes are observed on the Ref sample grown at 385 °C indicating nonuniform growth rate in different regions. Adding Bi results

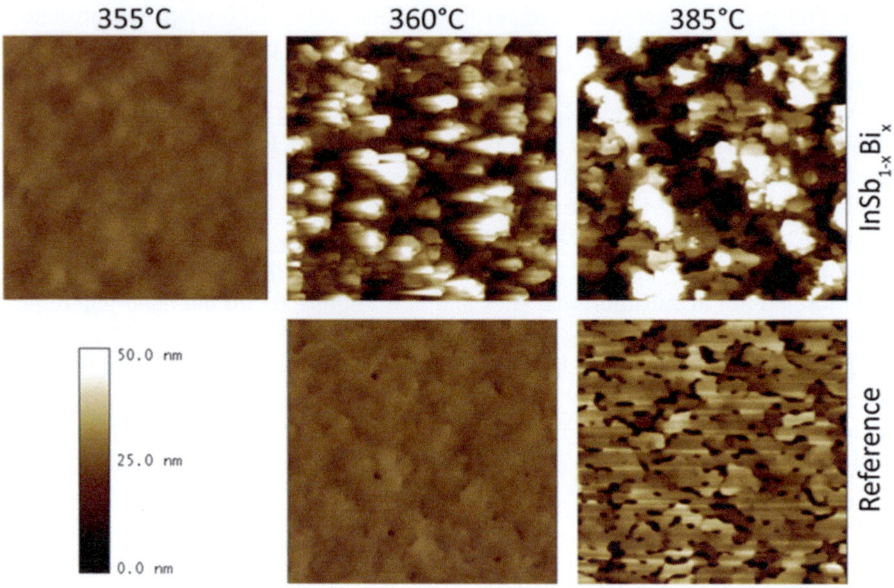

Fig. 1.8 10 × 10 μm² AFM images of $InSb_{1-x}Bi_x$ samples. The *first row* shows the $InSb_{1-x}Bi_x$ samples and the *second row* the InSb reference samples

in larger 3D structures with irregular shapes compared with the reference sample grown at the same temperature. The RMS roughness values measured from both 1 × 1 μm² and 10 × 10 μm² scans increase with growth temperature. The $InSb_{1-x}Bi_x$ has rougher surface than the InSb reference samples but the difference in the 1 × 1 μm² scans is small.

1.3.1.2 Bi Incorporation

RBS measurements are carried out to examine Bi incorporation and the results of four samples are summarized in Fig. 1.9a. Compared with the reference sample, a peak/step-like signal can be observed at the channel value of around 480 for all the $InSb_{1-x}Bi_x$ samples (not very obvious for the one grown at 355 °C). From the simulations, the deduced Bi concentrations are summarized in Fig. 1.9b. The highest measured Bi percentage is 2 % in the sample grown at 370 °C. RBS channeling measurements are performed for two $InSb_{1-x}Bi_x$ samples. By evaluating the yield difference, about 70 and 57 % of the incorporated Bi atoms are estimated to be at substitutional sites for the $InSb_{1-x}Bi_x$ layers grown at 360 and 385 °C, respectively. A simulated curve based on sharp interfaces is also shown by the green line in Fig. 1.9a. A clear trend can be found that the steps in the measured curves become blunt when the growth temperature increases, indicating interface roughening.

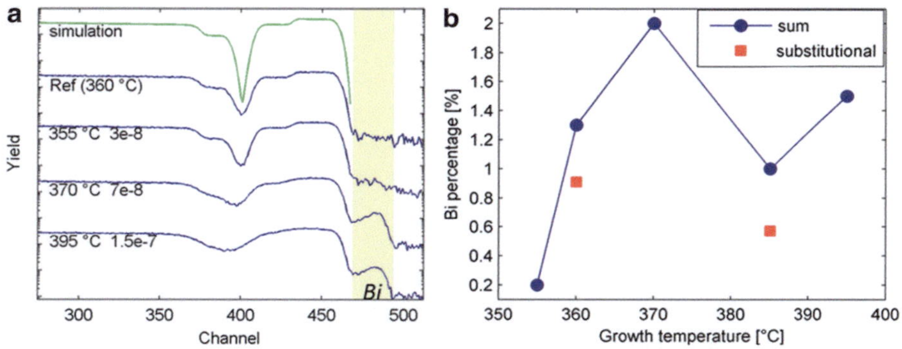

Fig. 1.9 (**a**) RBS spectra of $InSb_{1-x}Bi_x$ samples. The *green curve* is the simulation based on sharp interfaces without Bi. The *yellow belt marks* the signal region from Bi. (**b**) Bi concentrations in total (*blue circles*) and at substitutional sites (*red squares*)

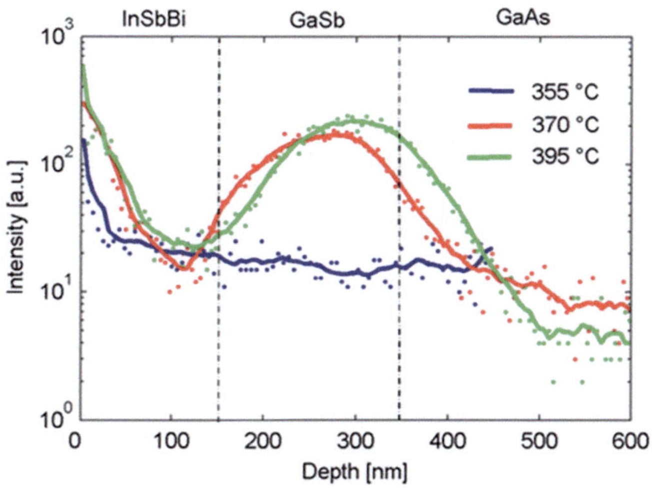

Fig. 1.10 Bi depth profiles of the $InSb_{1-x}Bi_x$ samples grown at 355, 370, and 395 °C measured by negative O_x ions in SIMS. The *dots* are measured data points and the *curves* are five-point smoothened results

Figure 1.10 shows the Bi profiles for $InSb_{1-x}Bi_x$ samples grown at 355, 370, and 395 °C measured by SIMS. Clear Bi incorporation is observed; however, the Bi distribution is not uniform. The Bi composition decreases within the first 100 nm thick $InSb_{1-x}Bi_x$ layer (only the data from the very top (~30 nm) of the sample can be attributed to the stabilization process of the SIMS equipment). Surprisingly a huge Bi signal in the GaSb layer is found in the samples grown at 370 and 395 °C. The reason is unclear at the moment and might be possibly related to Bi droplets etching through the $InSb_{1-x}Bi_x$ layer into the GaSb layer. From transmission electron microscopy measurements (not shown here), Bi droplets in μm-size are observed causing crater-like defects down to the GaSb layer.

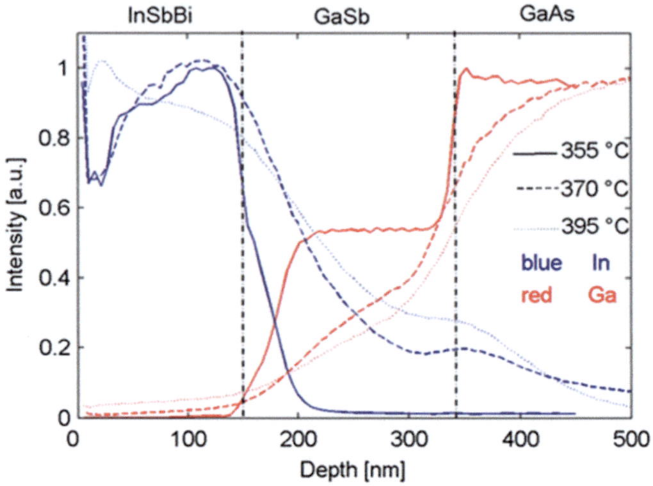

Fig. 1.11 Normalized depth profiles of In/Ga measured by negative O_x ions in SIMS for the $InSb_{1-x}Bi_x$ samples grown at 355, 370, and 395 °C

1.3.1.3 Bi-Induced In/Ga Intermixing

Bi-induced In/Ga intermixing in $InSb_{1-x}Bi_x$/InSb/GaSb was observed and found to be enhanced with increasing growth temperature as seen in Fig. 1.11. The $InSb_{1-x}Bi_x$ grown at 355 °C shows clear steps of both In and Ga profiles (normalized linear scale) at the $InSb_{1-x}Bi_x$/InSb/GaSb interfaces with small In/Ga intermixing. When the growth temperature is high, the steps become blunt indicating very strong In/Ga intermixing. When the growth temperature reaches 395 °C, a high tail of the Ga signal can be found even at the top of the $InSb_{1-x}Bi_x$ layer while In atoms can diffuse into the GaAs substrate. To the contrary, much less intermixing of group-V atoms is observed (not shown here).

1.3.2 $InSb_{1-x}Bi_x$ Grown by LPE

Das et al. reported the growth of $InSb_{1-x}Bi_x$ (4–5 μm thick) by LPE and results on absorption and Raman spectroscopy [33]. The samples were grown on InSb (100) substrates in a solvent of 95 % In and 5 % Sb and additional Bi up to 2.2 %. The maximum Bi incorporation is reported to be 0.4 %. Bi incorporation is confirmed by XRD as shown in Fig. 1.12. A sharp $InSb_{1-x}Bi_x$ peak is found at a larger diffraction angle than that of InSb substrate indicating lattice contraction for $InSb_{1-x}Bi_x$.

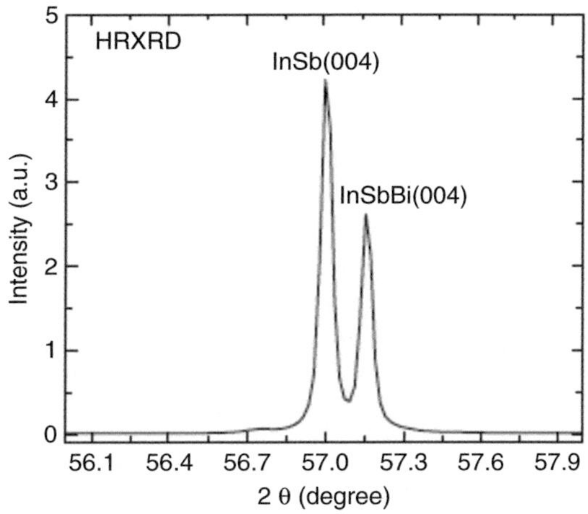

Fig. 1.12 XRD rocking curve for an LPE-grown $InSb_{1-x}Bi_x$ with 2.2 % Bi in the growth melt, reprinted with permission from Elsevier [33]

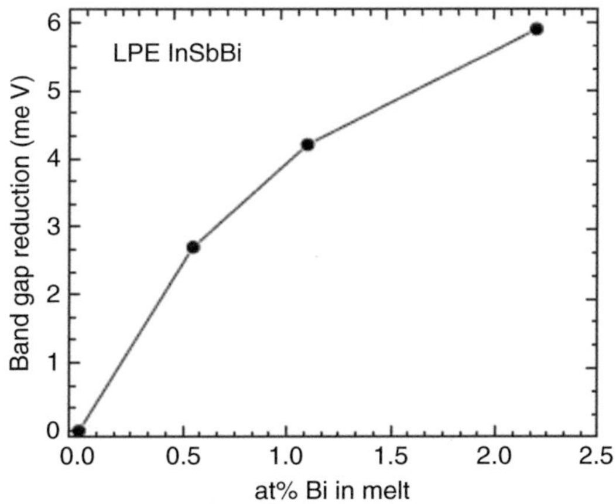

Fig. 1.13 Band-gap reduction in LPE-grown InSbBi plotted as a function of at.% Bi added to the growth melt, reprinted with permission from Elsevier [33]

Band-gap reduction of the $InSb_{1-x}Bi_x$ layers is observed in optical absorption measurements at room temperature and summarized in Fig. 1.13. It should be noted that the horizontal axis is the percentage of Bi in the growth melt rather than the Bi composition in the $InSb_{1-x}Bi_x$ layers. The band-gap reduction increases following the Bi content in the growth melt showing increasing Bi incorporation.

Figure 1.14 shows a Raman spectrum for one of the $InSb_{1-x}Bi_x$ samples. An explicit peak attributed to InBi transverse optical mode is found. Vibrational modes due to free Bi atoms are also presented in the spectrum. These results confirm that Bi atoms in the $InSb_{1-x}Bi_x$ layer reside at both substitutional and interstitial sites.

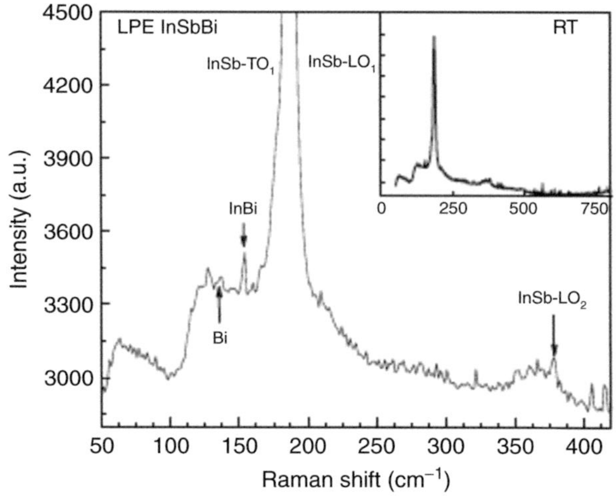

Fig. 1.14 Room temperature Raman spectrum for an LPE-grown InSbBi layer with 2.2 at.% Bi in the growth melt. *Inset* shows the full spectrum, reprinted with permission from Elsevier [33]

1.4 InAsBi

1.4.1 Surface Morphology

InAsBi alloys have been proposed for infrared detector materials operating in the wavelength range from 3 to 5 μm, which is the natural window in the atmosphere for free-space transmission. The theoretical eutectic temperature of InAsBi is as low as 110 °C, lower than the lowest possible crystal-growth temperature by MOVPE. Therefore, single phase InAsBi can only be obtained as a metastable alloy, and, if the growth condition is not appropriate, droplets or whiskers are formed on surface.

Ma et al. have grown InAsBi layers on InAs substrates at 400 °C by using MOVPE with the growth rates of 0.02 μm/min [40]. It was found that only a very narrow range of V/III ratio from 4.18 to 4.23 could yield smooth surface morphology. As shown in Fig. 1.15 for surface morphology, Bi droplets can be formed with increasing the TMBi partial pressure, and a higher TBAs partial pressure can suppress Bi droplet formation. The whiskers tend to appear in the region where the partial pressures of both TMBi and TBAs are high [41]. By SEM and EDX observations, the droplets were found mainly composed of Bi metal.

1.4.2 Bi Incorporation

Lowering the growth temperature is the most effective approach for increasing the Bi content in InAsBi alloys. For example, InAsBi samples with Bi concentrations as high as 6.1 % were successfully grown at a temperature of 275 °C [37, 42, 43].

Fig. 1.15 Surface morphology of MOVPE-grown InAsBi under various TBAs and TMBi pressures, reprinted with permission from Japan Society of Applied Physics [41]

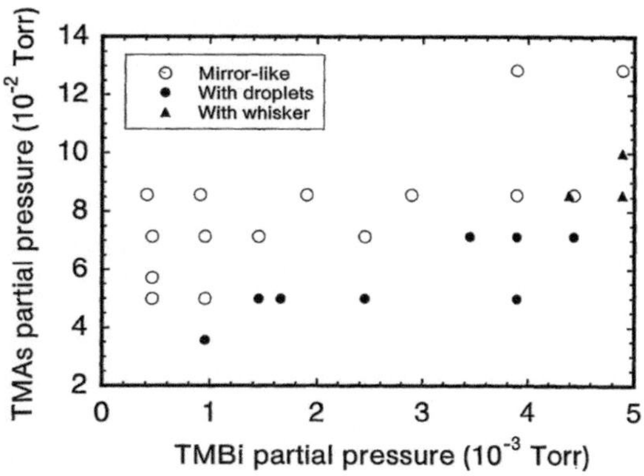

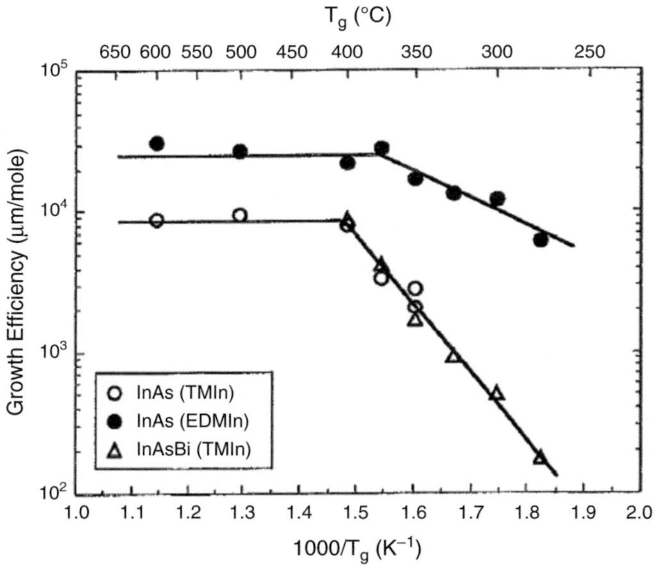

Fig. 1.16 Growth efficiency versus reciprocal temperature. The *lines* represent the best fits to the data points, reprinted with permission from American Institute of Physics [43]

The growth efficiency is a constant for temperatures above 400 °C and decreases exponentially with decreasing temperature with an activation energy of 24 kcal/mol as shown in Fig. 1.16, indicating the growth rate is diffusion-limited in this temperature regime.

For a specific growth temperature, the Bi composition of InAsBi is linearly proportional to the Bi/As ratio, in proportion to the Bi partial pressure and in inverse proportion to the As partial pressure [18, 41]. RBS channeling measurements along the [100], [110], and [111] directions confirm that 90 % of the Bi is incorporated into the InAs zinc-blende lattice sites [41].

The lattice constants of InAsBi samples are deduced from XRD and their variation with Bi composition follows the expression of $a = 6.058 + 0.966x$. An extrapolated lattice parameter for the hypothetical zinc-blende InBi is 7.024 Å [18].

1.4.3 *Electrical and Optical Properties*

Ma et al. reported that the background concentration of MOVPE-grown InAsBi measured by Hall-effect is n-type and increases from approximately 2.3×10^{16} to 10^{19} cm^{-3} as the growth temperature decreases from 600 to 325 °C, where carbon is the dominant impurity shown by SIMS results, thus carbon is supposed to be mainly a donor in InAsBi materials [43].

In the PL measurements of InAsBi samples, Fang et al. observed the peak energy decrease at a rate of 55 meV/%Bi with increasing Bi concentration in the range of 0–2.3 % [18, 44]. Okamoto et al. reported the dependence of InAsBi band-gap on the Bi content to be 42 meV/%Bi using absorption-coefficient and PL spectra [45]. Recently, a reduction of 38 meV/%Bi for MBE-grown InAsBi was reported, where the InBi lattice constant of 6.686 Å was used in the determination of Bi composition [46]. The temperature dependence of band-gap ($\Delta E_g/\Delta T$) of InAsBi decreases with increasing the Bi content, and 0.22 meV/K is obtained for $InAs_{0.963}Bi_{0.037}$ by optical-transmission measurement. Although it is widely recognized that a small E_g-material also shows a small $\Delta E_g/\Delta T$, the $\Delta E_g/\Delta T$ of 0.22 meV/K for $InAs_{0.963}Bi_{0.037}$ is smaller than that of InSb (0.29 meV/K) whose energy gap is smaller than that of InAsBi. This fact confirms the effect of Bi incorporation on decreasing the $\Delta E_g/\Delta T$. The optical properties have also been studied by Raman spectroscopy [47]. Bi- and As-related vibrational modes were observed. Bi-related vibrational modes indicate that the solubility of Bi is better in GaAs than in InAs. Furthermore, Raman spectra from InAsBi and GaAsBi samples show phonon–plasmon-coupled modes, which arise from the coupling between the LO phonon and the free charge carriers in the samples. The carbon-related vibrational modes are weaker in InAsBi samples than in GaAsBi samples, indicating the InAsBi samples have less carbon contamination compared to the GaAsBi samples.

1.5 InAsSbBi

1.5.1 *Surface Morphology*

In principle, InAsSbBi lattice matched to InSb substrates can reach the wavelength range from 8 to 12 μm. The tremendously large miscibility gap is the major factor limiting Bi incorporation into the InAsSb alloys. The InAs–InBi and InSb–InBi pseudo-binary phase diagrams were calculated and the critical transformation temperature was predicted to be 2,569 °C for the InAs–InBi system and 496 °C for the

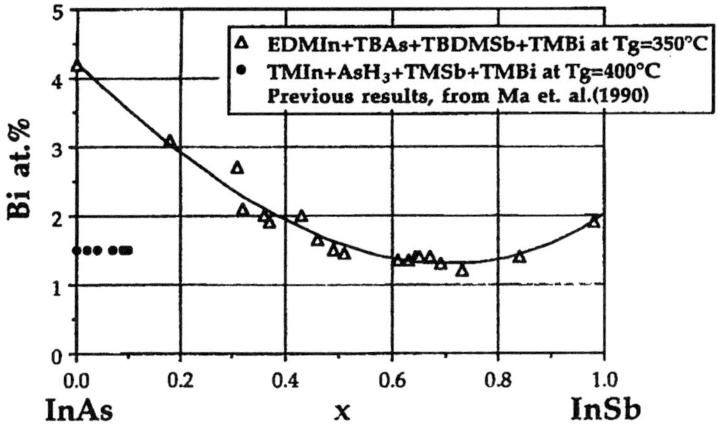

Fig. 1.17 Maximum Bi concentrations obtained without the presence of a second phase for $InAs_{1-x-y}Sb_xBi_y$ layers grown at 350 °C, reprinted with permission from American Institute of Physics [48]

InSb–InBi system [18]. Nevertheless, metastable InAsSbBi alloys have been grown with Bi concentrations far exceeding the solubility limit [18, 36]. $InAs_{0.889}Sb_{0.096}Bi_{0.015}$ layer was grown on InAs substrates with a thick graded transition layer, and a slightly crosshatched surface morphology was observed [40]. $InAs_{0.436}Sb_{0.52}Bi_{0.044}$ and $InAs_{0.47}Sb_{0.52}Bi_{0.01}$ were grown on GaAs substrates [36], and the surface morphology is dominated by the presence of precipitates, probably due to the polycrystalline phases of Bi.

1.5.2 Bi Incorporation

Huang et al. improved the precursors for the MOVPE growth of $InAs_{1-x-y}Sb_xBi_y$. The maximum Bi concentration that could be incorporated into the $InAs_{1-x-y}Sb_xBi_y$ is shown as a function of the Sb content, x, in Fig. 1.17. Above the maximum Bi concentration, which varies as the Sb concentration is changed, whiskers are formed on the surface and could not be avoided even under a careful optimization of the growth conditions [48]. SEM and EDX show that InBi is the major component in the liquid droplet at the tip of a whisker. It can be seen from Fig. 1.17 that the maximum Bi content is seen to approximately follow the band-gap of InAsSb.

InAsSbBi layers have also been synthesized by other methods, including flash-evaporation [49] and LPE [50] to overcome the problems in the cases of MBE or MOVPE employing low growth temperatures. In the flash-evaporation method, the obtained films have a number of inclusions composed mainly of InSbBi alloys located on the surface, whose density increases with the content of Bi in the vapor stream. They have a relatively small number of InBi inclusions and no In_2Bi

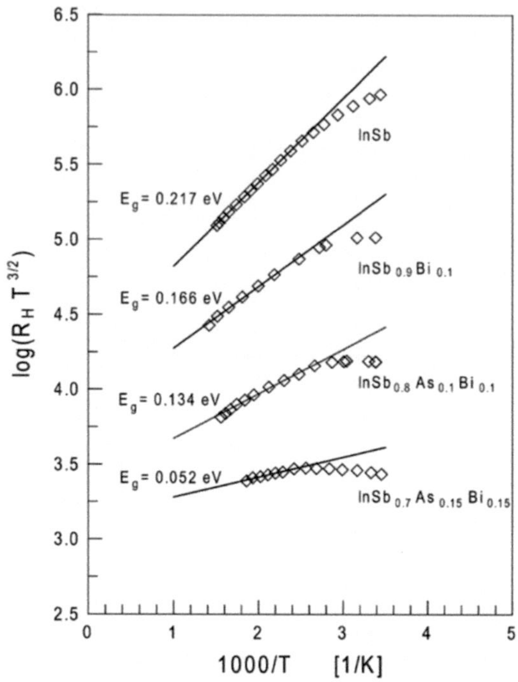

Fig. 1.18 The dependence of $\log R_H T^{3/2}$, which is related to the band-gap, on reciprocal temperature for InAsSbBi films, reprinted with permission from Wiley [49]

inclusions are detected in contrast to the InAsSbBi films grown by MBE or MOVPE. This may be because the Sb_4 molecules, promoting the formation of the InBi and In_2Bi phases, are not present in a large amount in the vapor stream of the flash-evaporation method [49].

1.5.3 Electrical and Optical Properties

The intrinsic InAsSbBi layers grown by MOVPE [35] and LPE [50] were both shown to be n-type, with the room temperature carrier concentration in the order of high 10^{16} to low 10^{17} cm^{-3}. Corresponding mobility values were in the 6,000–8,000 cm^2/V s range and around 3×10^4 cm^2/V s for MOVPE- and LPE-grown $InBi_{0.025}As_{0.105}Sb_{0.870}$ layer, respectively.

The PL measurements revealed a band-gap reduction rate of 46 meV/% Bi in $InAs_{1-x}Sb_x$ $(0.07<x<0.10)$. This implies that incorporation of only a few percent of Bi is required in $InAs_{0.35}Sb_{0.65}$ to achieve a band-gap of 0.1 eV, equivalent to a wavelength of 12 μm [18]. Oszwaldowski et al. have applied Hall coefficient R_H to estimate the band-gap of InAsSbBi samples by using the relationship of $E_g = 0.397(\Delta(\log R_H T^{3/2})/\Delta(1{,}000/T))$ [49]. The dependence of $\log(R_H T^{3/2})$ at different temperatures, which is related to the band-gap, is shown in Fig. 1.18.

It can be found that as the Bi content increases, the slope of the experimental $\log(R_H T^{3/2})$ decreases, indicating the reduced band-gap. Oszwaldowski et al. pointed out that the obtained band-gap values should be taken with care, as material regions with different band-gaps may coexist in InAsSbBi samples due to the alloy nonuniformity [49].

1.6 Optical Characterization of GaSb(Bi)/AlGaSb Quantum Wells

Excitation power- and magnetic field-dependent PL measurements are performed on the GaSb(Bi)/AlGaSb single QW (SQW) samples using the FTIR spectrometer-based PL technique [51], with the spectrometer in continuous- rather than step-scan mode [52, 53]. Ar+ laser at 514.5-nm line together with a laser power controller is used as pumping source for stable excitation power. For the magneto-PL measurements, a superconducting magneto-optical cryostat is employed in the Faraday configuration [54] for magnetic fields in a range of 0–10 T.

One $GaSb_{1-x}Bi_x/Al_{0.2}Ga_{0.8}Sb$ SQW sample grown at 380 °C and one GaSb/$Al_{0.2}Ga_{0.8}Sb$ SQW sample grown at 370 °C as a reference were investigated. The samples were grown on GaSb substrate by MBE with a structure of 5-nm-GaSb(Bi) well sandwiched by two 100-nm-$Al_{0.2}Ga_{0.8}Sb$ barriers, and finalized with a 10-nm-GaSb capping layer. Bi content was estimated to be about 0.7 % for the GaSbBi SQW sample, based on the data from the GaSbBi thin films grown at similar conditions.

Figure 1.19 shows PL spectra of GaSb and GaSbBi SQW samples at 11.5 K under a weak laser excitation of 4 mW. While the PL peak is relatively narrow for the GaSb SQW, that of the dilute-Bi SQW manifests an obvious red-shift as well as a significant broadening, indicating the real incorporation of Bi into the host GaSb QW layer.

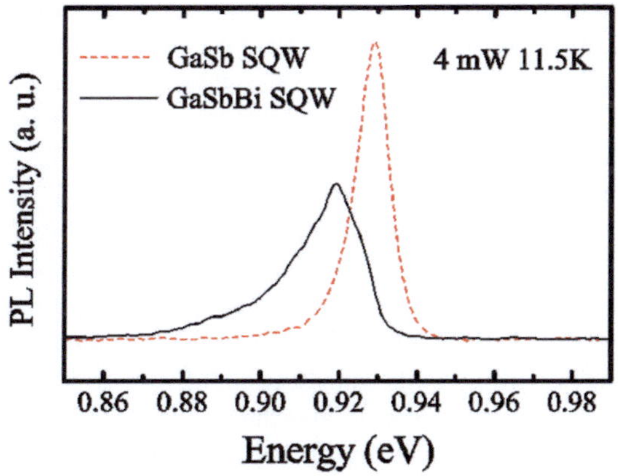

Fig. 1.19 PL spectra of GaSb and GaSbBi SQW samples at 11.5 K under a weak laser excitation of 4 mW

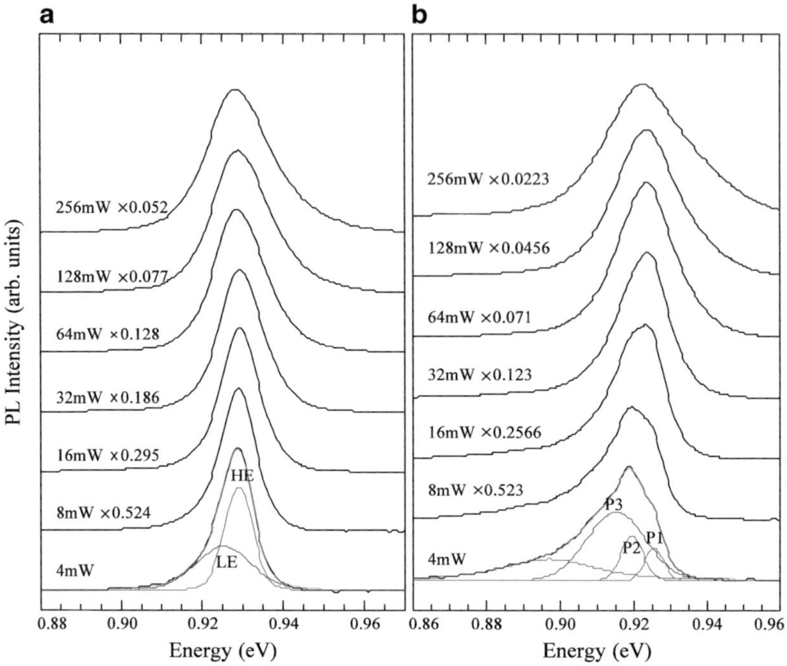

Fig. 1.20 PL spectra of GaSb and GaSbBi SQW samples at 11.5 K under different excitation power of 4–256 mW

To have a closer view of the PL difference, excitation power-dependent PL spectra are plotted in Fig. 1.20 for the two samples at 11.5 K. The PL peak of the GaSb SQW can be well fitted by two features, of which the high-energy (HE) feature dominates the intensity and gets enhanced with excitation power linearly, and can hence be ascribed to the interband transition, and the low-energy (LE) feature, similar to the typical QW PL band tail, may be related to defects and/or interfacial roughness-induced localized state. Meanwhile, the PL peak of the GaSbBi SQW can be reproduced by four features at low excitation levels, including a band tail similar to that of the GaSb SQW, a dominant broad feature P3, and two high-energy sharp features P1 and P2. As the excitation power gets higher, the features of P1 and P2 at about 0.92 eV become dominant and manifest clear blueshift, indicating the interband transition nature. The P3 feature should therefore point to the structural disorder-related transitions. At high excitation, the P1 and P2 features smear out due to the broadening by carrier–carrier scattering.

To evidence the origin of the P1 and P2 features, PL measurements of the two SQW samples are performed at 3.8 K under magnetic fields of 0–10 T, as illustrated in Fig. 1.21. Typical curve fittings are plotted for the PL spectra taken at 0 T. It is obvious that as the magnetic field gets stronger, the PL peak of the GaSbBi SQW sample gets narrower, and the P1 and P2 features become easily identifiable.

The peak-energies are depicted against magnetic field in Fig. 1.22 as dots for the HE feature of the GaSb SQW sample (left) and the P1 and P2 features of the GaSbBi

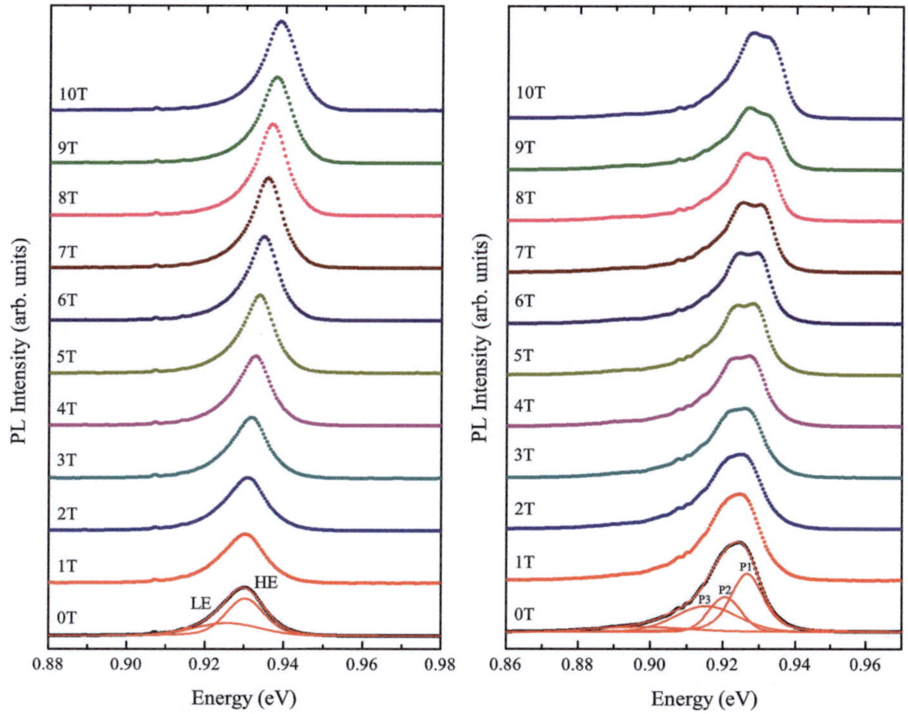

Fig. 1.21 PL spectra of GaSb and GaSbBi SQW samples at 3.8 K under different magnetic fields of 0–10 T

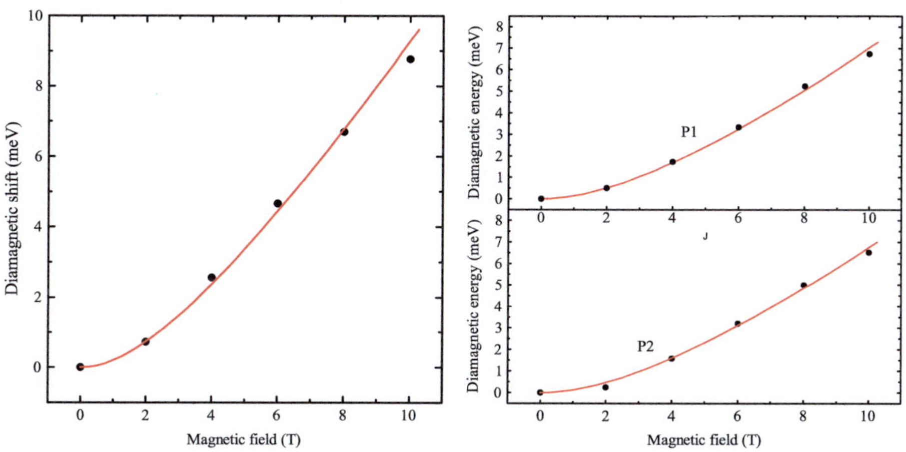

Fig. 1.22 Evolution of fitted PL feature energy with magnetic field for GaSb and GaSbBi SQW samples at 3.8 K

SQW sample (right), respectively. Solid lines are the fittings to the experimental data by a quasi-two-dimensional exciton effective-mass equation [54], suggesting the excitonic nature of the PL features. The P3 and bandtail features of the GaSbBi SQW sample, on the other hand, cannot be assumed to excitonic effect as the energies manifest a sharp fluctuation rather than a monotonic diamagnetic shift. The derived reduced effective mass and binding energy are, respectively, 0.0315 m_0 and 6.88 meV for the HE feature of the GaSb SQW, 0.0382 m_0 and 7.93 meV for the P2, and 0.0373 m_0 and 7.79 meV for P1 feature of the GaSbBi SQW sample. Correlating the HE behavior with that of the P1 or P2, it is obvious that the incorporation of Bi increases the reduced effective mass in the QW structure. On the other hand, the difference between the P1 and P2 features is only about 2.5 % in either reduced effective mass or exciton binding energy, indicating a similar mechanism for the two PL features. Detailed analysis suggests that the P1 and P2 are related to the band-to-band transition of QW with slightly different thicknesses, and this difference is estimated to be at a level of sub-nanometer, introduced possibly by the metallic Bi droplet at the GaSbBi/AlGaSb interfaces.

1.7 Summary

In this chapter we have reviewed the progresses of material growth and characterizations of $GaSb_{1-x}Bi_x$, $InSb_{1-x}Bi_x$, $InAs_{1-x}Bi_x$, and InAsSbBi films and GaSbBi QWs. To obtain high Bi incorporations and achieve high crystal quality is still very challenging. For $GaSb_{1-x}Bi_x$, the highest reported Bi incorporation is 0.7 % grown by MBE. The highest Bi incorporation in $InSb_{1-x}Bi_x$ is 14 %; however, most of the Bi atoms are at interstitial sites. Two close-positioned PL features identified in GaSbBi/AlGaSb SQW may hint the formation of metallic Bi droplets at the interfaces.

Acknowledgements The authors would like to thank Dr. Peixiong Shi from DANCHIP, Technical University of Denmark for SIMS measurements; Prof. Anders Hallen from ICT, Royal Institute of Technology for RBS measurements; Dr. Zonghe Lai and Ms. Ivy S. Roy from Chalmers University of Technology for TEM and AFM measurements; and Xiren Chen from Shanghai Institute of Technical Physics for PL measurements. The Swedish Research Council (VR) is acknowledged for financial support for this project.

References

1. Barnett, S.A.: Direct E_0 energy gaps of bismuth-containing III-V alloys predicted using quantum dielectric theory. J. Vac. Sci. Technol. A **5**, 2845–2848 (1987)
2. Francoeur, S., Seong, M.-J., Mascarenhas, A., Tixier, S., Adamcyk, M., Tiedje, T.: Band gap of $GaAs_{1-x}Bi_x$, $0<x<3.6\%$. Appl. Phys. Lett. **82**, 3874 (2003)
3. Alberi, K., Wu, J., Walukiewicz, W., Yu, K., Dubon, O., Watkins, S., Wang, C., Liu, X., Cho, Y.-J., Furdyna, J.: Valence-band anticrossing in mismatched III-V semiconductor alloys. Phys. Rev. B **75**, 045203 (2007)

4. Cooke, D.G., Hegmann, F.A., Young, E.C., Tiedje, T.: Electron mobility in dilute GaAs bismide and nitride alloys measured by time-resolved terahertz spectroscopy. Appl. Phys. Lett. **89**, 122103 (2006)
5. Oe, K., Okamoto, H.: New semiconductor alloy $GaAs_{1-x}Bi_x$ grown by metal organic vapor phase epitaxy. Jpn. J. Appl. Phys. **37**, L1283–L1285 (1998)
6. Tixier, S., Adamcyk, M., Tiedje, T., Francoeur, S., Mascarenhas, A., Wei, P., Schiettekatte, F.: Molecular beam epitaxy growth of $GaAs_{1-x}Bi_x$. Appl. Phys. Lett. **82**, 2245 (2003)
7. Fluegel, B., Francoeur, S., Mascarenhas, A., Tixier, S., Young, E.C., Tiedje, T.: Giant spin-orbit bowing in $GaAs_{1-x}Bi_x$. Phys. Rev. Lett. **97**, 67205 (2006)
8. Sweeney, S.J., Batool, Z., Hosea, T.J.C., Jin, S.R.: The potential of III-bismides for near- and mid-IR photonic devices. In: First International Workshop on Bismuth-Containing Semiconductors: Theory, Simulation, and Experiment, Michigan, USA (2010)
9. Germogenov, V.P., Otman, Y.I., Chaldyshev, V.V., Shmartsev, Y.V.: Width of the band gap in $GaSb_{1-x}Bi_x$ solid solutions. Sov. Phys. Semicond. **23**, 942–943 (1989)
10. Das, S.K., Das, T.D., Dhar, S., de la Mare, M., Krier, A.: Infrared photoluminescence of dilute GaSb:Bi alloys grown by liquid phase epitaxy. Infrared Phys. Technol. **55**, 156–160 (2012)
11. Vurgaftman, I., Meyer, J.R., Ram-Mohan, L.R.: Band parameters for III-V compound semiconductors and their alloys. J. Appl. Phys. **89**, 5815 (2001)
12. Kondow, M., Uomi, K., Hosomi, K., Mozume, T.: Gas-source molecular beam epitaxy of GaN_xAs_{1-x} using a N radical as the N source. Jpn. J. Appl. Phys. **33**, L1056–L1058 (1994)
13. Dutta, P., Bhat, H., Kumar, V.: The physics and technology of gallium antimonide: an emerging optoelectronic material. J. Appl. Phys. **81**, 5821–5870 (1997)
14. Hosoda, T., Kipshidze, G., Tsvid, G., Shterengas, L., Belenky, G.: Type-I GaSb-based laser diodes operating in 3.1- to 3.3 μm wavelength range. IEEE Photon. Technol. Lett. **22**, 718–720 (2010)
15. Zilko, J.L., Greene, J.E.: Growth of metastable $InSb_{1-x}Bi_x$ thin films by multitarget sputtering. Appl. Phys. Lett. **33**, 254 (1978)
16. Oe, K., Ando, S., Sugiyama, K.: $InSb_{1-x}Bi_x$ films grown by molecular beam epitaxy. Jpn. J. Appl. Phys. **20**, L303–L306 (1981)
17. Noreika, A.J., Takei, W.J., Francombe, M.H., Wood, C.E.C.: Indium antimonide-bismuth compositions grown by molecular beam epitaxy. J. Appl. Phys. **53**, 4932 (1982)
18. Fang, Z., Ma, K., Cohen, R., Stringfellow, G.: Photoluminescence of InAsBi and InAsSbBi grown by organometallic vapor phase epitaxy. J. Appl. Phys. **68**, 1187–1191 (1990)
19. Gladkov, P., Monova, E., Weber, J.: Liquid phase epitaxy and photoluminescence characterization of p-type GaSb layers grown from Bi based melts. J. Cryst. Growth **146**, 319–325 (1995)
20. Song, Y., Wang, S., Roy, I.S., Shi, P., Hallen, A.: Growth of $GaSb_{1-x}Bi_x$ by molecular beam epitaxy. J. Vac. Sci. Technol. B **30**, 02B114–02B117 (2012)
21. Danilewsky, A., Lauer, S., Meinhardt, J., Benz, K., Kaufmann, B., Hofmann, R., Dornen, A.: Growth and characterization of GaSb bulk crystals with low acceptor concentration. J. Electron. Mater. **25**, 1082–1087 (1996)
22. Janotti, A., Wei, S.-H., Zhang, S.B.: Theoretical study of the effects of isovalent coalloying of Bi and N in GaAs. Phys. Rev. B **65**, 115203 (2002)
23. Mohmad, A.R., Bastiman, F., Hunter, C.J., Ng, J.S., Sweeney, S.J., David, J.: The effect of Bi composition to the optical quality of $GaAs_{1-x}Bi_x$. Appl. Phys. Lett. **99**, 42107 (2011)
24. Moison, J.M., Guille, C., Houzay, F., Barthe, F., Van Rompay, M.: Surface segregation of third-column atoms in group III-V arsenide compounds: ternary alloys and heterostructures. Phys. Rev. B **40**, 6149 (1989)
25. Joukoff, B., Jean-Louis, A.M.: Growth of $InSb_{1-x}Bi_x$ single crystals by Czochralski method. J. Cryst. Growth **12**, 169–172 (1972)
26. Zilko, J., Greene, J.: Growth and phase stability of epitaxial metastable $InSb_{1-x}Bi_x$ films on GaAs. I. Crystal growth. J. Appl. Phys. **51**, 1549 (1980)
27. Zilko, J.L., Greene, J.E.: Growth and phase stability of epitaxial metastable $InSb_{1-x}Bi_x$ films on GaAs. II. Phase stability. J. Appl. Phys. **51**, 1560–1564 (1980)
28. Lee, J.J., Razeghi, M.: Investigation of novel InTlSb and InSbBi alloys for uncooled photodetector applications. In: Brown, G.J. (ed.) Photodetectors: Materials and Devices III, pp. 256–269. SPIE-INT Soc Optical Engineering, Bellinghan, USA (1998)

29. Lee, J.J., Razeghi, M.: Exploration of InSbBi for un cooled long-wavelength infrared photodetectors. Opto-Electr. Rev. **6**, 25–36 (1998)
30. Lee, J.J., Kim, J.D., Razeghi, M.: Room temperature operation of 8–12 μm InSbBi infrared photodetectors on GaAs substrates. Appl. Phys. Lett. **73**, 602–604 (1998)
31. Lee, J.J., Kim, J.D., Razeghi, M.: Exploration of novel InSbBi alloy for uncooled infrared photodetector applications. J. Korean Phys. Soc. **35**, S275–S278 (1999)
32. Wang, S., Song, Y., Roy, I.S.: Bismuth incorporation and lattice contraction in GaSbBi and InSbBi. In: Jaworski, M, Marciniak, M. (eds.) 2011 13th International Conference on Transparent Optical Networks (ICTON), IEEE, New York, USA (2011)
33. Das, S.C., Das, T.D., Dhar, S.: Infrared absorption and Raman spectroscopy studies of InSbBi layers grown by liquid phase epitaxy. Infrared Phys. Technol. **55**, 306–308 (2012)
34. Du, Q., Alperin, J., Wang, W.I.: Molecular beam epitaxial growth of GaInSbBi for infrared detector applications. J. Cryst. Growth **175–176**, 849–852 (1997)
35. Noreika, A.J., Greggi, J., Takei, W.J., Francombe, M.H.: Properties of MBE grown InSb and $InSb_{1-x}Bi_x$. J. Vac. Sci. Technol. A **1**, 558–561 (1983)
36. Humphreys, T., Chiang, P., Bedair, S., Parikh, N.: Metalorganic chemical vapor deposition and characterization of the In-As-Sb-Bi material system for infrared detection. Appl. Phys. Lett. **53**, 142 (1988)
37. Ma, K.Y., Fang, Z.M., Cohen, R.M., Stringfellow, G.B.: OMVPE growth and characterization of Bi-containing III–V alloys. J. Cryst. Growth **107**, 416–421 (1991)
38. Wagener, M.C.C., Botha, J.R., Leitch, A.W.R.: Characterization of secondary phases formed during MOVPE growth of InSbBi mixed crystals. J. Cryst. Growth **213**, 51–56 (2000)
39. Dixit, V.K., Keerthi, K.S., Bera, P., Bhat, H.L.: Growth of $InBi_xSb_{1-x}$ films on GaAs(0 0 1) substrates using liquid phase epitaxy and their characterization. J. Cryst. Growth **241**, 171–176 (2002)
40. Ma, K., Fang, Z., Jaw, D., Cohen, R., Stringfellow, G., Kosar, W., Brown, D.: Organometallic vapor phase epitaxial growth and characterization of InAsBi and InAsSbBi. Appl. Phys. Lett. **55**, 2420 (1989)
41. Okamoto, H., Oe, K.: Growth of metastable alloy InAsBi by low-pressure MOVPE. Jpn. J. Appl. Phys. **37**, 1608–1613 (1998)
42. Huang, K.T., Chiu, C.T., Cohen, R.M., Stringfellow, G.B.: InAsBi alloys grown by organometallic vapor phase epitaxy. J. Cryst. Growth **134**, 29–34 (1993)
43. Ma, K.Y., Fang, Z.M., Cohen, R.M., Stringfellow, G.B.: Investigation of organometallic vapor phase epitaxy of InAs and InAsBi at temperatures as low as 275 °C. J. Appl. Phys. **70**, 3940 (1991)
44. Ma, K.Y., Fang, Z.M., Cohen, R.M., Stringfellow, G.B.: Ultra-low temperature OMVPE of InAs and InAsBi. J. Electron. Mater. **21**, 143–148 (1992)
45. Okamoto, H., Oe, K.: Structural and energy-gap characterization of metalorganic-vapor-phase-epitaxy-grown InAsBi. Jpn. J. Appl. Phys. **38**, 1022–1025 (1999)
46. Svensson, S.P., Hier, H., Sarney, W.L., Donetsky, D., Wang, D., Belenky, G.: Molecular beam epitaxy control and photoluminescence properties of InAsBi. J. Vac. Sci. Technol. A **30**, 02B109 (2012)
47. Verma, P., Oe, K., Yamada, M., Harima, H., Herms, M., Irmer, G.: Raman studies on $GaAs_{1-x}$ Bi_x and $InAs_{1-x}$ Bi_x. J. Appl. Phys. **89**, 1657 (2001)
48. Huang, K., Chiu, C., Cohen, R., Stringfellow, G.: InAsSbBi alloys grown by organometallic vapor-phase epitaxy. J. Appl. Phys. **75**, 2857 (1994)
49. Oszwaldowski, M., Berus, T., Szade, J., Józwiak, K., Olejniczak, I., Konarski, P., Jozwiak, K.: Structural properties of InSbBi and InSbAsBi thin films prepared by the flash-evaporation method. Cryst. Res. Technol. **36**, 1155–1171 (2001)
50. Dixit, V.K., Keerthi, K.S., Bhat, H.L., Bera, P., Hegde, M.S.: Structural and compositional analysis of $InBi_xAs_ySb_{(1-x-y)}$ films grown on GaAs(0 0 1) substrates by liquid phase epitaxy. Appl. Surf. Sci. **220**, 321–326 (2003)
51. Shao, J., Lu, W., Lu, X., Yue, F., Li, Z., Guo, S., Chu, J.: Modulated photoluminescence spectroscopy with a step-scan Fourier transform infrared spectrometer. Rev. Sci. Instrum. **77**, 63104–63106 (2006)

52. Zhu, L., Shao, J., Lu, X., Guo, S., Chu, J.: Competition of compressive strain with substrate misorientation in CuPt-type ordered GaInP/AlGaInP quantum wells. J. Appl. Phys. **109**, 13506–13509 (2011)
53. Zhang, X., Shao, J., Chen, L., Lu, X., Guo, S., He, L., Chu, J.: Infrared photoluminescence of arsenic-doped HgCdTe in a wide temperature range of up to 290 K. J. Appl. Phys. **110**, 43503–43507 (2011)
54. Shao, J., Lu, X., Yue, F., Huang, W., Guo, S., Chu, J.: Magnetophotoluminescence study of $Ga_xIn_{1-x}P$ quantum wells with CuPt-type long-range ordering. J. Appl. Phys. **100**, 53522–53526 (2006)

Chapter 2
Bismide-Based Photonic Devices for Near- and Mid-Infrared Applications

S.J. Sweeney, I.P. Marko, S.R. Jin, K. Hild, Z. Batool, N. Hossain, and T.J.C. Hosea

Abstract Bismides are a new class of III–V semiconductor alloys which are gaining interest due to their many potential applications. In this chapter we show how the addition of bismuth atoms to III–V alloys gives rise to useful band structure properties, such as a large band gap bowing, making it possible to produce narrow gap (near- and mid-infrared) materials and devices on conventionally used III–V substrates such as GaAs and InP. In addition, bismuth-containing alloys can provide a large spin-orbit splitting energy which offers the potential to reduce or eliminate the important non-radiative Auger recombination and inter-valence band absorption processes which dominate the performance of near- and mid-infrared lasers and LEDs. Furthermore, we show that by adding nitrogen to the alloys, lattice-matched narrow band gap semiconductor heterostructures may be produced with the possibility of wide control of the conduction and valence offsets. Finally, we provide a discussion of the physics of devices based upon bismide alloys.

S.J. Sweeney (✉) • I.P. Marko • S.R. Jin • K. Hild • Z. Batool • N. Hossain
Advanced Technology Institute and Department of Physics, University of Surrey, Guildford, Surrey GU2 7XH, UK
e-mail: s.sweeney@surrey.ac.uk; i.marko@surrey.ac.uk

T.J.C. Hosea
Advanced Technology Institute and Department of Physics, University of Surrey, Guildford, Surrey GU2 7XH, UK

Ibnu Sina Institute for Advanced Studies, Faculty of Science, Universiti Teknologi Malaysia, 81310 Johor Bahru, Johor, Malaysia

H. Li and Z.M. Wang (eds.), *Bismuth-Containing Compounds*, Springer Series in Materials Science 186, DOI 10.1007/978-1-4614-8121-8_2,

2.1 Introduction

In a similar manner to the dilute nitrides, the incorporation of bismuth into III–V semiconductors such as GaAs causes a large band gap bowing which can be modelled using band anti-crossing (BAC) theory, albeit in the valence band (VB) for bismides vs. the conduction band (CB) for dilute nitrides [1]. In addition, due to the large size of bismuth atoms, the spin-orbit band moves down in energy and this, coupled with the upward movement of the valence band edge, gives rise to a very large spin-orbit splitting [2]. This opens up interesting new possibilities for efficient photonic devices, such as near- and mid-infrared lasers which are more thermally stable and less susceptible to losses compared to conventional InP-based devices. Since bismuth principally influences the valence band, whilst nitrogen influences the conduction band, combining bismuth and nitrogen in III–V alloys offers huge potential for engineering the conduction and valence band offsets, the strain, band gap and spin-orbit splitting, with wide scope for the design of photonic devices. Further applications exist in thermoelectrics and spintronics. In this chapter we review the interesting band structure properties of the bismides and discuss how such materials may be used to produce photonic devices with improved properties.

There is an increasing interest in incorporating bismuth into conventional III–V semiconductors, due to their potential to optimize the band structure and overcome several deficiencies in near- and mid-infrared light emitting devices. Substituting As with small amounts of Bi creates an energy level in the host valence band (VB) causing a VB anti-crossing (VBAC) interaction [1]. This strongly reduces the band gap, E_g, by ~80 and ~56 meV/Bi% in GaAsBi and InGaAsBi, respectively [3]. The VBAC interaction between the Bi level and the host III–V material system [4] is analogous to the conduction band (CB) anti-crossing widely reported for dilute nitride alloys [5]. As a result of this anti-crossing the VBs split into six valence sub-bands which may be grouped into the E_+ (HH^+, LH^+, SO^+) and E_- (HH^-, LH^-, SO^-) levels, where HH and LH refer to heavy and light holes, respectively, and SO to the spin-orbit split-off band [4]. As shown in Fig. 2.1 Bi enhances the SO splitting energy (Δ_{SO}) which is also interesting in the field of spintronics [6]. This occurs due to enhanced electron spin-orbital (SO) angular momentum interaction within the heavier Bi atoms together with the upward movement of the VB maximum (VBM) due to BAC. If Δ_{SO} can be made larger than E_g, this offers a route to reduce, or even eliminate, fundamental loss processes such as Auger recombination and inter-valence band absorption (IVBA), which dominate IR emitters such as 1.55 μm InGaAsP/InP lasers as illustrated in Fig. 2.2 [7–11]. Specifically, if Δ_{SO} can be made larger than E_g, the Auger process involving a conduction band electron, heavy-hole and spin split-off hole excited to heavy-hole band ("CHSH") [12] may be suppressed.

There have been a number of attempts to improve the performance of 1.55 μm lasers. The InGaAlAs/InP material system has been studied significantly owing to the fact that it helps reduce electron overflow because of its increased conduction band offset. However, it has only a small effect on Auger recombination. The more recently developed GaAs-based dilute nitride material system is attractive, but devices tend to

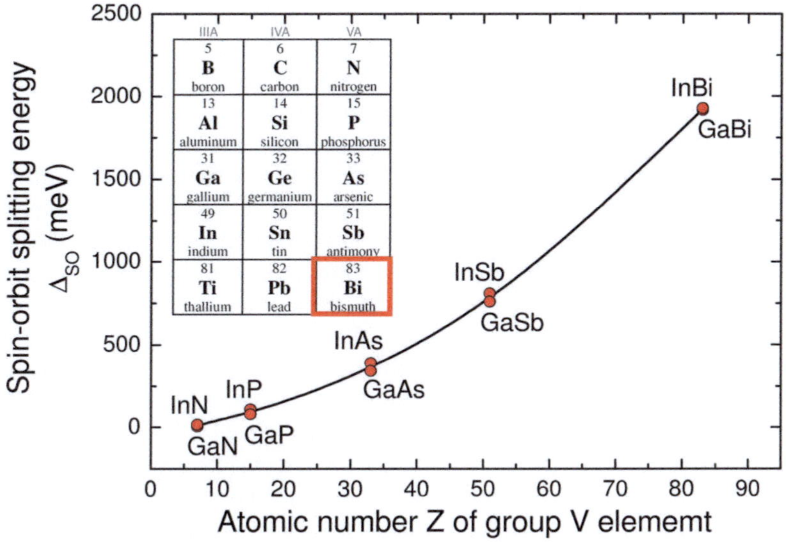

Fig. 2.1 Variation of the spin-orbit splitting energy as a function of group V atomic number for III–V binary compounds

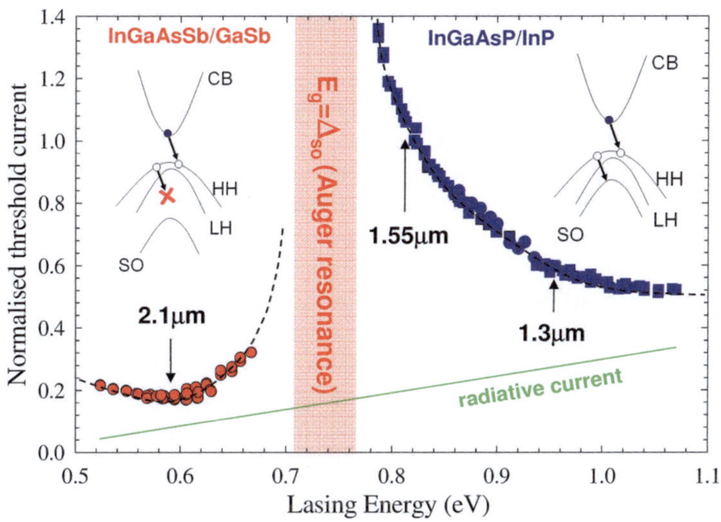

Fig. 2.2 Influence of Auger recombination on the threshold current of InP- and GaSb-based lasers in the near- and mid-infrared range

suffer from large defect-related recombination in addition to Auger recombination [13]. 1.3 μm InAs/GaAs [14] and 1.5 μm InAs/InP [15] quantum dot-based lasers have similarly been shown to suffer from Auger recombination and have comparable temperature sensitivity to conventional quantum well lasers unless p-doping is employed [16]. Moving into the mid-infrared, GaInAsSb/GaSb-based lasers operating between 2 and 3 μm have lower threshold current densities (also shown in Fig. 2.2) and a reduced sensitivity to temperature [17]. This is due to the large spin-orbit splitting (Δ_{SO}) in antimonide-alloys such that Δ_{SO} is larger than the band gap, E_g [18, 19]. This suppresses IVBA and CHSH-type Auger transitions via the spin-orbit band since energy and momentum can no longer be conserved. This leads to improved laser performance in the mid-infrared, an effect also observed in GaInAsSbP/InAs mid-infrared LEDs [20]. Achieving this in the near-infrared has proved troublesome owing to the difficulty in achieving $\Delta_{SO}>E_g$ at these larger band gaps. However, the unusual band structure properties of the bismide alloys provide a potential route to achieve this.

Tailoring both E_g and Δ_{SO} could make bismides potentially useful for improved operating efficiency of 1.3–1.6 μm laser diodes, optical amplifiers and optical modulators important for optical fibre communications [18, 19]. A thorough understanding of the fundamental physical properties of bismides is therefore crucial. Whilst there are reports in the past decade on the structural, optical [4, 6, 21], thermal [22, 23] and transport properties [24] of bismides, they remain a relatively underexplored family of alloys. In addition, similar to alloys such as InGaNAs, the temperature (T) dependence of the band gap, dE_g/dT, is expected to be weaker in bismides than in conventional III–Vs. Oe and Asai suggest that, because InBi and GaBi are semi-metallic, InGaAsBi structures with temperature-insensitive E_g may be possible [25]. Some reports indicate dE_g/dT is significantly weaker in GaAsBi [22], but there is controversy by how much [26, 27], and on the comparative influences of the VBAC interaction and material inhomogeneity.

There is relatively little information on dE_g/dT for InGaAsBi; Devenson et al. recently reported a *qualitatively* weaker dE_g/dT compared to InGaAs [28]; however, Marko et al. report relatively little influence of Bi content on dE_g/dT [29]. This is crucial to understand not only the physics but also the extent to which laser emission wavelength could be stabilized with temperature. Currently, the latter requires thermoelectric cooling, consumes significant power, reduces overall efficiency and adds complexity and cost. Together with suppression of non-radiative recombination, this could significantly improve the temperature stability of laser output power, with the added benefit of growth on well-established relatively inexpensive substrates: GaAsBi/GaAs is suitable for near-IR, particularly 1.5 μm telecomm devices; InGaAsBi or GaAsBiN/GaAs could span a wide spectral range from the near-IR out to >6 μm, whilst still maintaining tolerable strain levels on standard InP or GaAs substrates. The growth and processing of the GaSb-based materials is less advanced than for InP-based materials, which are being widely used commercially for optical fibre-based telecommunication applications. Furthermore, InP has higher thermal conductivity and lower electrical resistance than the quaternary compound AlGaAsSb, which shows advantage for room temperature continuous wave operation. Growth on the InP material system also benefits from well-known fabrication techniques and the availability of low-loss waveguides. The use of standard GaAs substrates for near- or mid-infrared applications is also beneficial from the point of view of well-established technology and possibility for the growth of lattice-matched Distributed Bragg Reflectors (DBRs) for Vertical Cavity Surface Emitting Lasers (VCSELs) and related devices.

In the next section, coming from a device and applications perspective, we focus on the band structure properties of bismide alloys for growth on GaAs substrates using simplified conduction and VBAC models. In addition we show how bismuth may be used to form alloys whereby the spin-orbit splitting energy (Δ_{SO}) is larger than the band gap (E_g) thereby providing a means of suppressing non-radiative CHSH (hot-hole producing) Auger recombination and IVBA, which is expected to improve the high-temperature performance and thermal stability of light emitting devices. It is also shown that preferential band offsets are achievable with GaAsBiN/GaAs, which makes this material system promising for photonic devices operating in the near- and mid-infrared.

2.2 Modelling of GaAsBiN/GaAs Structures for Device Applications

The unusually large band gap shrinkage due to the incorporation of Bi and N atoms in GaAs can be modelled using BAC theory [4, 30] as discussed in detail by Broderick, Usman and O'Reilly elsewhere in this book. It has been shown that the addition of N atoms into GaAs causes a BAC effect leading to the generation of E_+ and E_- conduction sub-bands. Consequently, the band gap (E_- to VBM splitting) decreases. This originates from the interaction between the nitrogen resonant level and the host conduction band due to the large difference in electronegativity of As and N [30]. Since the Bi level lies below the VBM of GaAs, the anti-crossing between the VBM and the Bi levels is similarly expected to occur, leading to band gap bowing via the formation of E_+ and E_- valence sub-bands and an effective upward shift of the VBM in GaAsBi [4]. Both interactions can be described well by using either the conduction BAC or VBAC model, which describes the interaction between the resonant level and the conduction (valence) band. In the simple approximation of this model the conduction band minimum (VBM) is given by [4, 30–32]

$$E_{C,V}[\mathrm{GaAs(N,Bi)}] = \frac{E_{C,V}(\mathrm{GaAs}) + E_{N,Bi} \mp \sqrt{[E_{C,V}(\mathrm{GaAs}) - E_{N,Bi}]^2 + 4C_{N,Bi}^2 y}}{2}, \quad (2.1)$$

where $E_{C,V}$ stands for the energy levels of the conduction band minimum (the VBM) of GaAs, $E_{N,Bi}$ and $C_{N,Bi}$ are the resonant nitrogen (bismuth) level in GaAs and the coupling parameters, respectively, and y is the composition of Bi or N. Other parameters of the quaternary alloys are calculated using the formulae and parameters from the literature [33, 34]. The relative alignment of the band edges between the quantum well and GaAs barrier is calculated using the model-solid theory [35], which describes the valence band edge by

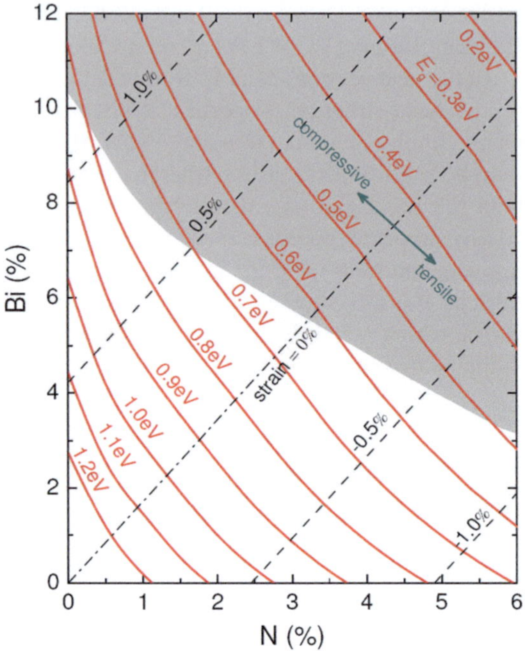

Fig. 2.3 Predicted band gap of GaAsBiN on GaAs. This quaternary offers wide scope for band structure engineering. The dependence of strain on N and Bi composition in GaAsBiN alloys on GaAs substrate is also included. The *shaded* region indicates where $\Delta_{SO} > E_g$

$$E_V = \begin{cases} E_{v,av} + \dfrac{\Delta_{SO}}{3} + \delta E_{hh}, & \text{for heavy holes} \\ E_{v,av} + \dfrac{\Delta_{SO}}{3} + \delta E_{lh}, & \text{for light holes} \end{cases} \tag{2.2}$$

where $E_{v,av}$ and Δ_{SO} are the average valence band edge energy [36] and the spin-orbit splitting energy, respectively. δE_{hh} (δE_{lh}) is the shift of the strained bulk valence band edge relative to the unstrained value. The shift of the conduction band edge due to the incorporation of Bi has been taken into account based on the data of Usman et al. [37] and experimentally verified by Batool et al. [2]. Figure 2.3 displays the predicted band gap of GaAsBiN as a function of Bi and N compositions at 300 K. It is shown that the quaternary alloys cover a wide energy range from 0.2 to 1.4 eV, i.e., covering the near and mid-infrared, for Bi up to 12 % and N up to 6 %. For such a range, the strain of GaAsBiN on GaAs remains within ±1.5 % (as shown in the inset of Fig. 2.4). The relatively small strain achieved with GaAsBiN alloys on GaAs is due to strain compensation between GaAsN and GaAsBi. As relatively smaller N atoms in GaAs cause tensile strain whilst larger Bi atoms in GaAs lead to compressive strain, the incorporation of both N and Bi compensates strain. Depending on N and Bi compositions, GaAsBiN can be flexibly designed under compressive or tensile strain on GaAs, which could also be used for producing polarization-insensitive semiconductor optical amplifiers by designing a series of layers in compressive and tensile strain [38]. In a similar way, a series of GaAsBi/GaAsN layer pairs may be used to develop type-II narrow gap super-lattice structures.

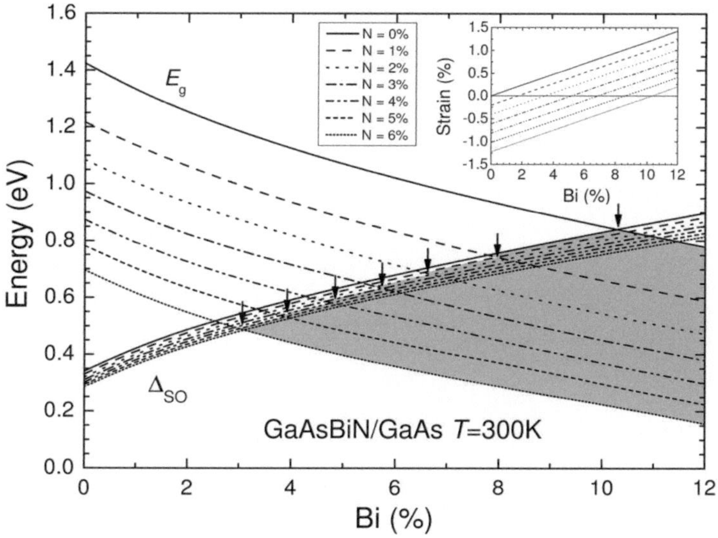

Fig. 2.4 Predicted band gap and spin-orbit splitting energy as a function of Bi and N compositions in GaAsBiN on GaAs. The crossing points of $\Delta_{SO}=E_g$ at various N compositions are marked using *arrows*. The *shaded* region indicates where $\Delta_{SO}>E_g$

These data clearly show the great potential of such alloys to operate in a wide spectral range whilst remaining lattice-matched to GaAs with applications in emitters, detectors and photovoltaics. VCSELs and related devices are particularly interesting due to the ease at which GaAs/AlGaAs high reflectivity DBRs may be fabricated. However, the use of a GaAsBiN active region provides a possibility to produce VCSELs in the mid-infrared, something which has hitherto been relatively complicated to realize.

In addition to the strong band gap reduction caused by the incorporation of either Bi or N, Fig. 2.4 shows the variation of Δ_{SO} as a function of Bi composition at various N fractions. This figure clearly shows how the addition of bismuth to GaAsN causes a strong increase in Δ_{SO}. The significant increase of Δ_{SO} with increasing Bi composition is attributed to the large atomic mass of bismuth which increases the interaction between the electron spin and orbital angular momentum. It has been shown that the CHSH (hot-hole producing) Auger process and IVBA dominate the threshold current and temperature sensitivity of InP- and GaAs-based lasers in the near-infrared [35]. Figure 2.1 shows that the spin-orbit splitting energy increases super-linearly with the cation atomic number of the group V element. Bismuth is the largest stable group V element and it has been theoretically predicted that III–V bismide compounds such GaBi, InBi and AlBi should have an extremely large Δ_{SO} of ~2 eV [39]. The data in Fig. 2.1 also largely explain the success of antimonides in the mid-infrared since it is relatively straightforward to achieve $\Delta_{SO}>E_g$ in such narrow gap alloys. However, achieving this in the near-infrared with antimonides is impossible due to the larger band gaps required to achieve higher energy optical transitions. Thus, the incorporation of bismuth offers the potential for narrow gap

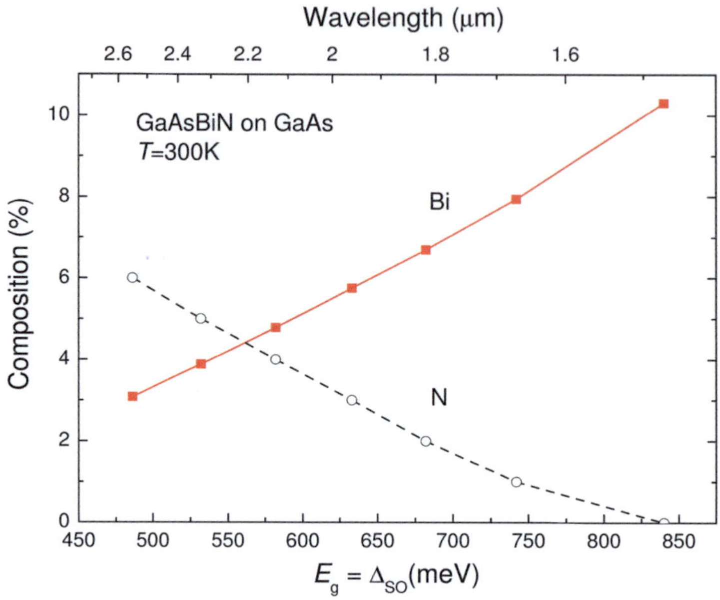

Fig. 2.5 Bi% required for achieving $\Delta_{SO}=E_g$ at various N% as a function of E_g (=Δ_{SO}) in GaAsBiN on GaAs

materials with large spin-orbit splitting. In contrast, nitrogen has little influence on the spin-orbit splitting due to its small size. Compared with the band gap and spin-orbit splitting energies of GaAsBiN, one can identify the composition region for which $\Delta_{SO}>E_g$ as indicated by the shaded region in Fig. 2.3.

In Fig. 2.4, arrows mark the crossing points where $\Delta_{SO}=E_g$ at various Bi and N compositions. Since the value of Δ_{SO} is primarily determined by the Bi fraction whilst the N fraction has little influence on Δ_{SO}, a higher Bi fraction is required to achieve $\Delta_{SO}>E_g$ for a larger band gap. The Bi fraction required for achieving $\Delta_{SO}=E_g$ at various N fractions is plotted as a function of E_g (=Δ_{SO}) in Fig. 2.5. For GaAsBiN/GaAs with Bi up to 12 % and N up to 6 %, which are within acceptable strain for photonic devices, the optimum band structure with $\Delta_{SO}>E_g$ for CHSH Auger and IVBA suppression is achievable for the energy range of ~450–850 meV (~1.5–2.7 μm). As bismuth-containing alloys offer a large Δ_{SO}, incorporation of bismuth offers the potential of narrow gap materials with a large spin-orbit splitting. It is evident that the optimum band structure (shaded region as shown in Figs. 2.3 and 2.4) for CHSH Auger and IVBA suppression ($\Delta_{SO}>E_g$) is possible over a wide range of band gaps raising the additional possibility of efficient GaAs-based lasers and LEDs operating in the mid-infrared.

Figure 2.6 displays the predicted band offsets of the conduction band (ΔE_C), heavy-hole (ΔE_{hh}) and light-hole (ΔE_{lh}) of GaAsBiN/GaAs as a function of Bi at various N compositions. Maintaining large band offsets is important to suppress electron or hole leakage. It can be seen that a type-I alignment of GaAsBi/GaAs occurs even in the nitrogen-free case due to the downward shift of the conduction

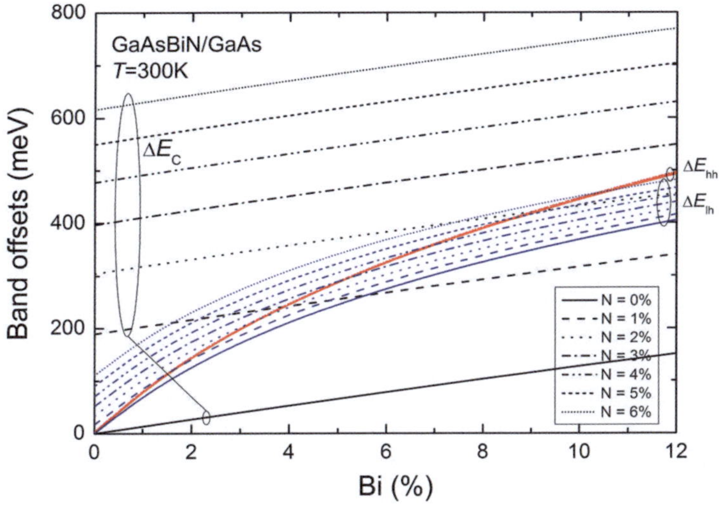

Fig. 2.6 Predicted band offsets of the conduction band, heavy-hole and light-hole bands as function of Bi at various N compositions in GaAsBiN on GaAs

band edge with increasing bismuth content [2, 37]. For GaAsBi/GaAs ΔE_C is smaller than both ΔE_{hh} and ΔE_{lh}. The incorporation of nitrogen in GaAs has been shown to lead to conduction BAC, creating a large ΔE_c and hence a strong band gap reduction [31, 32].

With the addition of N above 2 %, GaAsBiN/GaAs provides ΔE_C greater than ΔE_{hh} and ΔE_{lh}. The maximum value for the band offsets can reach ~0.77 eV (ΔE_C), ~0.50 eV (ΔE_{hh}) and ~0.48 eV (ΔE_{lh}) for Bi up to 12 % and N up to 6 %. A large band offset provides improved carrier confinement, which suppresses carrier leakage occurring due to thermal ionization of carriers from the active region. This should result in an improved high-temperature performance of light emitting devices. The incorporation of bismuth and nitrogen provides flexibility to design GaAsBiN/GaAs structures with strong control of ΔE_C and ΔE_V. This is also of value to detectors or photovoltaics where the carrier extraction efficiency is dependent upon the band offsets.

It should be noticed that the increase of ΔE_{hh} arises primarily from the increase of the bismuth composition, which is due to the upward shift of the valence band edge by adding bismuth atoms into GaAs. N has a small effect on ΔE_{hh} where we assume that this is entirely due to the effect of strain. The incorporation of nitrogen also slightly increases ΔE_{lh} due to the tensile strain-induced shift of the VBM relative to that of GaAs. The large valence band offset provided by Bi is also expected to suppress hole leakage effects, which can limit the maximum operating temperature of conventional GaInAsSb/GaSb mid-infrared lasers, particularly towards longer wavelengths [40]. GaAsBiN/GaAs therefore provides more flexibility in the optimization of laser structures with the desired conduction and valence band offsets to suppress leakage effects with the added advantage of growth on GaAs.

The device benefits may be achieved by incorporating nitrogen into GaAsBi to form the quaternary alloy GaAsBiN. Dilute nitrides such as Ga(In)AsN/GaAs have been widely explored for their potential applications in telecommunications devices such as VCSELs operating at 1.3 and 1.55 μm. Combining bismuth and nitrogen potentially offers increased scope for band structure design; this includes (a) significant band gap control, since both Bi and N lead to a large reduction in band gap; (b) strain control, since Bi and N have opposite effects on the strain in GaAsNBi and can together provide lattice matching to GaAs; (c) a large spin-orbit splitting due to bismuth, as described above and (d) control of both the conduction (ΔE_C) and valence ($\Delta E_{hh/lh}$) band offsets since N mainly influences the conduction band whilst Bi strongly influences the valence band. It is therefore possible to design structures by using GaAsBiN with $\Delta_{SO} > E_g$, to fully suppress CHSH-type Auger processes and also IVBA whilst providing high and controllable conduction and valence band offsets. Finally, since N and Bi primarily influence the conduction and valence band edges, respectively, type-II super-lattices based upon GaAsBi/GaAsN may open up still further possibilities in device design, particularly for longer wavelength detectors.

To summarize this section, the band parameters, such as the band gap, band offsets, spin-orbit splitting energy and strain, of GaAsBiN alloys on GaAs have been presented. It is shown that GaAsBiN alloys have the potential to cover a wide spectral range from near- to mid-infrared with flexible control of the band offsets and spin-orbit splitting. Furthermore, by using GaAsBiN with a spin-orbit splitting energy larger than the band gap and large band offsets, one may design potential Auger- and leakage-free devices for long-wavelength device applications. The next section considers experimental spectroscopic results for GaAsBi/GaAs structures.

2.3 Band Structure of GaAsBi/GaAs Layers Studied Using Photo-modulated Spectroscopy

In this section, the results of optical studies of $GaBi_xAs_{1-x}$ with bismuth concentrations of up to $x = 10.4$ % grown compressively strained on a GaAs substrate are discussed. In addition to a bowing of the band gap and spin-orbit splitting energies, a crossover in energy between the CB-to-HH^+ transition and the HH^+-to-SO^+ transition is obtained. The investigated samples are $GaBi_xAs_{1-x}$ epilayers, grown on undoped (100) GaAs by molecular beam epitaxy with bismuth concentrations of $x = 2.3$, 4.5, 8.5 and 10.4 %. The samples are all uncapped with the exception of the 2.3 % sample which is capped with 300 nm of GaAs. Further details about growth can be found in the paper of Lu et al. [41, 42]. The thicknesses of GaAsBi layers (30–40 nm) are all below the estimated critical thicknesses [21] and so are fully strained (pseudomorphic) to the GaAs substrate, as confirmed by earlier X-ray diffraction (XRD) studies of Bi concentrations up to 10 % [2, 21].

Figure 2.7 presents the result of room temperature photo-modulated reflectance (PR) spectroscopy on four $GaAs_{1-x}Bi_x$ (2.3 % $\leq x \leq$ 10.4 %) samples in the region of the fundamental band gap of GaAsBi. The figure shows the three PR

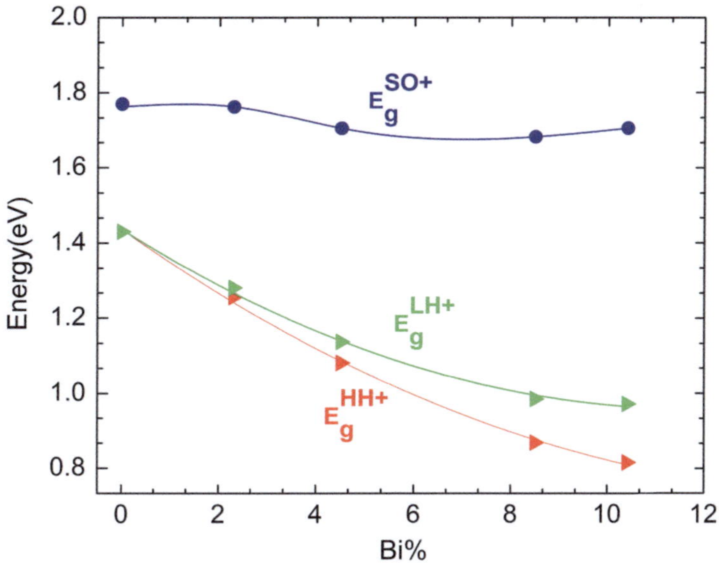

Fig. 2.7 Transition energy values $E_g^{HH^+}$, $E_g^{LH^+}$ and $E_g^{SO^+}$ obtained by fitting the room temperature experimental PR spectra [2]. The *curves* are only guides to the eye. The fitting uncertainties are within the symbol size

transition energies: $E_g^{SO^+}$, $E_g^{HH^+}$ and $E_g^{LH^+}$. Such a composition dependence of the interband transitions has been attributed to the BAC interaction of the Bi level with the VB of GaAs [4]. The $E_g^{HH^+}$ and $E_g^{LH^+}$ PR features are due to the strain-induced VB splitting in this alloy due to the lattice mismatch to GaAs [2].

By using the results of Fig. 2.7 one can plot the splittings $E_g^{SO^+} - E_g^{HH^+}$ and $E_g^{SO^+} - E_g^{LH^+}$ which are equivalent to the transition energy values from the SO^+ to HH^+ and LH^+ valence band edges, respectively. Figure 2.8 shows that $E_g^{SO^+} - E_g^{HH^+}$, the splitting between the SO^+ and the top of valence band (which is the heavy-hole in this case), becomes equal to the observed HH^+ transition energy $E_g^{HH^+}$ at a bismuth concentration of ~9.0 ± 0.2 % in these strained samples. This result has importance for device applications due to the possibility of reducing the CHSH Auger losses involving hole excitation into the spin-orbit band at higher bismuth concentrations.

Based on this data on the known strain in the GaAsBi layers on GaAs, the unstrained band gap energy for free-standing GaAsBi may be determined as shown in Fig. 2.9. From this data it is predicted that the energy crossover $\Delta_{SO} = E_g$ occurs at bismuth concentrations of 10.5 ± 0.2 % for such free-standing GaAsBi in good agreement with theory [37].

The relatively small change in the E_g^{SO}-transition in Fig. 2.7 indicates that the CB edge must also be moving down in energy with increasing Bi content. This, combined with the reduction in band gap and upward movement of the valence band edge owing to BAC, suggests that a type-I conduction/valence band alignment is

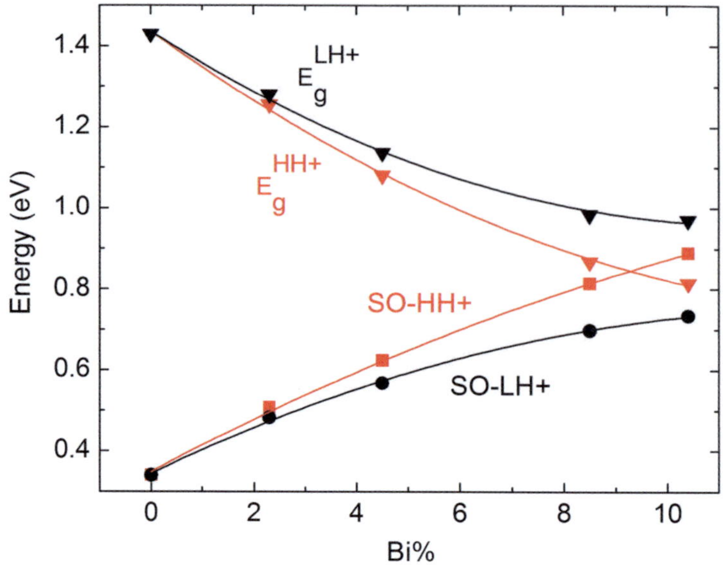

Fig. 2.8 Fitted transition energy values $E_g^{HH^+}$, $E_g^{LH^+}$ and the differences between these and the $E_g^{SO^+}$ results in Fig. 2.7, as a function of bismuth concentration. Fitting uncertainties are within the symbol size

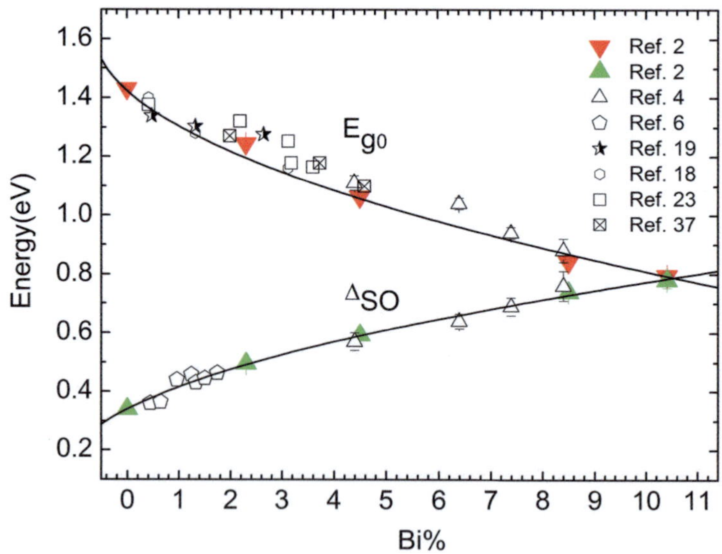

Fig. 2.9 The *filled triangles* show the composition dependence of the room temperature spin-orbit splitting energy Δ_{SO} and unstrained band gap energy E_{go} for free-standing GaAsBi. The *curves* are guides to the eye. The other symbols show the results of other authors [4, 6, 21, 22, 26, 52]

present for GaAsBi on GaAs. These conclusions are consistent with theoretical tight binding studies of the band structure of free-standing bulk GaAsBi which predict that the CB edge drops about five times faster than the SO edge with bismuth concentration, causing a net red-shift in the CB–SO transition energy of ~22 meV/%Bi [37].

To summarize, in this section the electronic band structure of the bismide alloy $GaBi_xAs_{1-x}$ in the energy regions of both the fundamental band gap and spin-orbit split-off feature was discussed. Room temperature photo-reflectance measurements on compressively strained $GaAs_{1-x}Bi_x$ layers on GaAs substrates, with Bi concentrations between 2.3 and 10.4 %, show that the energy of the strained heavy-hole band gap $E_g^{HH^+}$ is resonant with the difference in energy between the HH^+ and SO^+ band edges at a bismuth concentration of ~9.0 ± 0.2 %. Furthermore, this spectroscopic data suggests that the band alignment for GaAsBi grown on GaAs will be type-I, in line with theoretical predictions.

2.4 InGaAsBi/InP Structures

In this section the band parameters such as band gap, spin-orbit splitting energy, band offsets and strain of InGaAsBi on InP are discussed based on the recent theoretical work of Jin and Sweeney [43] and experimental data [29]. InGaAsBi is promising for near- and mid-infrared photonic devices operating in the wavelength range of 1.5–4 μm. A significant reduction of band gap due to the incorporation of Bi atoms in InGaAs can again be described using the relatively simple VBAC theory [1, 4, 31, 32]. It treats the interaction of the two types of states, i.e., the extended VBM and the localized Bi state, as a perturbation. Since the Bi level lies below the VBM of GaAs (0.4 eV relative to the VBM), the interaction of the localized Bi level and the extended valence band states leads to an anti-crossing of these two states, which is similar to that of GaAsN [30, 37]. In the approximation of this model the band gap of unstrained InGaAsBi can be expressed by [44]

$$E_g^{unstr}(\text{InGaAsBi}) = E_g(\text{InGaAs}) - \frac{\Delta E_{\text{VBM-Bi}}}{2}\left[\sqrt{1 + 4\left(\frac{C_{\text{Bi}}}{\Delta E_{\text{VBM-Bi}}}\right)^2 x} - 1\right], \quad (2.3)$$

where E_g(InGaAs) is the band gap of InGaAs, $\Delta E_{\text{VBM-Bi}}$ is the energy separation between the localized resonant bismuth level and the VBM of InGaAs (initially assumed equal to that for GaAs), C_{Bi} is the coupling parameter between the two type of states and x is Bi composition. As a first approximation, the coupling parameter C_{Bi} is taken as a constant. The valence band alignment between GaAs and InGaAs, however, varies with increasing In composition [45]. As shown in Fig. 2.10, the incorporation of In shifts the VBM upwards relative to that of GaAs. This leads to the increases of the separation between the VBM of InGaAs and Bi level and

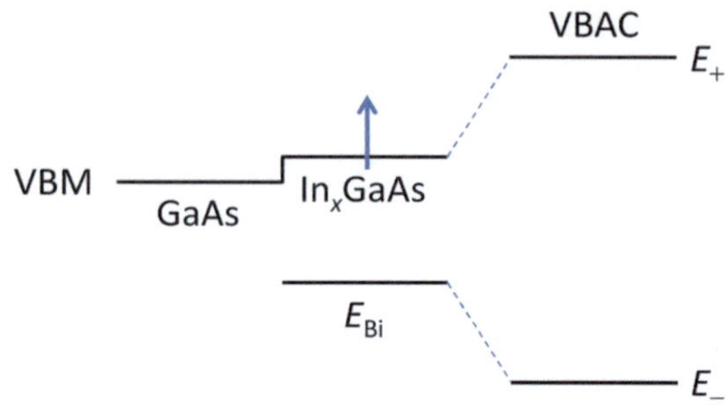

Fig. 2.10 The incorporation of indium in GaAs will shift the valence band maximum (VBM) upwards relative to that of GaAs, leading to a larger separation between the VBM of InGaAs and Bi level (E_{Bi}) and therefore a weaker anti-crossing interaction

therefore a weaker anti-crossing interaction, i.e., the reduction of E_g in InGaAsBi is slightly slower than that of GaAsBi with increasing Bi fraction.

The relative alignment of the band edges between InGaAsBi and GaAs barrier is calculated using the model-solid theory [35] as described in (2.2). $E_{v,av}$ of GaBi and InBi is estimated from the dependence of the valence band edge energies vs. lattice constant in Ga- and In-containing group III–V binaries, which gives $E_{v,av}$(GaBi)~−6.28 eV and $E_{v,av}$(InBi)~−6.15 eV. δE_{hh} (δE_{lh}) is the shift of the strained bulk valence band edge relative to the unstrained one, which leads to the valence band splitting. Both bismuth and strain-induced shifts of the conduction band minimum (determined using E_g from (2.3) and the calculated position of the VBM) and VBM are also taken into account [45]. In addition a super-linear increase of Δ_{SO} in GaAsBi with increasing bismuth composition observed recently is assumed in these calculations [45]. All band parameters of InGaAsBi quaternary alloys were calculated using the formulae and recommended band structure parameters including the so-called bowing parameters of the related binary and ternary alloys in the literature [33, 34].

Figure 2.11 displays the predicted band gap of InGaAsBi as a function of Bi composition at various In≤0.6 at 300 K. The experimental data of $In_{0.53}Ga_{0.47}As_{1-x}B$ i_x measured using different techniques including photo-reflectance (PR), photoluminescence (PL) and absorption at room temperature [29] are also shown in Fig. 2.11, which is consistent with theoretical predictions. It is shown that InGaAsBi covers the energy region from 0.3 to 0.8 eV, corresponding to the wavelength from 1.5 to 4 μm, suggesting their potential applications for near- and mid-infrared optoelectronics. In addition to the strong band gap reduction caused by the incorporation of Bi, Fig. 2.11 clearly shows how the addition of bismuth to InGaAsBi causes a significant increase in Δ_{SO}. The strong increase of Δ_{SO} with increasing Bi composition is attributed to the large atomic mass of bismuth which increases the interaction between the electron spin and orbital angular momentum.

Earlier work [46] has shown that the CHSH (hot-hole producing) Auger process and IVBA, both involving transitions to the spin split-off (SO) band, dominate the threshold current and temperature sensitivity of InP- and GaAs-based lasers in the near-infrared. These processes also significantly increase towards longer wavelengths in the near-infrared. However, moving into the mid-infrared, GaInAsSb/GaSb-based lasers operating between 2 and 3 μm have lower threshold currents and

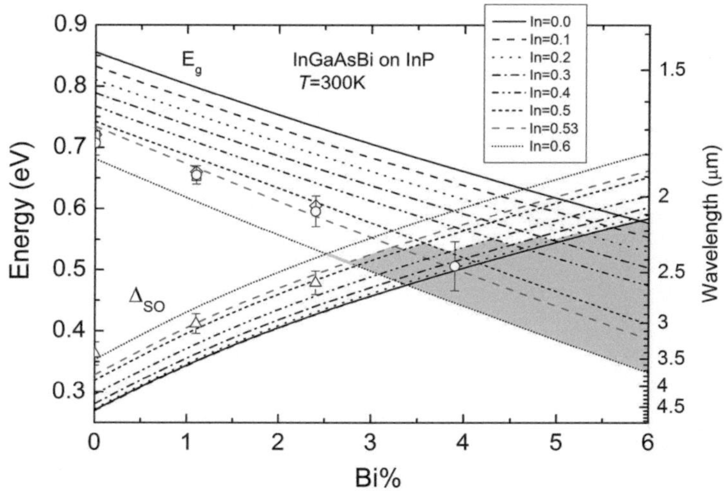

Fig. 2.11 Predicted band gap of InGaAsBi on InP at 300 K. This quaternary offers wide scope for band structure engineering from 0.4 to 0.8 eV (1.5–4 eV). The spin-orbit splitting energy of InGaAsBi alloys is also included. The *shaded* region indicates where $\Delta_{SO} \geq E_g$ for In composition of up to 60 %

a reduced sensitivity to temperature [45]. This is due to the large spin-orbit splitting energy Δ_{SO} in antimonide-alloys such that Δ_{SO} is larger than the band gap E_g. This situation inhibits IVBA and CHSH-type Auger transitions via the spin-orbit band since energy and momentum can no longer be conserved, leading to improved laser performance in the mid-infrared. Thus, the incorporation of bismuth in InGaAsBi alloys offers the potential for narrow gap materials with large spin-orbit splitting. In contrast, indium has relatively smaller influence on the spin-orbit splitting than that of Bi atoms. The composition region for which $\Delta_{SO} \geq E_g$ for indium content of up to 60 % is indicated by the shaded region in Fig. 2.11. From this figure, the Bi fraction required for achieving $\Delta_{SO} = E_g$ can be determined, which is shown in Fig. 2.12. It has shown that the shortest wavelength for achieving $\Delta_{SO} = E_g$ using InGaAsBi is ~2.15 μm (~0.58 eV). This implies that using higher In and Bi fractions InGaAsBi can provide larger $\Delta_{SO} > E_g$ in the mid-infrared region (>~2.15 μm), offering InGaAsBi as a novel material system for mid-infrared light emitting devices.

Figure 2.13 displays the strain of InGaAsBi on an InP substrate. Since $In_{0.53}Ga_{0.47}As$ is lattice-matched to InP, InGaAs with In < 0.53 is under tensile strain whilst In > 0.53 causes compressive strain. Using an In fraction of 40–60 %, the strain of InGaAsBi with Bi of up to 6 % on InP is within ±1 %. Hence InGaAsBi can be flexibly designed under compressive or tensile strain on InP depending on the In and Bi compositions, which could also be used for producing polarization-insensitive semiconductor optical amplifiers by designing a series of layers in compressive and tensile strain [38]. These data again clearly show the potential of such alloys to operate in a wide spectral range whilst remaining lattice-matched to InP.

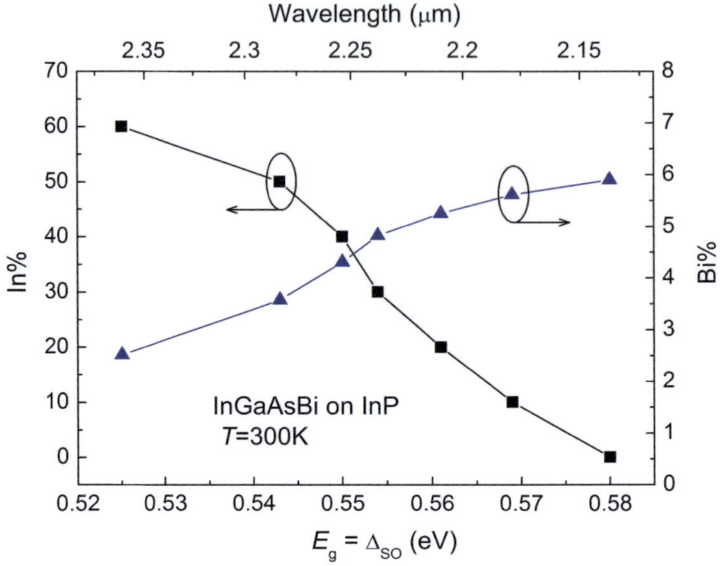

Fig. 2.12 Bi% required for achieving $\Delta_{SO} = E_g$ at various In% as a function of E_g ($= \Delta_{SO}$) in InGaAsBi on InP

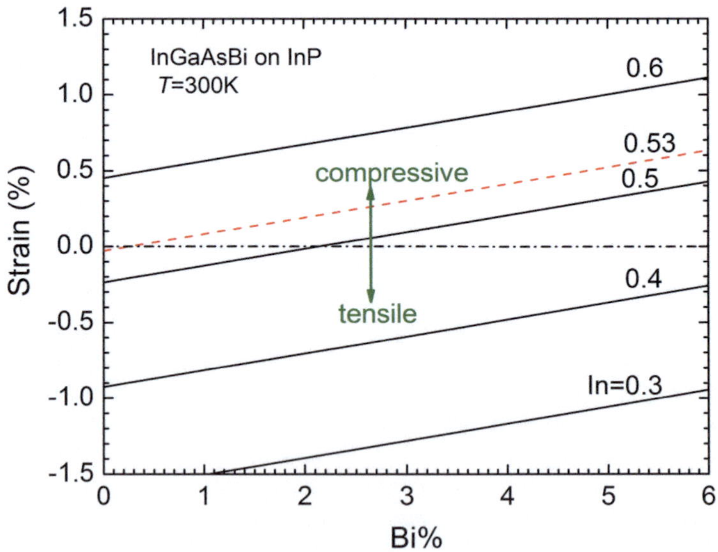

Fig. 2.13 Strain of InGaAsBi on InP at various In% as a function of Bi% at 300 K

Maintaining large band offsets is important to offer both good electron and hole confinement for suppressing electron and hole leakage in device designs. Figure 2.14 displays the predicted band offsets of the conduction band (ΔE_C), heavy-hole (ΔE_{hh}) and light-hole (ΔE_{lh}) of InGaAsBi/InP as a function of Bi at In compositions of In = 40–60 % at 300 K. The inset shows the band offsets of $In_{0.53}Ga_{0.47}AsBi$/InP.

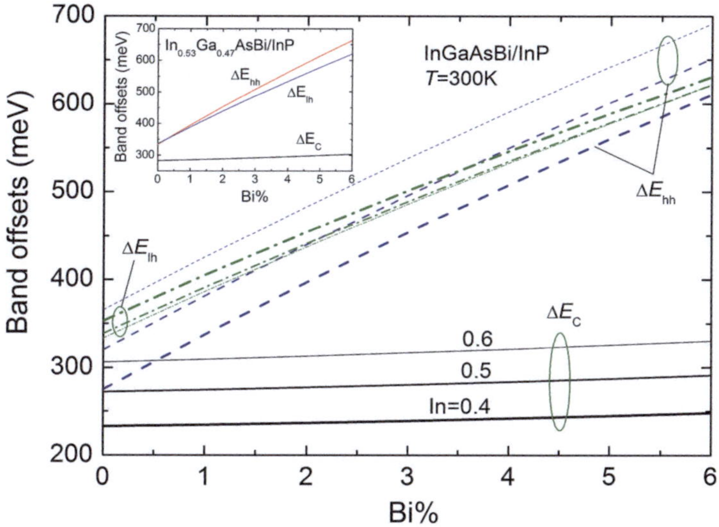

Fig. 2.14 Predicted band offsets of the conduction band, heavy-hole and light-hole bands as function of Bi at In compositions of 40, 50 and 60 % in InGaAsBi on InP at 300 K

It can be seen that with increasing bismuth composition the valence band offsets (ΔE_{hh} and ΔE_{lh}) rapidly increase, which is due to the strong upward shift of the top of the valence band with increasing Bi fraction. With increasing bismuth composition the calculated shift of the conduction band edge is smaller than that of the free-standing InGaAsBi predicted by theory [37], which can be attributed to the compressive strain-induced effects of the conduction band edge due to the addition of bismuth atoms. For InGaAsBi/InP, ΔE_C is smaller than both ΔE_{hh} and ΔE_{lh}. The large valence band offset provided by Bi is also expected to suppress hole leakage effects, which currently limit the maximum operating temperature of conventional GaInAsSb/GaSb mid-infrared lasers, particularly towards longer wavelengths [40]. It should be noticed that the increase of ΔE_C is strongly influenced by the indium composition. Thus, in order to achieve a higher conduction band offset, larger In compositions need to be used. For example, in the indium-free case (highly tensile strained) it is estimated that $\Delta E_C = 75$ meV, $\Delta E_{hh} = 411$ meV and $\Delta E_{lh} = 702$ meV for $GaAs_{0.94}Bi_{0.06}$/InP. For $In_{0.53}Ga_{0.47}As_{0.94}Bi_{0.06}$/InP, however, $\Delta E_C = 303$ meV, $\Delta E_{hh} = 660$ meV and $\Delta E_{lh} = 618$ meV, which shows both good electron and hole confinement energies.

A large band offset provides improved carrier confinement, which suppresses carrier leakage occurring due to thermal ionization of carriers from the active region. This should result in an improved high-temperature performance of light emitting devices. The incorporation of bismuth and indium provides flexibility to design InGaAsBi/InP structures with strong control of ΔE_C and ΔE_V. This is also of value to devices such as detectors or photovoltaics. The device benefits may be achieved by incorporating indium into GaAsBi to form the quaternary alloy InGaAsBi on InP. InGaAs/InP has been widely applied in telecommunications devices operating in near-infrared. Similar to the GaAsBiN/GaAs system, InGaAsBi/InP potentially offers increased scope for band structure design including (a) significant band gap control in the mid-infrared region, (b) strain control, (c) a large spin-orbit splitting due to bismuth, as described above and (d) control of both the conduction (ΔE_C) and valence (ΔE_{hh}) band offsets. It is therefore possible to design structures by using InGaAsBi with $\Delta_{SO} > E_g$, to fully suppress CHSH-type Auger processes and also IVBA whilst providing high and controllable conduction and valence band offsets to manage carrier capture and escape.

In summary, the band parameters, such as the band gap, band offsets, spin-orbit splitting energy and strain, of InGaAsBi alloys on InP have been presented. It is shown that InGaAsBi alloys are promising for near- and mid-infrared photonic devices with flexible control of the band offsets and spin-orbit splitting. Furthermore, by using InGaAsBi with a spin-orbit splitting energy larger than the band gap together with large band offsets, one may design potential Auger and leakage-free devices for long-wavelength device applications compatible with standard InP substrates.

2.5 GaAsBi/GaAs Device Studies

There has been relatively little research to date on the carrier recombination and temperature-dependent processes occurring in GaAsBi-based devices. The determination of the alignment of conduction and valence bands at the $GaAs_{1-x}Bi_x$/GaAs interface is critical for theoretical modelling and design optimization. Type-I band alignment provides an increased optical gain due to better electron–hole overlap. On the other hand, type-II band alignment could potentially allow access to longer wavelengths [47]. This section discusses carrier recombination processes of $GaAs_{1-x}Bi_x$/GaAs LEDs to aid in the design and optimization of device structures. Using high hydrostatic pressure and low temperature techniques, the processes limiting device performance are probed [27]. The pressure and current dependence measurements also provide evidence for the type-I band alignment at the $GaAs_{1-x}Bi_x$/GaAs interface.

The devices in this study were grown in a VG-V80H MBE system on n-doped (001) GaAs substrates. The active region consists of a 50 nm $GaAs_{0.986}Bi_{0.014}$ layer between two 25 nm GaAs spacer layers. The entire active region is sandwiched between a 1,000 nm p-doped and a 1,000 nm n-doped GaAs buffer layers. Further details of the growth and processing of similar devices can be found in ref [48].

The emission wavelength of the device is measured to be ~936 nm at 260 K. Figure 2.15 shows the temperature dependence of the EL peak, which shifts with temperature at the rate of -0.30 ± 0.01 meV/K (calculated by a linear fit) corresponding to a low wavelength change of 0.19 ± 0.01 nm/K over the temperature range 160–260 K. The temperature dependence of the fundamental energy gap of GaAs, E_g(GaAs), is obtained from the Varshni equation (given in Fig. 2.15) [49]

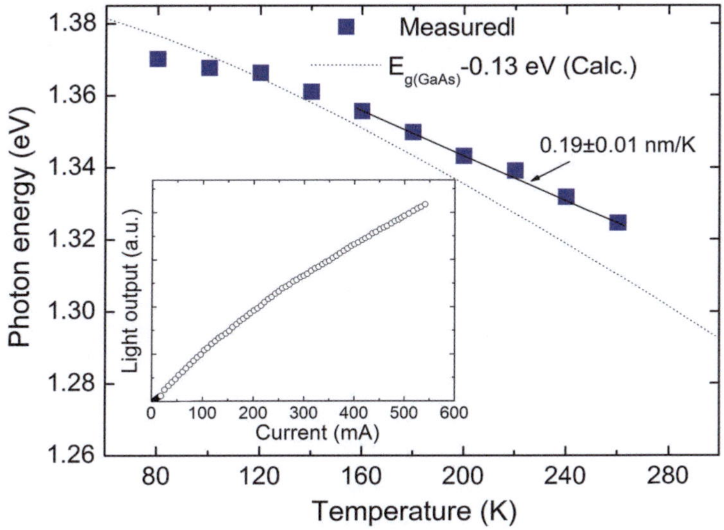

Fig. 2.15 Temperature dependence of the measured EL emission peak. *L–I* characteristics at 80 K (*inset*)

$$E_g(T) = E_0 - \alpha T^2 / (T + \beta), \tag{2.4}$$

where E_0 (1.519 eV [33, 34]) is the band gap at $T=0$ K, and α (0.54 meV/K [33, 34]) and β (204 K [33, 34]) are the Varshni parameters. The temperature dependence of the band gap of GaAs determined by the Varshni parameters is given in Fig. 2.15 (vertically offset for comparison purposes). It shows that the EL peak of the GaAsBi devices shifts less with increasing temperature than the band gap of GaAs. This behaviour is consistent with previous photoluminescence measurements [50] and may be attributed to BAC and/or localization effects [37].

Figure 2.16 shows the change in efficiency as a function of temperature for a forward bias current density of ~370 A/cm^2. A rapid decrease in efficiency with increasing temperature in these devices implies that some non-radiative loss mechanism(s) is (are) activated at higher temperature. The light–current characteristics show a sub-linear behaviour (inset of Fig. 2.15) even at temperatures as low as 80 K which indicates a presence of a loss mechanism with a stronger carrier density (n) dependence than the radiative current [51]. Since carrier leakage has an approximately exponential dependence on n, this may suggest that carrier leakage plays an important role in these devices, if, as expected, Auger recombination is negligible at these short wavelengths [27].

To further probe the recombination mechanisms, high pressure techniques were utilized. The application of hydrostatic pressure mainly affects the conduction band (CB) causing an increase in the direct band gap of III–V semiconductors and is, therefore, an ideal method to investigate the important band gap-dependent

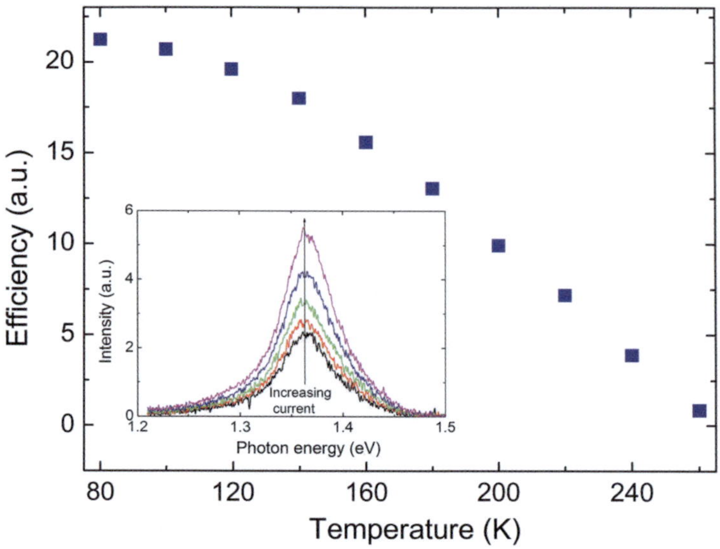

Fig. 2.16 Emission efficiency as a function of temperature. Current dependence of EL emission for current 20–100 mA with 20 mA step (*inset*)

non-radiative processes. Figure 2.17 shows the EL spectra (on a log scale) as a function of pressure for a forward bias current density of 124 A/cm^2 at 80 K. The decrease in overall EL emission intensity together with increased (relative to the GaAsBi) emission from the GaAs layers with increasing pressure (see Fig. 2.17) suggests the presence of a carrier leakage path in these devices, where the leakage activation energy (ΔE) decreases with increasing pressure (since dE/dP(GaAsBi) > dE/dP(GaAs), as shown in the inset of Fig. 2.17). The pressure (P) dependence of the leakage current density, J_{leak}, can be written simply as [52]

$$J_{\text{leak}}(P) = J_{\text{leak}}(0)\exp\left(-\frac{\mathrm{d}\Delta E}{\mathrm{d}P}\frac{P}{k_B T}\right), \tag{2.5}$$

where k_B is Boltzmann's constant, T is the absolute temperature and $J_{leak}(0)$ is the leakage current at atmospheric pressure. The pressure dependence of the higher energy emission peak corresponding to GaAs was obtained from a Gaussian fit to the measured EL spectra at 80 K (an example of such a fit is shown in the inset of Fig. 2.18). The integrated intensity (normalized at 0 kbar) of the GaAs emission as a function of pressure is shown in Fig. 2.18.

The pressure coefficient for the $GaAs_{0.986}Bi_{0.014}$ Γ minimum (taken from the EL spectra and shown in inset of Fig. 2.17) is measured to be $dE\Gamma$(GaAsBi)/dP = 11.8 ± 0.3 meV/kbar and the pressure coefficient of the Γ minimum of GaAs is well known ($dE\Gamma$(GaAs)/dP = 10.7 meV/kbar) [33, 34], from which $d\Delta E/dP = -1.1$ meV/kbar. Substituting this value into (2.5) yields the solid line shown in Fig. 2.18. As one can see, this provides an excellent agreement with the experimental data confirming that carrier leakage into GaAs barriers is occurring in these devices [27].

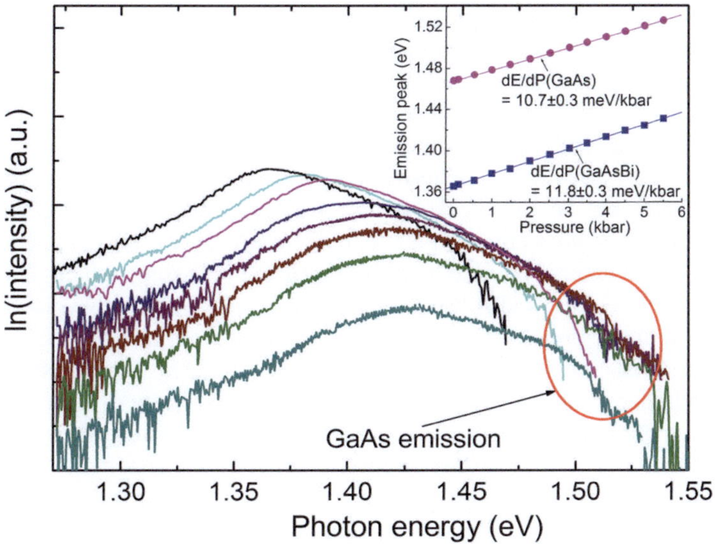

Fig. 2.17 Pressure dependence of the natural log of measured EL emission spectra for pressure 0–6 kbar at 80 K. It shows a relative increase in GaAs emission with increasing pressure compared to GaAsBi emission. GaAsBi (measured) and GaAs (fitted) emission peaks as a function of applied pressure (*inset*)

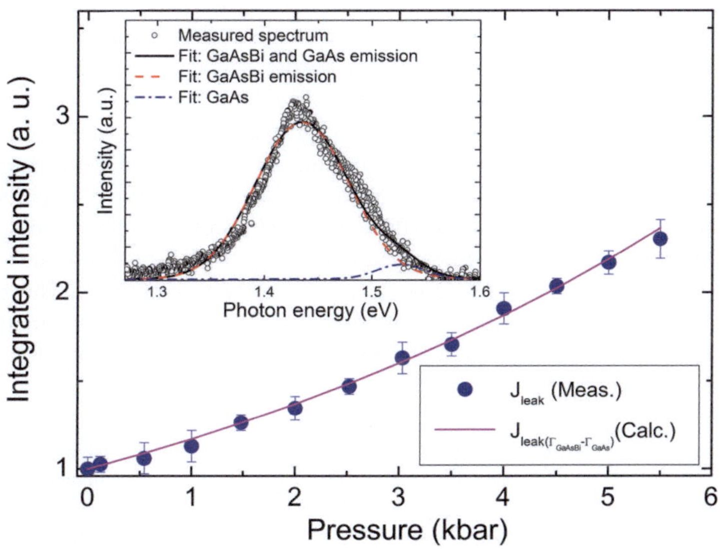

Fig. 2.18 The integrated intensity (normalized at 0 kbar) of fitted GaAs emission as a function of pressure. An example fit is shown in the *inset*

The GaAsBi emission energy has a stronger pressure dependence than that for GaAs but a lower temperature dependence of the band gap. Localization near the VB edge may play a role in the lower temperature dependence [23]. Since pressure mainly affects the CB, it is less influenced by Bi inhomogeneities. There is no evidence of a blue shift of the electroluminescence peak with increasing current injection (inset of Fig. 2.16) in this material at atmospheric temperature and pressure as would typically be the case for a type-II material. With increasing pressure, a relative increase in GaAs emission compared to GaAsBi emission suggests that the band offset for electrons reduces with increasing pressure, which leads to an increased carrier escape and subsequent radiative and non-radiative recombination outside the active region. Hence, it may be concluded that the higher pressure coefficient of GaAsBi compared to GaAs coupled with an increase in emission from the GaAs cladding region with increasing pressure is consistent with a type-I band alignment at the $GaAs_{1-x}Bi_x$/GaAs interface. This is consistent with the spectroscopic and theoretical calculations discussed earlier in this chapter.

2.6 Conclusions

This chapter has explored the potential of bismide alloys for applications in near- and mid-infrared photonic devices. It has been shown that alloys such as GaAsBi(N) offer the potential to obtain materials with band gaps covering the near- and mid-infrared ranges whilst also providing the beneficial condition that $\Delta_{SO} > E_g$ which would allow for the suppression of the major Auger and optical loss processes which plague lasers and LEDs at these wavelengths. Furthermore, owing to the wide degree of control of the band offsets, carrier leakage can effectively be suppressed. Growth of such alloys close to lattice match on GaAs substrates also opens up many opportunities in developing mid-infrared photonics on standard and inexpensive GaAs substrates, with significant potential for vertical geometry devices such as VCSELs and related devices. Similarly, InGaAsBi provides potential for close to lattice-matched mid-infrared devices on InP substrates. This again opens up the possibility to use standard InP-based telecoms laser processes to develop new devices in the mid-infrared. There is clearly significant potential for the bismides in efficient and temperature-stable photonic devices, and, with developments in the growth quality of such materials, devices such as these will hopefully one day become commonplace.

Acknowledgements We are grateful to all of the collaborative partners that we have worked with on bismide alloys, and, in particular, the groups of Prof. Tom Tiedje at the University of Victoria, Canada, and Prof. Joshua Zide at the University of Delaware, USA, for the supply of samples. We are very pleased to acknowledge financial support for this activity as follows: Engineering and Physical Sciences Research Council (EPSRC) grants EP/H005587/1 and EP/G064725/1, European Union Seventh Framework Programme (BIANCHO; FP7-257974), the Technology Strategy Board "ETOE2" project, the Islamic University of Bahawalpur FDP and the Kwan Trust for funding a studentship for Z.B.; UTM and MOHE GUP grant 01H55 for T.J.C.H.

References

1. Alberi, K., Wu, J., Walukiewicz, W., Yu, K.M., Dubon, O.D., Watkins, S.P., Wang, C.X., Liu, X., Cho, Y.-J., Furdyna, J.: Valence-band anticrossing in mismatched III–V semiconductor alloys. Phys. Rev. B **75**, 045203 (2007)
2. Batool, Z., Hild, K., Hosea, T.J.C., Lu, X.F., Tiedje, T., Sweeney, S.J.: The electronic band structure of GaBiAs/GaAs layers: Influence of strain and band anti-crossing. J. Appl. Phys. **111**, 113108 (2012)
3. Petropoulos, J.P., Zhong, Y., Zide, J.M.O.: Optical and electrical characterization of InGaBiAs for use as a mid-infrared optoelectronic material. Appl. Phys. Lett. **99**, 031110 (2011)
4. Alberi, K., Dubon, O.D., Walukiewicz, W., Yu, K.M., Bertulis, K., Krotkus, A.: Valence band anticrossing in $GaBi_xAs_{1-x}$. Appl. Phys. Lett. **91**, 051909 (2007)
5. Zhang, Y., Mascarenhas, A., Wang, L.W.: Similar and dissimilar aspects of III–V semiconductors containing Bi versus N. Phys. Rev. B **71**, 155201 (2005)
6. Fluegel, B., Francoeur, S., Mascarenhas, A., Tixier, S., Young, E.C., Tiedje, T.: Giant spin-orbit bowing in $GaAs_{1-x}Bi_x$. Phys. Rev. Lett. **97**, 067205 (2006)
7. Sweeney, S.J.: Bismide-alloys for higher efficiency infrared semiconductor lasers. In: Conference Digest—IEEE International Semiconductor Laser Conference, Kyoto, 111 (2010)
8. Phillips, A.F., Sweeney, S.J., Adams, A.R., Thijs, P.J.A.: The temperature dependence 1.3- and 1.5-μm compressively strained InGaAs(P) MQW semiconductor lasers. IEEE J. Sel. Top. Quant. Electron. **5**(3), 401–412 (1999)
9. Higashi, T., Sweeney, S.J., Phillips, A.F., Adams, A.R., O'Reilly, E.P., Uchida, T., Fujii, T.: Experimental analysis of temperature dependence in 1.3-m AlGaInAs–InP strained MQW lasers. IEEE J. Sel. Top. Quant. Electron. **5**(3), 413–419 (1999)
10. Sweeney, S.J., Jin, S.R.: Bismide-nitride alloys for mid-infrared photonics. In: 10th International Conference on Mid-Infrared Optoelectronics: Materials and Devices (MIOMD-X), Shanghai, China (2010)
11. Sweeney, S.J., Batool, Z., Hild, K., Jin, S.R., Hosea, T.J.C.: The potential role of bismide alloys in future photonic devices. In: 13th International Conference on Transparent Optical Networks, Mo.C6.3, Stockholm, Sweden, 26–30 June 2011
12. Loehr, J.P., Singh, J.: Effect of strain on CHSH Auger recombination in strained $In_{0.53+x}Ga_{0.47-x}$ As on InP. IEEE J. Quant. Electron. **29**, 2583 (1993)
13. Fehse, R., Jin, S., Sweeney, S.J., Adams, A.R., O'Reilly, E.P., Riechert, H., Illek, S., Egorov, A.Y.: Evidence for large monomolecular recombination contribution to threshold current in 1.3μm GaInNAs semiconductor lasers. Electron. Lett. **37**, 1518 (2001)
14. Marko, I.P., Adams, A.R., Sweeney, S.J., Mowbray, D.J., Skolnick, M.S., Liu, H.Y., Groom, K.M.: Recombination and loss mechanisms in low-threshold InAs-GaAs 1.3-μm quantum-dot lasers. IEEE J. Sel. Top. Quant. Electron. **11**, 1041 (2005)
15. Sayid, S.A., Marko, I.P., Sweeney, S.J., Barrios, P., Poole, P.J.: Efficiency limiting processes in 1.55μm InAs/InP-based quantum dots lasers. Appl. Phys. Lett. **97**, 161104 (2010)
16. Masse, N.F., Sweeney, S.J., Marko, I.P., Adams, A.R., Hatori, N., Sugawara, M.: Temperature dependence of the gain in *p*-doped and intrinsic 1.3μm InAs/GaAs quantum dot lasers. Appl. Phys. Lett. **89**, 191118 (2006)
17. O'Brien, K., Sweeney, S.J., Adams, A.R., Murdin, B.N., Salhi, A., Rouillard, Y., Joullie, A.: Recombination processes in midinfrared InGaAsSb diode lasers emitting at 2.37mm. Appl. Phys. Lett. **89**, 051104 (2006)
18. Sweeney, S.J.: Light-emitting semiconductor device. Patent WO 2010/149978 (2010)
19. Sweeney, S.J.: The potential of bismide alloys for efficient near- and mid-infrared semiconductor lasers. SPIE Photonics West 2010, paper [7616-11] (2010)
20. Cheetham, K.J., Krier, A. Marko, I.P., Aldukhayel, A., Sweeney, S.J.: Direct evidence for suppression of Auger recombination in GaInAsSbP/InAs mid-infrared light-emitting diodes. Appl. Phys. Lett. **99**, 141110 (2011)

21. Tixier, S., Adamcyk, M., Tiedje, T., Franceour, S., Mascarenhas, A., Wei, P., Schiettekatte, F.: Molecular beam epitaxy growth of $GaAs_{1-x}Bi_x$. Appl. Phys. Lett. **82**, 2245 (2003)
22. Yoshida, J., Kita, T., Wada, O., Oe, K.: Temperature dependence of $GaAs_{1-x}Bi_x$ band gap studied by photoreflectance spectroscopy. Jpn. J. Appl. Phys. **42**, 371 (2003)
23. Pettinari, G., Polimeni, A., Capizzi, M., Blokland, J.H., Christianen, P.C.M., Maan, J.C., Young, E.C., Tiedje, T.: Influence of bismuth incorporation on the valence and conduction band edges of $GaAs_{1-x}Bi_x$. Appl. Phys. Lett. **92**, 262105 (2008)
24. Cooke, D.G., Hegmann, F.A., Young, E.C., Tiedje, T.: Electron mobility in dilute GaAs bismide and nitride alloys measured by time-resolved terahertz spectroscopy. Appl. Phys. Lett. **89**, 122103 (2006)
25. Oe, K., Asai, H.: Proposal on a temperature-insensitive wavelength semiconductor laser. IEICE Trans. Electron. **E79-C**, 1751 (1996)
26. Francoeur, S., Seong, M.-J., Mascarenhas, A., Tixier, S., Adamcyk, M., Tiedje, T.: Band gap of $GaAs_{1-x}Bi_x$, $0<x<3.6\%$. Appl. Phys. Lett. **82**, 3874 (2003)
27. Hossain, N., Marko, I.P., Jin, S.R., Hild, K., Sweeney, S.J., Lewis, R.B., Beaton, D.A., Tiedje, T.: Recombination mechanisms and band alignment of $GaAs_{1-x}Bi_x$/GaAs light emitting diodes. Appl. Phys. Lett. **100**, 051105 (2012)
28. Devenson, J., Pačebutas, V., Butkutė, R., Baranov, A., Krotkus, A.: Structure and optical properties of InGaAsBi with up to 7% bismuth. Appl. Phys. Express **5**, 015503 (2012)
29. Marko, I.P., Batool, Z., Hild, K., Jin, S.R., Hossain, N., Hosea, T.J.C., Petropoulos, J.P., Zhong, Y., Dongmo, P.B., Zide, J.M.O., Sweeney, S.J.: Temperature and Bi-concentration dependence of the bandgap and spin-orbit splitting in InGaBiAs/InP semiconductors for mid-infrared applications. Appl. Phys. Lett. **101**, 221108 (2012)
30. Shan, W., WaluKiewiez, W., Ager, J.W., Haller, E.E., Geisz, J.F., Friedman, D.J., Olson, J.M., Kurtz, S.R.: Band anticrossing in GaInNAs alloys. Phys. Rev. Lett. **82**, 1221 (1999)
31. Tixier, S., Webster, S.E., Young, E.C., Tiedje, T., Francoeur, S., Mascarenhas, A., Wei, P., Schiettekatte, F.: Band gaps of the dilute quaternary alloys $GaN_xAs_{1-x-y}Bi_y$ and $Ga_{1-y}In_yN_xAs_{1-x}$. Appl. Phys. Lett. **86**, 112113 (2005)
32. Nacer, S., Aissat, A., Ferdjani, K.: Band gap and band offsets of GaNAsBi lattice matched to GaAs substrate. Opt. Quant. Electron. **40**, 677 (2008)
33. Vurgaftman, I., Meyer, J.R.: Band parameters for III–V compound semiconductors and their alloys. J. Appl. Phys. **89**, 5815 (2001)
34. Vurgaftman, I., Meyer, J.R.: Band parameters for nitrogen-containing semiconductors. J. Appl. Phys. **94**, 3675 (2003)
35. Lin, G., Lee, C.P.: Comparison of 1300nm quantum well lasers using different material systems. Opt. Quant. Electron. **34**, 1191 (2002)
36. Van de Walle, C.G.: Band lineups and deformation potentials in the model-solid theory. Phys. Rev. B **39**, 1871 (1989)
37. Usman, M., Broderick, C.A., Lindsay, A., O'Reilly, E.P.: Tight-binding analysis of the electronic structure of dilute bismide alloys of GaP and GaAs. Phys. Rev. B **84**, 245202 (2011)
38. Adams, A.R.: Strained-layer quantum-well lasers. IEEE J. Quant. Electron. **17**, 1364 (2011)
39. Carrier, P., Wei, S.H.: Calculated spin-orbit splitting of all diamondlike and zincblende semiconductors: Effects of $p_{1/2}$ local orbitals and chemical trends. Phys. Rev. B **70**, 035212 (2004)
40. Shterengas, L., Belenky, G.L., Kim, J.G., Martinelli, R.U.: Design of high-power room-temperature continuous-wave GaSb-based type-I quantum-well lasers with l>2.5mm. Semicond. Sci. Technol. **19**, 655 (2004)
41. Lu, X., Beaton, D.A., Lewis, R.B., Tiedje, T., Zhang, Y.: Composition dependence of photoluminescence of $GaAs_{1-x}Bi_x$ alloys. Appl. Phys. Lett. **95**, 041903 (2009)
42. Lu, X., Beaton, D.A., Lewis, R.B., Tiedje, T., Whitwick, M.B.: Effect of molecular beam epitaxy growth conditions on the Bi content of $GaAs_{1-x}Bi_x$. Appl. Phys. Lett. **92**, 192110 (2008)
43. Jin, S.R., Sweeney, S.J.: Unpublished
44. Sweeney, S.J., Jin, S.R.: Bismide-nitride alloys: Promising for efficient light emitting devices in the near- and mid-infrared. J. Appl. Phys. **113**, 043110 (2013)

45. Wei, S.H., Zunger, A.: Calculated natural band offsets of all II–VI and III–V semiconductors: Chemical trends and the role of cation d orbitals. Appl. Phys. Lett. **72**, 2011 (1998)
46. Sweeney, S.J., Adams, A.R., Silver, M., O'Reilly, E., Watling, J., Walker, A., Thijs, P.: Transport phenomena-dependence of threshold current on QW position and on pressure in 1.5mm InGaAs(P) lasers. Phys. Status Solidi B. **211**, 525 (1999)
47. Johnson, S.R., Guo, C.Z., Chaparro, S., Sadofyev, Y.G., Wang, J., Cao, Y., Samal, N., Xu, J., Yu, S.Q., Ding, D., Zhang, Y.-H.: GaAsSb/GaAs band alignment evaluation for long-wave photonic applications. J. Cryst. Growth **251**, 521–525 (2003)
48. Lewis, R.B., Beaton, D.A., Lu, X., Tiedje, T.: $GaAs_{1-x}Bi_x$ light emitting diodes. J. Cryst. Growth **311**, 1872–1875 (2009)
49. Eliseev, P.G., Perlin, P., Lee, J., Osinski, M.: "Blue" temperature-induced shift and band-tail emission in InGaN-based light sources. Appl. Phys. Lett. **71**, 569 (1997)
50. Mohmad, A.R., Bastiman, F., Ng, J.S., Sweeney, S.J., David, J.P.R.: Photoluminescence investigation of high quality $GaAS_{1-x}Bi_x$ on GaAs. Appl. Phys. Lett. **98**, 122107 (2011)
51. Chamings, J., Adams, A.R., Sweeney, S.J., Kunert, B., Volz, K., Stolz, W.: Temperature dependence and physical properties of Ga(NAsP)/GaP semiconductor lasers. Appl. Phys. Lett. **93**, 101108 (2008)
52. Huang, W., Oe, K., Feng, G., Yoshimoto, M.: Molecular-beam epitaxy and characteristics of $GaN_yAs_{1-x-y}Bi_x$. J. Appl. Phys. **98**, 053505 (2005)

Chapter 3
Theory of the Electronic Structure of Dilute Bismide Alloys: Tight-Binding and k · p Models

Christopher A. Broderick, Muhammad Usman, and Eoin P. O'Reilly

Abstract Dilute bismide alloys have been the focus of increasing research effort in recent years, due in large part to their novel electronic properties. In particular, they display significant potential for achieving highly efficient photonic devices operating at telecomm wavelengths (1.3–1.5 μm). However, despite substantial progress in the growth and characterisation of dilute bismides, there have been comparatively few theoretical investigations of this novel material system. We summarise here aspects of our theoretical work on the electronic and optical properties of dilute bismide alloys. We present tight-binding and $\mathbf{k} \cdot \mathbf{p}$ models for the electronic structure of (In)$GaBi_xAs_{1-x}$, in which the strong reduction of the band gap (E_g) and increase in the spin-orbit-splitting energy (Δ_{SO}) are explained in terms of a band-anticrossing interaction between the extended states of the host matrix valence band edge and Bi-related resonant impurity states lying in the valence band. Our results, which are in good agreement with the available experimental data, serve to elucidate the origins of the novel electronic properties of dilute bismide alloys and confirm the crossover to an $E_g < \Delta_{SO}$ regime in $GaBi_xAs_{1-x}$ for $x \gtrsim 11$ %, a condition which should lead to suppressed Auger recombination in long wavelength devices. The dilute bismide $\mathbf{k} \cdot \mathbf{p}$ model is applied to calculate the effect of Bi incorporation on the band structure and optical gain of dilute bismide quantum well structures, and some general trends relevant to laser operation are identified. We also extend our models to the quaternary dilute bismide–nitride alloy $GaBi_xN_yAs_{1-x-y}$ and show

C.A. Broderick • E.P. O'Reilly
Tyndall National Institute, Lee Maltings, Dyke Parade, Cork, Ireland

Department of Physics, University College Cork, Cork, Ireland
e-mail: chris.broderick@tyndall.ie; eoin.oreilly@tyndall.ie

M. Usman (✉)
Tyndall National Institute, Lee Maltings, Dyke Parade, Cork, Ireland
e-mail: usman@alumni.purdue.edu

H. Li and Z.M. Wang (eds.), *Bismuth-Containing Compounds*, Springer Series in Materials Science 186, DOI 10.1007/978-1-4614-8121-8_3,

how co-alloying of Bi and N offers broad scope for band structure engineering which should lead to the realisation of highly efficient GaAs-based long wavelength photonic devices.

3.1 Introduction

Highly mismatched semiconductor alloys such as GaN_yAs_{1-y} have attracted considerable interest in recent years, both from a fundamental perspective and also because of their potential device applications [1–3]. When a small fraction of As is replaced by N in GaAs, the band gap E_g initially decreases rapidly, by ~150 meV when 1 % of As is replaced by N [4].

Similar behaviour has been experimentally observed in $GaBi_xAs_{1-x}$, where E_g has been observed to decrease initially by ~90 meV [5, 6] per % of Bi replacing As. Additionally, recent experiments [7, 8] have also revealed the presence of a large bowing of the spin-orbit-splitting (SO) energy (Δ_{SO}), which increases strongly with increasing Bi composition.

Furthermore, it was shown recently using photo-modulated reflectance (PR) measurements that increasing the Bi composition in $GaBi_xAs_{1-x}$ leads to the onset of an $E_g < \Delta_{SO}$ regime in the alloy [8]. The onset of this $E_g < \Delta_{SO}$ regime in the dilute bismides raises some interesting possibilities to improve the operating efficiency of telecomm wavelength optoelectronic devices by providing the possibility to suppress the dominant loss mechanisms via band structure engineering.

Auger recombination is the dominant intrinsic loss mechanism that ultimately limits today's telecomm lasers. Two main Auger recombination processes are considered in telecomm lasers, namely the CHSH process (in which a **C**onduction electron and a **H**eavy-hole recombine across the energy gap, but, instead of emitting a photon, they excite into the **S**pin-split-off band a **H**ole from near the valence band maximum) and the CHCC process (in which a **C**onduction electron recombines with a **H**eavy-hole, exciting an electron from near the bottom of the **C**onduction band to a higher energy **C**onduction band state). The CHCC and CHSH processes are depicted schematically in Fig. 3.1 a, b, respectively. Previous experimental and theoretical analysis has shown that the CHSH process is dominant in conventional InP-based lasers, providing a current path which typically accounts for the majority of the Auger current [9–11].

In addition to Auger recombination, inter-valence band absorption (IVBA) also degrades laser performance. In the IVBA process, a photon is re-absorbed by an electron in the spin-split-off band which is excited to an empty (hole) state near the top of the valence band. Like the CHSH Auger process, IVBA is therefore also sensitive to the magnitude of the spin-orbit-splitting energy relative to that of the band gap. IVBA causes the threshold carrier density to increase superlinearly with temperature [12], further exacerbating Auger recombination and additionally degrading the differential quantum efficiency above threshold. The combination of

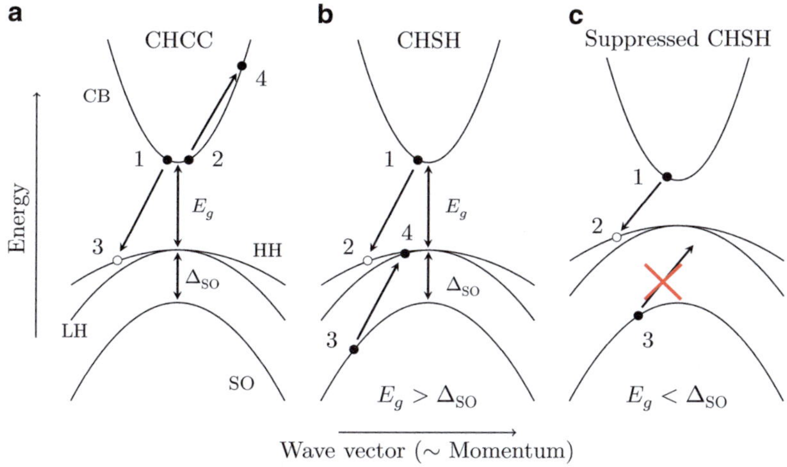

Fig. 3.1 (**a**) Schematic illustration of a CHCC Auger recombination process in which a conduction electron (1) recombines with a valence hole (3), with the released energy exciting a second conduction electron (2) to a vacant higher conduction state (4), such that energy and momentum are conserved. (**b**) Schematic illustration of a CHSH Auger recombination process in which a conduction electron (1) recombines with a valence hole (2), with the released energy exciting a valence hole (4) to the spin-split-off band (3), such that energy and momentum are conserved. (**c**) Schematic illustration of (**b**) in the case where the spin-orbit-splitting exceeds the band gap, $E_g < \Delta_{SO}$. For $E_g < \Delta_{SO}$ the CHSH process is forbidden by conservation of energy, since the energy required to excite a hole to the spin-split-off band exceeds the energy made available by the electron–hole recombination

Auger recombination and IVBA causes the output power of telecomm lasers to be strongly temperature sensitive over the typical operating temperature range [13].

It would therefore be highly beneficial if these loss mechanisms could be eliminated from telecomm lasers. This cannot be achieved using conventional III–V alloys based on (GaAlIn)(NPAsSb), where the energy gap E_g always exceeds the spin-orbit-splitting energy Δ_{SO} in materials that emit at 1.3 and 1.5 μm. However, just as the introduction of N gives the possibility to engineer the conduction band structure, the introduction of Bi then provides an opportunity to engineer the valence band structure potentially providing a route to produce telecomm and longer wavelength lasers with suppressed losses [14].

The $E_g < \Delta_{SO}$ regime is therefore of interest for the design of highly efficient and thermally stable optoelectronic devices since it opens up the possibility of suppression of the non-radiative CHSH Auger recombination process, a loss mechanism which dominates the threshold current and degrades temperature stability in III–V multinary lasers operating at telecommunication wavelengths [3, 14]. This strongly motivates the development of suitable models for the description of the electronic structure of dilute bismide alloys, which can then be applied to the study and prediction of Bi-based device characteristics.

The extreme band gap bowing observed in $In_{1-x}Ga_xN_yAs_{1-y}$ has been well explained in terms of a band-anticrossing (BAC) interaction between two levels, one at energy E_{CB} associated with the extended conduction band edge (CBE) states of the host (In)GaAs matrix and the second at energy E_N associated with the highly localised N-related resonant impurity states in the alloy [4].

Bi, being the largest stable group V element, is significantly larger and more electropositive than As. It should therefore be expected that any Bi-related impurity levels should lie in the valence band or above the valence band edge (VBE) and that, if an anticrossing interaction occurs in $GaBi_xAs_{1-x}$, it will occur between the Bi-related impurity levels and the VBE of the host (GaAs) matrix.

Recently the presence of such an interaction has been suggested [6, 15, 16], but there remains controversy [17] as to whether or not a BAC-like model can be applied in the case of the dilute bismides. This controversy is in part due to the absence of Bi-related features in the valence band in measured spectra of $GaBi_xAs_{1-x}$, which, based on previous studies of GaN_yAs_{1-y}, one might expect to be present in the case of a BAC interaction.

In [16] we presented a nearest-neighbour sp^3s^* tight-binding model to describe the electronic structure of dilute bismide alloys of GaAs and GaP. The sp^3s^* model agrees well both with experimental measurements and pseudopotential calculations [18] of the variation of E_g and Δ_{SO} with Bi composition in $GaBi_xAs_{1-x}$ over the investigated composition range.

Calculations based on this model reproduce the experimentally observed transition to an $E_g < \Delta_{SO}$ regime in the free-standing alloy for $x \approx 10.5\,\%$ [8], as well as agreeing with the conclusions of the pseudopotential calculations of [18], namely that an isovalent Bi impurity forms a resonant state below the GaAs VBE and that the alloy VBE is derived predominantly from the host VBE as opposed to a Bi bound state.

Analysis of the tight-binding calculations confirms the presence of a BAC interaction in the valence band in $GaBi_xAs_{1-x}$, occurring between the GaAs VBE and lower-lying Bi-related impurity states which are resonant with the GaAs VBE. In addition, our analysis shows that the failure to observe any Bi-related features below the VBE in PR spectra results from the broadening of the Bi resonant impurity states by the large density of host valence states with which they are resonant [16].

Using the results of the tight-binding calculations we derive a 12-band $\mathbf{k} \cdot \mathbf{p}$ model for $GaBi_xAs_{1-x}$ which includes BAC interactions between the GaAs VBE states and Bi resonant impurity states lying close in energy to the GaAs VBE. The model is shown to be in excellent agreement with full tight-binding supercell calculations and experimental measurements of the band gap and spin-orbit-splitting across a large composition range and, combined with a detailed tight-binding based analysis, demonstrates emphatically the validity of the BAC model as applied to $GaBi_xAs_{1-x}$.

The dilute bismide-nitride alloy $GaBi_xN_yAs_{1-x-y}$ has also been identified as being highly promising for the design of highly efficient optoelectronic devices since it opens up several avenues by which the band structure can be engineered. In particular, $GaBi_xN_yAs_{1-x-y}$ is predicted to have a giant band gap bowing, allowing for emission at 1.3–1.5 μm and longer wavelengths on a GaAs substrate, as well as retaining the large spin-orbit-splitting bowing that is characteristic of $GaBi_xAs_{1-x}$, thereby offering the possibility of Auger suppression [14]. Also, incorporation of Bi and N in GaAs introduces compressive and tensile strain, respectively, when grown on a GaAs substrate, so that a high degree of strain control should be possible in $GaBi_xN_yAs_{1-x-y}$, providing further opportunities for the manipulation of the electronic properties.

Based on an sp^3s^* tight-binding Hamiltonian for $GaBi_xN_yAs_{1-x-y}$ we show that the effects of Bi and N on the electronic structure of GaAs are largely independent of one another, both in ordered and disordered crystals [19, 20], so that a simple description of the electronic structure is possible. Based on this insight we derive a 14-band **k·p** model for $GaBi_xN_yAs_{1-x-y}$ [21], which accurately describes the near zone centre band dispersion of ordered $GaBi_xN_yAs_{1-x-y}$ crystals.

Since dilute bismide alloys are promising candidates for the design of optoelectronic devices with improved operating efficiency and reduced temperature sensitivity, we present calculations of the electronic and optical properties of a number of ideal Bi-containing quantum well structures using the 12-band **k·p** model. These calculations are used to identify some of the key effects of Bi incorporation on the electronic and optical properties, including the evolution of the gain and of inter-valence band absorption as the Bi composition passes beyond the point where $E_g = \Delta_{SO}$.

The remainder of this chapter is organised as follows. In Sect. 3.2 we briefly outline the tight-binding model used to study the electronic structure of dilute bismide alloys and review the theory of band-anticrossing due to impurity substitution in the framework of a tight-binding model. In Sect. 3.3 we use tight-binding calculations on ordered supercells to verify the BAC model for the valence band structure of $GaBi_xP_{1-x}$, $GaBi_xAs_{1-x}$ and $InBi_xAs_{1-x}$. In Sect. 3.4 we apply our tight-binding model to the study of the electronic structure of realistic $GaBi_xAs_{1-x}$ alloys by performing calculations on large, disordered supercells. Section 3.5 expands upon the discussions of Sect. 3.3, deriving a 12-band **k·p** Hamiltonian for dilute bismide alloys. To demonstrate the accuracy and utility of this **k·p** Hamiltonian, the model is applied to the study of $GaBi_xAs_{1-x}$, $InBi_xAs_{1-x}$ and $In_{1-x}Ga_xBi_yAs_{1-y}$ alloys and compared to the results of experimental measurements on these alloys. In Sect. 3.6 we extend our tight-binding and **k·p** models to the dilute bismide–nitride alloy $GaBi_xN_yAs_{1-x-y}$ and discuss some of the key features of the electronic structure of this novel material. In Sect. 3.7 we apply the 12-band **k·p** model to the study of the electronic and optical properties of a number of ideal Bi-containing quantum well structures and identify some important trends for dilute bismide-based device operation. Finally, in Sect. 3.8 we conclude and provide some outlook for future work in the area of dilute bismide alloys.

3.2 Tight-Binding Model for Dilute Bismide Alloys

3.2.1 Overview of the Tight-Binding Model

We presented in [16] a nearest-neighbour sp^3s^* tight-binding Hamiltonian, including spin-orbit-coupling, for $GaBi_xAs_{1-x}$. Here we describe how this Hamiltonian can be combined with sp^3s^* Hamiltonians for InAs, InBi and GaN to obtain sp^3s^* Hamiltonians for the quaternary dilute bismide alloy $In_{1-x}Ga_xBi_yAs_{1-y}$ and the bismide–nitride alloy $GaBi_xN_yAs_{1-x-y}$.

In our tight-binding model we begin by relaxing the atomic positions in a given supercell to their lowest energy configuration using a valence force field model based on the Keating potential [22, 23].

The on-site energies of the relaxed supercell Hamiltonian are taken to depend on the overall neighbour environment. The bond angle dependence of the two-centre integrals is determined using the expressions of Slater and Koster [24] and the inter-atomic interaction matrix elements are taken to vary with the relaxed nearest-neighbour bond length d as $\left(\frac{d_0}{d}\right)^{\eta}$, where d_0 is the equilibrium bond length in the equivalent binary compound and η is a dimensionless scaling parameter (the magnitude of which depends on the type of interaction being considered). Full details of the tight-binding model can be found in ref [16].

3.2.2 Tight-Binding Analysis of Valence Band-Anticrossing in Dilute Bismide Alloys

The BAC model explains the extreme band gap bowing observed in Ga(In)N_xAs_{1-x} in terms of an interaction between two levels, one at energy E_{CB} associated with the extended CBE state $|\psi_{c,0}\rangle$ of the Ga(In)As matrix and the other at energy E_N associated with the localised N impurity defect states $|\psi_N\rangle$, with the two states linked by an N-composition-dependent matrix element V_{Nc} describing the interaction between them [4]. The CBE energy of (In)GaN_yAs_{1-y} is then given by the lower eigenvalue, E_-, of the 2-band Hamiltonian

$$\begin{pmatrix} E_N & V_{Nc} \\ V_{Nc} & E_{CB} \end{pmatrix} \tag{3.1}$$

where, $1/6$.

A resonant feature associated with the upper eigenvalue, E_+, of Eq. (3.1) has been observed in PR measurements [4, 25, 26], appearing in GaN_yAs_{1-y} for $y \gtrsim 0.2\,\%$ and remaining a relatively sharp feature until $y \approx 3\,\%$, beyond which composition it broadens and weakens, when the resonant state becomes degenerate with the large L-related conduction band density of states [27].

To investigate the resonant state $|\psi_N\rangle$, and its behaviour, we used the sp^3s^* tight-binding Hamiltonian to calculate the electronic structure of ordered GaN_yAs_{1-y} supercells [27, 28]. By comparing the calculated host and alloy CBE states, $|\psi_{c,0}\rangle$ and $|\psi_{c,1}\rangle$, in large 1,728 atom supercells ($Ga_{864}N_1As_{863}$ and $Ga_{864}As_{864}$, respectively), we derived the nitrogen resonant state, $|\psi_N\rangle$, associated with an isolated N atom. In the BAC model, $|\psi_{c,1}\rangle$ is a linear combination of $|\psi_{c,0}\rangle$ and $|\psi_N\rangle$, with $|\psi_N\rangle$ then given by:

$$|\psi_N\rangle = \frac{|\psi_{c,1}\rangle - \alpha|\psi_{c,0}\rangle}{\sqrt{1-|\alpha|^2}} \tag{3.2}$$

where $\alpha = \langle\psi_{c,0}|\psi_{c,1}\rangle$. The evolution of the conduction band structure in an ordered GaN_yAs_{1-y} supercell can then be very well described by associating a localised resonant state with each N atom in the supercell [29].

In the dilute bismide case the BAC interaction is expected to occur in the valence band. Previous analysis [6] has included a BAC interaction with each of the three doubly degenerate valence bands, namely the heavy-hole (HH), light-hole (LH) and spin-split-off (SO) bands. We find below that there is a BAC interaction with the HH and LH bands, but that there is no calculated BAC interaction involving the SO band. For dilute bismides, the Hamiltonian analogous to (3.1) is therefore:

$$\begin{pmatrix} E_{\mathrm{HH}} & 0 & V_{\mathrm{Bi}} & 0 \\ 0 & E_{\mathrm{LH}} & 0 & V_{\mathrm{Bi}} \\ V_{\mathrm{Bi}}^{*} & 0 & E_{\mathrm{Bi}} & 0 \\ 0 & V_{\mathrm{Bi}}^{*} & 0 & E_{\mathrm{Bi}} \end{pmatrix} \tag{3.3}$$

where $E_{\mathrm{HH/LH}}$ are the (degenerate) host matrix VBE energies, E_{Bi} is the energy of the Bi-related resonant state energy and $V_{\mathrm{Bi}} = \beta_{\mathrm{Bi}} \sqrt{x}$ is the composition-dependent interaction matrix element linking the host matrix and Bi-related impurity states. For E_{Bi} above (below) the host matrix VBE the repulsive interaction described by (3.3) pushes down (up) the alloy VBE energy.

Because of the fourfold degeneracy of the VBE states, we need to modify the approach used in Eq. (3.2) to determine the BAC interaction in Bi-containing alloys.

In general, for an isolated Bi atom in GaAs, the impurity atom exhibits T_d symmetry so we can expect the impurity states$|\psi_{\mathrm{Bi}}\rangle$ to be fourfold degenerate, with energy E_{Bi}. These states$|\psi_{\mathrm{Bi},i}\rangle$ are calculated following a similar procedure to that in Eq. (3.2), generalised to take account of the fourfold degeneracy [16].

We can then calculate explicitly E_{Bi} and the interaction between$|\psi_{\mathrm{Bi},i}\rangle$ and the host matrix VBE, β_{Bi}, by calculating the appropriate matrix elements of the full TB Hamiltonian for the Bi-containing alloy, $\hat{H}$, for any one of the four degenerate Bi states:

$$E_{\mathrm{Bi}} = \langle \psi_{\mathrm{Bi}} | \hat{H} | \psi_{\mathrm{Bi}} \rangle \tag{3.4}$$

$$\beta_{\mathrm{Bi}} = \frac{V_{\mathrm{Bi}}}{\sqrt{x}} = \frac{\langle \psi_{\mathrm{Bi}} | \hat{H} | \psi_{v,0} \rangle}{\sqrt{x}} \tag{3.5}$$

By constructing the Bi-related impurity state in $\mathrm{Ga}_M\mathrm{Bi}_1\mathrm{As}_{M-1}$ using the generalisation of Eq. (3.2) we show that (i) it interacts with the $\mathrm{Ga}_M\mathrm{As}_M$ VBE via an anti-crossing interaction and (ii) that this approach can be used to explain the strong bowing of the band gap and spin-orbit-splitting present in $\mathrm{GaBi}_x\mathrm{As}_{1-x}$.

3.2.3 Calculation of Fractional Γ Character in Alloy Supercells

In the ideal BAC model of Eqs. (3.1) and (3.3), the host matrix band edge states are expected to mix in the alloy with the localised Bi-related defect states. In order to explore this BAC interaction, we calculate $G_\Gamma(E)$, which is the projection of the host matrix band edge states onto the full spectrum of levels in the alloy supercell. This allows us to readily examine the effects of Bi incorporation on the electronic structure of a chosen host matrix compound and provides valuable information on the Bi-induced mixing of bands.

In what follows, we use the subscripts $l,0$ and $k,1$ to index the unperturbed host and Bi-containing alloy states, respectively. The $G_\Gamma(E)$ spectrum is calculated by projecting the unperturbed X_MY_M (X = Ga, In; Y = P, As) supercell states at $\Gamma, |\psi_{l,0}\rangle$, onto the spectrum of the $X_MBi_LY_{M-L}$ alloy supercell, $E_k, |\psi_{k,1}\rangle$:

$$G_\Gamma(E) = \sum_k \sum_{l=1}^{g(E_l)} f_{\Gamma,kl}\,\delta(E_k - E) \tag{3.6}$$

$$f_{\Gamma,kl} = |\langle \psi_{k,1} | \psi_{l,0} \rangle|^2 \tag{3.7}$$

Here $\Delta\ (E_k - E)$ is the Dirac delta function centred on E_k, $g(E_l)$ is the degeneracy of the host matrix band having energy E_l at the Γ point in the Brillouin zone and $f_{\Gamma,k\,l}$ is the fractional Γ character of the alloy state $|\psi_{k,1}\rangle$. When plotting $G_\Gamma(E)$ below, the delta function is scaled so that the maximum plotted value of $G_\Gamma(E)$ equals 1 when a set of host matrix states are 100 % projected to states at a single energy in a given alloy supercell.

We use $G_\Gamma(E)$ firstly to confirm the presence of a BAC interaction in ordered $GaBi_xP_{1-x}$ and $GaBi_xAs_{1-x}$ alloys, and then to analyse the effects of Bi pair and cluster states, as well as the evolution of the BAC interaction in disordered structures. We also use $G_\Gamma(E)$ to explore the evolution of the conduction and SO band edges in $GaBi_xAs_{1-x}$ with Bi composition, showing that the variation in both cases is attributable to the conventional alloying effect. These calculations are discussed in detail in the next two sections.

3.3 Valence Band-Anticrossing in Dilute Bismide Alloys

We have shown previously that it is straightforward to identify and analyse a band-anticrossing interaction in ordered supercell structures where the localised defect state is in the energy gap, as is the case for GaN_yP_{1-y} [30]. It has also been shown experimentally that an isolated Bi atom acts as an impurity in GaP, giving rise to a defect level lying approximately 100 meV above the VBE [31]. We therefore start our calculations by investigating the electronic structure in a series of ordered, cubic $Ga_MBi_1P_{M-1}$ supercells containing $2M = 8N^3$ atoms, for $2 \leq N \leq 8$, in which a single substitutional Bi atom is inserted into the GaP matrix. Having established the main factors influencing the valence band structure in $GaBi_xP_{1-x}$, we then consider $GaBi_xAs_{1-x}$, using the phosphide results to guide and support the interpretation of the arsenide calculations.

Figure 3.2a plots the fractional Γ character, $G_\Gamma(E)$, associated with the GaP VBE states in the series of $Ga_MBi_1P_{M-1}$ supercells as the supercell size (Bi composition)

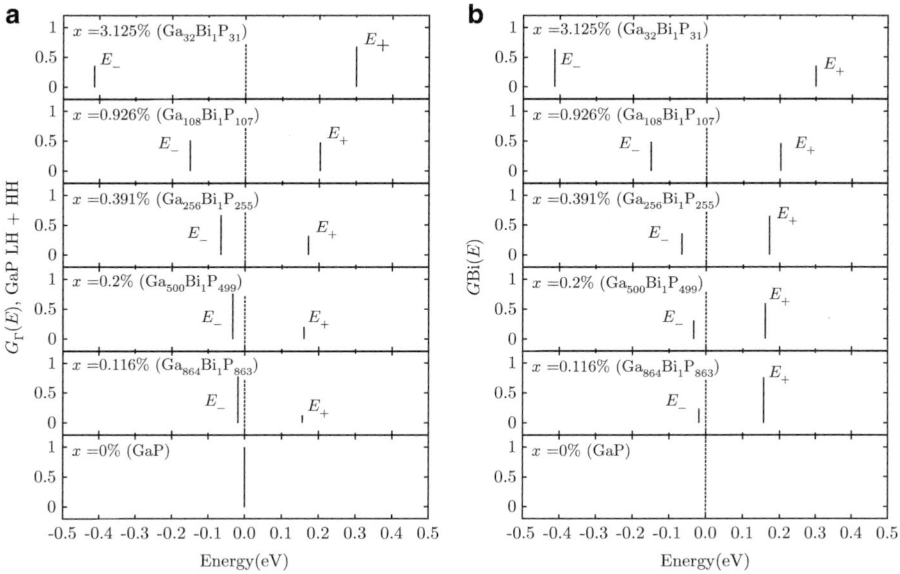

Fig. 3.2 Calculated character spectra of (**a**) combined GaP HH + LH Γ character and (**b**) Bi impurity state character, $G_{Bi}(E)$, for a series of ordered $Ga_MBi_1P_{M-1}$ supercells. In each case the zero of energy is taken at the GaP VBE and indicated by a *vertical dashed line*

decreases (increases) from $N=6$ ($x=0.116\,\%$) to $N=2$ ($x=3.125\,\%$). For reference, we also include in Fig. 3.2a the calculated $G_\Gamma(E)$ for a pure GaP supercell, in which 100 % of the HH/LH Γ character resides, by definition, on the VBE.

When we replace a single P atom by Bi in a 1,728 atom $Ga_{864}P_{864}$ supercell, we find that this gives rise to a fourfold degenerate defect level which lies above, and interacts with, the GaP VBE. The amplitude of the unperturbed GaP VBE states is distributed over two levels, corresponding to E_- and E_+, as expected from the band-anticrossing model of Eq. (3.3) and as confirmed by examining the projection of the $|\psi_{Bi}\rangle$ character $G_{Bi}(E)$ in Fig. 3.2b. We calculate using Eqs. (3.4) and (3.5) that the isolated Bi impurity level lies approximately 0.12 eV above the GaP VBE, and that each impurity state is highly localised, with 31 % of its probability density localised on the Bi atom and the four nearest-neighbour Ga sites. We also calculate that the band-anticrossing interaction V_{Bi} between the Bi-related defect levels at energy E_{Bi} and the GaP VBE is of magnitude $V_{Bi} = \beta_{Bi}\sqrt{x}$, with $\beta_{Bi} = 1.40$ eV.

The calculated magnitude of the interaction parameter β_{Bi} in $GaBi_xP_{1-x}$ is then comparable to that which we have calculated previously [1] for GaN_yP_{1-y} (β_N = 1.74 eV) and for GaN_yAs_{1-y} (β_N = 2.00 eV).

Having identified that an anticrossing interaction does indeed occur between Bi-related defect levels and the GaP VBE in $GaBi_xP_{1-x}$, we now use this understanding to perform our analysis of the valence band structure of $GaBi_xAs_{1-x}$. We present evidence for the presence of a similar BAC interaction in $GaBi_xAs_{1-x}$ by considering

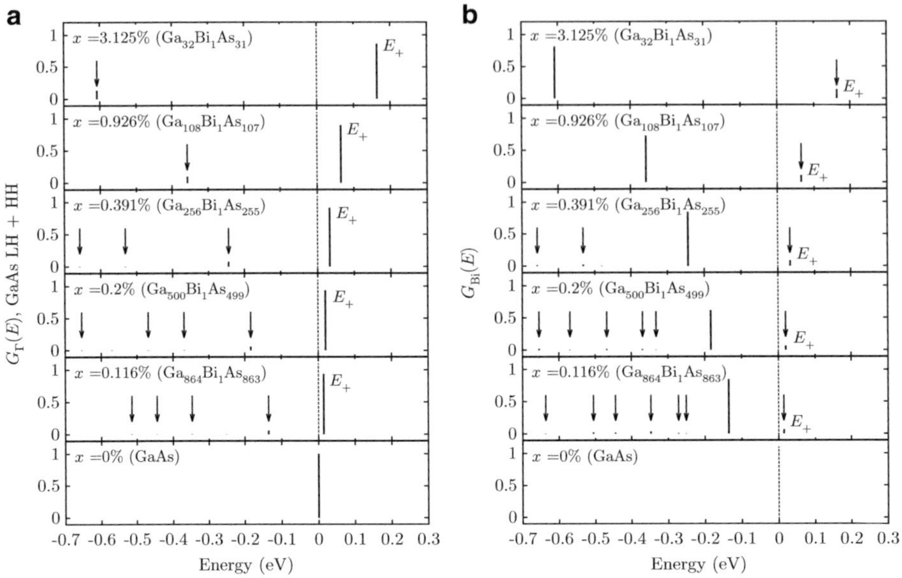

Fig. 3.3 Calculated character spectra of (**a**) combined GaAs HH + LH Γ character and (**b**) Bi impurity state character, $G_{Bi}(E)$, for a series of ordered $Ga_MBi_1As_{M-1}$ supercells. In each case the zero of energy is taken at the GaAs VBE and indicated by a *vertical dashed line*; *arrows* highlight $Ga_MBi_1As_{M-1}$ valence states with (small) non-zero Γ character

the $G_Γ(E)$ spectra of the GaAs VBE and the impurity state associated with a substitutional Bi atom in a series of ordered, cubic $Ga_MBi_1As_{M-1}$ supercells containing $2M = 8N^3$ atoms, for $2 \leq N \leq 8$.

We proceed by inserting a single substitutional Bi impurity in a series of cubic $2M$ atom Ga_MAs_M supercells and calculate the $G_Γ(E)$ spectrum in each case by projecting the wave functions of the fourfold degenerate Ga_MAs_M VBE onto the full spectrum of the Bi-containing alloy [16].

We can see from Fig. 3.3a that the GaAs VBE character in each supercell resides predominantly (>70 %) on the alloy VBM, with the remainder distributed over a series of lower-lying valence levels. This is in agreement with the pseudopotential calculations of Zhang et al. [18], namely that the $GaBi_xAs_{1-x}$ VBE is derived mainly from that of GaAs, with Bi forming a resonant impurity state lying below the GaAs VBE in energy.

We calculate that the state which mixes with each of the GaAs VBE states is a localised, fourfold degenerate Bi resonant impurity state similar in character to that found in $GaBi_xP_{1-x}$ [16]. We also find that with decreasing supercell size (increasing Bi composition) the strength of the interaction between the resonant state and the GaAs VBE increases, pushing the alloy VBE, E_+, upwards in energy and mixing more of the calculated $G_Γ(E)$ spectrum into the valence band as more of the GaAs VBE fractional Γ character transfers to the lower BAC level(s), E_-. This can clearly be seen in Fig. 3.3a, which plots the calculated $G_Γ(E)$ spectra obtained by projecting the Ga_MAs_M VBE wave functions onto the spectrum of the series of ordered $Ga_MBi_1As_{M-1}$ supercells.

In each supercell we see that the Γ character associated with the GaAs VBE is distributed over several supercell valence levels, with the broadening of the calculated $G_\Gamma(E)$ increasing with supercell size (see Fig. 3.3a).

Figure 3.3b shows the calculated $G_{\mathrm{Bi}}(E)$ spectra for the same set of $Ga_MBi_1As_{M-1}$ supercells as in Fig. 3.3a, calculated by replacing $|\psi_{l,0}\rangle$ by $|\psi_{\mathrm{Bi},i}\rangle$ in Eq. (3.6) and thereby projecting one of the Bi impurity states onto the full spectrum of the Bi-containing supercell. We see that the Bi-related states are spread over several supercell valence levels, reflecting that $|\psi_{\mathrm{Bi},i}\rangle$ is resonant with the GaAs valence states [16, 18]. We attribute this broadening of the Bi impurity state to the large density of zone centre host valence states with which the $|\psi_{\mathrm{Bi}}\rangle$ are resonant [32, 33].

Comparing Fig. 3.3a and b we see that the Bi-related impurity states and the Γ character of the GaAs VBE states intermix in the Bi-containing supercells, so that the alloy VBE is comprised of an admixture of the host VBE and the Bi resonant states, consistent with the presence of a BAC interaction between them.

In GaN_yAs_{1-y} the presence of an upper N-related BAC feature has been observed for $y \lesssim 3\,\%$, beyond which composition it broadens and weakens [25] as the Γ character of the N-related states becomes spread over a large number of upper conduction levels, when E_{N} becomes degenerate with the large density of conduction states in the L-valley [27]. The observed distribution of Γ character in Fig. 3.3a is then analogous to the GaN_yAs_{1-y} case for $y \gtrsim 3\,\%$. The failure to observe any feature in measured spectra of $GaBi_xAs_{1-x}$ associated with transitions to a resonant state in the valence band can be understood in terms of the broadening and delocalisation of the Bi resonant states, due to the large density of host valence states.

By examining the character of the impurity states $|\psi_{\mathrm{Bi},i}\rangle$ associated with an isolated Bi impurity in increasingly large supercells, we approach the dilute doping limit. Our results show that a substitutional Bi atom in $Ga_MBi_1P_{M-1}$, $Ga_MBi_1As_{M-1}$ or $In_MBi_1As_{M-1}$ forms a set of fourfold degenerate resonant states at energy E_{Bi}, which are highly localised about the Bi site, and that the strength of the interaction between these states and the host VBE varies with Bi composition x as $V_{\mathrm{Bi}} = \beta_{\mathrm{Bi}}\sqrt{x}$ for large supercells.

Figure 3.4 shows the calculated values of the energy separation between the host VBE and Bi resonant state, $\Delta E_{\mathrm{Bi}} = E_{\mathrm{Bi}} - E_{v,0}$, as well as the BAC coupling parameter β_{Bi}, as a function of supercell size for a series of ordered $Y_MBi_1As_{M-1}$ (Y = Ga, In) supercells. We observe a strong variation of ΔE_{Bi} and β_{Bi} with Bi composition at larger x, with the trends stabilising as x approaches the dilute doping limit in the large ($\gtrsim$2,000 atom) supercells. This variation is primarily due to the increased number of host valence band states which fold back to the Γ point as the supercell size is increased, which makes more states available to construct $|\psi_{\mathrm{Bi}}\rangle$ in the larger cells [16]. Therefore, we conclude that large supercell calculations are required when investigating the behaviour of a substitutional Bi impurity in (In)GaAs alloys.

It is interesting to note here the similar character of the Bi-related states in $Ga_MBi_1As_{M-1}$ and in $In_MBi_1As_{M-1}$: the impurity states associated with isolated Bi impurities in GaAs and InAs lie within 0.1 eV of one another in all supercells

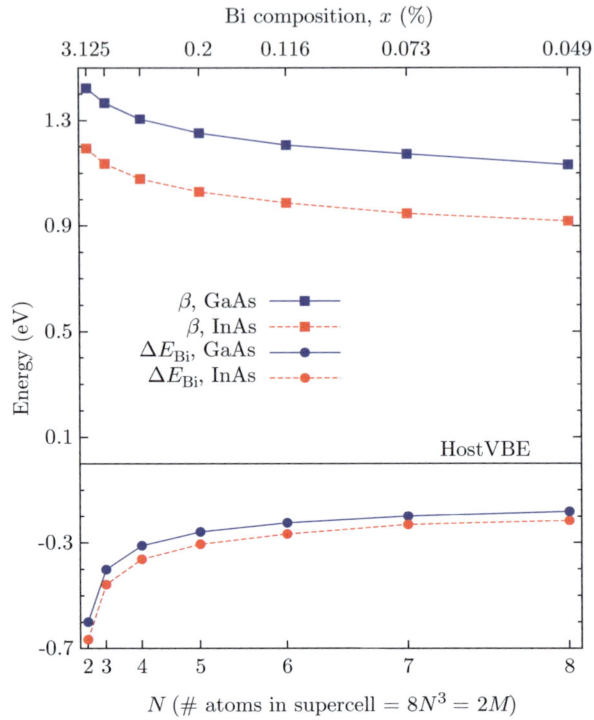

Fig. 3.4 Calculated separation between the Bi impurity state and host VBE, ΔE_{Bi} (*circles*), *lower panel*, and BAC coupling parameter, β_{Bi} (*squares*), for substitutional Bi incorporation in a series of ordered, cubic $Y_M Bi_1 As_{M-1}$ (Y = Ga, In) supercells, *upper panel*. As a guide to the eye, *solid* (*dashed*) *lines* have been inserted to highlight the evolution of the BAC parameters with Bi composition in the $Ga_M Bi_1 As_{M-1}$ ($In_M Bi_1 As_{M-1}$) cases. The zero of energy for ΔE_{Bi} is taken in each case at the host VBE and is denoted by a *horizontal line* at 0 eV

considered. This is in stark contrast to the case of a substitutional nitrogen (N) atom in GaAs or in InAs, where the N-related impurity states lie approximately 0.19 eV and 0.97 eV above the host conduction band edge, respectively [1]. In the dilute nitride case this different behaviour arises primarily due to differences in the GaAs and InAs conduction band structures. Since Bi primarily affects the valence band in (In)$GaBi_xAs_{1-x}$, we attribute the similarity of the Bi-related states in $GaBi_xAs_{1-x}$ and in $InBi_xAs_{1-x}$ to the relative similarity of the valence band structure of GaAs and InAs [34].

We list in Table 3.1 the calculated values of ΔE_{Bi} and β_{Bi} in 64 ($N=2$) and 4,096 ($N=8$) atom $GaBi_xP_{1-x}$, $GaBi_xAs_{1-x}$ and $InBi_xAs_{1-x}$ supercells containing a single substitutional Bi atom. Our calculations predict that a substitutional Bi impurity produces a resonant state lying approximately 0.18 eV (0.22 eV) below the GaAs (InAs) VBE in the dilute doping limit.

In summary, our tight-binding calculations have shown that the VBE states in ordered crystals of $GaBi_xP_{1-x}$ and (In)$GaBi_xAs_{1-x}$ are well described in terms of a BAC interaction between the host matrix VBE and resonant impurity states associated with substitutional Bi atoms. We now turn to consider the electronic structure of disordered alloys, following which we will derive in Sect. 3.5 a 12-band $\mathbf{k} \cdot \mathbf{p}$ Hamiltonian to describe the band structure of $GaBi_xAs_{1-x}$ and $InBi_xAs_{1-x}$ crystals.

Table 3.1 Calculated energy of the Bi impurity state relative to the host matrix VBE (ΔE_{Bi}) and BAC coupling parameter (β_{Bi}) for Bi incorporation in cubic 64 atom ($N=2$) and 4,096 atom ($N=8$) $GaBi_xP_{1-x}$, $GaBi_xAs_{1-x}$ and $InBi_xAs_{1-x}$ supercells, each of which contain $8N^3$ atoms in total, with a single substitutional Bi atom on a group V site

Supercell	x (%)	ΔE_{Bi} (eV)	β_{Bi} (eV)
$Ga_{32}Bi_1P_{31}$	3.125	−0.16	1.92
$Ga_{2048}Bi_1P_{2047}$	0.049	0.12	1.40
$Ga_{32}Bi_1As_{31}$	3.125	−0.60	1.42
$Ga_{2048}Bi_1As_{2047}$	0.049	−0.18	1.13
$In_{32}Bi_1As_{31}$	3.125	−0.67	1.19
$In_{2048}Bi_1As_{2047}$	0.049	−0.22	0.92

3.4 Tight-Binding Analysis of the Electronic Structure of Disordered $GaBi_xAs_{1-x}$ Alloys

3.4.1 *Evolution of $GaBi_xAs_{1-x}$ Valence States with* x

Figure 3.5 shows the calculated evolution of $G_\Gamma(E)$ for the highest valence states in a series of $Ga_{2048}Bi_MAs_{2048-M}$ supercells, where the supercell energy spectrum is, in each case, projected onto the unperturbed GaAs valence band edge HH and LH wave functions (see Eq. (3.7)). For reference, the lowest panel, (a), shows the calculated $G_\Gamma(E)$ for an unperturbed $Ga_{2048}As_{2048}$ supercell, which is simply a single peak of unit amplitude, centred at the energy of the $Ga_{2048}As_{2048}$ HH/LH energy levels.

We observe in Fig. 3.5 that the calculated VBE energy shifts upwards with increasing Bi composition, x, and that the overall Γ character of the valence band maximum state decreases with increasing x, as expected from the BAC model of Eq. (3.3). The solid circles and triangles in Fig. 3.6 show how the calculated fractional Γ character, $f_{\Gamma,v} = |\langle \psi_{v,0} | \psi_{v,1} \rangle|^2$, of each of the two highest valence states decreases with increasing Bi composition in the supercells considered.

The solid line in Fig. 3.6 shows the expected variation in Γ character using Eq. (3.3) with $E_{HH/LH} = 1.03x$ eV, $E_{Bi} = -0.18$ eV and $V_{Bi} = \beta_{Bi}\sqrt{x}$, with $\beta_{Bi} = 1.13$ eV. It can be seen that a good fit is obtained at low x, with the quality of the fit decreasing at higher values of x, as we move into a more disordered regime in the alloy.

A resonant feature associated with the upper eigenvalue, E_+, of the Hamiltonian of Eq. (3.1) has been observed in PR measurements [25, 26] on GaN_yAs_{1-y} for $x>0.2\%$, giving a relatively sharp feature until x reaches approximately 3 %, beyond which composition the feature broadens and weakens, when the resonant state becomes degenerate with the L-related CB levels [27]. It has not proved possible to observe such a feature at any composition in $GaBi_xAs_{1-x}$, and this has led

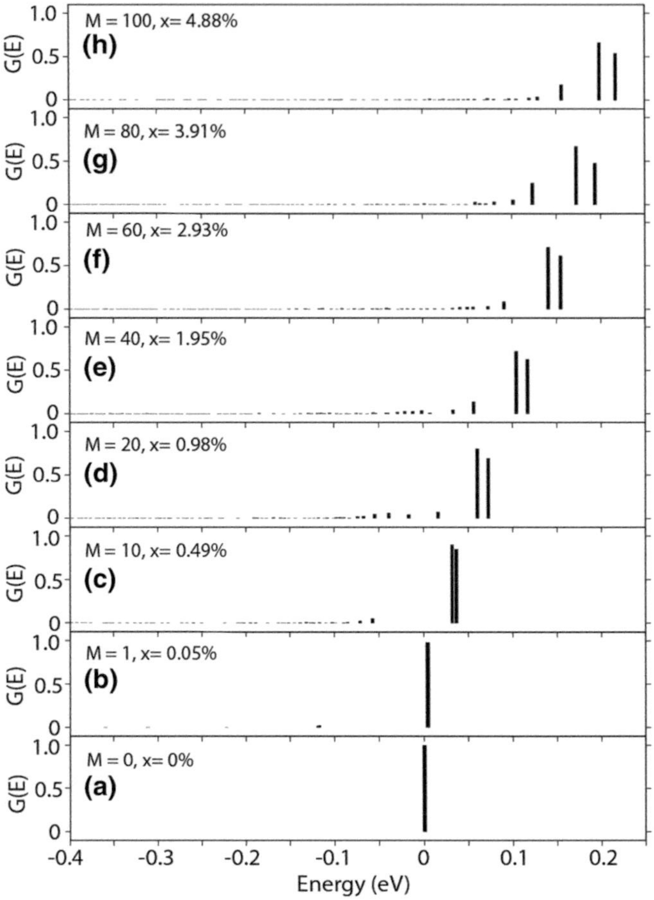

Fig. 3.5 Combined fractional HH/LH Γ character for a series of disordered $Ga_{2048}Bi_MAs_{2048-M}$ supercells for several Bi compositions, $x = \frac{M}{2048}$. The zero of energy is taken at the GaAs VBE. (Reprinted with permission from [16]. Copyright of the American Physical Society (2011))

to controversy as to whether there is a VB-BAC interaction present in the alloy [17]. It can be seen from Fig. 3.5 that the GaAs VBE character in $GaBi_xAs_{1-x}$ is mixed into a wide range of the alloy valence states so that, unlike in the $GaBi_xP_{1-x}$ case, no single state can be identified as the E_+ level in $GaBi_xAs_{1-x}$. This strong mixing is to be expected, because the width of the resonant level is proportional to the density of states [32, 33]. The large density of valence band states then explains the failure to observe an E_+ level in measured $GaBi_xAs_{1-x}$ spectra.

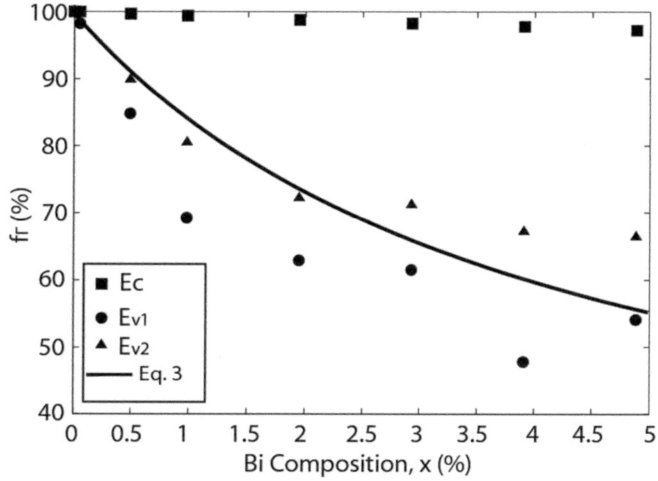

Fig. 3.6 Fractional Γ character (cf. Eq. (3.7)) of the lowest conduction band (E_c) and the highest two valence band (E_{v1} and E_{v2}) edges as a function of Bi composition, x. The *solid line* shows the valence band Γ character calculated using parameters in Eq. (3.3) extracted from the ordered supercell calculations. (Reprinted with permission from [16]. Copyright of the American Physical Society (2011))

3.4.2 Band Gap and Spin-Orbit-Splitting Energies Versus x in $GaBi_xAs_{1-x}$

Figure 3.7 shows how the fractional Γ character of the SO band edge evolves with increasing Bi composition. As the Bi composition increases, the Γ character spreads out over several states and it becomes unclear to choose a particular energy for E_{SO}. The values of the E_{SO} energies from panels (b) to (g) of Fig. 3.7 are obtained by taking a weighted average of the energies with their corresponding Γ character as well as by picking the energies with the highest Γ character. The values from the two methods come out to be very close (typically within 0.1 meV of each other) and we deduce that both methods are equally good in this case. The Bi-induced disorder leads to a broadening of the projected Γ character, $G_Γ(E)$, with increasing Bi composition, but no anticrossing behaviour is observed in this broadened spectrum. Hence, the shift and broadening of the SO state projection is attributable to conventional alloying and disordering effects. For this reason we conclude that our calculations do not support the existence of an additional Bi resonant state interacting with the SO band, as introduced in the $\mathbf{k} \cdot \mathbf{p}$ Hamiltonian of Alberi et al. [6].

Figure 3.8 shows the calculated variation of the lowest conduction, the two highest valence, and the spin-split-off band edge energies with increasing Bi composition x in the disordered 4,096 atom $GaBi_xAs_{1-x}$ supercells. It can be seen that the calculated conduction band edge energy shifts down linearly with increasing Bi composition by about 28 meV per % of Bi replacing As. The solid squares in Fig. 3.6

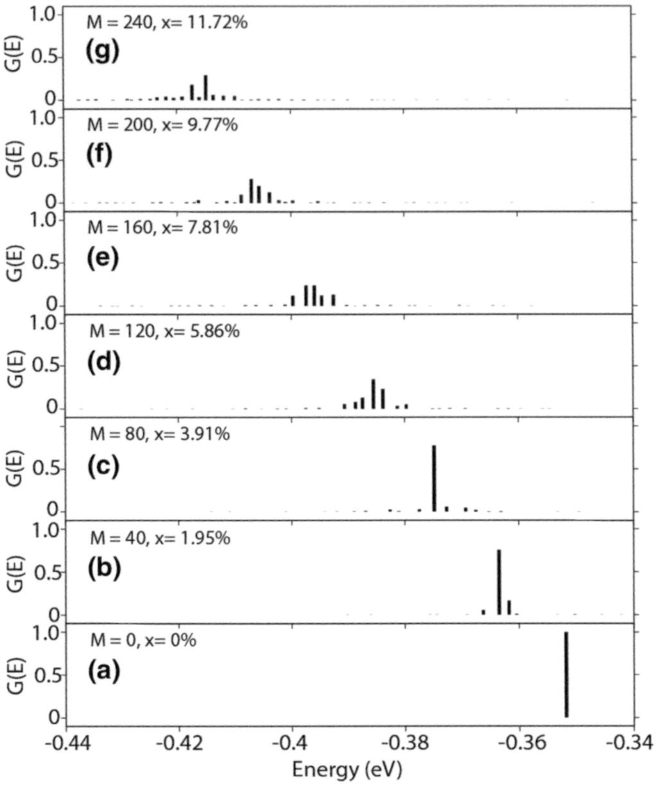

Fig. 3.7 Fractional SO Γ character for a series of disordered $Ga_{2048}Bi_MAs_{2048-M}$ supercells for several Bi compositions, $x = \frac{M}{2048}$. (Reprinted with permission from [16]. Copyright of the American Physical Society (2011))

show that E_c retains $\geq 97\,\%$ GaAs conduction band Γ character for $x \leq 5\,\%$. With a linear decrease in the energy of E_c and the Γ character remaining close to 100 %, we conclude that the evolution of the conduction band states is also best described using a conventional alloy model.

An important feature of $GaBi_xAs_{1-x}$ alloys is that, as the Bi composition increases, the decreasing band gap $E_g = E_c - E_{v1}$ and the increasing spin-orbit-splitting energy $\Delta_{SO} = E_{v1} - E_{SO}$ result in a crossover of these two energies, with $E_g < \Delta_{SO}$ at higher Bi compositions. It has been suggested that this should lead to a suppression of the dominant CHSH Auger recombination mechanism in $GaBi_xAs_{1-x}$ at higher Bi compositions, which could be highly beneficial for the design of high efficiency lasers and semiconductor optical amplifiers (SOAs) operating at telecommunication and at longer wavelengths [3, 14]. Recent PR measurements [8] have confirmed this crossover of the energy gap and spin-orbit-splitting energies in bulk $GaBi_xAs_{1-x}$ alloys for compositions above approximately $x = 10.5\,\%$.

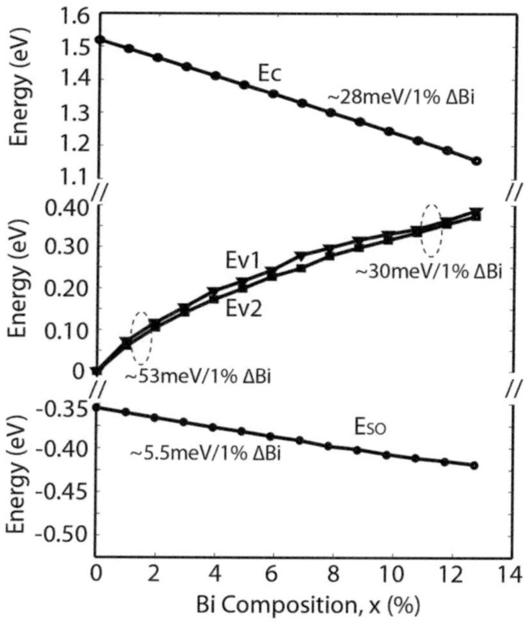

Fig. 3.8 Variation of the lowest conduction band (E_c), highest two valence band (E_{v1} and E_{v2}) and spin-split-off band (E_{SO}) edge energies as a function of Bi composition x in 4,096 atom disordered $GaBi_xAs_{1-x}$ supercells. The conduction band edge E_c decreases with a linear slope, changing by about 28 meV for every 1 % increase in the Bi composition. The highest two valence band edges E_{v1} and E_{v2} exhibit a non-linear variation with Bi composition. For small Bi composition, the variation is larger—about 53 meV per % Bi. For higher Bi compositions, this decreases to roughly 30 meV per % Bi. The spin-split-off band E_{SO} also decreases linearly, by 5.5 meV per % Bi. (Reprinted with permission from [16]. Copyright of the American Physical Society (2011))

Figure 3.9 plots the calculated band gap E_g and spin-orbit-splitting energy Δ_{SO} as a function of Bi composition in 4,096 atom disordered $GaBi_xAs_{1-x}$ supercells. The calculations were undertaken using the low temperature $GaBi_xAs_{1-x}$ band parameters presented here (E_g(GaAs) = 1.519 eV), and the CBE energy has then been shifted down by 92 meV for all Bi compositions, to fit to the GaAs energy gap at room temperature. Each data point shown in the figure is calculated by taking the average of three different random distributions of Bi atoms on anion sites in the $GaBi_xAs_{1-x}$ supercell. The results presented indicate that the decreasing band gap and increasing spin-orbit-splitting do indeed lead to a crossover of E_g and Δ_{SO} at approximately 11 % Bi composition, in agreement with PR measurements [8]. For a broader comparison with other experiments, we have also included several sets of experimentally measured band gaps and spin-orbit-splitting energies as indicated in Fig. 3.9 along with our calculated values. Our results exhibit a close match with the experimental data points over the entire composition range considered, confirming the validity of our model.

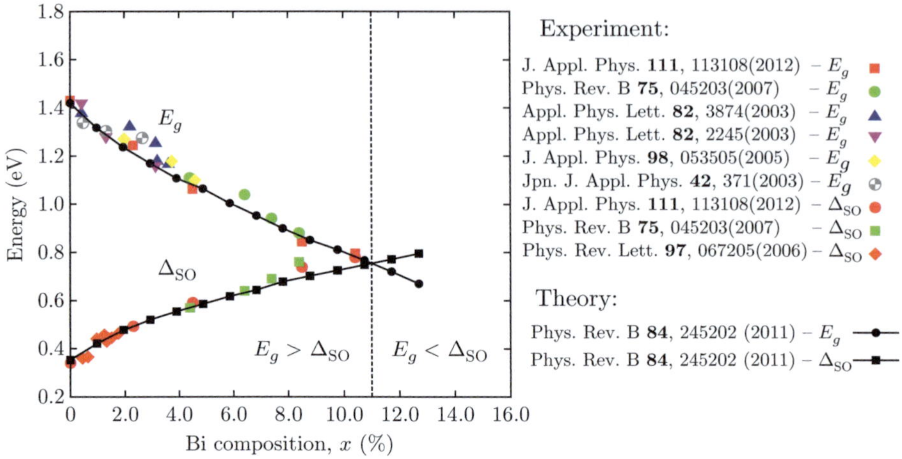

Fig. 3.9 Room temperature variation of the band gap (E_g) and spin-orbit-splitting energy (Δ_{SO}) as a function of Bi composition (x) in $GaBi_xAs_{1-x}$. *Filled black squares* (*circles*) denote the variation of E_g (Δ_{SO}) with x, obtained from sp^3s^* tight-binding calculations on a series of large, disordered $GaBi_xAs_{1-x}$ supercells; the *solid black lines* have been added as a guide to the eye. Various symbols: spectroscopic measurements of E_g and Δ_{SO} using a variety of techniques. The *vertical dashed line* at $x \approx 11$ % denotes the Bi composition at which a crossover to $E_g < \Delta_{SO}$ occurs in the alloy

3.5 12-Band k · p Hamiltonian for Dilute Bismide Alloys

In this section, we apply the tight-binding model to strengthen the case for the valence band BAC interaction by demonstrating that modification of an 8-band $\mathbf{k}\cdot\mathbf{p}$ Hamiltonian for GaAs to include a Bi composition-dependent BAC interaction between the GaAs VBE and lower-lying Bi-related resonant states allows us to describe the band structure of ordered $GaBi_xAs_{1-x}$ supercells, in the vicinity of the band edges, with a high degree of accuracy. The validity of the model is then reinforced by considering its application to disordered $GaBi_xAs_{1-x}$ alloys.

A previous attempt by Alberi et al. [6] to create a $\mathbf{k}\cdot\mathbf{p}$ model for $GaBi_xAs_{1-x}$ was based on an estimation of the energy of the Bi impurity state, E_{Bi}, which relied on the experimentally observed value of approximately 0.1 eV above the GaP VBE in GaP:Bi. The valence band offset of 0.5 eV between GaAs and GaP was then invoked to give a value of E_{Bi} lying 0.4 eV below the GaAs VBE. Such an approach does not account fully for the chemical and size differences between a P and a Bi atom, and between an As and a Bi atom.

The model reported by Alberi et al. [6] also included an additional BAC interaction between the GaAs SO band and a Bi-related band lying approximately 2 eV below the GaAs VBE, resulting in a 14-band Hamiltonian. The BAC interaction parameter β_{Bi} was then obtained by fitting the resulting variation of the $GaBi_xAs_{1-x}$ valence and SO bands to match the experimentally observed reduction (increase) of the band gap (spin-orbit-splitting energy) with increasing Bi composition.

Our detailed analyses presented in Sect. 3.4 indicate that the variation of the SO band edge in $GaBi_xAs_{1-x}$ can be described using a conventional alloy model, without the need to include the SO-BAC interaction of Alberi et al. The observed strong bowing of Δ_{SO} with Bi composition is instead accounted for both by the BAC induced upward shift of the VBE energy and a small downward shift of the SO band edge energy, attributable to the conventional alloying effect.

Furthermore, explicit calculation of E_{Bi} and β_{Bi} is possible in the framework of the tight-binding model [16], eliminating the need to fit the BAC parameters to experimental data. Our calculations suggest that the Bi impurity states lie approximately 180 meV below the GaAs VBE, as opposed to the estimated 400 meV of [6].

Based on the insights gained from tight-binding calculations, we derive here a 12-band $\mathbf{k}\cdot\mathbf{p}$ model for $GaBi_xAs_{1-x}$ which includes BAC interactions between the GaAs VBE and Bi impurity states lying below the VBE. The model is shown to be in excellent agreement with full tight-binding calculations and experimental measurements of the band gap and spin-orbit-splitting energy across a large composition range. Combined with a detailed tight-binding based analysis it demonstrates emphatically the validity of the BAC model as applied to $GaBi_xAs_{1-x}$.

3.5.1 Derivation of the 12-Band Model from Supercell Calculations

Since tight-binding calculations indicate that dilute Bi incorporation is to be considered as impurity-like, we begin in the case of $GaBi_xAs_{1-x}$ with a GaAs host matrix and parametrise the corresponding 8-band model directly from the tight-binding Hamiltonian [35, 36]. Then, guided by tight-binding calculations on ordered $Ga_MBi_1As_{M-1}$ supercells, we extend to a 12-band model by including BAC interactions, V_{Bi}, between the GaAs VBE and the fourfold degenerate Bi-related resonant impurity states at energy E_{Bi}.

We begin by diagonalising the Hamiltonian of the equivalent Bi-free Ga_MAs_M supercell at Γ, from which we obtain the 8-band basis states $|u_1\rangle,\ldots,|u_8\rangle$. This is possible since the Ga_MAs_M zone centre states form a complete set which can be used as basis states to calculate the Ga_MAs_M and $Ga_MBi_1As_{M-1}$ band structure. Following this approach we construct an 8-band parameter set for GaAs directly from the FTB calculations which leads to an 8-band model that is able to reproduce the FTB band structure of a given Ga_MAs_M supercell to a high degree of accuracy in the vicinity of the zone centre.

The derived set of GaAs parameters represent a typical tight-binding fit with, e.g., the calculated value of the Kane interband momentum matrix element ($P=11.1$ eV Å) being about 75 % of its measured value [35, 37]. We note that this does not affect the validity of the conclusions to be drawn from the model and recommend that the Bi-related parameters presented in Tables 3.1 and 3.2 be combined with a standard set of GaAs $\mathbf{k}\cdot\mathbf{p}$ parameters, from, e.g., ref. [37], when using the 12-band Hamiltonian.

Table 3.2 Calculated virtual crystal contributions to the band edge energy shifts due to Bi incorporation in $GaBi_xAs_{1-x}$ and $InBi_xAs_{1-x}$

Host matrix	α_{Bi} (eV)	γ_{Bi} (eV)	κ_{Bi} (eV)
GaAs	2.82	0.55	1.01
InAs	2.60	0.01	1.01

Using the basis states obtained from the FTB calculations we evaluate the dependence of the band edge energies on Bi composition as $\langle u_i | \hat{H}(x) | u_i \rangle, i = 1,\ldots,8$. We calculate that the band edge energies vary with Bi composition as:

$$E_{\text{CB}}(x) = E_{\text{CB},0} - \alpha_{\text{Bi}} x \tag{3.8}$$

$$E_{\text{HH}}(x) = E_{\text{LH}}(x) = E_{\text{VB},0} + \kappa_{\text{Bi}} x \tag{3.9}$$

$$E_{\text{SO}}(x) = E_{\text{SO},0} - \gamma_{\text{Bi}} x \tag{3.10}$$

where the conduction and spin-split-off band edges vary linearly with Bi composition and are well described by Eqs. (3.8) and (3.10). The calculated variations of the band edge energies in $GaBi_xAs_{1-x}$ and $InBi_xAs_{1-x}$ are listed in Table 3.2.

Based on the analysis of Sect. 3.3 we can account for the effect of Bi incorporation on the GaAs VBE via a BAC interaction with coupling strength β_{Bi} between the GaAs VBE and lower-lying Bi-related resonant impurity levels at energy E_{Bi}. We can choose the fourfold degenerate states at energy E_{Bi} to have LH and HH symmetry, so that the pairs of impurity states then interact solely with the GaAs LH and HH states, respectively. The inclusion of these impurity states into the $\mathbf{k}\cdot\mathbf{p}$ basis gives a 12-band $\mathbf{k}\cdot\mathbf{p}$ Hamiltonian for $GaBi_xAs_{1-x}$:

$$\begin{pmatrix}
E_{\text{CB}}(x) & -\sqrt{3}T_+ & \sqrt{2}U & -U & 0 & 0 & 0 & 0 & -T_- & -\sqrt{2}T_- & 0 & 0 \\
 & E_{\text{HH}}(x) & \sqrt{2}S & -S & V_{\text{Bi}} & 0 & 0 & 0 & -R & -\sqrt{2}R & 0 & 0 \\
 & & E_{\text{LH}}(x) & Q & 0 & V_{\text{Bi}} & T_+^* & R & 0 & \sqrt{3}S & 0 & 0 \\
 & & & E_{\text{SO}}(x) & 0 & 0 & \sqrt{2}T_+^* & \sqrt{2}R & -\sqrt{3}S & 0 & 0 & 0 \\
 & & & & E_{\text{Bi}} & 0 & 0 & 0 & 0 & 0 & 0 & 0 \\
 & & & & & E_{\text{Bi}} & 0 & 0 & 0 & 0 & 0 & 0 \\
 & & & & & & E_{\text{CB}}(x) & -\sqrt{3}T_- & \sqrt{2}U & -U & 0 & 0 \\
 & & & & & & & E_{\text{HH}}(x) & \sqrt{2}S^* & -S^* & V_{\text{Bi}} & 0 \\
 & & & & & & & & E_{\text{LH}}(x) & Q & 0 & V_{\text{Bi}} \\
 & & & & & & & & & E_{\text{SO}}(x) & 0 & 0 \\
 & & & & & & & & & & E_{\text{Bi}} & 0 \\
 & & & & & & & & & & & E_{\text{Bi}}
\end{pmatrix} \tag{3.11}$$

where $V_{Bi} = \beta_{Bi}\sqrt{x}$ and the band edge energies are taken to vary as in Eqs. (3.8)–(3.10). The Bi impurity states are taken to be dispersionless. We work here within the framework of the 8-band $\mathbf{k}\cdot\mathbf{p}$ model, as described for instance in [38], where the definitions of the matrix elements E_{CB}, E_{HH}, E_{LH}, E_{SO}, P, Q, R, S, $T_{\pm}$ and U can also be found. Since the Hamiltonian is a Hermitian matrix, we neglect to show the below diagonal entries; they can be obtained from the above diagonal matrix elements by Hermitian conjugation.

Using Eqs. (3.4) and (3.5) we calculate the values of E_{Bi} and β_{Bi} in a series of $Ga_MBi_1As_{M-1}$ supercells. This, combined with an appropriate 8-band parameter set for GaAs, as well as the Bi compositional dependence of the band edges given in Eqs. (3.8)–(3.10), fully parametrises the 12-band Hamiltonian.

In order to compare the results of the 12-band model with the full tight-binding (FTB) calculations we begin by first establishing 8-band $\mathbf{k}\cdot\mathbf{p}$ models for GaAs and InAs. The zone centre (In)GaAs states obtained from the FTB calculation at Γ provide the basis states to parametrise the 8-band GaAs and InAs Hamiltonians directly from their respective sp^3s^* Hamiltonians [39]. The 8-band $\mathbf{k}\cdot\mathbf{p}$ models obtained from this approach for GaAs and InAs are compared to the FTB band structures in Fig. 3.10a and b, where we observe excellent agreement between both sets of calculations in the vicinity of the Γ point.

Using these 8-band models as starting points, we construct the 12-band Hamiltonian for a given Bi-containing supercell by firstly using Eqs. (3.4) and (3.5) to calculate E_{Bi} and β_{Bi} for the supercell. These parameters are then combined with the 8-band parameters of the corresponding Bi-free supercell as well as the Bi composition dependence of the band edge energies [21] to give the appropriate $\mathbf{k}\cdot\mathbf{p}$ Hamiltonian for the $Y_MBi_1As_{M-1}$ supercell.

Figure 3.10c and d compare the band structures calculated using this approach with the FTB calculations of the band structure for the 64 atom, $Ga_{32}Bi_1As_{31}$ and $In_{32}Bi_1As_{31}$ supercells ($x=3.125\,\%$). The calculations demonstrate excellent agreement between the sp^3s^* and 12-band models. The presence of a BAC interaction in the valence band is confirmed by the existence of tight-binding bands that correspond in both cases to the lower energy Bi-related impurity bands (E_-) of the 12-band Hamiltonian. These exemplar calculations confirm the validity of the BAC approach to describe the valence band structure of dilute bismide alloys.

It is, however, important to note that the direct correspondence observed between the Bi impurity bands in the calculations presented in Fig. 3.10c and d occurs because the highest folded valence bands in the 64 atom $Y_{32}As_{32}$ (Y = Ga, In) supercells lie over 1 eV below the host VBE. This results in a low density of host valence states in the proximity of the Bi-related impurity states, when compared to larger supercells in which there is a higher density of folded host matrix states, leading to the Bi impurity forming a resonant state, which is spread over several of the host levels. This energy broadening of the lower-lying impurity bands accounts for the inability to resolve any Bi-related features lying below the VBE in spectroscopic measurements [16].

In summary, the overall agreement of the 12-band $\mathbf{k}\cdot\mathbf{p}$ model with the FTB calculations confirms that the BAC model accurately describes the main effects of Bi on the band structure of ordered (In)$GaBi_xAs_{1-x}$ structures. In the next section, we apply the 12-band $\mathbf{k}\cdot\mathbf{p}$ model to compare the calculated variations of the energy

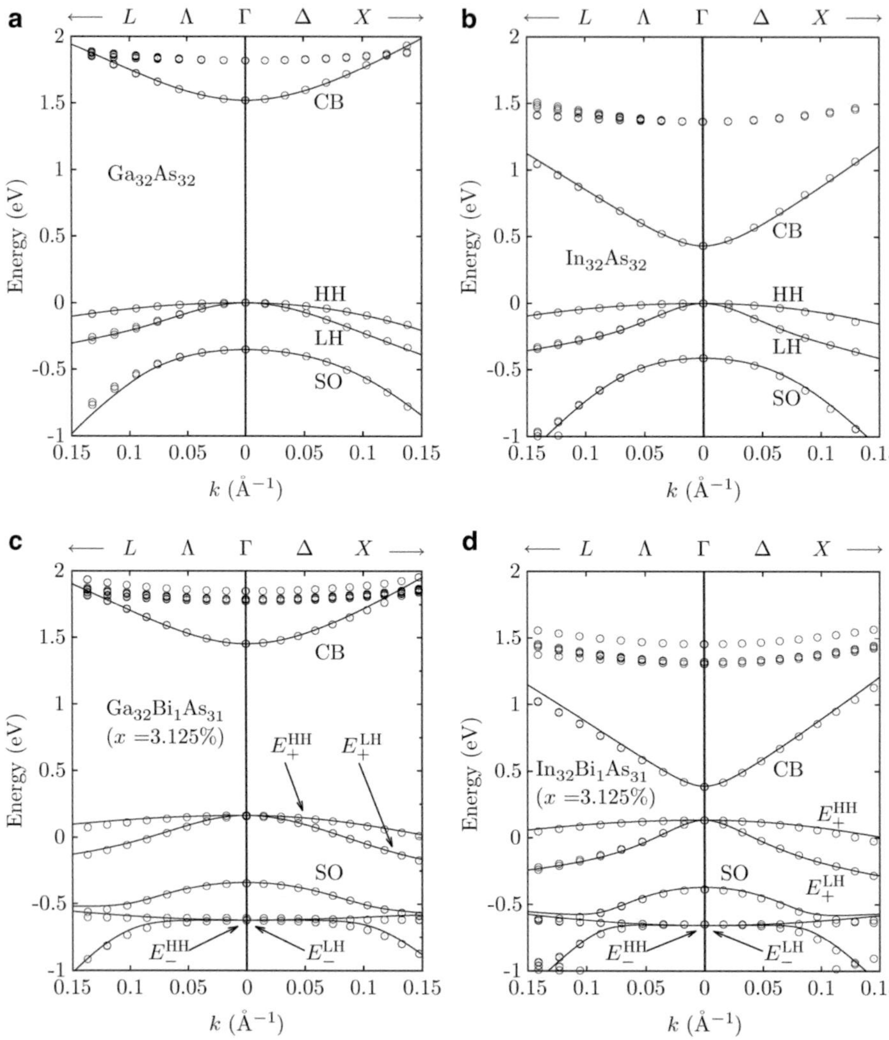

Fig. 3.10 Band structure of (**a**) $Ga_{32}As_{32}$ and (**b**) $In_{32}As_{32}$ calculated using sp^3s^* tight-binding (*open circles*) and 8-band $\mathbf{k}\cdot\mathbf{p}$ (*solid lines*) Hamiltonians. Band structure of (**c**) $Ga_{32}Bi_1As_{31}$ and (**d**) $In_{32}Bi_1As_{31}$ (x=3.125 %) calculated using sp^3s^* tight-binding (*open circles*) and 12-band $\mathbf{k}\cdot\mathbf{p}$ (*solid lines*) Hamiltonians. The band structures are shown here along the Λ and Δ directions, close to the Γ point in the Brillouin zone. Note that the band labels refer to the character of each band at k=0; these labels do not retain their meaning at finite k when bands start to mix

gap, E_g and spin-orbit-splitting energy Δ_{SO} with spectroscopic measurements on a series of $In_{0.53}Ga_{0.47}Bi_xAs_{1-x}$/InP epilayers.

The BAC model therefore provides a valid and accurate description of the modified valence band structure of ordered $GaBi_xAs_{1-x}$ crystals, but with the caveat that the lower-lying Bi impurity bands should not be interpreted as actual material bands—they exist in the model to describe the net effect of the Bi-related impurity states that interact with the host matrix VBE.

The validity of the BAC model is reinforced by showing that the description of the variation of E_g and Δ_{SO} with Bi composition obtained from the 12-band model is in excellent agreement with FTB calculations and experimental measurements across the investigated composition range. In the next section it will be demonstrated that this interpretation remains valid in the case of a disordered $GaBi_xAs_{1-x}$ alloy, where the quality of the fit to the FTB calculated band dispersion using the 12-band model decreases somewhat due to a loss of both short- and long-range order in the crystal.

3.5.2 *Application of the 12-Band Model to $GaBi_xAs_{1-x}$ and $In_{1-x}Ga_xBi_yAs_{1-y}$ Alloys*

As discussed in [16], the inclusion of multiple impurity atoms in a given $GaBi_xAs_{1-x}$ supercell lowers the symmetry and lifts the degeneracy of the associated resonant states, $|\psi_{Bi}\rangle$. As we move towards a disordered alloy, the GaAs VBE interacts with a full distribution of impurity states related to isolated Bi atoms, similar to those observed for N atoms in GaN_yAs_{1-y} [29] and GaN_yP_{1-y} [30], as well as with Bi pairs and clusters, where a Bi pair is formed when a Ga atom has two Bi neighbours. As predicted by the BAC model, the calculated alloy VBE energy shifts upwards and experiences an overall reduction in GaAs Γ character with increasing Bi composition.

Further detailed analysis based on tight-binding calculations has indicated that as the Bi composition increases in $GaBi_xAs_{1-x}$ the resulting disorder and random formation of Bi pairs and clusters gives rise, in addition to the resonant states associated with isolated Bi atoms, to pair and cluster-related impurity states which lie predominantly below the GaAs VBE energy. These states are strongly broadened due to the large density of GaAs valence states with which they are resonant. The GaAs VBE Γ character is then also spread over a wide range of alloy valence states, due to the broad distribution of resonant state energies and the large valence band density of states [16].

In order to demonstrate that the 12-band $\mathbf{k}\cdot\mathbf{p}$ model presented in the previous subsection is nevertheless applicable to the realistic case of a disordered $GaBi_xAs_{1-x}$ alloy, we apply it here to investigate disordered supercells. We use it to calculate E_g and Δ_{SO} in a series of large (4,096 atom) $GaBi_xAs_{1-x}$ supercells in which the Bi atoms are randomly distributed, thereby introducing also pairs and clusters in a statistically random manner.

Figure 3.11a compares the room temperature band gap and spin-orbit-splitting energies calculated from FTB calculations on large, disordered supercells with those calculated using the 12-band model. We see that the two sets of calculations are in excellent agreement, confirming that the BAC model is capable of

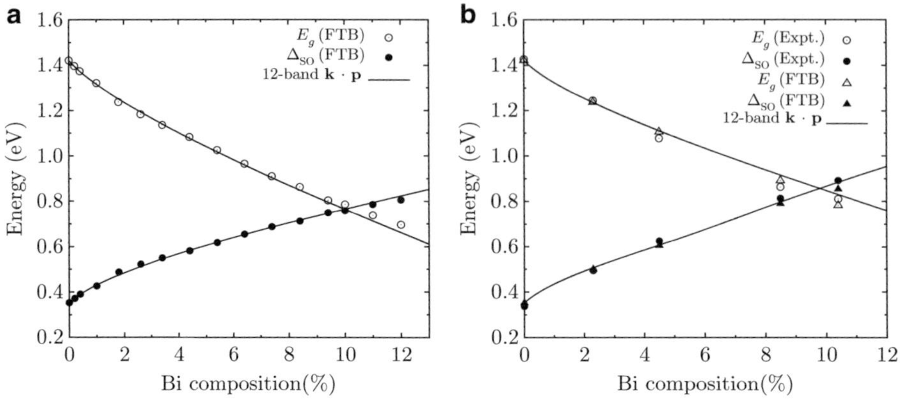

Fig. 3.11 (**a**) Room temperature band gap (E_g) and spin-orbit-splitting energy (Δ_{SO}) as a function of Bi composition in free-standing $GaBi_xAs_{1-x}$, calculated using the 12-band $\mathbf{k}\cdot\mathbf{p}$ model (*solid lines*) and full tight-binding (FTB) calculations on large, disordered supercells (*open and closed circles*). (**b**) *Open symbols*: photo-reflectance measured (*open circles*) and FTB calculated (*open triangles*) values of E_g as a function of Bi composition in a series of strained $GaBi_xAs_{1-x}$/GaAs layers. *Closed symbols*: photo-reflectance measured (*closed circles*) and FTB calculated (*closed triangles*) values of Δ_{SO} as a function of Bi composition in a series of strained $GaBi_xAs_{1-x}$/GaAs layers. *Solid lines*: variation of E_g and Δ_{SO} in bulk $GaBi_xAs_{1-x}$ lattice-matched to GaAs, calculated using the 12-band $\mathbf{k}\cdot\mathbf{p}$ model

reproducing the variations of the band edge energies in an accurate manner with Bi composition over a significant composition range.

Furthermore, we have shown in [16] that the sp^3s^* tight-binding model we have developed for $GaBi_xAs_{1-x}$ is in very good agreement with a range of experimental measurements of E_g and Δ_{SO} in $GaBi_xAs_{1-x}$, across the full composition range for which experimental data exists in the literature. We therefore conclude that, since the 12-band $\mathbf{k}\cdot\mathbf{p}$ model is in close agreement with the sp^3s^* tight-binding model, that the 12-band model presented here provides a realistic and accurate description of the critical energies in the $GaBi_xAs_{1-x}$ band structure.

In order to confirm this, Fig. 3.11b compares the calculated variation of E_g and Δ_{SO} from the 12-band model with Bi composition for $GaBi_xAs_{1-x}$ grown epitaxially on GaAs against PR measurements on a series of MBE-grown $GaBi_xAs_{1-x}$/GaAs layers [8, 40, 41]. We see that the 12-band model is in very good agreement with experiment over the full composition range considered, confirming its usefulness for materials and device analysis.

Some limitations of the model should, however, be noted. In a disordered $GaBi_xAs_{1-x}$ alloy, the loss of short- and long-range order introduces localised distortions of the crystal lattice. It also gives rise to localised states associated with Bi pairs and clusters, which can lie close to the GaAs VBE energy. Both of these effects perturb the band structure significantly, the first by reduction of symmetry and the second by strong perturbation of the host valence band structure due to hybridisation of the band structure with this distribution of localised states. While we have shown that the BAC model provides an accurate description of the

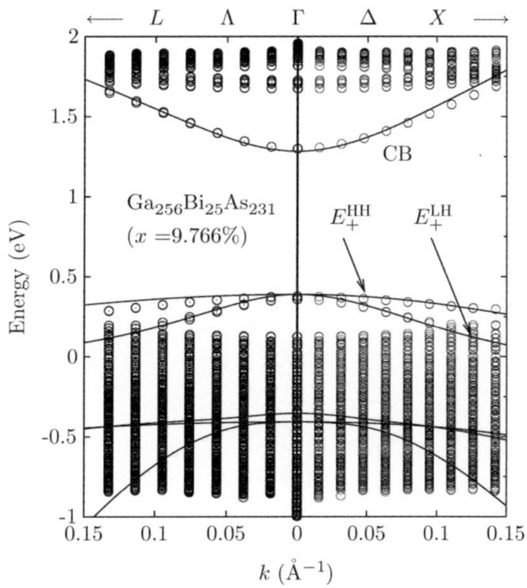

Fig. 3.12 Calculated band dispersion of a disordered $Ga_{256}Bi_{25}As_{231}$ (x=9.766%) supercell, along the Λ and Δ directions, close to the centre of the Brillouin zone, using the sp^3s^* (*open circles*) and 12-band $\mathbf{k} \cdot \mathbf{p}$ (*solid lines*) Hamiltonians. Note that the band labels refer to the character of each band at k=0; these labels do not retain their meaning at finite k when bands start to mix

band edge energies at Γ, we might expect that the quality of the description of the band dispersion should degrade somewhat compared to the ordered case of Fig. 3.10c and d—this indeed turns out to be the case. Figure 3.12 compares the calculated FTB band dispersion close to the zone centre of a disordered $Ga_{256}Bi_{25}As_{231}$, $x \sim 9.8\,\%$, supercell with that of the 12-band model. We see that in line with the above discussion, the band edge energies at Γ are well described by the 12-band model but that the quality of the description of the band dispersion reduces more rapidly with increasing wave vector than in the ordered case.

We now extend the 12-band $\mathbf{k} \cdot \mathbf{p}$ model outlined in Sect. 3.5.1 for $GaBi_xAs_{1-x}$ and $InBi_xAs_{1-x}$ to the quaternary dilute bismide alloy $In_{1-x}Ga_xBi_yAs_{1-y}$ [42]. Using the derived $\mathbf{k} \cdot \mathbf{p}$ Hamiltonian, we demonstrate the validity of the approach outlined in Sects. 3.3 and 3.5.1 by comparing the calculated variations of the room temperature band gap and spin-orbit-splitting energies with Bi composition against spectroscopic measurements on a series of molecular beam epitaxy (MBE) grown $In_{1-x}Ga_xBi_yAs_{1-y}$ samples, grown pseudomorphically on InP:Fe substrates. Details of the sample growth and spectroscopic measurements can be found in refs. [43] and [44].

In order to extend the 12-band models of Sect. 3.5.1 to the quaternary $In_{1-x}Ga_xB$ i_yAs_{1-y} alloy we linearly interpolate between the BAC parameters and Bi compositional dependence of the band offsets for $GaBi_xAs_{1-x}$ and $InBi_xAs_{1-x}$ derived from large supercell tight-binding calculations. The $In_{1-x}Ga_xBi_yAs_{1-y}$ layer is taken as being uniformly strained onto an InP substrate. We use Végard's law to interpolate between the lattice and elastic constants of the constituent binary compounds in order to obtain those of the quaternary alloy. We also use linear interpolation for the conduction and valence band deformation potentials.

Figure 3.13 shows the variation of E_g and Δ_{SO} as a function of x for $In_{0.53}Ga_{0.47}Bi_xAs_{1-x}$ grown pseudomorphically on InP, calculated using the 12-band

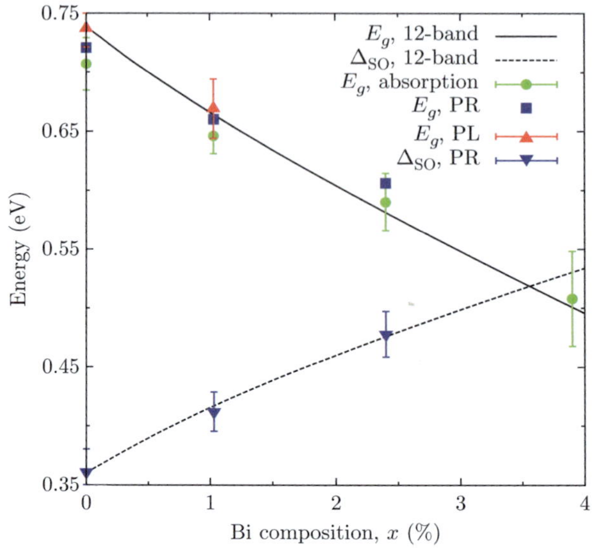

Fig. 3.13 Measured and calculated variation of the band gap (E_g) and spin-orbit-splitting energy (Δ_{SO}) as a function of Bi composition (x) for $In_{0.53}Ga_{0.47}Bi_xAs_{1-x}$ grown pseudomorphically on InP. *Solid and dashed lines*: calculation of E_g and Δ_{SO} using the 12-band $\mathbf{k} \cdot \mathbf{p}$ model. Various symbols: values of E_g and Δ_{SO} obtained from optical absorption, photo-modulated reflectance (PR) and photo-luminescence (PL) measurements on a series of MBE-grown $In_{0.53}Ga_{0.47}Bi_xAs_{1-x}$/InP samples [43, 44]

$\mathbf{k} \cdot \mathbf{p}$ model, compared to optical absorption, PR and photo-luminescence (PL) measurements of the samples discussed in [42–44]. Overall, we note the excellent agreement between the calculated and measured values of E_g and of Δ_{SO} for the different compositions, confirming that the 12-band $\mathbf{k} \cdot \mathbf{p}$ model provides an accurate description of the valence band structure of dilute bismide alloys. We conclude that the 12-band $\mathbf{k} \cdot \mathbf{p}$ model should therefore be well suited to analyse the electronic and optical properties of dilute bismide materials and device structures, as discussed further in Sect. 3.7.

3.6 Tight-Binding and $\mathbf{k} \cdot \mathbf{p}$ Models for the Electronic Structure of Dilute $GaBi_xN_yAs_{1-x-y}$

Co-alloying N and Bi in GaAs offers further potential for band structure engineering to design highly efficient and temperature insensitive photonic components operating at telecomm and longer wavelengths. The large variation in the GaN_yAs_{1-y} and $GaBi_xAs_{1-x}$ band gaps with nitrogen and bismuth composition enables the $GaBi_xN_yAs_{1-x-y}$ material system to span a wide wavelength range, even when grown on a GaAs substrate. Since replacing As by N and Bi in GaAs introduces, respectively, tensile and compressive strain, the N and Bi compositions in $GaBi_xN_yAs_{1-x-y}$ can be chosen to lattice-match the alloy to a GaAs substrate, while allowing access to the 1.3 μm and 1.5 μm wavelength ranges and beyond. Theoretical calculations have shown that $GaBi_xN_yAs_{1-x-y}$ retains the large upward bowing of

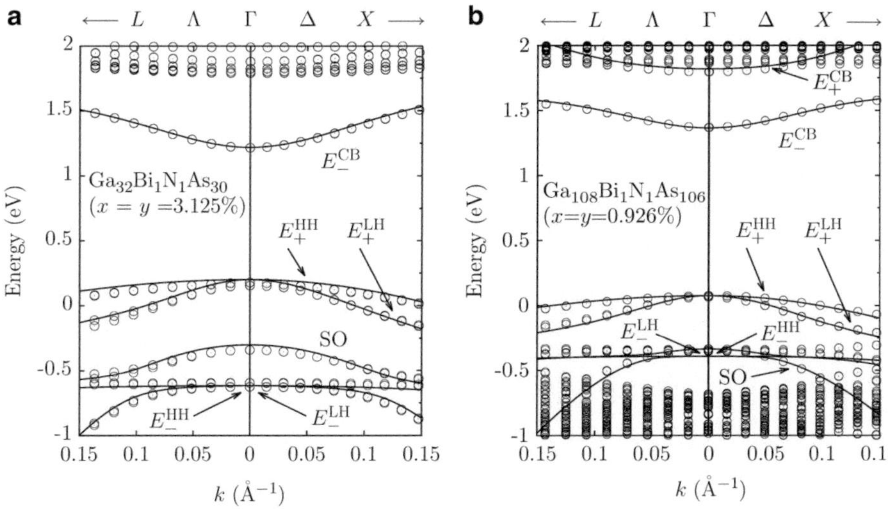

Fig. 3.14 Calculated band dispersion of ordered, cubic (**a**) $Ga_{32}Bi_1N_1As_{30}$ ($x = y = 3.125$ %) and (**b**) $Ga_{108}Bi_1N_1As_{106}$ ($x = y = 0.926$ %) supercells, along the **Λ** and *Δ* directions, close to the centre of the Brillouin zone, using the sp^3s^* (*open circles*) and 14-band **k·p** (*solid lines*) Hamiltonians. Note that the N-related impurity band in (*a*) is out of scale. Note that the band labels refer to the character of each band at k=0; these labels do not retain their meaning at finite k when bands start to mix

the spin-orbit-splitting energy present in $GaBi_xAs_{1-x}$ [20]. Because co-alloying Bi with N causes a giant reduction in the band gap, this offers the possibility to eliminate the CHSH Auger recombination mechanism in lasers operating across a wide wavelength range, from 1.5 μm through mid-infrared wavelengths [45].

By examining in detail the effects of co-alloying N and Bi in a series of ordered $GaBi_xN_yAs_{1-x-y}$ supercells using the tight-binding model, we showed that the effects of Bi and N on the GaAs electronic structure are largely decoupled from each other [19–21]. This indicates that the $GaBi_xN_yAs_{1-x-y}$ electronic structure admits analysis in terms of separate N- and Bi-induced BAC interactions occurring in the conduction and valence bands, respectively [21].

In order to derive a 14-band Hamiltonian for $GaBi_xN_yAs_{1-x-y}$ from the tight-binding calculations we first examine the effects of Bi and N when both are doped into GaAs in dilute quantities. We examine a series of ordered, cubic $Ga_MBi_1N_1As_{M-2}$ supercells (in which the N-Bi separation is chosen to be as large as possible) using an sp^3s^* tight-binding model of $GaBi_xN_yAs_{1-x-y}$. In this manner we determine that, even for the smallest supercell considered, the effects of isolated N and Bi atoms on the electronic structure are decoupled in the ordered case and hence that the electronic structure of ordered $GaBi_xN_yAs_{1-x-y}$ crystals admits a simple interpretation in terms of separate N- and Bi-related BAC interactions in the conduction and valence bands, respectively.

Guided by this, we modify the 12-band Hamiltonian of Eq. (3.11) to account for the presence of N in the usual way [46]. We begin with a $Ga_MBi_1N_1As_{M-2}$ supercell and calculate the Bi (N) related BAC parameters in the equivalent N (Bi) free supercell.

Figure 3.14a and b shows the calculated band dispersion close to the zone centre of 64 and 216 atom $GaBi_xN_yAs_{1-x-y}$ supercells using the sp^3s^* and 14-band **k · p**

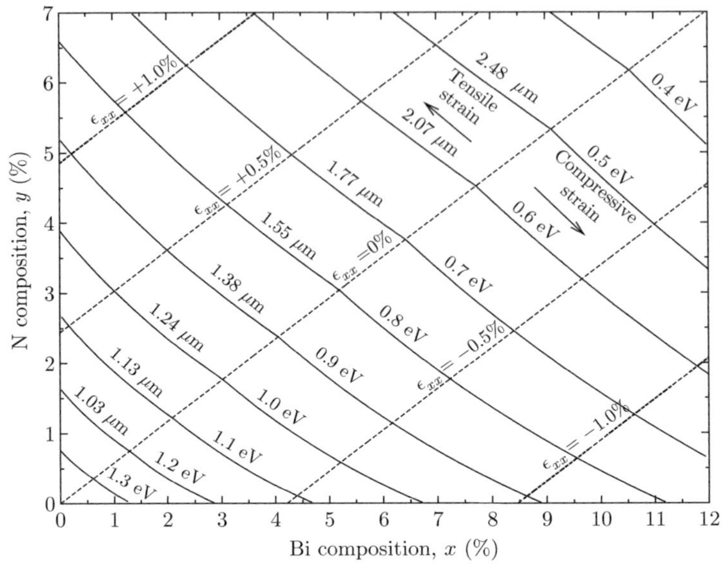

Fig. 3.15 Calculated room temperature variation of the band gap, E_g, as a function of Bi and N composition, x and y, for $GaBi_xN_yAs_{1-x-y}$ epitaxially grown on GaAs. *Solid lines* denote paths in the composition space along which E_g is constant; *dashed lines* denote paths along which the strain is constant. The giant band gap bowing in this material system allows access to a very wide wavelength range with only moderate strain on a GaAs substrate

Hamiltonians, from which we see that the extension to a 14-band model accounts accurately for N and Bi co-alloying, by reproducing accurately the band edges energies and modified band dispersion of the FTB calculations.

Figure 3.15 shows the variation of the band gap as a function of Bi and N composition for $GaBi_xN_yAs_{1-x-y}$ epitaxially grown on GaAs, calculated using the 14-band $\mathbf{k}\cdot\mathbf{p}$ model. This shows the wide wavelength range accessible using $GaBi_xN_yAs_{1-x-y}$ grown on a GaAs substrate. Figure 3.16 shows the calculated difference between the spin-orbit-splitting energy and the band gap energy over the same composition range, showing the regions in which $\Delta_{SO} > E_g$ and suppression of CHSH Auger recombination can be expected to occur.

3.7 Consequences of Bi Incorporation for Optoelectronic Device Operation

Using the 12-band $\mathbf{k}\cdot\mathbf{p}$ model of Sect. 3.5 we turn now to investigate the influence of Bi incorporation on the optoelectronic properties of dilute bismide quantum wells (QWs). We calculate in this section the band structure and material gain of two prototypical Bi-containing QWs using the 12-band $\mathbf{k}\cdot\mathbf{p}$ model to investigate general trends that should be expected in dilute bismide-based devices.

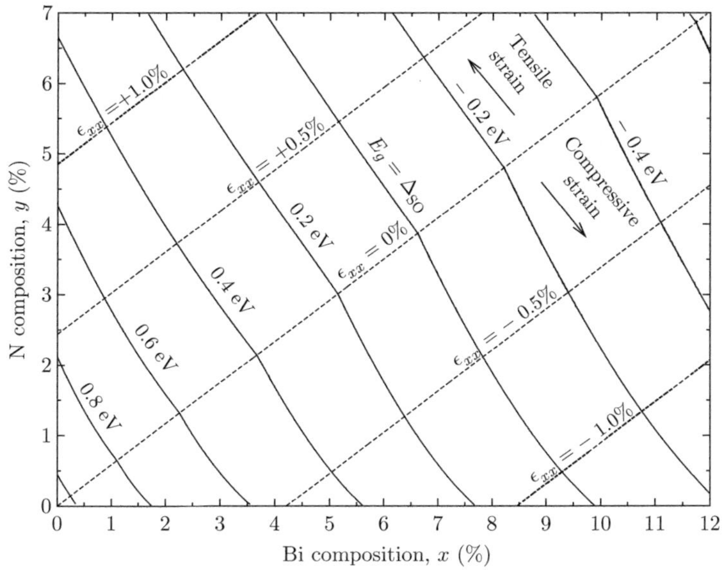

Fig. 3.16 Calculated room temperature variation of the difference between the band gap and spin-orbit-splitting energy ($E_g - \Delta_{SO}$), as a function of Bi and N composition, x and y, for $GaBi_xN_yAs_{1-x-y}$ epitaxially grown on GaAs. *Solid lines* denote paths in the composition space along which $E_g - \Delta_{SO}$ is constant; *dashed lines* denote paths along which the strain is constant. Regions in which $E_g < \Delta_{SO}$ (to the right of the $E_g = \Delta_{SO}$ contour) indicate alloys in which suppression of CHSH Auger recombination and inter-valence band absorption can be expected to occur

The QW calculations are performed in the envelope function approximation using an efficient plane wave expansion method [47] to numerically calculate the QW band structure and envelope functions. We use a gain model based on density matrix theory [48] to evaluate the influence of the modified QW band structure and density of states on the material gain in the QWs. The optical matrix elements used in the calculation of the material gain have been evaluated using the general formalism of Szmulowicz [49]. As a reference we also consider an equivalent Bi-free QW, so that we can identify the main influences of Bi incorporation on the optoelectronic properties of the QWs. The QW structures considered are listed in Table 3.3. QW 1 is an ideal $GaBi_xAs_{1-x}$/GaAs QW containing 3.5 % Bi and is based on a recent MOVPE-grown multiple quantum well structure [50].

The calculated conduction band offset, including the effects of compressive strain, for the 3.5 % Bi QW (QW 1) is 48 meV, compared to 118 meV for the InGaAs/GaAs case (QW 2). This results in very weak electron confinement in the $GaBi_{0.035}As_{0.965}$/GaAs QW. Figure 3.17a–c shows the calculated valence band structures for the different QW structures. The smaller strain in QW 1 results in a reduced splitting between the HH and LH valence band edge states compared to the splitting in QW 2. The calculated HH (LH) band offset for QW 1 is 209 meV (183 meV), as compared to 64 meV (−10 meV) in QW 2 (there are no confined light-hole states in

Table 3.3 QW structures for which the band structure, optical matrix elements and material gain have been calculated. QWs 1 and 3 are $GaBi_xAs_{1-x}$ QWs containing 3.5 % and 10 % Bi, respectively, while QW 2 is an $In_yGa_{1-y}As$ QW with the In composition chosen to have the same bulk conduction band to heavy-hole energy gap as QW 1 at the zone centre. All calculations are performed at $T=300$ K using a basis of 61 plane waves in a supercell with a total length of 21 nm

QW structure	Well	Well width (nm)	Barrier
QW 1	$GaBi_{0.035}As_{0.965}$	7	GaAs
QW 2	$In_{0.18}Ga_{0.82}As$	7	GaAs
QW 3	$GaBi_{0.1}As_{0.9}$	7	GaAs

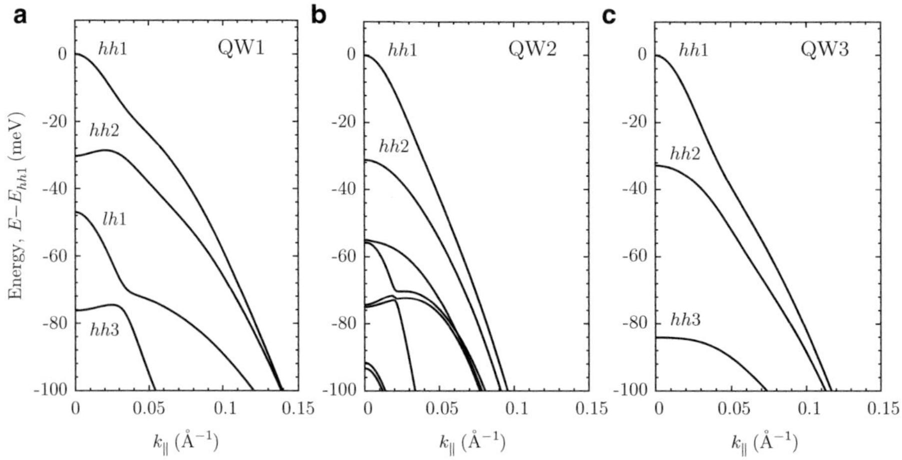

Fig. 3.17 Calculated valence band structure of the QW structures described in Table 3.3: (**a**) QW 1, (**b**) QW 2 and (**c**) QW 3. The 12-band **k** · **p** Hamiltonian of Sect. 3.5 was used to calculate the band structures of QWs 1 and 3, and an 8-band **k** · **p** Hamiltonian was used to calculate the band structure of QW 2. In each case the zero of energy is taken at the energy of the *hh*1 bound state, in order to facilitate comparison between the different valence band structures. The predominant character of the bound states at the zone centre ($k_{\|}=0$) is indicated next to bands containing bound states. The calculated QW band gaps $E_{e1}-E_{hh1}$ for QWs 1, 2 and 3 are 1.20 eV (1.03 μm), 1.30 eV (0.95 μm) and 0.95 eV (1.31 μm), respectively

QW 2). This leads to an increased density of states near the valence band maximum in the $GaBi_{0.035}As_{0.965}$ structure compared to either of QWs 2 or 3.

Figure 3.17c shows the calculated valence band structure of QW 3, the QW structure containing 10 % Bi. Because the conduction band offset increases with increasing Bi composition, we find an improved electron confinement in this structure, resulting also in an increased overlap between the confined electron and hole states. Furthermore, the increased compressive strain in the 10 % Bi structure also leads to an increased HH–LH splitting compared to the 3.5 % Bi case (QW 1), resulting in a reduced density of states at the valence band maximum. Both of these factors lead to an increase in the calculated material gain as a function of carrier density compared to the 3.5 % case, as discussed further below.

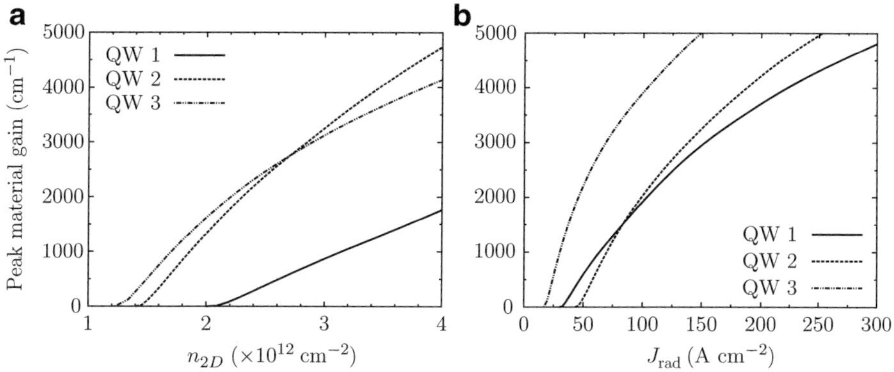

Fig. 3.18 Calculated variation of the peak material gain as a function of (**a**) sheet carrier density n_{2D} and (**b**) radiative current density $J_{\rm rad}$, for the QW structures described in Table 3.3

Figure 3.18a and b shows the calculated variation of the peak material gain as a function of sheet carrier density and of radiative current density, respectively, in QWs 1, 2 and 3. We calculate a strong reduction of the material gain in the 3.5 % Bi-containing QW (QW 1) compared to the InGaAs QW (QW 2). There are three main reasons for this reduction in the material gain. Firstly, as discussed above, the reduced conduction band offset in QW 1 compared to QW 2 results in a weak confinement of the single bound electron state ($e1$) in QW 1. This results in a reduction of the $e1$-$hh1$ spatial overlap, which decreases the calculated value of the $e1$-$hh1$ optical matrix element in QW 1 compared to QW 2. Secondly, Bi-induced mixing between the host matrix VBE states and localised Bi-related resonant impurity states in the valence band further reduces the $e1$-$hh1$ optical matrix element in QW 1 compared to QW 2. Finally, the increased hole effective mass in QW 1 compared to QW 2—due to the BAC interaction between the host matrix VBE states and localised Bi-related resonant impurity states—results in an increase in the density of states at the VBE in QW 1 compared to QW 2, which further reduces the material gain.

This analysis suggests that it should be expected that laser devices based upon QW structures such as that presented in [50] should perform poorly compared to their Bi-free counterparts. However, the primary aspect of the $GaBi_xAs_{1-x}$ electronic structure that can be exploited for the design of highly efficient photonic devices with suppressed Auger losses and reduced temperature sensitivity is the transition to an $E_g < \Delta_{SO}$ regime for $x \gtrsim 11\,\%$. It is for this reason that the 10 % Bi structure is of interest, since at this composition we approach the $E_g < \Delta_{SO}$ crossover and reach a QW band gap of approximately 1.3 μm, which lies at the shorter wavelength telecomm band.

Because the conduction band offset increases with increasing Bi composition, we find an improved electron confinement in the 10 % Bi structure, resulting also in an increased overlap between the confined electron and hole states. The increased compressive strain in the 10 % Bi structure also leads to an increased HH–LH splitting compared to the 3.5 % Bi case, resulting in a reduced density of states at the VBE for QW 3 compared to QWs 1 and 2. Both of these factors lead to an increase in the calculated material gain as a function of carrier density for QW 3 compared to the QW 1 case.

Our initial analysis here suggests that structures with higher Bi compositions are ideal candidate structures for laser emission since the gain and differential gain in a $GaBi_xAs_{1-x}$/GaAs-based QW laser designed to suppress Auger recombination will be comparable to that which can be achieved in an InGaAs/GaAs QW structure. The initial outlook on the prospect of designing an efficient dilute bismide-based laser is therefore promising, but we note that our calculations still require us to quantify the effects of Bi incorporation on intra-valence band absorption and Auger recombination in dilute bismide-based QW laser structures, which is the subject of ongoing research we are undertaking.

3.8 Conclusions

In summary, this chapter reports on an sp^3s^* tight binding model including a valence force field relaxation scheme, which we use to calculate the electronic structure of large dilute bismide alloy supercells. Supercell calculations confirm that Bi interacts via a band-anticrossing interaction with the valence band states of the host material, similar to the well-established band-anticrossing interaction of N with the conduction band states. The experimentally measured large reduction in the $GaBi_xAs_{1-x}$ band gap with increasing x is explained via a BAC induced shift in the valence band edge energy and a conventional alloy type shift in the conduction band edge energy. We further derived 12- and 14-band $\mathbf{k}\cdot\mathbf{p}$ models from the tight-binding calculations for $(In)GaBi_xAs_{1-x}$ and $GaBi_xN_yAs_{1-x-y}$ alloys by adding Bi- and N-related states into an 8-band $\mathbf{k}\cdot\mathbf{p}$ model for GaAs. Our models accurately reproduce the experimentally observed crossover to $E_g < \Delta_{SO}$ in $GaBi_xAs_{1-x}$ at $x \approx 11\,\%$, and in $In_{0.53}Ga_{0.47}Bi_xAs_{1-x}$/InP at $x \approx 3.5\,\%$.

Our theoretical analysis of the electronic structure of the quaternary bismide–nitride alloy $GaBi_xN_yAs_{1-x-y}$ indicates that Bi and N largely influence the electronic structure independently of one another, introducing large modifications of the valence and conduction bands via independent band-anticrossing interactions in the respective bands. This offers a large degree of control over the band structure while simultaneously allowing lattice-matching to a GaAs substrate at telecomm wavelengths.

Finally, our analysis of the band structure and material gain in ideal $GaBi_xAs_{1-x}$ quantum wells indicates that in high Bi-composition structures, where Auger recombination is expected to be suppressed, the material gain is comparable to that of a standard Bi-free quantum well. We therefore conclude that, at least in principle, dilute bismide alloys offer a pathway to good quality photonic devices operating at telecomm wavelengths.

Overall, our theoretical analysis provides detailed insight into the electronic structure of dilute bismide alloys and confirms the significant promise of these alloys to realise telecomm and infrared wavelength photonic devices operating at higher efficiencies and with reduced temperature sensitivity.

Acknowledgements C.A. Broderick acknowledges financial support from the Irish Research Council under the Embark Initiative (RS/2010/2766). M. Usman and E.P. O'Reilly acknowledge financial support from the European Union Seventh Framework Programme (BIANCHO; FP7-257974) and Science Foundation Ireland (10/IN.1/I299). M. Usman acknowledges the use of computational resources from the National Science Foundation funded Network for Computational Nanotechnology through http://www.nanohub.org. The tight binding calculations were performed using the *N*ano*E*lectronic *MO*deling (NEMO 3-D) simulator [51, 52].

References

1. O'Reilly, E.P., Lindsay, A., Klar, P.J., Polimeni, A., Capizzi, M.: Semicond. Sci. Technol. **24**, 033001 (2009)
2. Henini, M.: Dilute Nitride Semiconductors. Elsevier, Oxford (2005)
3. Broderick, C.A., Usman, M., Sweeney, S.J., O'Reilly, E.P.: Semicond. Sci. Technol. **27**, 094011 (2012)
4. Shan, W., Walukiewicz, W., Ager III, J.W., Haller, E.E., Geisz, J.F., Friedman, D.J., Olson, J.M., Kurtz, S.R.: Phys. Rev. Lett. **82**, 1221 (1999)
5. Tixier, S., Webster, S.E., Young, E.C., Teidje, T., Francoeur, S., Mascarenhas, A., Wei, P., Schiettekatte, F.: Appl. Phys. Lett. **86**, 112113 (2005)
6. Alberi, K., Wu, J., Walukiewicz, W., Yu, K.M., Dubon, O.D., Watkins, S.P., Wang, C.X., Liu, X., Cho, Y.J., Furdyna, J.: Phys. Rev. B **75**, 045203 (2007)
7. Fluegel, B., Francoeur, S., Mascarenhas, A., Tixier, S., Young, E.C., Tiedje, T.: Phys. Rev. Lett. **97**, 067205 (2006)
8. Batool, Z., Hild, K., Hosea, T.J.C., Lu, X., Tiedje, T., Sweeney, S.J.: J. Appl. Phys. **111**, 113108 (2012)
9. Higashi, T., Sweeney, S.J., Phillips, A.F., Adams, A.R., O'Reilly, E.P., Uchida, T., Fujii, T.: IEEE J. Sel. Top. Quant. Electron. **5**, 413 (1999)
10. Sweeney, S.J., Adams, A.R., Silver, M., O'Reilly, E.P., Watling, J.R., Walker, A.B., Thijs, P.J.A.: Phys. Stat. Sol. B **525**, 211 (1999)
11. Silver, M., O'Reilly, E.P., Adams, A.R.: IEEE. J. Quant. Electron. **33**, 1557 (1998)
12. Adams, A.R., Heasman, K.C., Hilton, J.: Semicond. Sci. Technol. **2**, 761 (1987)
13. Sweeney, S.J., Thijs, P.J.A.: 16th IEEE LEOS meeting, Tucson, Arizona, USA, (2003)
14. Sweeney, S.J., Batool, Z., Hild, K., Jin, S.R., Hosea, T.J.C.: In: Proceedings of the 13th International Conference on Transparent Optical Networks, Stockholm, Sweden, 2011
15. Alberi, K., Dubon, O.D., Walukiewicz, W., Yu, K.M., Bertulis, K., Krotkus, A.: Appl. Phys. Lett. **91**, 051909 (2007)
16. Usman, M., Broderick, C.A., Lindsay, A., O'Reilly, E.P.: Phys. Rev. B **84**, 245202 (2011)
17. Deng, H.X., Li, J., Li, S.S., Peng, H., Xia, J.B., Wang, L.W., Wei, S.H.: Phys. Rev. B **82**, 193204 (2010)
18. Zhang, Y., Masarenhas, A., Wang, L.-W.: Phys. Rev. B **71**, 155201 (2005)
19. Broderick, C.A., Usman, M., O'Reilly, E.P.: In: Proceedings of the 13th International Conference on Transparent Optical Networks, Stockholm, Sweden, 2011
20. Usman, M., Broderick, C.A., O'Reilly, E.P.: An atomistic understanding of the electronic structure of GaAs-bismides-nitrides (in preparation, 2013)

21. Broderick, C.A., Usman, M., O'Reilly, E.P.: 12- and 14-band **k** ·**p** Hamiltonians for dilute bismide and bismide-nitride semiconductor alloys (submitted, 2013)
22. Keating, P.N.: Phys. Rev. **145**, 637 (1966)
23. Lazarenkova, O.L., von Allmen, P., Oyafuso, F., Lee, S., Klimeck, G.: Appl. Phys. Lett. **85**, 4193 (2004)
24. Slater, J.C., Koster, G.F.: Phys. Rev. **94**, 1498 (1954)
25. Perkins, J.D., Mascarenhas, A., Zhang, Y., Geisz, J.F., Friedman, D.J., Olson, J.M., Kurtz, S.R.: Phys. Rev. Lett. **82**, 3312 (1999)
26. Klar, P.J., Grüning, H., Heimbrodt, W., Koch, J., Höhnsdorf, F., Stolz, W., Vicente, P.M.A., Camassel, J.: Appl. Phys. Lett. **76**, 3439 (2000)
27. Lindsay, A., O'Reilly, E.P.: Phys. E **21**, 901 (2004)
28. O'Reilly, E.P., Lindsay, A., Tomić, S., Kamal-Saadi, M.: Semicond. Sci. Technol. **17**, 870 (2002)
29. Lindsay, A., O'Reilly, E.P.: Phys. Rev. Lett. **93**, 196402 (2004)
30. Harris, C., Lindsay, A., O'Reilly, E.P.: J. Phys. Condens. Matter **20**, 295211 (2008)
31. Trumbore, F.A., Gershenzon, M., Thomas, D.G.: Appl. Phys. Lett. **9**, 4 (1966)
32. Fano, U.: Phys. Rev. **124**, 1866 (1961)
33. Fahy, S., Lindsay, A., Ouerdane, H., O'Reilly, E.P.: Phys. Rev. B **74**, 035203 (2006)
34. Janotti, A., Wei, S.H., Zhang, S.B.: Phys. Rev. B **65**, 115203 (2002)
35. Lindsay, A., O'Reilly, E.P.: Phys. Stat. Sol. B **216**, 131 (1999)
36. Benchamekh, R., Nestoklon, M., Jancu, J.M., Voison, P.: Chapter 2, Semiconductor Modelling Techniques. Springer Series in Material Science, vol. 159. Springer, Heidelberg (2012)
37. Vurgaftman, I., Meyer, J.R., Ram-Mohan, L.R.: J. Appl. Phys. **89**, 5815 (2001)
38. Meney, A.T., Gonul, B., O'Reilly, E.P.: Phys. Rev. B **50**, 10893 (1994)
39. Lindsay, A., Tomić, S., O'Reilly, E.P.: Solid State Electron. **47**, 443 (2003)
40. Lu, X., Beaton, D.A., Lewis, R.B., Tiedje, T., Zhang, Y.: Appl. Phys. Lett. **95**, 041903 (2009)
41. Lu, X., Beaton, D.A., Lewis, R.B., Tiedje, T., Whitwick, M.B.: Appl. Phys. Lett. **92**, 192110 (2008)
42. Broderick, C.A., Usman, M., O'Reilly, E.P.: Phys. Stat. Sol. B **250**, 773 (2013)
43. Petropoulos, J.P., Zhong, Y., Zide, J.M.O.: Appl. Phys. Lett. **99**, 031110 (2011)
44. Marko, I.P., Batool, Z., Hild, K. Jin, S.R., Hossain, N., Hosea, T.J.C., Petropoulos, J.P. Zhong, Y., Dongmo,P.B., Zide, J.M.O., Sweeney, S.J.: Appl. Phys. Lett. **101**, 221108 (2012)
45. Sweeney, S.J., Jin, S.R.: J. Appl. Phys. **113**, 043110 (2012)
46. Tomić, S., O'Reilly, E.P., Fehse, R., Sweeney, S.J., Adams, A.R., Andreev, A.D., Choulis, S.A., Hosea, T.J.C., Riechert, H.: IEEE J. Sel. Top. Quant. Electron. **9**, 1228 (2003)
47. Healy, S.B., O'Reilly, E.P.: IEEE J. Quant. Electron. **42**, 608 (2006)
48. Ahn, D., Chuang, S.L.: IEEE J. Quant. Electron. **26**, 13 (1990)
49. Szmulowicz, F.: Phys. Rev. B **51**, 1613 (1995)
50. Ludewig, P., Knaub, N., Stolz, W., Volz, K.: J. Cryst. Growth (10.1016/j.jcrysgro.2012.07.002, 2012)
51. Klimeck, G., Ahmed, S.S., Bae, H., Kharche, N., Clark, S., Haley, B., Lee, S., Naumov, M., Ryu, H., Saied, F., Prada, M., Korkusinski, M., Boykin, T.B., Rahman, R.: IEEE Trans. Electron. Dev. **54**, 2079 (2007)
52. Klimeck, G., Ahmed, S.S., Kharche, N., Korkusinski, M., Usman, M., Prada, M., Boykin, T.B.: IEEE Trans. Electron. Dev. **54**, 2090 (2007)

Chapter 4
Dilute Bismuthides on an InP Platform

Yujun Zhong, Pernell Dongmo, and Joshua Zide

Abstract Incorporating a small amount of bismuth into conventional III–V compounds (GaAs, InGaAs) creates materials known as dilute bismuthides. They have attracted increasing attention over recent years due to their unique optical and electrical properties. They are potential candidates for mid-infrared (mid-IR) devices because of the reduced band gap resulting from valence band anticrossing (VBAC). Meanwhile, they serve as good thermoelectric materials due to the combination of high thermoelectric power factor inherited from InGaAs and the reduced thermal conductivity caused by the incorporation of bismuth. One advantage of InGaBiAs on an InP platform over GaBiAs on GaAs is its possibility to grow a film lattice-matched to the substrate, which is more desirable in optoelectronics. The growth conditions of InGaBiAs on an InP platform by molecular beam epitaxy (MBE) are discussed in details. Similar to GaBiAs growths, low growth temperature and moderate Bi/As ratio are beneficial for bismuth to incorporate. The compositions of InGaBiAs samples are studied by high resolution X-ray diffraction (HRXRD) and Rutherford backscattering spectrometry (RBS). The results from reciprocal space mapping (RSM) indicate that most of the samples are nearly 100 % strained. The band gaps of InGaBiAs are measured by spectrophotometry and modeled by VBAC theory. The good agreement between the experimental and simulation shows an effective band gap reduction due to the incorporation of bismuth. Photo reflectance (PR) and contactless electroreflectance (CER) studies confirm the results from spectrophotometry and indicate the band gap between conduction band and spin orbit is not affected by the variation of bismuth concentration. Unintentionally doped InGaBiAs samples exhibit promising electrical properties

Y. Zhong • P. Dongmo • J. Zide (✉)
Department of Materials Science and Engineering, University of Delaware, Newark, DE 19716, USA
e-mail: yjzhong@udel.edu; pdongmo@udel.edu; zide@udel.edu

H. Li and Z.M. Wang (eds.), *Bismuth-Containing Compounds*, Springer Series in Materials Science 186, DOI 10.1007/978-1-4614-8121-8_4,

comparable to InGaAs and expected lower thermal conductivity. N-type InGaBiAs:Si films with different doping levels show potential application in thermoelectrics and highly conductive materials are promising as contact materials for heterojunction bipolar transistor (HBT) and in other applications.

4.1 Introduction

Research into III–V compounds started in the 1960s and made phenomenal progress in the 1970s through 1990s [1]. The major application areas for GaAs and InP-based materials are high speed electronic and photonic devices due to their high electron mobility and direct band gap. They also possess high radiation hardness and wide operation temperature range, which make them of special interest for space and military use. GaAs-based materials are mostly utilized in microwave devices, high speed digital integrated circuits, and as substrates for light emitting diodes (LED), laser diodes (LD), lasers, and so on. For InP, the most common use is to grow ternary or quaternary alloys which are lattice-matched to the substrates. These materials are particularly interesting for fiber optical communication systems due to their 1.55 μm wavelength fitting right in the low optical loss window. In the 1980s, fast development of optical communication technology has brought forth a variety of InP-based transistors, LEDs, lasers, and photodiodes in the long wavelength region. Hence, the technique of InP bulk crystal growth, InP-based material deposition, and InP-based device processing are becoming increasingly mature. Accordingly, the cost has decreased over the years.

Highly mismatched alloys (HMAs) have become increasingly attractive in recent years due to their interesting properties resulting from the hybridization between the extended energy states of the matrix material and the localized states of the minority component [2]. The minority component is always isovalent with one element in the matrix material but has a large electronegativity mismatch. The incorporation of the minority component will lead to a strong band restructuring which usually generates peaks in the density of states (DOS) or even anticrossing near conduction band or valence band. Novel optical and electrical properties could result from this band restructuring and be applied to new areas like mid-IR detectors and thermoelectrics. Currently, HMAs have been mostly studied in group IV, III–V, and II–V include $Ge_{1-x}Sn_x$ [3], $Ga(In)N_{1-x}As_x$ [4, 5], and $ZnTe_{1-x}Se_x$ [6].

Dilute bismuthides, which are one class of HMAs, are formed by incorporating a small amount of bismuth into III–V alloys such as GaAs and InGaAs. Previous research has been devoted to the growth and properties of $GaBi_xAs_{1-x}$, with incorporation of Bi up to 13 % [7–10]. Recently, there has been limited effort put into the growths of InGaBiAs [8, 11]. This chapter will focus on a detailed exploration of the growths and properties of InGaBiAs on InP substrates, which are similar in many ways to GaBiAs on GaAs but different in several aspects. The incorporation of bismuth will lead to valence band anticrossing (VBAC) because of the overlapping of Bi 6*p* energy level and the top of GaAs or InGaAs valence band, as shown in Fig. 4.1. There has been some skepticism about VBAC, but its use is justified as

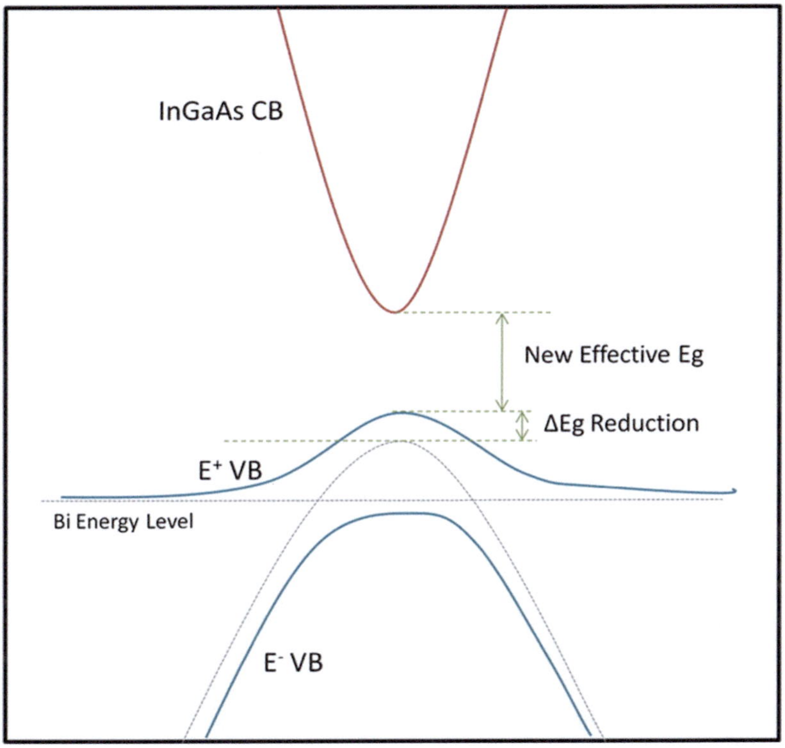

Fig. 4.1 VBAC diagram for InGaBiAs

a practical description of observed phenomena. According to Pauli exclusion principle, electrons with the same quantum number cannot occupy the same energy level; the top valence band will therefore be forced to split into two separate bands E+ and E−. The E+ energy band is shifted towards the conduction band, leading to a reduced band gap. Unlike in dilute nitride, in which the conduction band of GaAs or InGaAs is strongly perturbed, VBAC hardly affects the electronic properties of the conduction band, which gives dilute bismuthide a significant advantage. Growing quaternary alloy InGaBiAs has an edge over GaBiAs in terms of the possibility of depositing films that are lattice-matched to the InP substrate. Because of this, it is theoretically possible to make the cut-off long wavelength limit up to 6 μm with zero strain, which is very desirable for optical and electronic materials.

Another application of InGaBiAs is thermoelectric materials. Thermoelectric materials convert a temperature difference into an electrical voltage or vice versa. They could be used either in power generation (Seebeck effect) or cooling (Peltier effect) applications, and are promising for a variety of sustainable energy applications, including waste-heat recovery. A simple scheme of a thermoelectric device is shown in Fig. 4.2, as a p-type and n-type thermoelectric materials are connected electrically in series and thermally in parallel. The figure of merit, expressed in (4.1), is a crucial factor to judge whether the material is a good thermoelectric material.

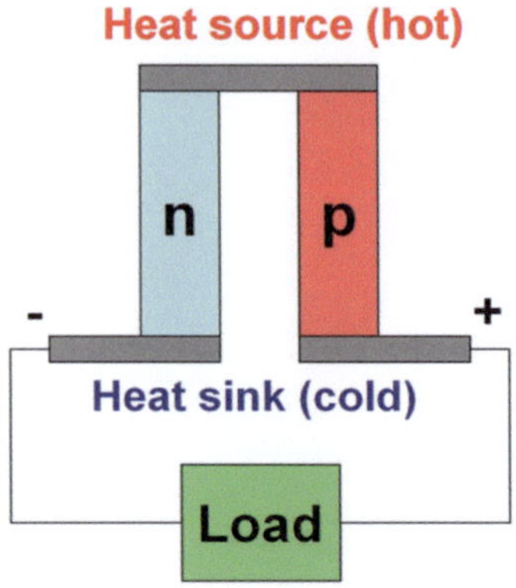

Fig. 4.2 Diagram of a basic thermoelectric component

$$Z = \frac{S^2\sigma}{\kappa} \tag{4.1}$$

Z, defined as the product of electrical conductivity σ, square of Seebeck coefficient *S*, and the reciprocal of thermal conductivity κ, is the figure of merit. *ZT* is a dimensionless coefficient where *T* is the average temperature of the two ends. A larger *ZT* represents a faster energy conversion efficiency which is considered better for thermoelectric materials. The product $S^2\sigma$ is defined as thermoelectric power factor (TPF).

Good thermoelectric materials require high TPF and low thermal conductivity, and InGaBiAs satisfies both. They inherit both a high TPF from InGaAs and a reduced thermal conductivity due to the increased scattering of electrons and phonons lead by the incorporation of large bismuth atoms. Si-doped InGaBiAs have been experimentally proven to be decent n-type thermoelectric materials at room temperature, and highly promising thermoelectric materials at high temperatures. They are also expected to be good p-type thermoelectric materials due to the peak of DOS near Bi 6*p* energy level. Similar to Tl-doped PbTe which is a high ranking p-type thermoelectric material [12], an abrupt increase of DOS below the Fermi level could greatly increase the Seebeck coefficient *S*, as shown in Fig. 4.3. In the end, they are compatible with other good thermoelectric materials such as nanoparticles embedded in III–V compounds.

4.2 Background of MBE

MBE is a well-established technique to grow epitaxial structures for both scientific research and industrial production. It is known for its relatively well-controlled growth conditions and its ability to produce very high quality materials.

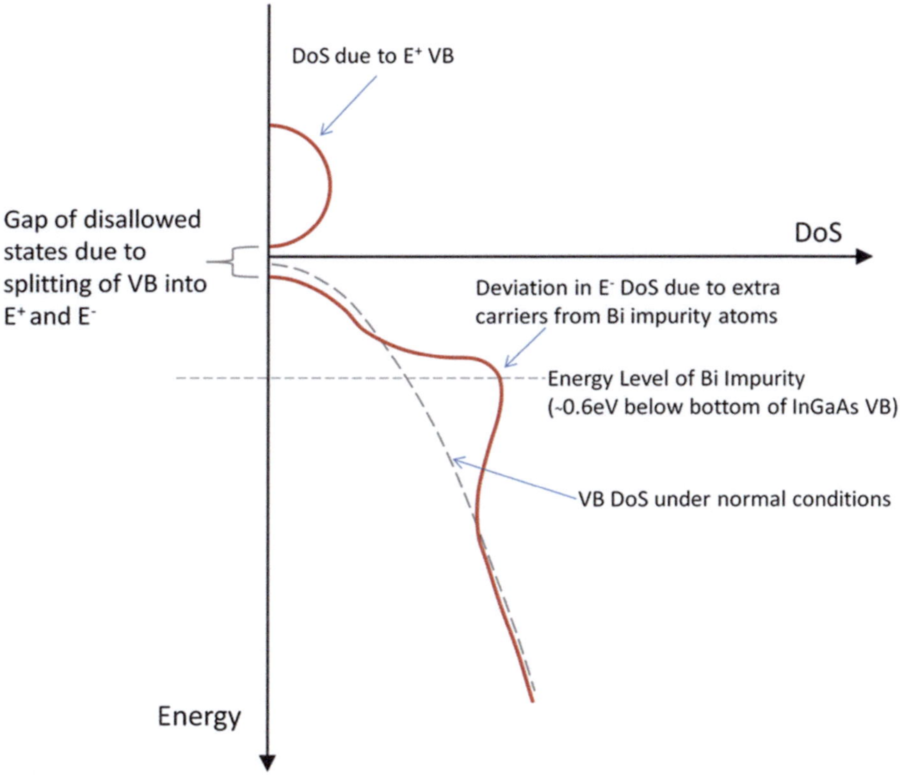

Fig. 4.3 A distorted DOS in the valence band of InGaBiAs due to the presence of Bi impurity level

The thickness of the deposition layer could be precisely modulated at atomic scale. MBE is kinetically dominated in nature, allowing the creation of some metastable alloys or structures under thermodynamically immiscible conditions. There has been work on growing bismuth containing materials by metalorganic vapor phase epitaxy (MOVPE) [13], but the films have quality issues. In the past decade, researchers have focused primarily on MBE-grown dilute bismuthides [7, 8, 14–16].

Band edge (BE) thermometry, which is used to read the substrate temperature, is important in bismuth containing material growths because growth must occur at fairly low growth temperature (around 300 °C). BE thermometry has advantages over a conventional pyrometer because the effectiveness of conventional pyrometers falls quickly when the temperature is below 400 °C. The principle of BE thermometry is that the band gap of a semiconductor is a function of temperature; BE thermometry measures the position of the band edge of the substrate, thereby determining its temperature.

4.3 MBE Growths of InGaBiAs

Recently, we have reported systematic study of the growth conditions of InGaBiAs on InP:Fe substrates by MBE [16]. We have grown both lattice-mismatched and nominally lattice-matched samples. Typically, lattice-mismatched samples are grown by incorporating bismuth into $In_{0.53}Ga_{0.47}As$, while nominally lattice-matched samples are grown by decreasing the In to Ga ratio and make the quaternary alloy lattice-matched to the substrate. Generally, InGaBiAs require relatively low growth temperature (around 300 °C), high As to Bi beam equivalent pressure (BEP) ratio (around 100), and strict stoichiometric III–V ratio. Specifically, two variables in our set of experiments are independently changed: growth temperature and Bi to As BEP ratio.

The growth conditions of dilute bismuthides are far from the regular GaAs or InGaAs growth. First, all growths are carried under relatively low growth temperature because it has been proven that Bi evaporates during a substrate heating from 300 to 400 °C [17]. Second, Bi/As BEP ratio is also critical in determining whether Bi will incorporate successfully. According to literature, Bi works as a very good surfactant [11]. However, if there are too many Bi atoms arriving at one time, they tend to stay on the surface to form Bi droplets, and few will incorporate. In addition, due to the favorable binding energy between group III and As atoms, too many As atoms will impede Bi incorporation as well. The window for Bi to incorporate is therefore very narrow.

$In_xGa_{1-x}Bi_yAs_{1-y}$ samples are grown on (001) oriented InP:Fe substrates in a III–V solid source MBE system, equipped with effusion cells for In, Ga, and Bi, and a two-zone valved cracker source for As_2. The substrate temperature is monitored by measuring the temperature-dependence of the substrate band edge, and BEP is measured using an ionization gauge. The wafers are prebaked in prep chamber at 325 °C for 2 h and then moved to the growth chamber for another bake at 400 °C for 20 min. The native oxide on the InP substrate is desorbed under an As overpressure of 1.0×10^{-5} Torr at about 565 °C until a clear RHEED 2×4 pattern appears. The substrate temperature is then reduced to 250–330 °C and the As BEP is reduced to 2.0×10^{-6} Torr. The Ga BEP ranges from 7.5×10^{-8} to 2.1×10^{-7} Torr, and the BEP of In falls between 2.69×10^{-7} and 5.67×10^{-7} Torr. The flux ratio between Ga and In is fixed, and various BEP values determine different growth rates, which are varying between 450 and 800 nm/h. A 50–70 nm buffer layer $In_{0.53}Ga_{0.47}As$ is first deposited onto the substrate to provide a clean epitaxial surface. Then, the film layer $In_{0.53}Ga_{0.47}Bi_yAs_{1-y}$ with various Bi BEPs (ranging from 1.0×10^{-8} to 3.0×10^{-8} Torr) is grown. During all growths, RHEED displays clear 2×4 patterns.

Figure 4.4a shows Bi concentration as a function of growth temperature. The two sets of points represent different Bi/As ratio, while each set has a constant Bi/As ratio. Generally, Bi content increases with decreased growth temperature and reaches a plateau when the growth temperature is below 285 °C. Figure 4.4b indicates Bi concentration dependence on various Bi BEPs. Bi concentration gradually increases with higher Bi BEPs, and the highest Bi content (average value because this sample has Bi concentration gradient) reaches 6.75 %.

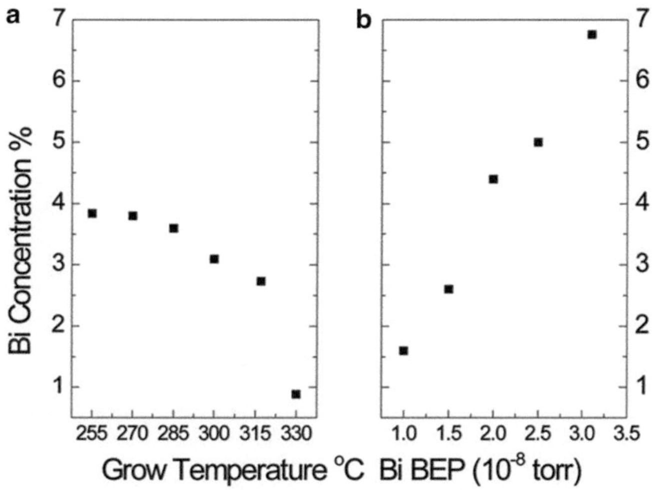

Fig. 4.4 Effects of growth conditions on bismuth incorporation in InGaBiAs. (**a**) Bi content dependence on growth temperature. (**b**) Bi content dependence on Bi BEP

4.4 Characterization and Properties

We have conducted a series of measurements including composition, strain and relaxation, surface morphology, optical, electrical, and thermoelectrical properties of InGaBiAs samples. From these measurements, we conclude that InGaBiAs are promising for both optoelectronic and thermoelectric applications.

4.4.1 Composition

The composition measurements are performed using two methods: high resolution X-ray diffraction (HRXRD) and Rutherford backscattering spectrometry (RBS). The first method is good at sensing the small variation of In/Ga ratio but less accurate to differentiate bismuth concentration, while the latter is great to determine the bismuth concentration but not very sensitive to In/Ga ratio change. The combination of the two methods can be used to reliably determine the composition of each sample.

XRD is a common technique in thin film characterization. It analyzes the X-ray diffraction pattern and follows Bragg equation. For highly perfect crystals, the width of the rocking curve is only a few arc seconds while typical ordinary XRD has a divergence around 500 arc sec. Therefore, a system with high angular resolution and sufficient monochromatic source with adequate scattered intensity is required for HRXRD. In this measurement, the data is collected with coupled omega-2theta

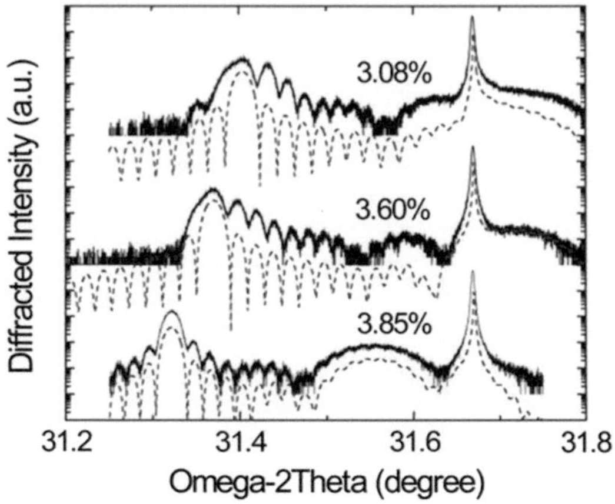

Fig. 4.5 HRXRD (004) omega-2theta scans for $In_xGa_{1-x}Bi_yAs_{1-y}$ epilayers on InP (*solid lines*) with Bi content of 3.08, 3.60, and 3.85 % (RBS). The *dash lines* are simulations (Color online)

scans along the (004) direction by HRXRD. In Fig. 4.5, the film peak is to the left of the substrate peak, which is at 31.6682°. This indicates a lattice constant increase caused by the successful incorporation of large bismuth atoms. The buffer layer $In_{0.53}Ga_{0.47}As$, which is much thinner than the film, modulates the whole spectrum by a larger period leading to the broad bump between the substrate peak and the film peak. The existence of thickness fringes indicates very smooth epitaxial interfaces and reasonably good crystalline quality. The deviation of the buffer layer peak position indicates an indium concentration variation. The simulation results further confirm that In% varies by approximately ±2 % over the whole set of samples (only some are included in this figure). Therefore the determination of Bi content from omega-2theta scans directly is not ideal. In addition, the lattice constant of InGaAs matrix units near the Bi atoms will dilate due to the incorporation of large Bi atoms, rendering the theoretical lattice constants of InAs, GaAs, InBi, and GaBi (6.058, 5.653, 6.686, and 6.324 A [18]) to be no longer applicable.

RBS is ideal to analyze Bi concentration because of its high sensitivity of heavy elements like bismuth. It can be used to quantitatively measure the composition, film thickness, and depth profiling of individual elements. The principle of RBS is that a beam of high energy particles (usually alpha particles, He^{2+}, less commonly, protons, in the MeV range) bombards the sample, and some are backscattered by the sample's atomic nuclei. These backscattered ions are then collected by the energy-sensitive detector. The collected ions with larger residue energy, which are backscattered by heavier atoms, will appear at higher channel number. In addition to composition measurements, RBS can also be used to study the crystal structure by channeling scan. When the incident beam is aligned with rows of atoms in certain direction, the bombarding ions will only be backscattered from the first few monolayers of the atoms (under which the rest of the atoms are buried) which will lead to a much lower backscattering intensity compared to the incidence at a nonaligned angle. However, interstitial defects in the sample will increase the backscattering in the aligned direction, which leads to a peak in the channeling data. Therefore, channeling scan is a good way to indicate the quality of the films.

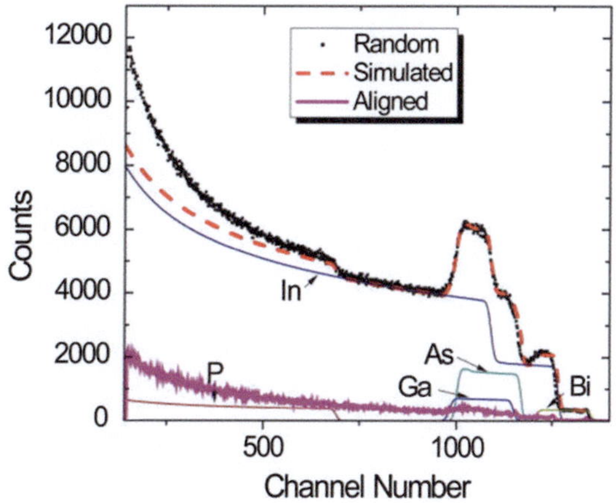

Fig. 4.6 Randomly oriented (*dots*), simulated (*dash line*), and aligned (*solid line*) RBS spectrum for $In_{0.49}Ga_{0.51}Bi_{0.03}As_{0.97}$ epilayers. Components of In, Ga, Bi, As, and P are indicated based on simulation (Color online)

Figure 4.6 shows randomly oriented and aligned RBS spectra and the simulation for one InGaBiAs sample. The individual contributions from each element are specified. Bi concentration can easily be determined from the height of the Bi signal, which is clearly separated from the others. Relatively high crystalline quality for this sample is demonstrated by a much lower yield in the aligned spectrum than the random spectrum. No observation of obvious peaks in the aligned spectrum confirms the lack of interstitial defects and indicates the substitutional incorporation nature for almost all Bi atoms.

4.4.2 *Strain and Relaxation*

In terms of consideration of optical and electrical properties, the degree of relaxation for lattice-mismatched alloys is an important concern because dislocations and defects resulting from relaxation can significantly affect the material quality. Reciprocal space mapping (RSM), which records the intensity at different coordinates in reciprocal space corresponding to independently varied omega-2theta values in real space, is a common tool to study strain and relaxation. Figure 4.7 shows an asymmetric RSM scan along (224) direction for one InGaBiAs sample, where the horizontal and vertical axes represent the in-plane (110) and cross-plane (001) directions. The upper and lower peaks (larger and smaller coordinates in (001) direction) correspond to InP substrate and $In_{0.520}Ga_{0.480}Bi_{0.036}As_{0.964}$ film, respectively, and the series of peaks in between originate from thickness fringes as appeared in Fig. 4.5 as well. The same horizontal coordinate indicates the film is totally strained, while a larger lattice constant in cross-plane direction means the film is under compressive strain. It is worth noting that the broadening of the film peak along ω scan direction rather than the in-plane direction is either due to dislocation-induced tilt/twist or wafer curvature [19] other than the finite thickness

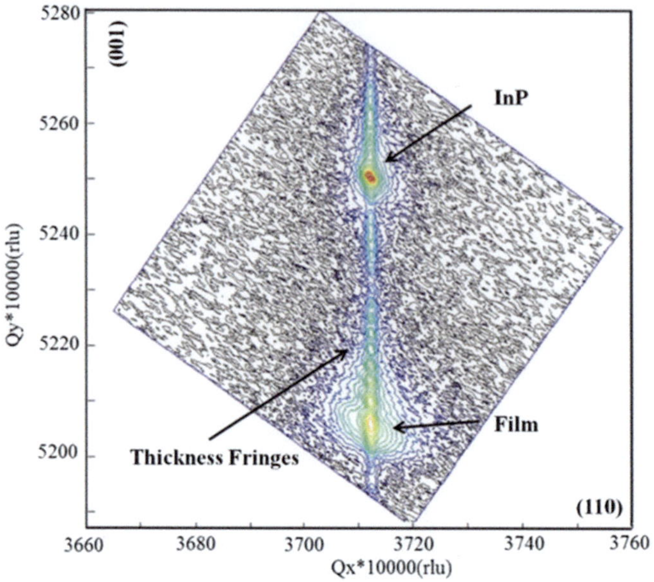

Fig. 4.7 A typical (224) reciprocal space map for $In_{0.52}Ga_{0.48}Bi_{0.036}As_{0.964}$. The Q_x and Q_y represent the in-plane and cross-plane directions. Intensities are on a log scale (Color online)

in the cross-plane direction. Interestingly, as observed in other low-temperature-grown systems as well [20], this fully strained sample's thickness has greatly exceeded the theoretical critical thickness predicted by a simplified Matthews–Blakeslee model [21] which assumes the epilayer and substrate are isotropic elastic materials (280 vs. 69 nm). One possible explanation is the dislocations being isolated from each other [22]. Because of the low growth temperature, the dislocations lack sufficient thermal energy to move. As a result, relaxation cannot be formed. One sample with 6.7 % bismuth is annealed at 550 °C for 30 min, and RSM results show no sign of relaxation, indicating there should be sufficient thermal budget for processing of devices.

4.4.3 Surface Morphology Study

As stated in the previous section, low growth temperature and Bi/As BEP ratio from 1/200 to 3/200 are the optimum growth window for high quality InGaBiAs depositions. However, even within this optimum growth regime, Bi droplets started to form on the surface once the growth temperature is above 285 °C or Bi BEP is above 2.0×10^{-8} Torr, indicating the difficulty of Bi incorporation. All bismuth droplets exhibit clear phase segregation from the brightness contrast shown in the scanning electron microscope (SEM) image in Fig. 4.8, and the energy dispersive

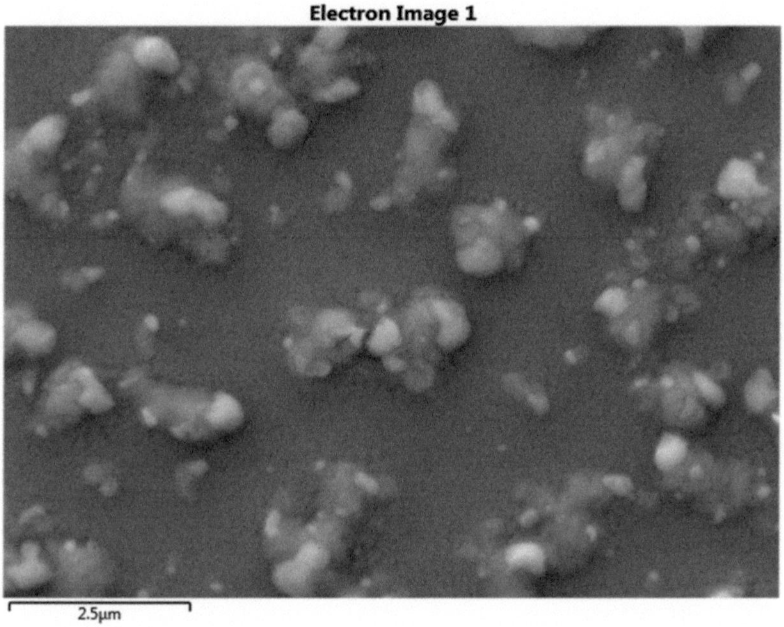

Fig. 4.8 SEM picture for the surface of InGaBiAs with clear observation of droplets. The *brightness* variation indicates the droplets are phase-segregated

spectroscopy (EDS) results suggest one side is bismuth richer, which agrees with the previous report [23]. The density and size of the droplets will increase if the growth temperature or Bi/As BEP ratio increases until a full floating layer is formed on the surface, thereby hindering new adatom incorporation. An extreme case is encountered when the Bi BEP reached 3.1×10^{-8} Torr. There is a bismuth concentration gradient increasing in the growth direction, demonstrated by the inclined shape of Bi signal in the RBS spectrum shown in Fig. 4.9. A rough linear fitting for 0.0625 % Bi%/nm (Bi concentration ranging from 1.4 to 14.4 % in the growth direction) displays the bismuth concentration variation rate due to the limitation of the resolution of RBS measurements, though the actual changing pattern could be more complex. The bismuth concentration gradient suggests a thin bismuth floating layer and it is very close to the highest Bi limit in the given growth conditions. In general, the growth conditions for InGaBiAs agree qualitatively with $GaBi_xAs_{1-x}$ growths [7, 24] though some specific parameters differ.

A preliminary study of the bulk film by high angle annular dark field (HAADF) scanning transmission electron microscopy (STEM) has been conducted. These images are formed by collecting scattered electrons with an annular-shaped dark-field detector which avoids the main beam that is passed through an objective aperture but collects electrons from an annulus around the beam. This is advantageous in terms of signal collection efficiency. This technique, commonly called *z*-contrast STEM, is highly dependent on the atomic number difference between the lattice

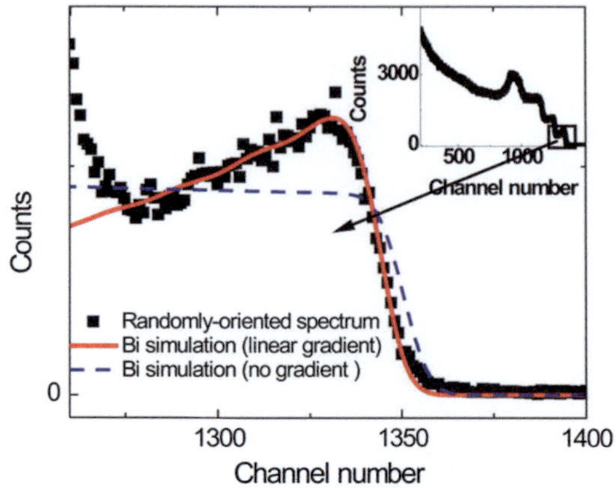

Fig. 4.9 Detail of the large-mass edge of the randomly oriented (*dots*) RBS spectrum for $In_{0.54}Ga_{0.46}Bi_{0.0675}As_{0.9325}$, the whole spectrum is in the inset. *Solid line* is the simulation for Bi signal with an approximately linear Bi concentration gradient of about 0.0625 %/nm in the growth direction. *Dash line* is a conventional fitting with a constant Bi concentration which clearly does not fit these data (Color online)

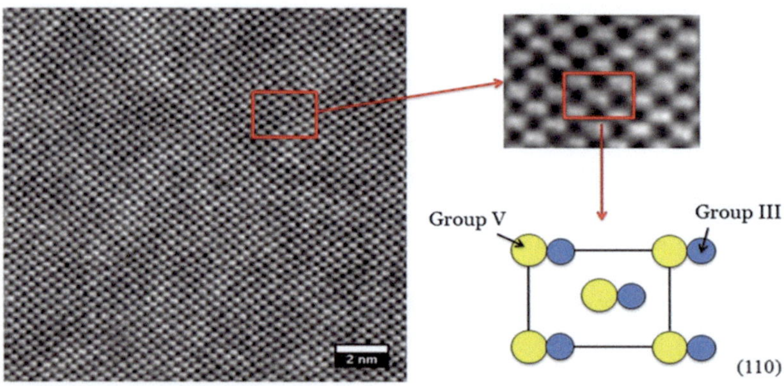

Fig. 4.10 HAADF-STEM picture of the bulk of the film InGaBiAs. The *top right* is a magnification of a region on the *left* side. Each *dot* has a dumbbell shape and is composed of two columns of atoms (group V on the *left* and group III on the *right*) as shown in the *bottom right* (Color online)

atoms, and a brighter region represents more electrons scattered by heavier atoms in the column. The applications of this technique include quantitative elemental mapping, compositional analysis for semiconductor, and observation of dislocations or grain boundaries.

Figure 4.10 shows the HAADF-STEM image taken from (110) direction for InGaBiAs. Inset 1 is a magnification of a localized region, while inset 2 indicates the scheme of image. Each dot has a dumbbell shape and is composed by two columns of atoms. The left side belongs to group III atoms and the right side represents group V. The brightness of each side of the dot indicates an average composition of the columns, where a higher brightness is intuitively related with more heavy elements due to the higher efficiency to scatter electrons. By comparing the contrast between the brightness of the right columns, we can, in principle, conclude whether bismuth clusters form in the bulk film. Preliminary data are inconclusive, but suggest clusters might be forming.

4.4.4 Optical Properties

4.4.4.1 E_0 Band Gap Measurement

The E_0 band gap (band gap between conduction band and valence band) is measured at room temperature by spectrophotometry, which quantitatively measures the reflection and transmission light intensity variation as a function of the source light's wavelength. In our case, transmitted light intensity is utilized for analysis since it is strongly dependent on the E_0 band gap where the transmission curve will display a clear edge once the incident wavelength is resonant with E_0.

Absorption coefficient, describing how absorbent this material behaves for light, is given from Beer's law in the following function:

$$\alpha = -(1/d)\ln(I_{\text{norm}}) \tag{4.2}$$

where α is the absorption coefficient, I_{norm} is the normalized transmission intensity, and d is the thickness of the film. The absorption coefficient could be expressed in another equation as a function of wavelength as well,

$$\alpha = 4\pi n / \lambda \tag{4.3}$$

with n being the refractive index and λ the wavelength of the material, given as

$$\lambda = 1.24 / E_{\text{g}} \tag{4.4}$$

for λ in microns. For direct band gap materials, such as GaAs, InGaAs, and InGaBiAs, the square of absorption coefficient is proportional to the band gap energy E_0. The band gap energy could be determined from the start point of the linear region of α^2 vs. varied band gap energies curve.

Theoretical values are calculated using a simple VBAC model where two interacting energy levels are considered: one is the localized Bi impurity level and the other is the extended states of InGaAs matrix. By applying the perturbation theory, we could conclude the following eigenvalue problem:

$$\begin{vmatrix} E - E_{\text{M}} & V_{\text{MB}} \\ V_{\text{MB}} & E - E_{\text{B}} \end{vmatrix} = 0 \tag{4.5}$$

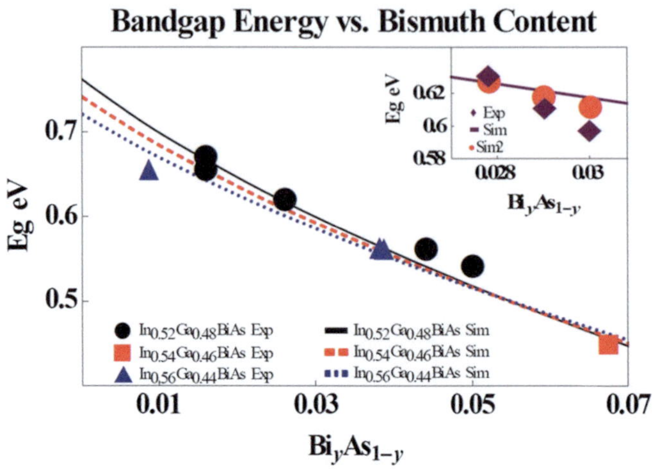

Fig. 4.11 *Lines* represent simulation results (predicted from VBAC model) and *marks* represent experimental gap energies of $In_xGa_{1-x}Bi_yAs_{1-y}$ as a function of Bi concentration. *Solid line/circle*, *dash line/square*, and *dot line/triangle* indicate different In concentration which are shown in the legend. *Inset* shows results of nominally lattice-matched samples band gap values. *Line* represents calculation results with composition of InGaBiAs perfect lattice-matched to InP substrate; *dots* are calculation with actual composition of our samples; *diamonds* are measured band gaps from experiments (Color online)

where E_M and E_B are the energies of the InGaAs matrix valence band edge and of the Bi level relative to the top of the conduction band, respectively, and V_{MB} is the matrix element describing the interaction strength between InGaAs valence band and Bi energy level, which indicates there is a mixing and anticrossing of these states. The two solutions of the equation above can be given as

$$E_{\pm} = \frac{(E_B + E_M \pm [(E_B - E_M)^2 + 4V_{MB}^2]^{1/2})}{2} \tag{4.6}$$

E_M dependent on composition is given by

$$E_0 = 0.356 + 0.7x + 0.4x^2 \tag{4.7}$$

where x is gallium concentration [25]. And E_B is assumed to constantly stay 0.4 eV [26] below the top of the valence band of InGaAs and will not be affected by the composition variation. We have not yet taken the strain effect into account due to the shortage of information on the effects of strain in these materials, though the effect is clearly important given the strain and relaxation results from RSM. Both the experimental results and theoretical calculation display an obvious band gap reduction in Fig. 4.11 due to the incorporation of bismuth. At low bismuth concentration, the band gap decreases approximately linearly at a rate of ~56 meV/%Bi, which is a moderate rate compared to 88 meV/Bi% for GaBiAs [27] and 12 meV/In% for

InGaAs [28]. The long wavelength limit of InGaBiAs from experiments thus far is 2,755 nm, corresponding to an average bismuth concentration of 6.75 %. There are small variations in the In/Ga ratio which can be observed from the HRXRD and RBS results. Their effect on the actual band gap by the three sets of simulations with varied In/Ga ratios was excluded. By comparing the experimental results and simulations, they agree reasonably well with each other; the scatter in the data could result from the presence of bismuth cluster which could affect the effective band gap in an anomalous way [29, 30] or some unexpected defects.

4.4.4.2 Spin-Orbit Split (Δ_{SO}) Transitions

For InP-based III–V alloys, (typically InGa(Al)As and InGaAsP with small strain), when the fundamental energy band gap (E_0) is larger than spin-orbit split (Δ_{SO}) transitions, the performances of InP platform lasers will be hindered due to the Auger processes in the valence band. However, when the E_0 band gap becomes smaller than the Δ_{SO} transition in mid-IR wavelength alloys, the Auger process will be suppressed. Therefore, the efficiency of the laser would be greatly enhanced. In this context, InGaBiAs would be a perfect example to study the energy band gap and the spin-orbit splitting.

Kudrawiec et al. first report their experimental measurements of the E_0 and Δ_{SO} for InGaBiAs by CER [31]. CER is a noninvasive technique to study the optical transitions in semiconductor thin films, low dimensional structures, and real devices [32]. The principle of this technique is to measure the derivative of the reflectance spectrum with respect to modulating electric field.

Figures 4.12 and 4.13 show CER spectra measured at room temperature as well as the simulation for the E_0 and $E_0 + \Delta_{SO}$ transitions, respectively. The results from the spectrophotometry measurements confirm the E_0 transitions moving to a smaller band gap due to the incorporation of bismuth. They also confirm the $E_0 + \Delta_{SO}$ transitions remaining unchanged in spite of the variation of bismuth concentration within the experimental error, which indicates the Δ_{SO} transition energy increases at equal rate as the E_0 transition reduction rate. The CER data are expressed in thick grey line in Fig. 4.12 and fitted by Aspnes formula:

$$\frac{\Delta R}{R}(E) = \mathrm{Re}[Ce^{i\theta}(E - E_j + i\Gamma)^{-m}] \tag{4.8}$$

where $(\Delta R / R)(E)$ is the energy dependence of CER signal, C and θ are the amplitude and phase of the line, and E_j and Γ are the energy and the broadening parameter of the optical transition, respectively. The term m could be assumed to be 3 for an inhomogeneous system. The moduli of CER resonances, expressed in dashed lines, can be obtained from the following equation with the same parameters from above.

$$\Delta\rho\,(E) = \frac{|C|}{[(E - E_j)^2 + \Gamma^2]^{m/2}} \tag{4.9}$$

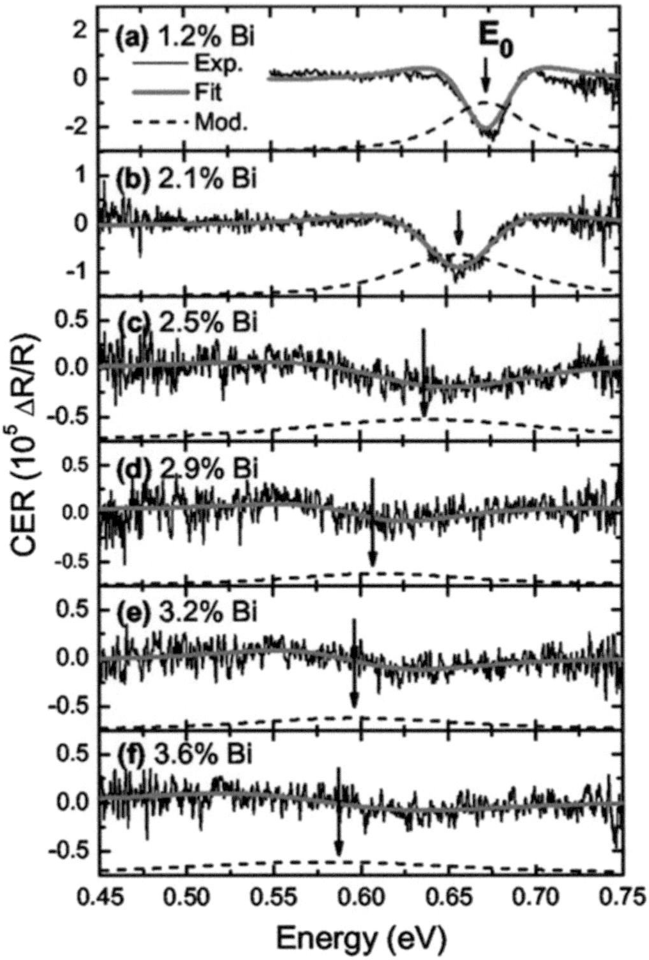

Fig. 4.12 CER measurements (*solid lines*) as well as the prediction from model (*dash lines*) for the E_0 transition in InGaBiAs

Figure 4.14 shows the energies of E_0, Δ_{SO}, and $E_0 + \Delta_{SO}$ transitions extracted from equations above. The first two energies follow opposite trends with the increase of bismuth concentration, and they begin to become comparable with each other when the Bi% is over 3 %. This is greatly encouraging because the similar trend does not happen until Bi% > 10 % in GaBiAs, where incorporating bismuth becomes really difficult because of past experience. Figure 4.14 also shows that the decrease of E_0 equals the increase of Δ_{SO} within the experimental error. This indicates that the shift of the band mainly happens in the heavy/light hole band while the conduction and spin-orbit split band are not affected by the VBAC, as shown in Fig. 4.15. This is in good agreement with the common anion rule applied in III–V compounds where

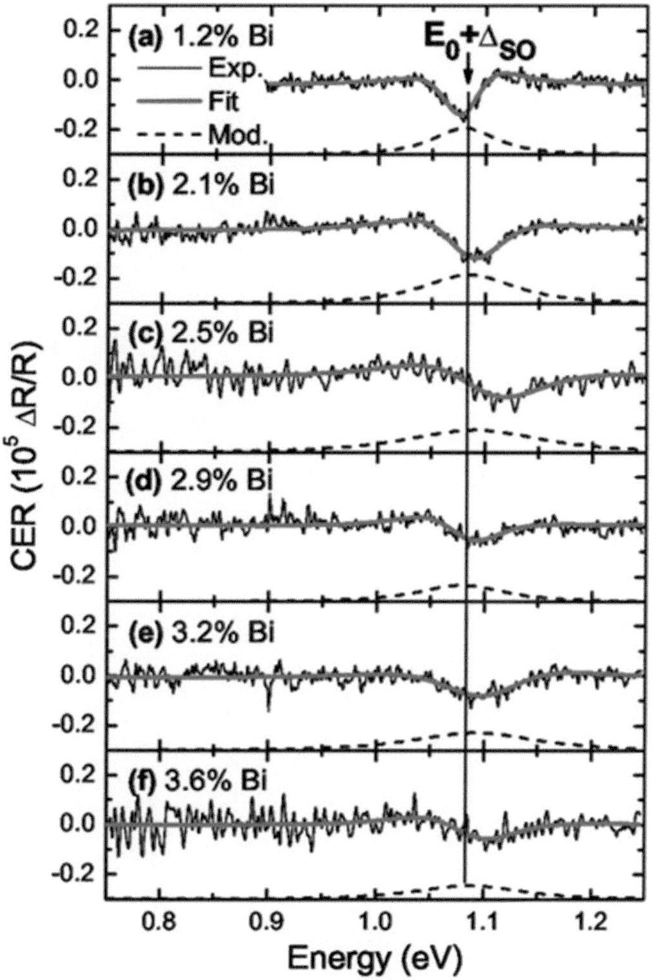

Fig. 4.13 CER measurements (*solid lines*) as well as the prediction from model (*dash lines*) for the $E_0+\Delta_{SO}$ transition in InGaBiAs

conduction band is dominated by the cation group III while valence band is governed by anion group V atoms. The energy gap E_0 of InGaBiAs decreases when heavy anions Bi replace lighter As. The spin-orbit splitting has the opposite trend of E_0 where it increases with heavier anion groups. For example, the spin-orbit splittings in Ga-V (GaP, GaAs, GaSb, and GaBi) are 0.08, 0.34, 0.76 [33], and 2.1 eV [31], while in In-V (InP, InAs, InSb, and InBi), the spin-orbit splitting are 0.11, 0.39, 0.81 [33], and 2.1 eV [31]. Therefore, the Δ_{SO} transition for InGaBiAs increases with higher bismuth concentration as the above III–V alloys.

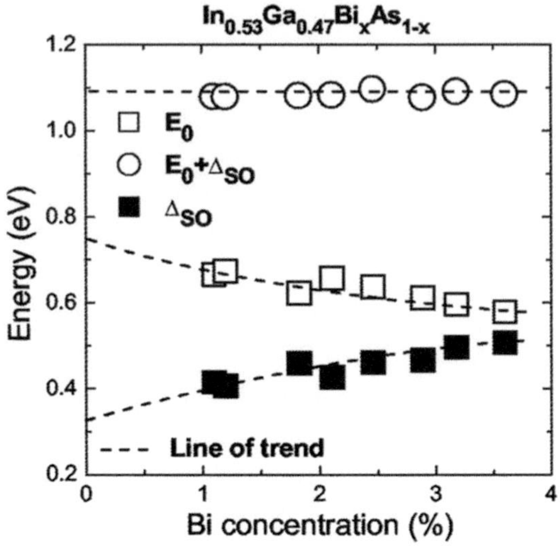

Fig. 4.14 Comparison of E_0 (*open squares*), $E_0+\Delta_{SO}$ (*open circles*), and Δ_{SO} (*closed squares*) transition energies

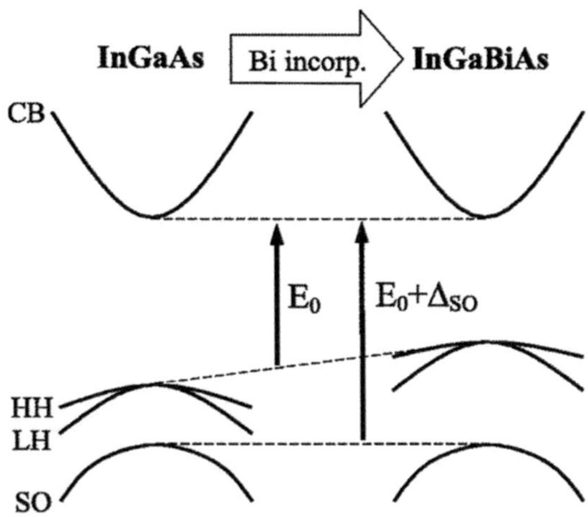

Fig. 4.15 Diagram of energy transition change due to the incorporation of bismuth. The E_0 transition is smaller with Bi incorporation, while the $E_0+\Delta_{SO}$ transition remains unchanged

The E_0 band gap reduction rate given by CER is around 50 meV/Bi%, which is close to the result measured by spectrophotometry. Similar to conduction band anticrossing in dilute nitride [34], this value is smaller than expected considering the ab initio self-consistent calculations for III-Bi compounds [17], which indicates a simple virtual crystal approximation may not apply for this case.

Broderick et al. apply a 12 band k.p model based on an sp^3s^* tight-binding theory to calculate the band structures of InGaBiAs [35]. The model treats bismuth energy level as impurity-like, as well as the interaction between the Bi-related resonate states and extended states of matrix material as band anticrossing (BAC). They compare the calculation results with experimental measurements via multiple methods and exhibit an overall good agreement. This theoretically confirms that the E_0

transition will decrease with the higher concentration of bismuth; Δ_{SO} transition will follow an opposite trend, which indicates InGaBiAs as a potentially good candidate for a mid-IR laser material.

4.4.5 *Electrical Properties*

4.4.5.1 Electrical Properties of Unintentionally Doped InGaBiAs Samples

Dilute bismuthides are comparable to dilute nitrides in many ways but have one distinct advantage over the latter: VBAC will hardly affect the electron electrical properties of the film. The electrical properties of InGaBiAs, shown in Fig. 4.16, are measured and characterized by Hall-effect and four-point probe measurements in the van der Pauw geometry [14]. The unintentionally doped InGaBiAs samples display n-type as opposed to the observation of p-type GaBiAs, which is probably due to the Bi_{III} antisite defects [36]. The sheet mobility of InGaBiAs films are generally high, ranging from 7,400 to 350 cm²/Vs. For the same bismuth concentration, InGaBiAs have higher mobility as compared to GaBiAs, which is mainly due to the better mobility inherited from the matrix material InGaAs as opposed to GaAs. The mobility of InGaBiAs is decreased by the increase of bismuth concentration, which is expected because of the enhanced scattering resulting from the incorporation of

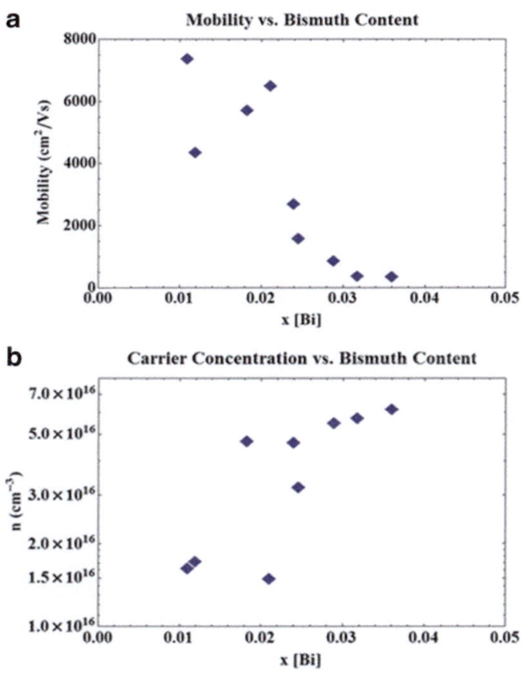

Fig. 4.16 Mobility and carrier concentration for unintentionally doped InGaBiAs based on different bismuth concentration (Color online)

large bismuth atoms. The bulk carrier concentration ranges from 1.5×10^{16} to 6.0×10^{16} cm^{-3} and increases with the increase of bismuth concentration. The scattering in the data could be due to the variation in terms of the sample quality.

4.4.5.2 Electrical Properties of Si-Doped n-Type InGaBiAs Samples

In practice, doped materials are of particular interest for thermoelectric applications. We study Si-doped n-type InGaBiAs grown by MBE on InP:Fe [37]. The growth temperature is fixed at 300 °C, as well as a constant As BEP at 2×10^{-6} Torr and Bi BEP at 1×10^{-8} Torr, corresponding to1.6 % Bi incorporation; while the silicon doping level varies by raising the cell temperature from 1,100 to 1,390 °C. We have also tested samples with higher Bi concentrations: 2.6 and 4.4 % corresponding to Bi BEP at 1.5×10^{-8} and 2.0×10^{-8} Torr, respectively, and with three different silicon cell temperatures of 1,200, 1,215, and 1,245 °C. The samples are grown along (001) direction on InP:Fe substrates, starting with a 70 nm InGaAs buffer layer intentionally lattice-matched to substrate and being followed by adding both bismuth and silicon for a 300 nm deposition.

The carrier concentration and mobility for each sample are measured by the same equipment as stated before. The conductivity is simply calculated by the following equation:

$$\sigma = e\mu n \tag{4.10}$$

where σ is the conductivity, e is the electron charge, μ is the mobility, and n is the carrier concentration. The squares and the solid line in Fig. 4.17a indicate the carrier concentration increases exponentially as a function of silicon cell temperature before it saturates, which is in good agreement with theory, and when the Si temperature is above 1,300 °C, some of the doping atoms start to become electrically inactive. When Bi concentration is at 1.6 %, the carrier concentration stays above InGaAs:Si; but at higher Bi concentration 2.6 % or above, the carrier concentrations drop slightly below InGaAs:Si, which suggests the maximum carrier concentration could be achieved between a bismuth concentration of 0 and 2.6 %. At low bismuth concentrations, a recent report [38] hypothesizes that bismuth may create a large amount of step edges at a microscopic scale, which greatly increases the possibilities for silicon adsorption. However, at higher bismuth concentrations, with increased bismuth presence, it may impede silicon incorporation, leading to a small reduction in carrier concentration. In general, the carrier concentration level in this material system is high compared to other III–V alloys. This is partially due to the growth condition, especially the low growth temperature, as well as the effect of bismuth. The near-stoichiometric growth permits low-temperature growth without the incorporation of As antisites, as observed in materials such as low temperature grown GaAs or LT-GaAs [39]. The largest carrier concentration measured is 6.4×10^{19} cm^{-3} for 1.6 % Bi. The relationship between electrical conductivity and carrier concentration is given in Fig. 4.17b. Generally, the conductivity increases with carrier concentration which indicates the mobility falls at a much smaller rate than the carrier concentration increases. In addition, the conductivity in this material system is very high as a semiconductor contributed by the extremely high carrier concentration and relatively large mobility. The highest electrical conductivity measured is 4,850 S/cm, which is almost one half of metal mercury and very promising as contact material in devices such as HBT, where contact resistance is critical to high-frequency performance. Some scattering has been observed for samples with high carrier concentration. This is mainly due to the reduced mobility which is probably brought by the lower material quality. Another possible explanation is intervalley scattering in the L-valley. A plot of mobility vs. carrier concentration is shown in Fig. 4.17c. As expected, the mobility is affected by the increased carrier concentration as well as the enhanced bismuth concentration due to the added scattering centers. However, the mobility still remains relatively high and drops slowly at low bismuth concentration. This explains the high conductivity of this material system.

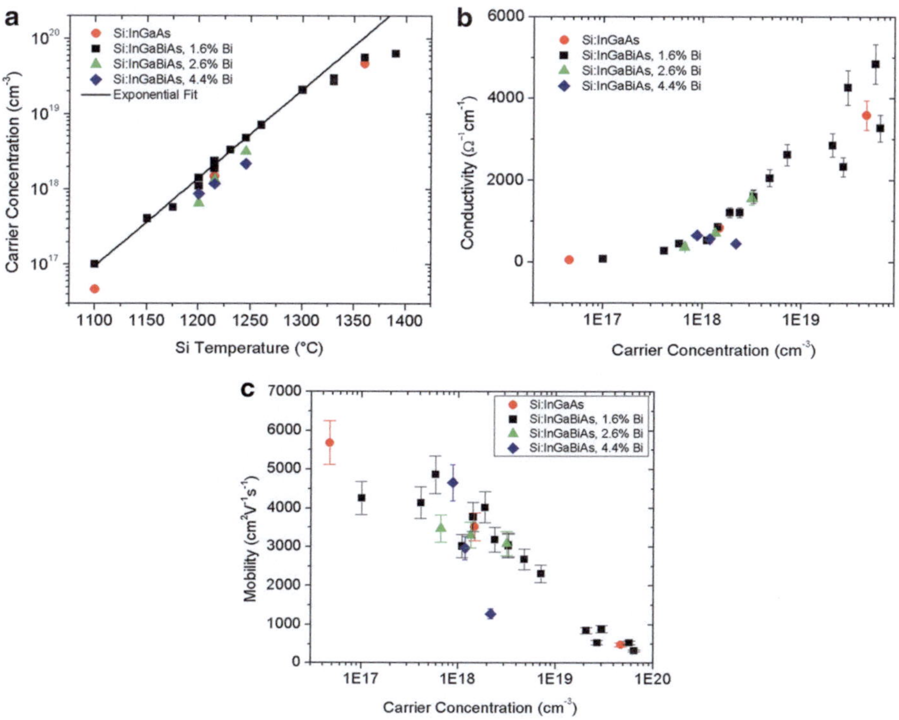

Fig. 4.17 Carrier concentration vs. Si temperature (**a**), along with conductivity (**b**) and mobility (**c**) vs. carrier concentration for Si-doped InGaBiAs. The exponential fit in Fig. 4.18a (*solid line*) represents the expected relationship between cell temperature and carrier concentration, modeled by the Hertz-Knudsen and Clausius–Clapeyron equations. The points are expected to deviate from the fit *line* because Si begins to incorporate on group V sites and starts to cancel out the electron contribution from Si sitting on a group III site (Color online)

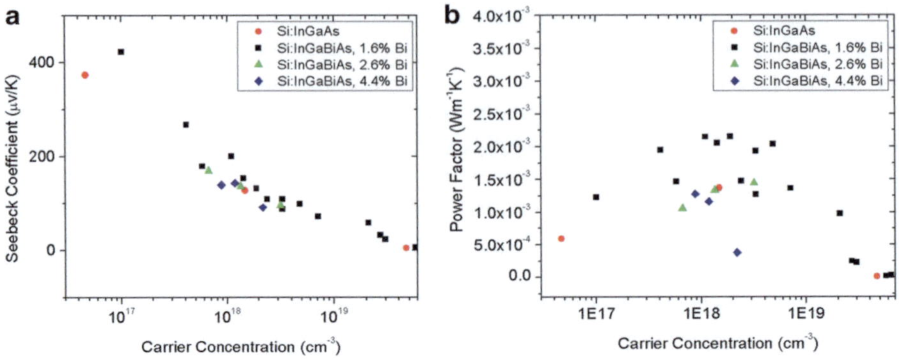

Fig. 4.18 Seebeck coefficient (**a**) and thermoelectric power factor (**b**) vs. carrier concentration for InGaAs:Si and InGaBiAs:Si (Color online)

4.4.6 Thermoelectric Properties

As introduced in the first section, Seebeck coefficient and thermal conductivity are two other important components other than electrical conductivity in terms of thermoelectric properties. The Seebeck coefficient measurement is started by cleaving samples to a rectangular shape with a dimension of 6×15 mm. Then a strip of indium is soldered on each end with an approximate thickness of 1 mm for electrical contacts. Then the sample bar is pasted onto two Peltier modules with one end heated under a controllable power supply and the other is cooled at room temperature. The temperature difference ΔT between the two ends are measured by type K thermocouples, while the voltage is obtained by a digital multimeter through two copper clips. The experiment begins by changing the output power at the heated end, follows by recording the voltage and temperature difference simultaneously, and ends by making a plot of V vs. ΔT linear curve. The slope of the curve is obtained as Seebeck coefficient.

Figure 4.18a, b shows the Seebeck coefficient and thermal power factor (TPF) mentioned in Sect. 4.1 as functions of carrier concentration. The Seebeck coefficient decreases gradually with the increase of carrier concentration which is expected given the increase in Fermi level relative to the bottom of the conduction band. TPF displays a broad peak over a moderate carrier concentration range from 3×10^{17} to 4×10^{18} cm^{-3} which is a balanced result from the product $S^2\sigma$. Comparing the bismuth concentration at 1.6–0 %, both Seebeck coefficient and TPF are improved, especially from low to moderate carrier concentrations (at a carrier concentration of ~1.4E–18 cm^{-3}, a 49.6 % improvement in TPF is observed). The enhancement is mainly contributed by a larger Seebeck coefficient. A possible explanation for the Seebeck improvement would be a more complex conduction band profile in InGaBiAs, contradicting with the popular view that dilute bismuthides share a similar conduction band structure with their matrix material. Moreover, there has been some recent theoretical work suggesting the conduction band of GaAs is affected by the presence of Bi atoms in GaBiAs [40].

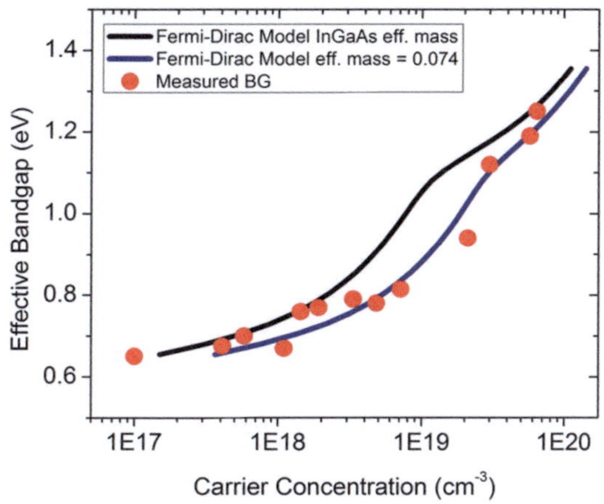

Fig. 4.19 Effective band gap vs. carrier concentration for our samples (*red dots*) relative to the Fermi–Dirac model of lattice-matched InGaAs. The *black line* is calculated from the Fermi–Dirac model using the InGaAs effective mass, and the *blue line* is calculated using a larger effective mass. The model with the larger effective mass strongly agrees with measured samples, suggesting that the effective mass in the conduction band is increased with bismuth incorporation (Color Online)

We model the Burstein–Moss shift in effective band gap of these materials (as detailed below), which suggests a possible modification of the conduction band to InGaAs occurs due to the incorporation of bismuth. Both the Seebeck coefficient and TPF have little to no improvement with higher bismuth concentration at 2.6 and 4.4 %.

There are three main sources of error in Seebeck measurements. The largest source of error (±1 K) comes from the temperature measurement due to the thermal contact between the sample and the indium metal. The error in acquiring the Seebeck coefficient is very small because the R^2 for the linear V vs. ΔT fitting is close to 1. At room temperature, the error contribution from the conduction between the film and substrate is negligible; however, it may turn out to be a big concern at high temperature measurements where the substrate begins to become conductive.

As stated above, a Burstein–Moss shift is observed in the InGaBiAs:Si films. Using the Fermi–Dirac equation for degenerate semiconductors, Fig. 4.19 shows a plot of effective band gap vs. carrier concentration for the InGaBiAs:Si films containing 1.6 % Bi and the Fermi–Dirac model of lattice-matched InGaAs. From the plot, there is an obvious deviation between the model and InGaBiAs:Si films, which suggests that the InGaBiAs:Si conduction band differs from that of InGaAs:Si. However, if the effective mass in the Γ valley of the Fermi–Dirac model is increased, then the model has a better fit to the experimental data. This suggests that effective mass is increased when Bi is incorporated; this would also explain the increase in Seebeck coefficient at relevant concentrations.

The last parameter in the equation of figure of merit is thermal conductivity. There are several approaches to measure thermal conductivity such as the three omega, frequency domain thermoreflectance (FDTR), and time domain

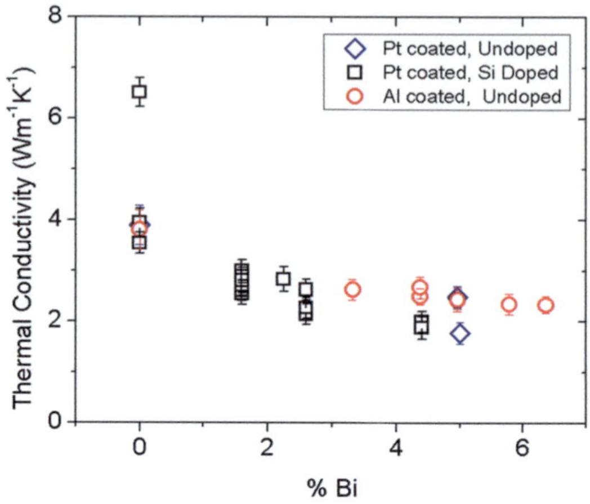

Fig. 4.20 Thermal conductivity vs. Bi concentration on undoped, InGaAs:Si and InGaBiAs:Si (Color online)

thermoreflectance (TDTR) methods [41]. Compared to the other two, TDTR has the advantage of being effective for films that are electrically conductive or semiconducting, and it is the method we adopted. TDTR is a pump-probe technique. In this setup, the pump laser is modulated at 11 MHz by a Ti:Sapphire oscillator and linearly amplified with a sinusoid, the probe laser is given a ~6 ns delay with respect to the pump pulses by a mechanical delay stage, and the ratio of the in-phase to out-of-phase signal of the probe beam is monitored by a lock-in amplifier. The samples are first deposited with Al or Pt film transducer before TDTR measurement. A multilayer thermal model [42–44] is used to fit the TDTR data. The heat capacity of the metal transducer is obtained from literature, and the thermal conductivity of it is approximated from electrical resistivity measurements [45–47]. There is a relatively high tolerance to the variation of thermal conductivity of the metal transducer over the short time delay during these TDTR measurements. The heat capacity of the substrate [45–47] is adopted from literature. The heat capacity of the sample is calculated by averaging the value of $In_{0.55}Ga_{0.45}As$ [48] and Bi [47]. To determine the thermal conductivity of the bismuthide samples, the TDTR data are fitted by varying the thermal conductivity of the bismuthide films as well as slightly adjusting the thermal boundary conductance between the metal and the bismuthide samples. This system is much more sensitive to small changes in thermal conductivity than the thermal boundary conductance due to the thickness of the samples and their relatively low thermal conductivity given the thermal penetration depth of our laser at 11 MHz; thereby the uncertainty in determining the thermal conductivity of the alloy films is reduced.

Thermal conductivity measurements are performed over a wide range of bismuthide samples including both unintentionally doped and silicon-doped with varied bismuth concentration. A plot of thermal conductivity versus bismuth concentration is given in Fig. 4.20. In general, thermal conductivity is reduced due to the increase of bismuth incorporation which is a direct result of alloy scattering. One InGaAs:Si sample has a relatively high thermal conductivity, and this is due to its high electrical conductivity. According to a calculation by Wiedemann–Franz Law, roughly 40 % of the total thermal conductivity was from electronic contribution; the thermal conductivity of that sample would be comparable with the other samples if we deduct the electronic contribution. The electronic contribution is much smaller for the rest InGaAs:Si samples indicating by their moderate conductivity. The sources of error bars come from the metal film thickness measurements and the standard deviation about the mean value through multiple measurements.

4.5 Conclusion

In this work, we systemically study the growth conditions and properties of the new material InGaBiAs. We find that low growth temperature, strict As/Bi and III/V ratio, and an appropriate growth rate are critical in Bi incorporation. Generally, a lower growth temperature or a higher Bi/As ratio increases Bi incorporation, but there is a limit point for both conditions. We measure the composition properties of InGaBiAs by HRXRD and RBS for both lattice-mismatched and nominally lattice-matched samples. The results indicate our samples have good crystalline properties and are almost free of interstitial defects. The strain and relaxation study, performed by RSM, show our lattice-mismatched films are compressively strained and greatly exceed the theoretical critical thickness due to the special growth condition. The optical properties were measured by spectrophotometry and CER; both show the E_0 band gap energies decrease with increased Bi concentration, and at low Bi concentration, the band gap drops at a rate about 56 meV/%Bi, which agrees with the VBAC model. CER measurements also state that the $E_0 + \Delta_{SO}$ remains the same for samples with varied bismuth concentrations, which exhibits a promising way to suppress the Auger process in valence band and greatly enhance the laser efficiency. All of our unintentionally doped samples display n-type properties with relative high mobility and moderate carrier concentration comparable to InGaAs. The mobility decreases as Bi concentration increases, while the carrier concentration has an opposite trend. The silicon-doped films exhibit very high conductivity as a semiconductor material due to the special growth conditions and the help of bismuth, which are potential candidates for contact materials in electronic devices. They show a slightly higher Seebeck coefficient with low concentration of bismuth compared to InGaAs:Si on the same doping level. This is probably due to a more complex conduction band structure indirectly confirmed by a modified Burstein–Moss shift model. TDTR measurements show a decreasing trend in thermal conductivity with increasing Bi concentration. For moderate doped samples, a peak ZT at room temperature of 0.23 was recorded; however, better performance at higher temperature are expected. Lastly, InGaBiAs is promising for both optoelectronics and thermoelectric materials.

Acknowledgements The authors wish to acknowledge our collaborators: Prof. Robert Kudrawiec and his group from Wroclaw University of Technology and Prof. Eoin O'Reilly and his group from Tyndall National Institute for the help of Sect. 4.4.4.2. We also thank our collaborator: Prof. Patrick Hopkins and his group from University of Virginia with the helpful measurements of thermal conductivity. In addition, we thank Prof. James LeBeau and his group from North Carolina State University for the HAADF-STEM picture. Finally, we acknowledge the US Office of Naval Research for financial support, primarily through the Young Investigator Program.

References

1. Osamu Wada, H.H.: InP-Based Materials and Devices: Physics and Technology, p. 1. Wiley, New York (1999)
2. Lee, J.-H., Wu, J., Grossman, J.C.: Enhancing the thermoelectric power factor with highly mismatched isoelectronic doping. Phys. Rev. Lett. **104**, 016602 (2010)
3. Alberi, K., Blacksberg, J., Bell, L.D., Nikzad, S., Yu, K.M., Dubon, O.D., Walukiewicz, W.: Band anticrossing in highly mismatched Sn_xGe_{1-x} semiconducting alloys, Phys. Rev. B. **77**, 073202 (2008)
4. Vurgaftman, I., Meyer, J.R.: Band parameters for nitrogen-containing semiconductors, J. Appl. Phys. **94**, 3675 (2003)
5. Wu, J., Walukiewicz, W., Yu, K.M., Denlinger, J.D., Shan, W., Ager III, J.W., Kimura, A., Tang, H.F., Kuech, T.F.: Valence band hybridization in N-rich $GaN_{1-x}As_x$ alloys. Phys. Rev. B. **70**, 115214 (2004)
6. Walukiewicz, W., Shan, W., Yu, K., Ager, J., Haller, E., Miotkowski, I., Seong, M., Alawadhi, H., Ramdas, A.: Interaction of localized electronic states with the conduction band: band anticrossing in II-VI semiconductor ternaries. Phys. Rev. Lett. **85**, 1552 (2000)
7. Lu, X., Beaton, D.A., Lewis, R.B., Tiedje, T., Whitwick, M.B.: Effect of molecular beam epitaxy growth conditions on the Bi content of $GaAs_{1-x}Bi_x$.pdf. Appl. Phys. Lett. **92**, 192110 (2008)
8. Feng, G., Yoshimoto, M., Oe, K., Chayahara, A., Horino, Y.: New III-V Semiconductor InGaAsBi Alloy Grown by Molecular Beam Epitaxy. Jpn. J. Appl. Phys. **44**, L1161 (2005)
9. Norman, A.G., France, R., Ptak, A.J.: Atomic ordering and phase separation in MBE $GaAs_{1-x}Bi_x$. J. Vac. Sci. Technol. B. Microelectron. Nanometer. Struct. **29**, 3 (2011)
10. Kini, R.N., Bhusal, L., Ptak, A.J., France, R., Mascarenhas, A.: Electron Hall mobility in GaAsBi. J. Appl. Phys. **106**, 043705 (2009)
11. Feng, G., Oe, K., Yoshimoto, M.: Temperature dependence of Bi behavior in MBE growth of InGaAs/InP. J. Cryst. Growth. **301–302**, 121 (2007)
12. Heremans, J.P., Jovovic, V., Toberer, E.S., Saramat, A., Kurosaki, K., Charoenphakdee, A., Yamanaka, S., Snyder, G.J.: Enhancement of thermoelectric efficiency in PbTe by distortion of the electronic density of states. Science **321**, 554 (2008)
13. Oe, K.: Characteristics of Semiconductor Alloy $GaAs_{1-x}Bi_x$. Jpn. J. Appl. Phys. **41**, 2801 (2002)
14. Petropoulos, J.P., Zhong, Y., Zide, J.M.O.: Optical and electrical characterization of InGaBiAs for use as a mid-infrared optoelectronic material. Appl. Phys. Lett. **99**, 031110 (2011)
15. Tixier, S., Adamcyk, M., Tiedje, T., Francoeur, S., Mascarenhas, A., Wei, P., Schiettekatte, F.: Molecular beam epitaxy growth of $GaAs_{1-x}Bi_x$. Appl. Phys. Lett. **82**, 2245 (2003)
16. Zhong, Y., Dongmo, P.B., Petropoulos, J.P., Zide, J.M.O.: Effects of molecular beam epitaxy growth conditions on composition and optical properties of $InxGa_{1-x}BiyAs_{1-y}$. Appl. Phys. Lett. **100**, 112110 (2012)
17. Laukkanen, P., Pakarien, J., Ahola-Tuomi, M., Kuzmin, M., Perala, R.E., Vayrynen, I.J.: Structural and electronic properties of Bi-adsorbate-stabilized reconstructions on the InP(100) and $GaAs_xN_{1-x}$(100) surfaces. Phys. Rev. B. **74**, 155302 (2006)

18. Janotti, A., Wei, S.-H., Zhang, S.B.: Theoretical study of the effects of isovalent coalloying of Bi and N in GaAs. Phys. Rev. B. **65**, 115203 (2002)
19. Moram, M.A., Vickers, M.E.: X-ray diffraction of III-nitrides. Rep. Prog. Phys. **72**, 036502 (2009)
20. Hauenstein, R.J., Clemens, B.M., Miles, R.H., Marsh, O.J.: Strain relaxation kinetics in $Si_{1-x}Ge_x$/Si heterostructures. J. Vac. Sci. Technol. B. Microelectron. Nanometer. Struct. **7**, 767 (1989)
21. Matthews, J.W., Blaskeslee, A.E.: Defects in epitaxial multilayers. J. Cryst. Growth. **27**, 118 (1974)
22. Tsao, J., Dodson, B., Picraux, S., Cornelison, D.: Critical stresses for Si_xGe_{1-x} Strained-Layer Plasticity. Phys. Rev. Lett. **59**, 2455 (1987)
23. Vardar, G., Warren, M.V., Kang, M., Jeon, S., Goldman, R.S.: Mechanisms of droplet formation during ga(in)asbi molecular beam epitaxy growth, presented at the 28th North American Molecular Beam Epitaxy Conference, San Diego, CA, 2011. North American molecular beam epitaxy (2012)
24. Yoshimoto, M., Murata, S., Chayahara, A., Horino, Y., Saraie, J., Oe, K.: Metastable GaAsBi alloy grown by molecular beam epitaxy. Jpn. J. Appl. Phys. **42**, L1235 (2003)
25. Swaminathan, V., Macrander, A.T.: Materials Aspects of GaAs and InP Based Structures, p. 27. Prentice Hall, Englewood Cliffs, NJ (1991)
26. Alberi, K., Dubon, O.D., Walukiewicz, W., Yu, K.M., Bertulis, K., Krotkus, A.: Valence band anticrossing in $GaBi_xAs_{1-x}$. Appl. Phys. Lett. **91**, 051909 (2007)
27. Francoeur, S., Seong, M.-J., Mascarenhas, A., Tixier, S., Adamcyk, M., Tiedje, T.: Band gap of $GaAs_{1-x}Bi_x$, $0<x<3.6\%$. Appl. Phys. Lett. **82**, 3874 (2003)
28. Nahory, R.E., Pollack, M.A., Johnston Jr., W.D., Barns, R.L.: Band gap versus composition and demonstration of Vegard's law for $In_{1-x}Ga_xAs_yP_{1-y}$ lattice matched to InP. Appl. Phys. Lett. **33**, 659 (1978)
29. Ciatto, G., Young, E.C., Glas, F., Chen, J., Mori, R.A., Tiedje, T.: Spatial correlation between Bi atoms in dilute $GaAs_{1-x}Bi_x$: From random distribution to Bi pairing and clustering. Phys. Rev. B. **78**, 035325 (2008)
30. Ciatto, G., Thomasset, M., Glas, F., Lu, X., Tiedje, T.: Formation and vanishing of short range ordering in $GaAs_{1-x}Bi_x$ thin films. Phys. Rev. B. **82**, 201304(R) (2010)
31. Kudrawiec, R., Kopaczek, J., Misiewicz, J., Petropoulos, J.P., Zhong, Y., Zide, J.M.O.: Contactless electroreflectance study of E_0 and $E_0+\Delta_{SO}$ transitions in $In_{0.53}Ga_{0.47}Bi_xAs_{1-x}$ alloys. Appl. Phys. Lett. **99**, 251906 (2011)
32. Misiewicz, J., Kudrawiec, R.: Contactless electroreflectance spectroscopy of optical transitions in low dimensional semiconductor structures. Opto-Electron. Rev. **20**, 101 (2012)
33. Vurgaftman, I., Meyer, J.R., Ram-Mohan, L.R.: Band parameters for III–V compound semiconductors and their alloys. J. Appl. Phys. **89**, 5815 (2001)
34. Shan, W., Walukiewicz, W., Ager III, J.W., Haller, E.E., Geisz, J.F., Friedman, D.J., Olson, J.M., Kurtz, S.R.: Band anticrossing in GaInNAs alloys. Phys. Rev. Lett. **82**, 1221 (1999)
35. Broderick, C.A., Usman, M., Reilly, E.P.O.: 12-band k . p model for dilute bismide alloys of (In)GaAs derived from supercell calculations. Phys. Status Solidi B **1** (2012)
36. Kini, R., Ptak, A., Fluegel, B., France, R., Reedy, R., Mascarenhas, A.: Effect of Bi alloying on the hole transport in the dilute bismide alloy $GaAs_{1-x}Bi_x$. Phys. Rev. B. **83**, 075307 (2011)
37. Dongmo, P.B., Zhong, Y., Attia, P., Bomberger, C., Cheaito, R., Ihlefeld, J.F., Hopkins, P.E., Zide, J.M.O.: Enhanced Room Temperature Electronic and Thermoelectric Properties of Dilute Bismuthides. J. Appl. Phys. **112**, 093710 (2012)
38. Duzik, A., Thomas, J.C., Millunchick, J.M., Lång, J., Punkkinen, M.P.J., Laukkanen, P.: Surface structure of bismuth terminated GaAs surfaces grown with molecular beam epitaxy, Surf. Sci. (2012). doi:10:1016/j.susc.2012.03.021
39. Warren, A.C., Woodall, J.M., Freeouf, J.L., Grischkowsky, D., Mclnturff, D.T., Melloch, M.R., Otsuka, N.: Arsenic precipitates and the semi-insulating properties of GaAs buffer layers grown by low temperature molecular beam epitaxy. Appl. Phys. Lett. **57**, 1331 (1990)

40. Pettinari, G., Polimeni, A., Capizzi, M., Blokland, J.H., Christianen, P.C.M., Maan, J.C., Young, E.C., Tiedje, T.: Influence of bismuth incorporation on the valence and conduction band edges of $GaAs_{1-x}Bi_x$. Appl. Phys. Lett. **92**, 262105 (2008)
41. Cahill, D.G., Goodson, K., Majumdar, A.: Thermometry and thermal transport in micro/nanoscale solid-state devices and structures. J. Heat. Transfer. **124**, 223 (2002)
42. Schmidt, A.J., Chen, X., Chen, G.: Pulse accumulation, radial heat conduction, and anisotropic thermal conductivity in pump-probe transient thermoreflectance. Rev. Sci. Instrum. **79**, 114902 (2008)
43. Cahill, D.G.: Analysis of heat flow in layered structures for time-domain thermoreflectance. Rev. Sci. Instrum. **75**, 5119 (2004)
44. Hopkins, P.E.: Influence of inter- and intraband transitions to electron temperature decay in noble metals after short-pulsed laser heating. J. Heat. Transfer. **132**, 122402 (2010)
45. Touloukian, Y.S.: Thermophysical Properties of matter-Thermal Conductivity: Nonmetallic Solids, vol. 2. Plenum, New York (1970)
46. Touloukian, Y.S.: Thermophysical Properties of Matter-Specific Heat: Nonmetallic Solids, vol. 5. Plenum, New York (1970)
47. Touloukian, Y.S., Buyco, E.H.: Thermaophysical Properties of Matter-Specific Heat: Metallic Elements and Alloys, vol. 4. Plenum, New York (1970)
48. Adachi, S.: Physical Properties of III-V Semiconductor Compounds: InP, InAs, GaAs, GaP, InGaAs and InGaAsP. Wiley, Weinheim (2004)

Chapter 5
Atmospheric-Pressure Metalorganic Vapor Phase Epitaxy of GaAsBi Alloy on GaAs Substrate

Hédi Fitouri, Ahmed Rebey, and Belgacem El Jani

Abstract This chapter deals with mainly the growth and the structural and optical characteristics of GaAsBi alloys. Firstly, the growth and the structural characteristic of GaAsBi alloys grown on GaAs substrate by conventional metalorganic vapor phase epitaxy technique have been investigated in detail and described in section "MOVPE Growth and Characterization of GaAsBi." Secondly, a novel method to epitaxy GaAsBi layer on (001) GaAs substrate is presented in section "Growth and Characterization of GaAsBi Alloy Under Alternated Bismuth Flows." Thirdly, the effects of thermal annealing on $GaAs_{1-x}Bi_x$ alloys are described in section "Effects of Thermal Annealing on $GaAs_{1-x}Bi_x$ Alloys." Trimethyl gallium, trimethyl bismuth (TMBi), and arsine were used as precursor sources. Growth parameters such as growth temperature, TMBi flow and V/III ratio were checked on a wide range. The growth was monitored in situ by laser reflectometry using a 632.8 nm beam. High-resolution X-ray diffraction, secondary ion mass spectroscopy, photoluminescence, and photoreflectance spectroscopy have been used to characterize the obtained GaAsBi layers. In analyzing the surface morphology, atomic force microscopy and scanning electron microscopy were used to qualify films properties. Result shows that the solid composition of GaBi content in the $GaAs_{1-x}Bi_x$ alloy reaches a maximum value of about 4 %. A significant redshift in the emission of GaAsBi of ~42 meV per percent Bi is observed. The surface shows a Bi island formation on all the samples. The density and size of these islands depend greatly on the growth conditions and growth method.

H. Fitouri (✉) • A. Rebey • B. El Jani
Faculté des Sciences de Monastir, Université de Monastir-Unité de Recherche sur les Hétéro-Epitaxies et Applications, Monastir 5019, Tunisia
e-mail: Hedi.Fitouri@fsm.rnu.tn; ahmed.rebey@fsm.rnu.tn

H. Li and Z.M. Wang (eds.), *Bismuth-Containing Compounds*, Springer Series in Materials Science 186, DOI 10.1007/978-1-4614-8121-8_5,

5.1 Introduction

GaAsBi has recently received considerable attention as an attractive candidate for developing applications in long-wavelength optoelectronics [1–3]. The addition of small amounts of Bi produces a large reduction in the GaAs band gap [4–11], and the temperature insensitivity of the band gap for $GaAs_{1-x}Bi_x$ was confirmed by photoluminescence (PL) measurement [2, 12] and photoreflectance (PR) spectroscopy [5]. Although Bi is isoelectronic with As, this atom behaves more like impurities than conventional alloying elements. The band gap reduction in the Bi alloy is believed to be due to a resonant interaction with the top of the valence band. The band gap temperature dependence of GaAsBi was fairly small (as small as 0.1 meV/K for 2.6 % of Bi) compared to that of GaAs [5]. This property would bode extremely well for using GaAsBi material as the active region in semiconductor lasers for communication systems where emission wavelengths are required to remain constant under ambient temperature variations. Recent theoretical and spectroscopic studies show that only the valence band of GaAs is affected by alloying it with Bi [10]. Since Bi has a strong tendency for surface segregation, the metalorganic vapor phase epitaxy (MOVPE) of GaAsBi alloys is not easy. Then, special growth conditions have been reported to achieve Bi incorporation [13–15]. The high incorporation of Bi atoms introduces the degradation of optical quality due to the carrier localization or the formation of deep levels in the midgap. The large size of Bi atoms supports on the one hand, the formation of metallic droplets, which are often observed on the surface during epitaxial growth, the droplet sizes and densities are reported to be dependent on the Bi flux and growth temperature [15]. On the other hand, the large atomic size of Bi gives rise to a large spin-orbit splitting. Many promising device applications of Bi containing compounds have already been recognized [8]. In future, the Bi containing compounds research can be divided into different areas including materials science, electronic properties, and devices. The large increase in spin-orbit splitting for electrons is attractive to make spintronic devices feasible [8].

5.2 MOVPE Growth and Characterization of GaAsBi

In this section, growth and characterization of GaAsBi are discussed. The atmospheric-pressure metalorganic vapor phase epitaxy (AP-MOVPE) optimization of GaAsBi alloy is studied. The growth parameters such as growth temperature, trimethylbismuth (TMBi) flow, and V/III ratio were checked on a wide range.

5.2.1 Conventional MOVPE Growth Method

The experimental procedure of the GaAsBi growth by conventional MOVPE technique and the sample characterization are presented. The epitaxial layers were grown in horizontal geometry AP-MOVPE reactor on exactly (100) Zn-doped GaAs

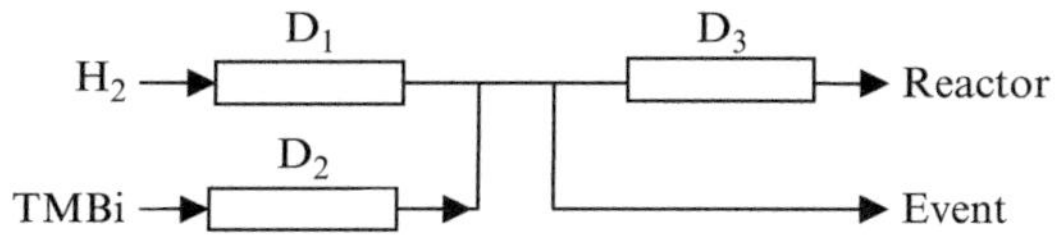

Fig. 5.1 Schematic representation of the double dilution equipment. D_1, D_2, and D_3 denote the mass flow controllers

epiready substrates. Trimethylgallium (TMG), TMBi, and pure arsine (AsH_3) were used as precursors. TMG and TMBi bubblers were housed in constant temperature baths kept at 5 °C and 0 °C, respectively. In order to obtain low TMBi flux with a better accuracy, we have used a double dilution system. Three mass flow controllers were connected as shown in Fig. 5.1.

The TMBi flow in the reactor D_{TMBi} is calculated from the following relation: $D_{TMBi} = D_2\ D_3\ /\ (D_1 + D_2)$ where (i=1, 2, 3) are the flows in the mass flow controllers. The carrier gas was Pd-diffused H_2. The total flow is fixed to 3.2 L/min. The substrate temperature was measured by a thermocouple inserted into the graphite susceptor. The substrate was ramped in a mixture flow of H_2 and AsH_3 at 650 °C for 10 min to remove the native surface oxide layer before growth. The temperature was decreased till 420 °C and stabilized. Then a GaAs buffer layer was grown (~80 nm). This was followed by the growth of GaAsBi epilayer at the chosen temperature. In situ reflectivity data were recorded using a 632.8 nm laser beam. Laser light is directed through one of the two reactor quartz windows with an angle of 30° to the stationary wafer. In order to avoid the signal fluctuations, the reflected beam was normalized by laser signal.

High resolution X-ray diffraction (HRXRD) characterizations coupled $\omega/2\theta$ measurements of (004) and (115) plane reflection were preformed systematically with copper target ($\lambda_{CuK}\alpha_1 = 1.54056$ Å) radiation from a high power X-ray generator Discover D8 (40 kV, 55 mA). The lattice parameter of the epilayer was evaluated and the bismuth content was determined using Vegard's law. Surface analysis was performed using a Philips XL-30 scanning electron microscope (SEM) and atomic force microscopy (AFM).

The Bi atom is approximately 25 % larger than Ga and As and tends to surface segregate during growth [16], making it a good surfactant [17]. When increasing Bi content, the band gap energy of GaAsBi alloy grown by low pressure MOVPE is reduced (~46 meV/%Bi) and shows a temperature insensibility [2, 5]. For this intention, we have tried to maximize the bismuth concentration by changing the following three important growth parameters: growth temperature, bismuth flow, and V/III ratio.

5.2.1.1 Growth Temperature Effect

In MOVPE technique, the growth of GaAsBi alloys is not straightforward [14]. The growth temperature is very important as it directly affects the growth process. Figure 5.2 shows the reflectivity variation during the growth of GaAsBi layers for different temperatures. We remark that the reflectivity signal decreases

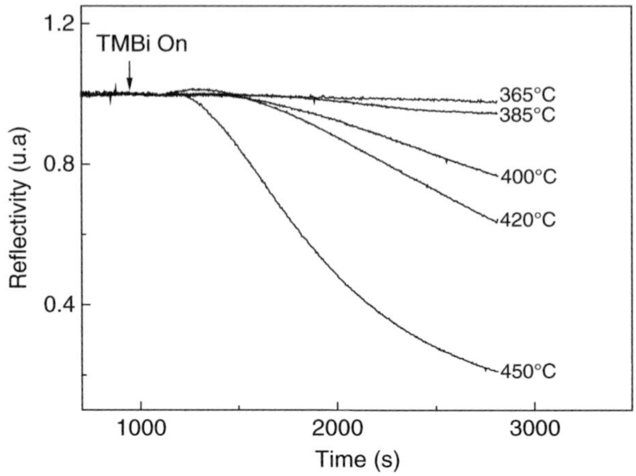

Fig. 5.2 Reflectivity variation versus growth time of GaAsBi layers elaborated at different temperatures

when temperature increases from 365 to 450 °C. This can be explained by the presence of a surface roughness which becomes more significant at elevated temperatures (450 °C).

Three-dimensional nucleation starts easily at elevated temperatures, and an important segregation of bismuth atoms occurs on the surface. These suggestions are confirmed by SEM observations [15]. It should be noted that these observations were seen by many groups [14, 18]. Indeed, Oe [14] has seen whisker crystals with different sizes randomly distributed on the most part of the substrate surface. On the other hand, it is well known that for lower temperatures (~350 °C) the decomposition rate of TMG and arsine is lower than 20 % [19]. So growing GaAsBi layers at lower temperatures and under atmospheric pressure becomes more difficult. For a temperature growth of 420 °C, we remark a decrease in islands' size (~0.25 μm) accompanied with an increase of their density (~10^8 cm^{-2}). The chemical composition of these defects is examined by SEM-EDX analysis and compared to that of flat surface. Figure 5.3a shows an SEM image of a surface layer containing island defects, and the inset Fig. 5.3b, c show EDX spectra measured at the top of the island and in the flat surface, respectively. It should be noted that islands are rich of Bi metal. The origin of the Bi island formation has been discussed by many authors [14, 20]. Indeed, similar observations were reported in case of LP-MOVPE of InAsBi [21] if the growth conditions are not appropriate. Probably Bi could form droplets on the surface. These droplets can also be seen when annealing GaAs substrate under AsH_3 and TMBi mixture flow at a temperature lower than 680 °C [16].

Figure 5.4 shows the diffraction pattern in the $\theta-2\theta$ scan of X-ray diffraction measurement for a $GaAs_{1-x}Bi_x$ layer grown at 420 °C on GaAs substrate. The spectrum shows three resolved peaks located at 31.627, 66.048, and 65.528° which are associated to the plane diffraction of (002) and (004) GaAs and (004) $GaAs_{1-x}Bi_x$, respectively (detail is shown in the inset figure). No other phases were identified.

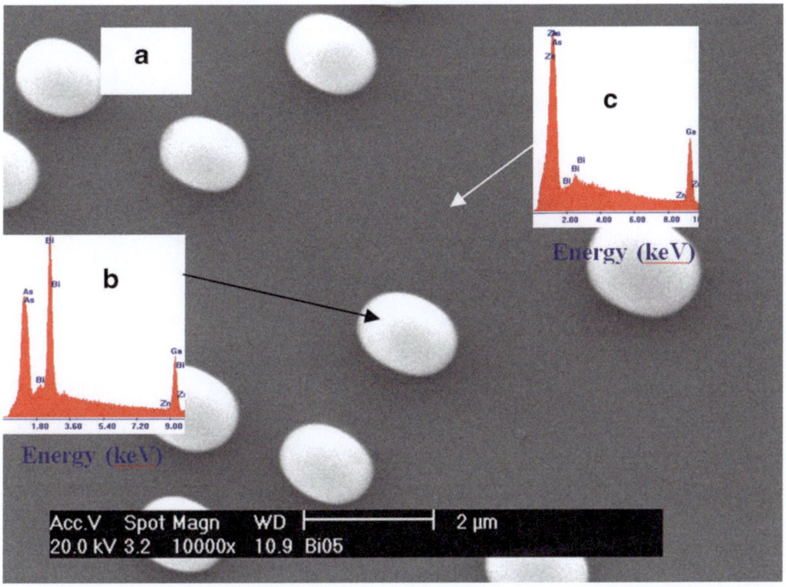

Fig. 5.3 (**a**) SEM image showing island defects on GaAsBi layer. (**b**) SEM-EDX spectrum measured at the top of an island. (**c**) SEM-EDX spectrum measured at the flat surface

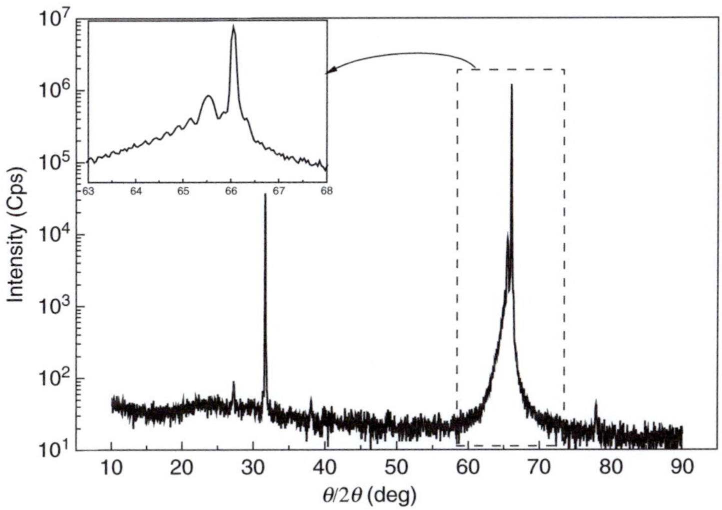

Fig. 5.4 X-ray $\theta-2\theta$ spectrum obtained for the diffraction of GaAsBi layer. The *inset* shows a zoom between $2\theta=63°$ and $68°$

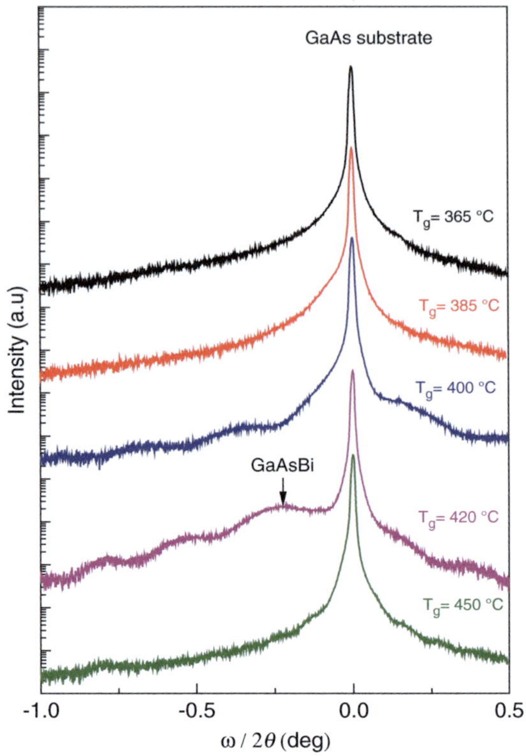

Fig. 5.5 High resolution X-ray diffraction (004) experimental $\omega/2\theta$ scans for five samples grown at different temperatures. All substrate peeks are adjusted on 0°

In order to study the crystallinity of the grown layers, (004) HRXRD measurements were performed on these samples. Figure 5.5 shows the HRXRD curves for different temperatures. For the samples grown at 385 and 400 °C, a small shoulder due to the diffraction from the grown layer can be seen at the lower-angle side of the substrate peak. A very weak shoulder is detected without Pendellösung oscillations for the sample grown at 450 °C. No diffraction peak or shoulder other than the substrate peak can be observed for the sample grown at 365 °C. On the other hand, a diffraction peak with Pendellösung oscillations are clearly seen in the curve of sample grown at 420 °C indicating a good crystalline quality of the GaAsBi epilayer. These results, which are in a good agreement with in situ laser reflectivity measurements and SEM observations, have guided us to choose 420 °C as an optimum growth temperature.

5.2.1.2 TMBi Flow Effect

Under this growth temperature, we have tried to grow GaAsBi using different TMBi flows. According to our experience, we have remarked that the density of islands decreases by increasing the substrate temperature, but islands' dimension increases.

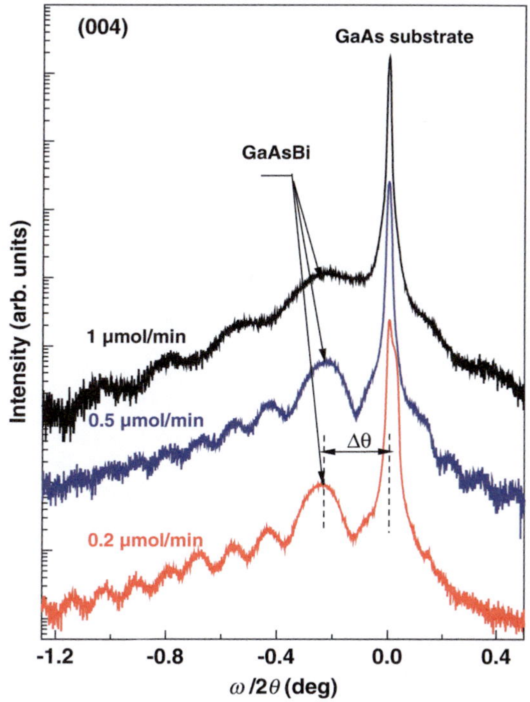

Fig. 5.6 HRXRD (004) experimental $\omega/2\theta$ scans for GaAsBi layers grown at 420 °C under different TMBi flow

In addition, this density increases with the growth duration and the flow rate of TMBi. We report in Fig. 5.6 HRXRD curves of three samples of GaAsBi alloy elaborated under 0.2, 0.5, and 1 μmol/min of TMBi flow at 420 °C. The variation of angular peak separation ($\Delta\theta_B$) between GaAs and $GaAs_{1-x}Bi_x$ as a function of Bi molar fraction is reported in Fig. 5.7. $\Delta\theta_B$ increases with increasing Bi molar fraction and saturates up to a TMBi flow of 0.2 μmol/min. It is possible that this saturation is originated from the formation of a liquid Bi phase on the surface limiting Bi incorporation.

In order to estimate the lattice parameter of GaAsBi and then Bi content, HRXRD measurements of symmetric (004) and asymmetric (115) plane diffractions were performed as shown in Fig. 5.8. Besides the two peaks relative to the layer and substrate diffractions, curves show clear Pendellösung oscillations, seen in the wings of peaks, indicating the presence of a smooth and coherent interface. These oscillations are extremely useful for layer thickness (t) determination, as the period ($\Delta\omega$) is independent of scattering power of layer composition. A differential form of Bragg's law gives: $t = \lambda\gamma_H / \Delta\,\omega\,\sin(2\,\theta_B)$ where $2\theta_B$ is the scattering angle, λ is the wavelength of the X-ray radiation, and $\gamma_H = \sin(\theta_B + \varphi)$. φ is the angle between the diffraction planes and the surface.

From the peak separation ($\Delta\theta_B$) between GaAs and $GaAs_{1-x}Bi_x$, we have also determined the lattice constants in growth direction ($a_\perp$) and in the plane ($a_{/\!/}$) of $GaAs_{1-x}Bi_x$ layer. Indeed, using a symmetric reflection (004), $a_\perp$ is deduced from

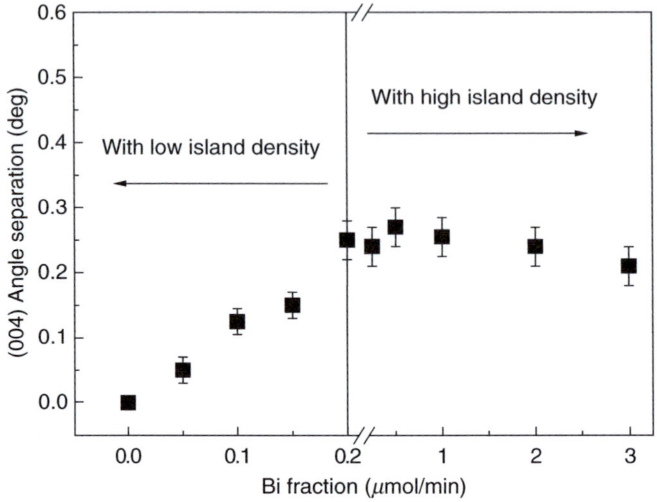

Fig. 5.7 Plots of (004) peak separation $\Delta\theta_B$ as a function of Bi molar fraction

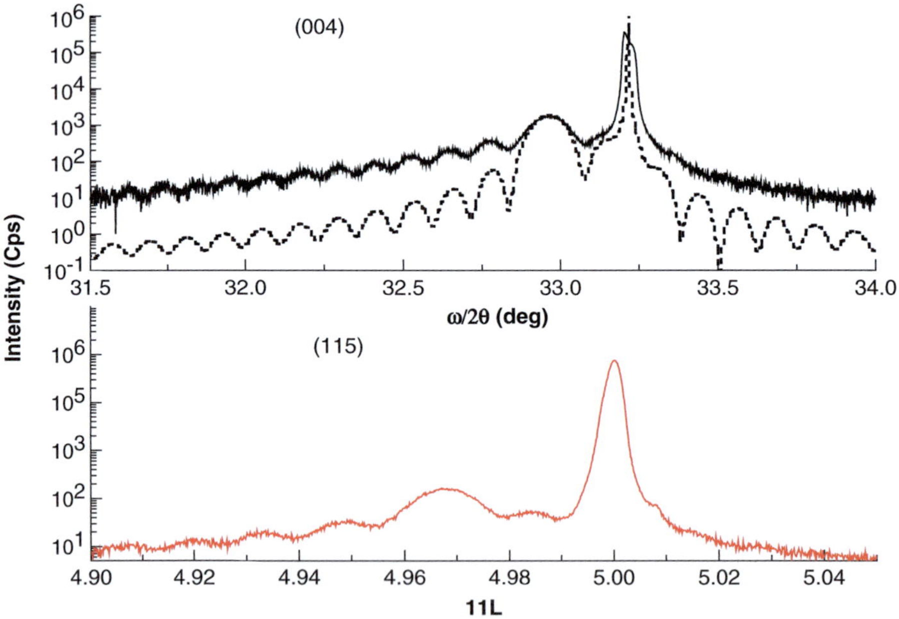

Fig. 5.8 In the *top*: HRXRD (004) experimental $\omega/2\theta$ curve (*solid line*) and simulated curve (*dotted line*) of $GaAs_{0.963}Bi_{0.37}$ layer. In the *bottom*: optimized scan of HRXRD (11*L*) reflection plane. *L* is changed around 5

the spacing of the layer (004) plane (d_{004}) in the form of $a_{\perp}=4d_{004}$. From asymmetric reflection (115), $a_{//}$ is deduced from the spacing of the layer (115) plane (d_{115}), in the form of

$$a_{\|} = \sqrt{2}\left[\frac{1}{d_{115}^2} - \frac{25}{a_{\perp}^2}\right]^{\left(-\frac{1}{2}\right)}$$

The lattice constant (a_0) of unstrained cubic $GaAs_{1-x}Bi_x$ is assumed from

$$a_{\perp} = a_{\|} + (a_0 - a_{\|})\frac{(C_{11} + 2C_{12})}{C_{11}}$$

where $C_{11}=11.92$ and $C_{12}=5.99$, the elastic constants for $GaAs_{1-x}Bi_x$, are supposed to be equal to those of GaAs. The Bi composition x is estimated from a_0 using Vegard's law:

$$a_0 = xa_{\text{GaBi}} + (1-x)a_{\text{GaAs}}$$

where a_{GaBi}, the lattice constant of hypothetical zinc-blend GaBi, is assumed to be 6.324 Å [22], and $a_{\text{GaAs}}=5.653$ Å is the lattice constant of GaAs.

Thickness homogeneity and layer composition were checked by recording (004) X-ray $\omega/2\theta$ curves for several positions in the layer surface. Small changes of 0.001 and 0.002/mm have been observed for $\Delta\omega$ and $\Delta\theta_B$, respectively. For all the three layers, the Bi content is found to be 3.7 % with a relative change of about 0.05 % per mm. Growth time of samples elaborated under 0.2 and 0.5 μmol/min of TMBi flow is greater than that of 1 μmol/min, which makes clear the difference in the oscillations period. Compared to GaAs, growing GaAsBi exhibits lower growth rates. It is known that using Bi as surfactant decreases the growth rate [17]. It should be noted that when TMBi flow was varied between 0.2 and 1 μmol/min no significant decrease of the growth rate was observed. The calculated layer thickness values were also compared to those deduced from simulated X-ray patterns using the dynamical theory depicted by dotted lines on Fig. 5.8. The tetragonal distortion induced by the epitaxy is taken into account using the elastic constants of GaAs. We found that there is a good agreement between experimental and simulated curves for a Bi content of 3.6 %. However, the simulated intensity is corrected by introducing a surface roughness of 10 nm.

5.2.1.3 The V/III Ratio Effect

The third parameter having a direct influence on the growth of GaAsBi alloy is the V/III ratio. We have checked its effect at the optimum growth temperature 420 °C and under a TMBi flow of 1 μmol/min. The experimental variation of V/III ratio was done by keeping constant the pressure of element III (TMG) and varying that of

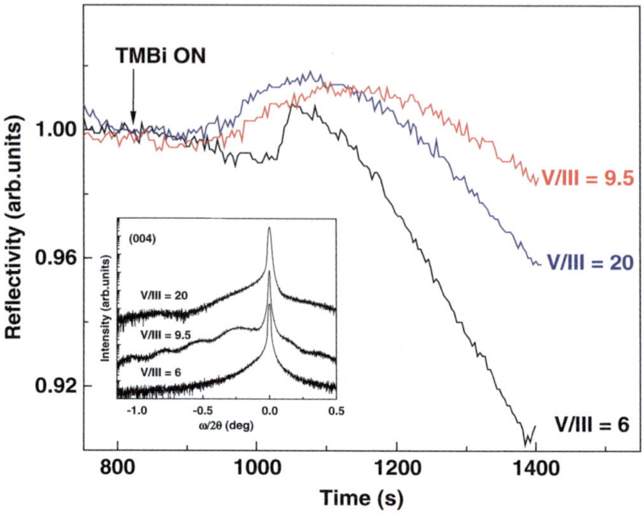

Fig. 5.9 Reflectivity variation during growth of GaAsBi layers at 420 °C under different V/III ratios. The *inset* shows HRXRD (004) experimental $\omega/2\theta$ scans of these samples

element V (AsH_3). We report in Fig. 5.9 the signal reflectivity variations for three V/III ratios. We can see that when TMBi is introduced into the reactor, the reflectivity signal takes a small increase for all samples. Indeed, compared to GaAs substrate, several groups [15, 23–25] have observed an increase in the refractive index of GaAsBi layer. On the other hand, the measured extinction coefficient of Bi films show very high values [26]. So, the small reflectivity increase and the lack of oscillations are probably the consequence of the elevated value of the extinction coefficient of GaAsBi films. After that, the reflectivity signal decreases more or less rapidly according to V/III ratio. Inset Fig. 5.9 shows HRXRD curves of these samples. The only diffraction peak clearly seen of GaAsBi layer is obtained for a V/III ratio of 9.5. A shoulder at the lower-angle side of the substrate peak was obtained for a V/III ratio of 20 indicating a low Bi composition. No GaAsBi diffraction peak was detected for a V/III ratio of 6. According to HRXRD analysis, in situ reflectivity measurements and SEM images, we can conclude that the maximum bismuth incorporation with the best surface quality is obtained in the sample elaborated at 420 °C with a V/III ratio of 9.5 and a TMBi flow of 0.2 μmol/min.

5.2.2 Photoreflectance Characterization of GaAsBi

The optical properties of these layers are investigated by photoreflectance (PR) spectroscopy. The PR measurements were carried out employing a standard setup with the 514.5 nm line of an argon laser as the pump light. The probe light was

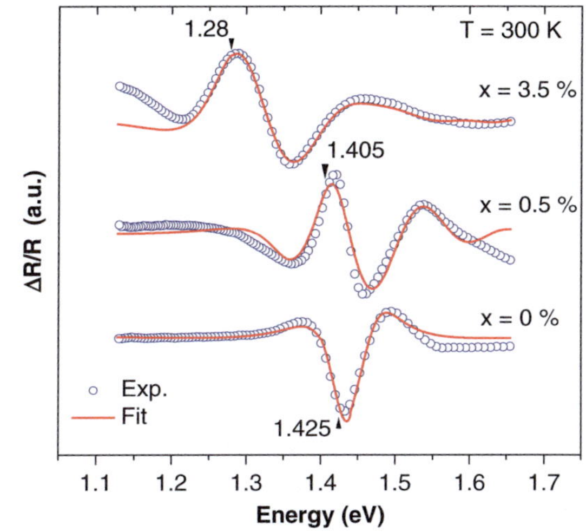

Fig. 5.10 PR spectra of $GaAs_{1-x}Bi_x$ for different Bi content. The *solid lines* represent adjustments by a multilayer model and the *arrows* show the position in energy of the band gap

obtained from a tungsten-halogen lamp analyzed with a 500 mm focal length monochromator. Figure 5.10 shows the PR spectra of $GaAs_{1-x}Bi_x$ layers for $x=0$, 0.5 and 3.5 % measured at 300 K. In spite of the fact that GaAsBi layers were grown on highly doped *p*-type GaAs:Zn substrates, no manifest Frank Keldish-type oscillation (FKO) related to the onset of the interband transition is clearly seen in the PR spectra. This may be explained by the presence of the GaAs buffer layer for which the band gap features coincide with GaAsBi FKOs. In general, the fast dumping of FKOs is due to the broadening effect and/or small electric field. Further, a nonuniform electric field in the space charge region may considerably affect the line shape of FKO [27]. In order to determine the band gap energy values, PR spectra can be analyzed by employing the third derivative functional form (TDFF) available for the low-field regime [28]. We have observed that experimental data cannot be well reproduced by a TDFF model. In fact, our PR spectra are better fitted using a model that takes into account the effects of the buffer layer and substrate, i.e., a multilayer model. Compared to the TDFF model, the multilayer model has the advantage to take into consideration optical interference effects as well as electric field inhomogeneity between different layers forming the structure. The solid lines in Fig. 5.1 show the line ship modeling of the experimental data with the model of equation: $\Delta R / R = (R_{\text{laser on}} - R_{\text{laser off}}) / R_{\text{laser off}}$. For Bi concentration $x=0$, 0.5 and 3.7 %, the best fit is obtained for band gap energies of 1.425, 1.405, and 1.28 eV, respectively (indicated by arrows). These values are consistent with the HRXRD measurements and show a net reduction of the band gap of GaAs with Bi incorporation.

The band gap energies of GaAsBi layers deduced from PR spectra are shown in Fig. 5.11 as a function of Bi content. A linear fit to the data gives a band gap reduction coefficient of 42 meV/% Bi. This value agrees well with 46 meV/% Bi reported earlier by Oe et al. [2], but disagree with those reported by Tixier et al. [9] (84 meV/% Bi) and Huang et al. [29] (62 meV/% Bi). One possible reason of this discrepancy

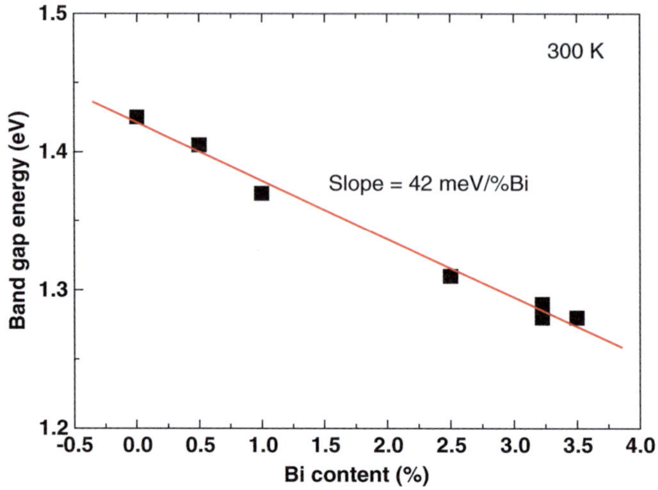

Fig. 5.11 Band gap energy of $GaAs_{1-x}Bi_x$ deduced from PR spectra versus Bi content. The *solid line* is a linear fit to the data

is the disparity in the band gap energy values for the same Bi content. Indeed, in PL measurements, the PL peak energy does not necessarily correspond to the band gap energy, whereas in PR measurements, estimation of the band gap energy depends strongly on the model used for spectra adjustment. The second and principal reason is the method used by these authors to measure bismuth concentration, i.e., Rutherford back scattering (RBS). Yoshimoto et al. [30] have established a phenomenological formula relating GaBi mole fraction x to the angular spacing 2θ of the (004) diffraction peak, expressed by $x(\%) = 6.93\ 2\theta(\text{deg})$. Applied to our results, this relation gives Bi composition values lower than those determined by X-ray diffraction measurements. Indeed, we found in this case a gap reduction of 79 meV/% Bi. This value agrees well with the slope of 84 meV/%Bi measured by Tixier et al. [9] on GaAsBi layers grown by molecular beam epitaxy.

5.3 Growth and Characterization of GaAsBi Alloy Under Alternated Bismuth Flows

5.3.1 Growth of GaAsBi Under Alternated Bismuth Flows

In the case of conventional MOVPE growth described in Sect. 5.2, the alloy epilayers could be grown within a very narrow range of growth conditions. Low growth temperature, V/III, and Bi/Ga ratios were used [13, 15]. Some discouraging results were obtained such as the small Bi islands were visible on the surface of the GaAsBi layer after growth, and the density of these islands increased with growth time. Therefore, islands should be obtained if the grown layer is thicker. However this did prevent further to apply the GaAsBi alloy layer to laser diodes and quantum well structure in which the layer thickness is superior to some nanometers. In this section, we report a successful growth method of GaAsBi by atmospheric pressure MOVPE.

We used in this method the same sources as the conventional growth method. The growth was carried out on (001) Zn-doped GaAs substrate. The sample was grown in horizontal geometry AP-MOVPE reactor. The precursor bubblers of TMG and TMBi were housed in constant temperature baths kept at 5 °C and 0 °C, respectively. The carrier gas was Pd-diffused H_2. Prior to starting growth the GaAs substrate was heated to 750 °C under a mixture flow of H_2 and AsH_3 to remove native surface oxide layer. The temperature was decreased till 420 °C and stabilized. At first, a GaAs buffer layer was grown onto the substrate at 420 °C for 150 s. AsH_3 and TMG fluxes were 446 and 22.8 μmol/min, respectively. We have kept these flows on, and TMBi supply (0.5 μmol/min) was switched on for about 10 s. After that, GaAs growth continues without TMBi flow during 10 min. This procedure was repeated four times, and finally the grown sample was capped by GaAs layer. The As/Ga, and the Bi/Ga ratios are about 20 and 0.02, respectively. These growth parameters are largely different from those for conventional AP-MOVPE growth of GaAsBi. It should be noted that no epitaxial GaAsBi layers on GaAs substrate are obtained under these condition growth by conventional AP-MOVPE process. The GaAsBi growth conditions are critical. The alloy epilayers could be grown within a very narrow range of growth conditions.

5.3.2 Characterization of GaAsBi Grown Under Alternated Bismuth Flows

Figure 5.12 shows the changes of reflectivity signal measured during the complete growth process of GaAsBi layer. Temperature profile during growth process is also inserted in the same figure. In fact, the grown layer was elaborated with several runs of alternated TMBi flows. For the first two runs of TMBi flow, the reflectivity signal increases. According to the surface roughness caused by Bi droplets, a decrease in the reflectivity signal is expected. This behavior may be attributed to the deposition of the first thin Bi flat islands. At this growth stage, the surface should present small Bi islands and no epitaxial GaAsBi layer can be obtained. These observations were also seen when annealing GaAs substrates under bismuth flow [16]. At the point where TMBi flow is switched off, a drastic drop of the reflectivity signal is observed indicating the beginning of GaAs growth. A similar reflectivity variation was observed for the second TMBi run. However, the reflectivity changes significantly when TMBi flow is again turned on into the MOVPE growth chamber. Indeed, the amplitude value decreases below that of GaAs substrate, indicating a persistence of three-dimensional growth mode.

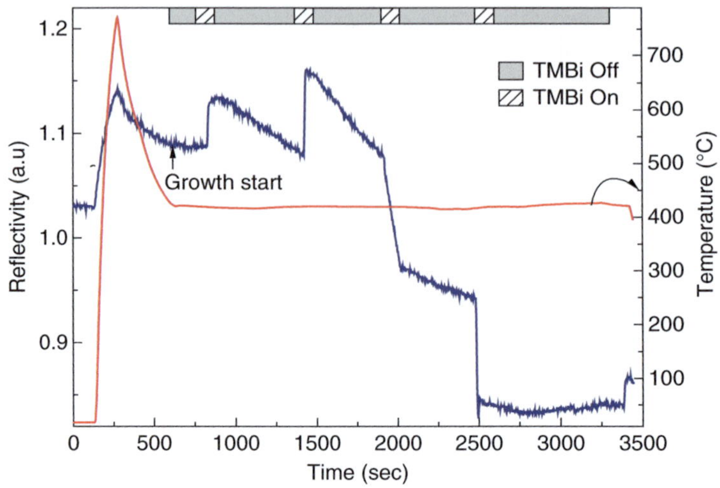

Fig. 5.12 Reflectivity signal (*left axis*) and growth temperature (*right axis*) recorded during the complete growth process of GaAsBi layer on GaAs substrate consisting of four deposits bismuth droplets and four GaAs

In order to analyze the crystalline quality of the sample grown after the first pair Bi/GaAs (till 1,400 s), (004) HRXRD measurement were performed (Fig. 5.13a). Other than GaAs substrate peak, a small shoulder at the lower-angle side was observed. This shoulder is attributed to the diffraction of the grown GaAsBi thin layer. On the other hand, to see Bi profile in the grown layer after the whole growth process, SIMS analysis was preformed.

Figure 5.14 shows the SIMS depth profile of the sample for the elements Ga and Bi. Results show no Bi periodicity. In fact, Bi atoms concentration decreases exponentially in the GaAsBi layer. So, our growth method provides a $GaAs_{1-x}Bi_x$-graded composition layer and not an alternated Bi/GaAs heterostructure. Indeed, Bi surface segregation at 420 °C is too high to allow its incorporation into the grown layer, and the multiquantum wells' structure formation. In our previous work on the optimization of conventional GaAsBi growth, Bi islands with different sizes randomly distributed on the substrate surface was observed at different growth temperatures [15]. SIMS results confirm the possibility of Bi atoms substitution by As atoms also in the interrupted growth process. At this time, this is a very tentative assignment and will require further investigation to ascertain its veracity. Nevertheless, the formation of a GaAsBi interfacial layer is supported by these SIMS result. Alternatively, a similar estimation of Bi content can be obtained from HRXRD measurements. The (004) HRXRD recorded pattern of the sample elaborated under the entire growth process is shown in Fig. 5.13b. The GaAsBi layer peak is clearly observed on the left-hand side of the substrate peak (the most intense feature). However, the broad feature observed in the left-hand side of the layer peak is attributed to a graded layer in GaAsBi/GaAs interface. The peak separation ($\Delta\theta$), measured for (004) reflection, between GaAs substrate and GaAsBi layer gives a Bi content of about 3 %. It also appears that in this pattern, no interference fringes were observed, and

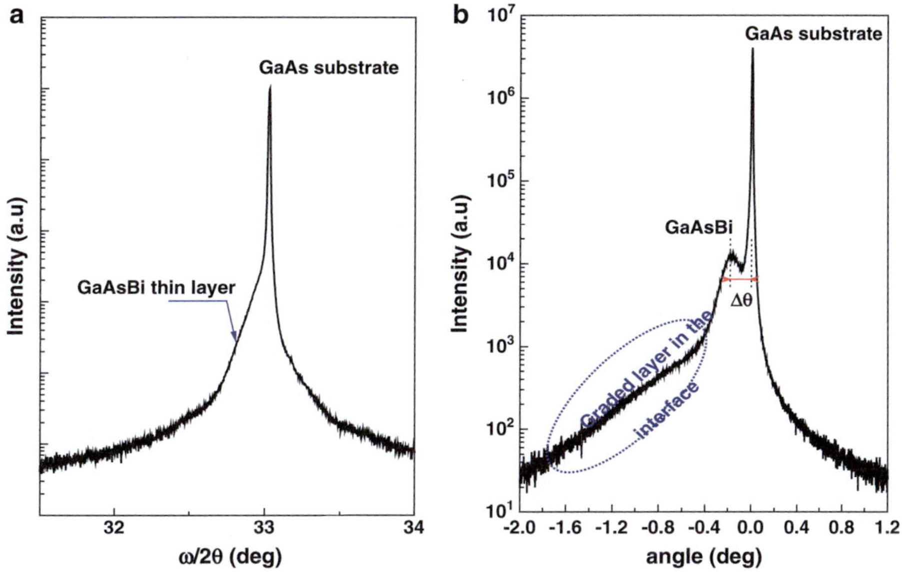

Fig. 5.13 HRXRD (004) experimental $\omega/2\theta$ curve of GaAsBi thin layer grown on (001) GaAs substrate. (**a**) After the first pair Bi/GaAs (till 1,400 s). (**b**) After the whole growth process

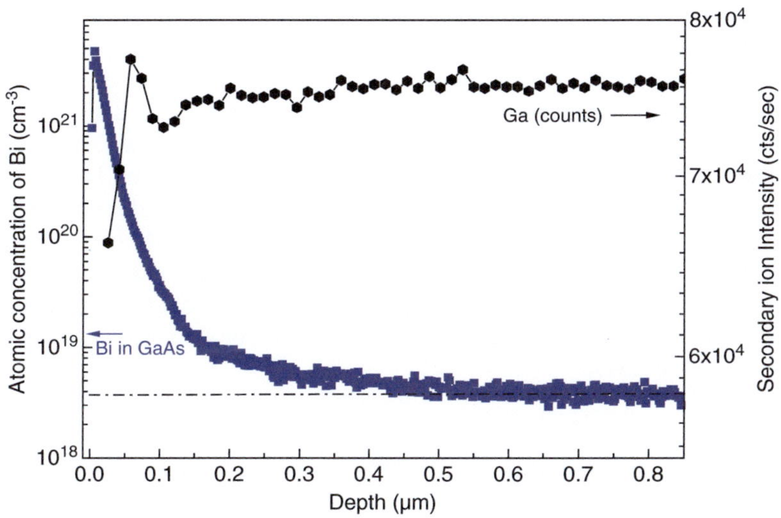

Fig. 5.14 SIMS profiles of Bi in the GaAsBi layer grown on GaAs substrate

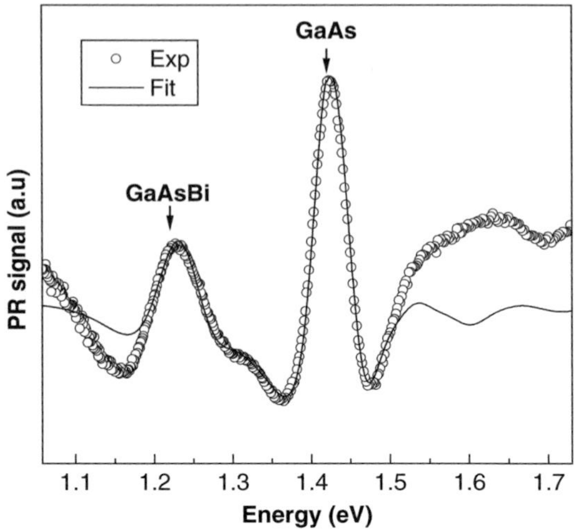

Fig. 5.15 Room temperature PR spectrum of the GaAsBi layer. The *solid line* represents adjustment by the multilayer model

the diffraction peak sits on wide diffuse scattering peak. The epilayer thickness of this sample, obtained by SEM in cross section, is ~280 nm. The Bi content homogeneity was checked by recording (004) HRXRD patterns in several positions of the layer surface. After the whole growth process, Bi islands are still present on the sample surface. However, compared to GaAsBi conventional growth, this growth method offers a low Bi island density on the sample surface.

The optical property of this layer is investigated by PR spectroscopy. The room temperature PR spectrum is shown in Fig. 5.15. It reveals the band edge transition in GaAsBi layer together with the band edge transition in GaAs layer (sharp peak at 1.42 eV). A fitting procedure based on a multilayer model, similar to that used in ref. [4], was used to adjust the PR spectrum. Analysis of PR spectrum by this model allows for the determination of the band gap energy and gives an estimation of both broadening parameter and built-in electric fields in the structure. Best fit to the data gives for the GaAsBi layer band gap energy of 1.22 eV, a broadening parameter of 60 meV and a built-in electric field of 59 kV/cm. The Bi content deduced from the energy shift (200 meV), using the band gap reduction coefficient of 42 meV/% Bi [4], is about 4.8 %. This value disagree with this estimated by HRXRD. The principal reason of this discrepancy is related to the method used to measure Bi content. Indeed, for such gradual layer it seems to be reasonable to find this difference associated to the ability of each method to estimate in depth the Bi content in the layer.

5.4 Effects of Thermal Annealing on $GaAs_{1-x}Bi_x$ Alloys

In this section, a systematic study on the effects of annealing on the structural and optical properties of MOVPE grown $GaAs_{1-x}Bi_x$ alloys will be reported. The thermal annealing was performed in the growth reactor for 15 min at different

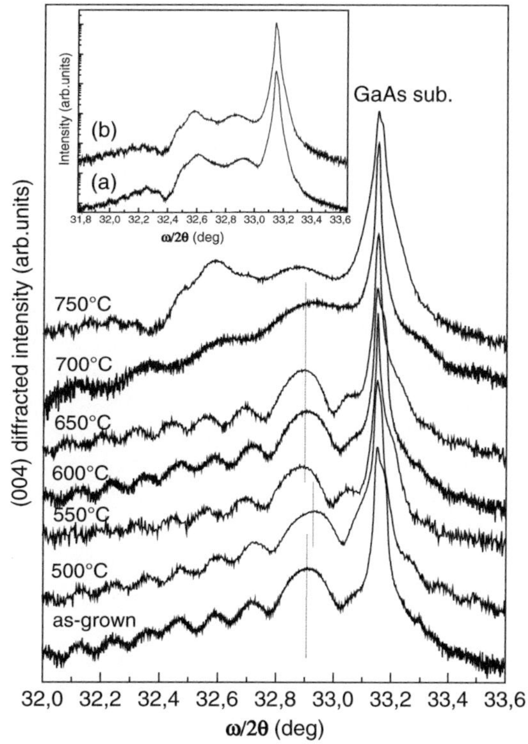

Fig. 5.16 HRXRD (004) $\omega/2\theta$ scans of as-grown and annealed $GaAs_{0.963}Bi_{0.037}$ samples. Annealed samples at 700 and 750 °C are 25 nm thick, and the others are 50 nm thick. *Inset* shows a comparison between the diffraction profiles of a layer grown under high Bi flow (**a**) and the annealed sample at 750 °C (**b**)

temperatures ranging from 500 to 750 °C. To prevent surface decomposition, samples are annealed under arsine flow. The samples discussed in this section were grown by conventional MOVPE method. Each sample consists of a 25–50 nm GaAsBi epilayer. Samples used for annealing were cleaved pieces of two GaAsBi layers having the same Bi content (3.7 %) and two different thicknesses: 25 and 50 nm. For annealing temperatures lower than 700 °C, layer thickness is 50 and 25 nm for higher ones (700 and 750 °C). Bi content and epilayer thicknesses were determined from analysis of HRXRD curves using (004) and (115) reflection.

Figure 5.16 shows the HRXRD patterns of as-grown and annealed $GaAs_{0.963}Bi_{0.037}$ layers. It is clear from this figure that HRXRD patterns have no detectable changes till an annealing temperature of 700 °C. A narrow peak and clear Pendellösung fringes were observed in all samples, indicating that GaAsBi/GaAs interface is fairly flat and GaAsBi layers are uniform and of a good crystalline quality. Compared with the others, we can see that only sample annealed at 500 °C shows a very small shift of the epilayer diffraction peak. However, the full width at half maximum (FWHM) of the substrate diffraction peak of this sample is smaller than that of the others. So, the narrowing of the main GaAs peak possibly causes the shift of the adjacent $GaAs_{0.963}Bi_{0.037}$ peak as an artifact. This small shift may also be related to vertical lattice shrinkage at this annealing temperature. The structural properties of the $GaAs_{0.963}Bi_{0.037}$ alloy are unaffected by thermal annealing till 700 °C indicating a high stability for device processing. For an annealing temperature of 750 °C, more diffraction peaks other than that of the GaAs substrate can be clearly seen. Pendellösung fringes disappear at this annealing temperature, indicating that the abrupt interface was destroyed. A comparison between HRXRD patterns of unannealed sample, grown under higher Bi flow rate (3 μmol/min) and reported in our study [13, 31], and that of annealed one at 750 °C, surprisingly shows the same diffraction profile (inset Fig. 5.1). According to the reproducibility of our measurements, the annealing effect on the crystalline quality is similar to that induced by a higher Bi flow. The presence of more peaks in the diffraction profiles can be attributed to a phase separation in GaAsBi alloy [2], or the presence of "micro regions" with different Bi content [32], or a difference in the strain state at different depths [33].

Figure 5.17 shows AFM images of $GaAs_{0.963}Bi_{0.037}$ samples annealed at 550, 700, and 750 °C. It is well known that when growing GaAsBi, bismuth islands start easily on the surface where density and size depend on the growth conditions [9, 15]. For an annealing temperature of 550 °C, bismuth islands are randomly distributed on the surface (Fig. 5.17a). Their diameter and height are about 160 and 20 nm, respectively. When annealing temperature is increased to 700 °C, a change in the surface morphology is observed. Islands density is clearly decreased making larger their average size: 500 nm diameter and 90 nm height. These observations were also seen when annealing high index GaAs substrates under bismuth flow [16]. For sample annealed at 750 °C, an important change in the surface morphology occurs. Surface sample contains pits with different sizes (Fig. 5.17c). Indeed, Bi islands were completely removed and each removed one creates a pit on the surface. From AFM observations, 700 °C is obviously the annealing temperature over which $GaAs_{0.963}Bi_{0.037}$ surface is no longer stable.

Low temperature PL (10 K) spectra of the $GaAs_{0.963}Bi_{0.037}$ alloy before and after annealing at different temperatures are shown in Fig. 5.18. The experimental traces presented in this figure are normalized to their maximum values of the *p*-type GaAs substrate emission located at 1.48 eV. For temperatures lower than 100 K, only one peak located at 1.34 eV was detected. This peak seems to be that of emission of localized states. Indeed, at lower temperatures, the PL of $GaAs_{1-x}Bi_x$ alloys is usually dominated by carriers localized at Bi pairs and clusters [3, 6, 34, 35]. These localized states, found in GaAsN [36, 37] and other materials [38, 39], are responsible for complex emission spectra. Recently, for a low Bi concentration ($x = 0.4$ %), Francoeur et al. [6] have demonstrated that the PL of localized centers in $GaAs_{1-x}Bi_x$ is very sensitive to temperature. These localized centers are thermally ionized when temperature exceeds 100 K, and the PL spectrum is dominated by free exciton emission only, leading to the measurement of the $GaAs_{1-x}Bi_x$ band gap. In our study, for higher temperatures (100 K), the PL intensities of all $GaAs_{0.963}Bi_{0.037}$ samples annealed at different temperatures are below the detection limit. So, from this temperature value, all bound excitonic states seem to disappear and no other transition showing free exciton contribution was measured. On the other hand, it was reported that samples grown by molecular beam epitaxy with a Bi concentration up to 3.1 %

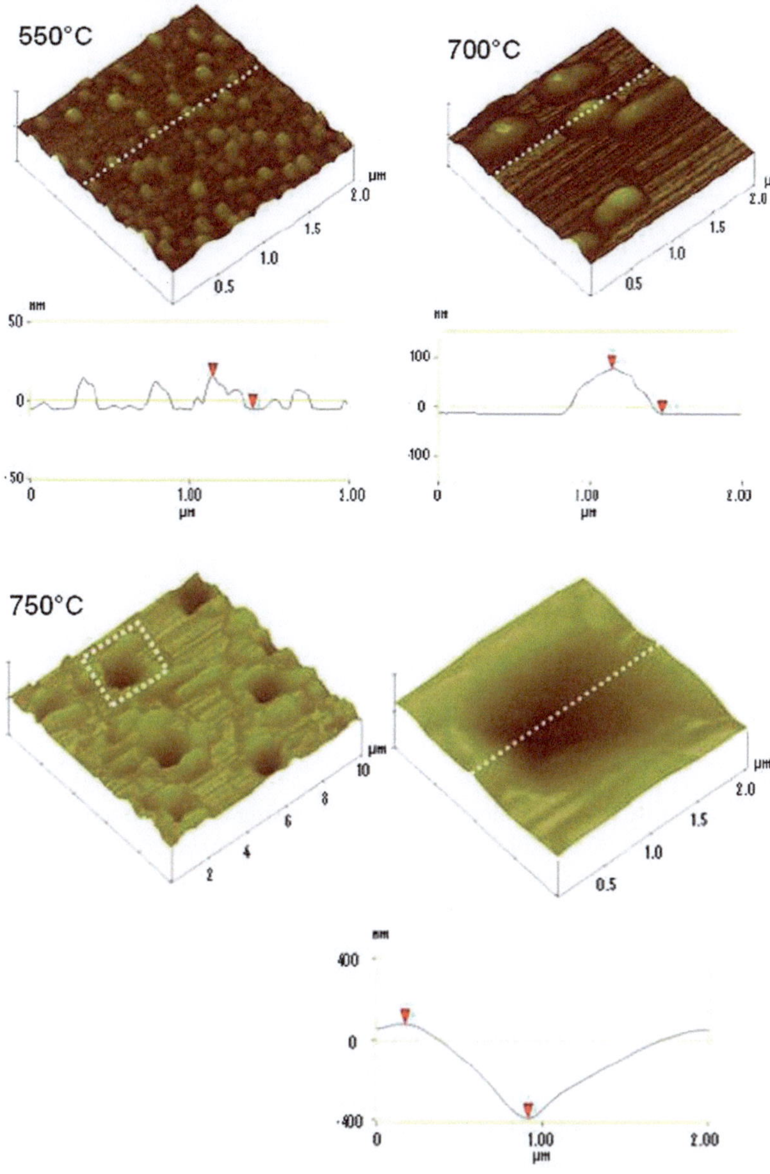

Fig. 5.17 AFM images of $GaAs_{0.963}Bi_{0.037}$ layers after annealing at different temperatures (**a**) 550°C, (**b**) 700 °C, and (**c**) 750 °C. AFM cross sections are presented below each image

have shown PL emission at room temperature [9]. This discrepancy is probably due to the growth techniques. An as-grown sample shows a poor PL spectrum, which can be ascribed to the weak carrier confinement.

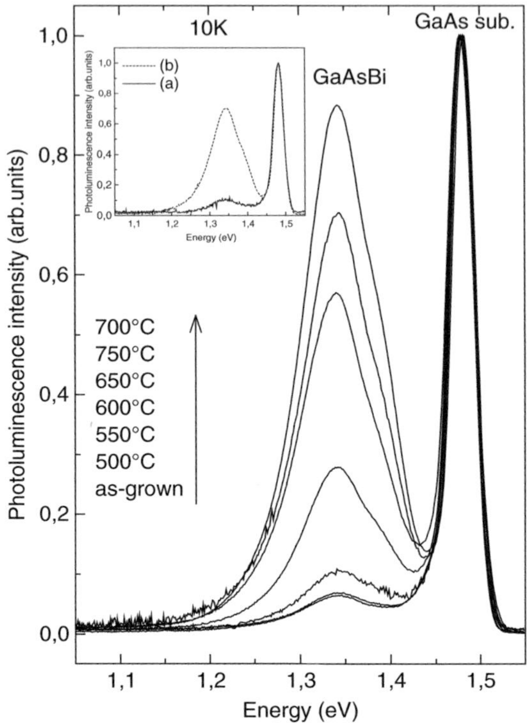

Fig. 5.18 Low temperature (10 K) PL spectra of as-grown and annealed $GaAs_{0.963}Bi_{0.037}$ samples. *Inset* shows a comparison between the PL spectra of a layer grown under high Bi flow (**a**) and the annealed sample at 750 °C (**b**)

The PL intensity is clearly increased with increasing annealing temperature. We note that for 25 and 50 nm thick $GaAs_{0.963}Bi_{0.037}$ layers annealed at 650 °C, the PL intensities give the same spectrum (not shown here). According to our Bi concentration (3.7 %), the energy change of the PL emission located at 1.34 eV and measured at 10 K is ~48 meV/% Bi. Since at low temperatures the PL spectra are dominated by localized excitons, we cannot evaluate the band gap change. It is obvious that a better understanding of this PL behavior needs accurate investigations. For all annealed samples, the PL peak energy of the $GaAs_{0.963}Bi_{0.037}$ layer remains unchanged. However, the only available study showing the annealing effect on the PL of $GaAs_{0.968}Bi_{0.032}$ layers provides a small blue shift [40]. The inset of Fig. 5.18 shows low temperature PL of an as-grown GaAsBi layer grown under high Bi flow and that of the sample annealed at 750 °C. In spite of the presence of more diffraction peaks of these two samples, the PL spectra show only one peak located at 1.34 eV. Figure 5.19 shows the tendency of the integrated intensity and the FWHM of the PL against annealing temperature. The PL intensity is improved significantly with increasing annealing temperature up to 700 °C. This increase is more than one order of magnitude compared with the as-grown sample. It turns out that thermal annealing is very efficient in decreasing the number of nonradiative recombination centers in GaAsBi. The annealing probably provides atomic rearrangement in the

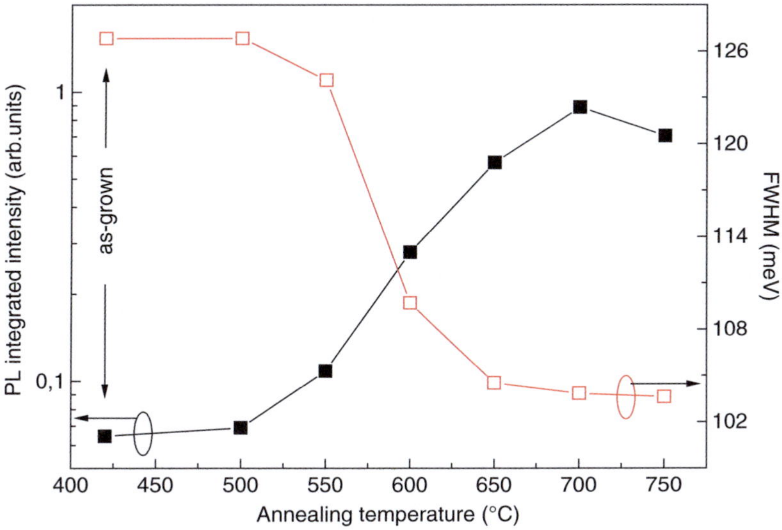

Fig. 5.19 Annealing temperature dependence of PL integrated intensity (*closed squares*) and the FWHM of the PL (*open squares*)

GaAsBi lattice, like nitrogen in GaAsN [41]. The FWHM decreases with annealing temperature, but remains relatively large. Its value is comparable to that found by Tixier et al. [9] in $GaAs_{0.969}Bi_{0.031}$. This can be attributed to the structural disorder caused by the low growth temperature and low V/III ratio. On the other hand, an exception can be seen for the PL intensity of the annealed sample at 750 °C. Indeed, the PL intensity decreases and the FWHM remains constant. The HRXRD pattern and AFM images of this sample, shown in Figs. 5.16 and 5.17, also present special behavior.

5.5 Conclusion

It has been established from our experimental results that the optimum AP-MOVPE growth conditions of GaAsBi alloys are further away from standard parameters. The effect of growth conditions such as growth temperature, TMBi flow, and V/III ratio was studied. According to in situ reflectivity measurements, SEM observations, and XRHRD analysis, we conclude that 420 °C, 0.2 µmol/min, and 9.5 are the optimized values of growth temperature, TMBi flow, and V/III ratio, respectively. These conditions allow GaAsBi layers of a good crystalline quality with a Bi concentration up to 3.7 %. We have also reported a successful growth method for growing a GaAsBi layer under alternated bismuth flows. The Bi islands were deposited using TMBi flashes during short growth times. We reported SIMS profile without Bi periodicity and then deduced the GaAsBi-graded composition layer. HRXRD curve shows a diffraction peak that can be attributed to the GaAsBi layer. With this method we minimized Bi island on the surface. The effect of thermal annealing on the structural and optical properties of the GaAsBi alloys was investigated. For annealing temperatures lower than 700 °C, high crystalline quality and remarkable stability of the $GaAs_{0.963}Bi_{0.037}$ alloy were found, while for higher temperature, 750 °C, HRXRD and AFM analysis shows a weakening of the structural quality. After annealing, the increase of integrated PL intensity was clearly observed indicating the improvement of the optical properties of $GaAs_{0.963}Bi_{0.037}$ samples at lower temperatures (100 K). The optimal annealing temperature was found to be 700 °C.

References

1. Francoeur, S., Seong, M.J., Mascarenhas, A., Tixier, S., Adamcyk, M., Tiedje, T.: Appl. Phys. Lett. **82**, 3874 (2003)
2. Oe, K.: Jpn. J. Appl. Phys. **41**, 2801 (2002)
3. Fluegel, B., Francoeur, S., Mascarenhas, A., Tixier, S., Young, E.C., Tiedje, T.: Phys. Rev. Lett. **97**, 067205 (2006)
4. Chine, Z., Fitouri, H., Zaied, I., Rebey, A., El Jani, B.: Semicond. Sci. Technol. **25**, 065009 (2010)
5. Yoshida, J., Kita, T., Wada, O., Oe, K.: Jpn. J. Appl. Phys. **42**, 371 (2003)
6. Francoeur, S., Tixier, S., Young, E., Tiedje, T., Mascarenhas, A.: Phys. Rev. B **77**, 085209 (2008)
7. Bertulis, K., Krotkus, A., Alekseienko, G., Pacebutas, V., Adomavicius, R., Molis, G., Marcinkevicius, S.: Appl. Phys. Lett. **88**, 201112 (2006)
8. Oe, K., Okamoto, H.: Jpn. J. Appl. Phys. **37**, L1283 (1998)
9. Tixier, S., Adamcyk, M., Tiedje, T., Francoeur, S., Mascarenhas, A., Wei, P., Schiettekatte, F.: Appl. Phys. Lett. **82**, 2245 (2003)
10. Alberi, K., Dubon, O.D., Walukiewicz, W., Yu, K.M., Bertulis, K., Krotus, A.: Appl. Phys. Lett. **91**, 051909 (2007)
11. Henini, M., Ibanez, J., Schmidbauer, M., Shafi, M., Novikov, S.V., Turyanska, L., Molina, S.I., Sales, D.L., Chisholm, M.F., Misiewicz, J.: Appl. Phys. Lett. **91**, 251909 (2007)
12. Pacebutas, V., Butkutė, R., Cechavicius, B., Kavaliauskas, J., Krotkus, A.: Thin Solid Films **520**, 6415 (2012)
13. Fitouri, H., Moussa, I., Rebey, A., Fouzri, A., El Jani, B.: J. Cryst. Growth **295**, 114 (2006)
14. Oe, K.: J. Cryst. Growth **237–239**, 1481 (2002)
15. Moussa, I., Fitouri, H., Rebey, A., El Jani, B.: Thin Solid Films **516**, 8372 (2008)
16. Fitouri, H., Moussa, I., Rebey, A., El Jani, B.: J. Cryst. Growth **300**, 347 (2007)
17. Wixom, R.R., Rieth, L.W., Stringfellow, G.B.: J. Cryst. Growth **265**, 367 (2004)
18. Takehara, Y., Yoshimoto, M., Huang, W., Saraie, J., Oe, K., Chayahara, A., Horino, Y.: Jpn. J. Appl. Phys. **45**, 67 (2006)
19. Larsen, C.A., Li, S.H., Buchan, N.I., Stringfellow, G.B., Brown, D.W.: J. Cryst. Growth **102**, 126 (1990)
20. Jiang, W.Y., Liu, J.Q., So, M.G., Myrtle, K., Kavanagh, K.L., Watkins, S.P.: J. Cryst. Growth **277**, 85 (2005)
21. Okamoto, H., Oe, K.: Jpn. J. Appl. Phys. **37**, 1608 (1998)
22. Janotti, A., Wei, S.H., Zhang, S.B.: Phys. Rev. B **65**, 115203 (2002)
23. Yamashita, K., Yoshimoto, M., Oe, K.: Phys. Status Solidi C **3**, 693 (2006)
24. Tumėnas, S., Karpus, V., Bertulis, K., Arwin, H.: Phys. Status Solidi C **9**, 1633 (2012)

25. Sedrine, N.B., Moussa, I., Fitouri, H., Rebey, A., El Jani, B., Chtourou, R.: Appl. Phys. Lett. **95**, 011910 (2009)
26. de Sande, J.C., Missana, T., Afonso, C.N.: J. Appl. Phys. **80**, 7023 (1996)
27. Shen, H., Dutta, M.: J. Appl. Phys. **78**, 2151 (1995)
28. Aspnes, D.E.: Surf. Sci. **37**, 418 (1973)
29. Huang, S.H., Balakrishnan, G., Khoshakhlagh, A., Jallipalli, A., Dawson, L.R., Huffaker, D.L.: Appl. Phys. Lett. **88**, 131911 (2006)
30. Yoshimoto, M., Murata, S., Chayahara, A., Horino, Y., Saraie, J., Oe, K.: Jpn. J. Appl. Phys. **42**, L1235 (2003)
31. Fitouri, H., Moussa, I., Rebey, A., El Jani, B.: Microelectron. Eng. **88**, 476 (2011)
32. El-Masry, N.A., Piner, E.L., Liu, S.X., Bedair, S.M.: Appl. Phys. Lett. **72**, 40 (1998)
33. Pereira, S., Correia, M.R., Pereira, E., O'Donnell, P., Aves, E., Sequeira, A.D., Franco, N.: Appl. Phys. Lett. **79**, 1432 (2001)
34. Pacebutas, V., Bertulis, K., Aleksejenko, G., Adomavicius, R., Molis, G., Krotkus, A.: Opt. Mater. **30**, 756 (2008)
35. Pettinari, G., Polimeni, A., Capizzi, M., Blokland, J.H., Christianen, P.C.M., Maan, J.C., Young, E.C., Tiedje, T.: Appl. Phys. Lett. **92**, 262105 (2008)
36. Polimeni, A., Bissiri, M., Augieri, A., Baldassari Hoger von Hogersthal, G., Capizzi, M., Gollub, D., Fischer, M., Reinhardt, M., Forchel, A.: Phys. Rev. B **65**, 235325 (2002)
37. Alberi, K., Wu, J., Walukiewicz, W., Yu, K.M., Dubon, O.D., Watkins, S.P., Wang, C.X., Liu, X., Cho, Y.-J., Furdyna, J.: Phys. Rev. B **75**, 045203 (2007)
38. Ouali, A.A., Yip, R.F., Brebner, J.L., Masut, R.A.: J. Appl. Phys. **83**, 3153 (1998)
39. Dudiy, S.V., Zunger, A., Felici, M., Polimeni, A., Capizzi, M., Xin, H.P., Tu, C.W.: Phys. Rev. B **74**, 155303 (2006)
40. Feng, G., Oe, K., Yoshimoto, M.: Jpn. J. Appl. Phys. **46**, L764 (2007)
41. Spruytte, S.G., Coldren, C.W., Harris, J.S., Wampler, W., Krispin, P., Ploog, K.H., Larson, M.C.: J. Appl. Phys. **89**, 44014 (2001)

Chapter 6
Group III–V Bismide Materials Grown by Liquid Phase Epitaxy

Sunanda Dhar

Abstract Bi containing III–V compound semiconductors have become important in the last few years for many potential device applications. Molecular beam epitaxy and metalorganic vapor phase epitaxy techniques have mostly been used for the growth of these compounds. We review here the application of the simple liquid phase epitaxy (LPE) technique for the growth of some members of the III–V-Bi series. Due to the restrictions of limited solubility of Bi in III–V solids, use of LPE has so far been limited to the growth of InSbBi and GaSbBi. We describe here details of the growth procedure and the characterization of these two ternaries by different groups with special emphasis on the structural, luminescence, and band gap reduction properties. Theoretical model based on diffusion of Bi across an LPE growth model is described with a view to optimize the growth parameters. Reference is also made to the melt growth of bulk crystals of these compounds and to the LPE growth of InPBi and InAsSbBi epitaxial layers.

6.1 Introduction

Group III–V bismide alloys have attracted considerable attention in the last few years and are being regarded as a new class of materials for applications in long wavelength optoelectronic devices. Incorporation of small quantities of bismuth to III–V alloys results in a large reduction in the band gap energy whereas the corresponding increase in lattice constant is rather small enabling better lattice match with the substrate than by the addition of In or Sb to the alloy. To date, most studies

---

S. Dhar (✉)
Department of Electronic Science, University of Calcutta,
92, A. P. C. Road, Kolkata 700009, India
e-mail: sdelc@caluniv.ac.in

H. Li and Z.M. Wang (eds.), *Bismuth-Containing Compounds*, Springer Series in Materials Science 186, DOI 10.1007/978-1-4614-8121-8_6,

on III–V-Bi materials focused on $GaAs_{1-x}Bi_x$ [1–4] for applications in optical communication and $InSb_{1-x}Bi_x$ [5–7] for devices operating in the 8–12 μm region. The III–V bismides have striking similarities with the well-known dilute III–V nitrides [8] in the sense that in both cases a drastic reduction in the band gap of the parent III–V compound is observed after the addition of small quantities of either Bi or N. However, there are some important differences between III–V-N and III–V-Bi material systems which have been detailed in references [9, 10]. In III–V nitrides, nitrogen forms an isolated impurity level within the conduction band which is resonant with the bottom of the conduction band, together with a number of N–N pair states. The resonant interaction of the nitrogen level with the conduction band minimum introduces large band gap bowing and a lowering of the conduction band minimum [11]. Bi, in contrast, forms an isolated energy level which is resonant with the valence band maximum of the III–V host resulting in an upward movement of the valence band and this has been predicted theoretically [9] and verified through experiments [12]. Like dilute nitrides, the band gap reduction property in dilute bismides is most successfully explained by the band anti-crossing model [3]. Further, the band gap reduction of GaAs due to Bi is reported to be smaller (88 meV/% Bi) than that due to nitrogen (about 200 meV/%N). On the other hand, bismides offer two distinct advantages over dilute nitrides. One is a relatively temperature-independent band gap [13] which may help to get laser diodes with temperature-insensitive emission wavelengths. The other is that, in contrast to dilute nitrides, where the electron mobility of the material is greatly degraded by N, the degradation of mobility with Bi addition is only marginal [14, 15]. Molecular beam epitaxy (MBE) and metalorganic vapor phase epitaxy (MOVPE) techniques are being used for the growth of GaAsBi [1, 16–19], InAsBi [20], and InSbBi [5–7]. MBE growth of the quaternary compound InGaAsBi, lattice-matched to InP substrates, has been reported [21] for the fabrication of temperature-insensitive lasers for fiber-optic communications.

6.2 Growth of III–V Bismides from Melt

Non-equilibrium growth techniques like MBE and MOVPE have the advantage of incorporating Bi in a much larger quantity than is allowed by the equilibrium solid solubility of the element in the parent III–V compound. Thus Bi incorporation up to 10 % has been achieved in GaAsBi [22], 6 % in InAsBi [23], and up to 7 % in InGaAsBi [24]. However, larger amounts of Bi in the solid are reported to reduce the luminescence from the material [22]. Growth of III–V bismide epilayers from liquid melts under thermal equilibrium, such as using the well-known liquid phase epitaxy (LPE) technique [25], appears to be a simpler alternative for growing the materials with high structural and luminescence properties. Thermodynamic calculations to assess the feasibility of LPE growth of some III–V-Bi compounds have been recently done by Kuznetsov et al. [26]. Further, applicability of the growth of bismides from solution is limited by the equilibrium solid solubility of Bi in the solid. Bi has an extremely small solubility in GaAs with an effective segregation coefficient of 1×10^{-4} [27]. Hence, although there are reports on growth of high purity GaAs layers from Bi solutions, growth of GaAsBi could not be realized. Similarly, Bi has a solubility as low as 0.02 at.% in InAs [28] and hence bulk crystal or LPE growth of InAsBi is ruled out. The most promising candidate for the melt growth seems to be InSbBi since Bi has an appreciable solubility up to 2.6 at.% in InSb [28]. Reports on the growth of bulk crystals of InSbBi date back to 1969 where Bi concentrations up to 2 mol% were reported by Jean-Louis et al. [29, 30]. A more comprehensive study was presented by the same group in 1972 [27] wherein single crystals of InSbBi were grown by a Czochralski pulling technique. The crystals had nonuniform Bi concentration along the length with a maximum Bi incorporation of 2.4 at.% at the central region. The study also showed that good crystals could be obtained for Bi concentration in the melt up to 30 %. Vertical Bridgman growth of InSbBi reported by Mohan et al. [31] produced crystals with uniform and smooth surface as long as the Bi content in the material was within 2 % whereas higher Bi concentration in the melt resulted in precipitates due to the formation of InSb–InBi eutectics. The authors also demonstrated a reduction of band gap with increase in Bi content, through infrared absorption measurements. These data are being reproduced in Fig. 6.1 which shows a near saturation of InSbBi band gap for Bi content above 0.2 and is evidently a consequence of the limited solubility of Bi in InSb.

6.2.1 LPE Growth of InSbBi

Phase diagram for the Bi–In–Sb system was calculated and reported by Cui and co-workers in 2002 [32]. Successful growth of InSbBi by LPE technique was reported earlier in 1999 by Gao and Yamaguchi [33]. The layers were grown in a horizontal

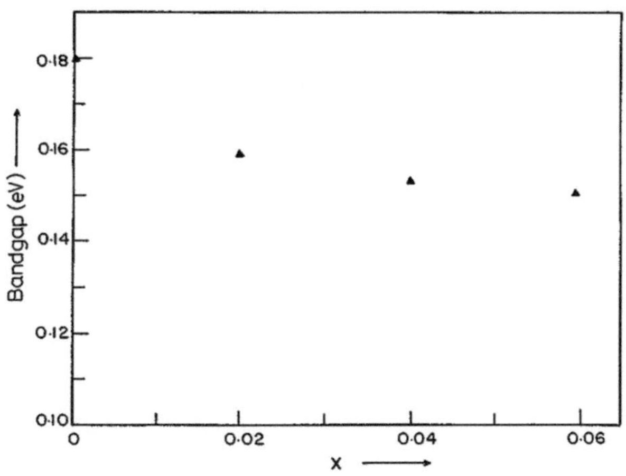

Fig. 6.1 Band gap of InSbBi plotted against Bi content x in the crystal. Reprinted from Mohan et al. J. Mater. Sc. Lett. **20**, 241 (2001) with permission from Springer Science + Business Media B. V.

sliding boat LPE reactor using In, Bi, or Sn as the solvent. Growth was done on InSb substrates at 350 °C under a melt supercooling of 10 °C and a cooling rate varied between 0.2 and 3 °C/min. Bi incorporation in the solid, measured by EPMA technique, was found to depend on the solvent and was 1.4 %, 1.26 %, and 0.85 % for growth from Sn, In, and Bi solvent, respectively. Bi incorporation from Sn + Bi solvent was found to be strongly influenced by the melt cooling rate during growth.

Raczynska et al. [34] reported LPE growth of InSbBi on (111)B surfaces at 400 °C using pure Bi solvents. The same group also performed an XPS analysis of the compositional changes on the surface of LPE-grown InSbBi as a result of low energy Ar ion bombardment and obtained the binding energy of Bi in InSbBi [35].

Dixit et al. [36] developed an LPE process for the growth of InSbBi on (001) GaAs substrates. The growth melt in this case was prepared by dissolving InSbBi crystal and InSb in In, keeping the ratio of In to Sb + Bi fixed at 1.86. Typical growth was done at 351 °C under an extremely low cooling rate of 0.2 °C/h. The authors reported a Bi composition of 2.18 at.% for films grown on InSb substrates and an enhanced Bi incorporation of 3.96 at.% when growth was done over GaAs substrates. Further studies are needed here to understand (a) the mechanism for the growth of highly mismatched InSbBi layers on GaAs substrates and (b) the incorporation of Bi greatly exceeding the thermodynamically possible solubility limit [27, 37] in InSb.

We have recently reported LPE growth of InSbBi layers from a melt containing very minute amounts of Bi [38]. The melt for the growth was prepared by saturating In with Sb by placing it over an undoped single crystal InSb wafer at about 360 °C. Up to 2.2 at.% pure Bi was added to this melt. Growth was typically done at 340–360 °C for 8–10 min under a melt supercooling of 10–12 °C and a continuous cooling rate of 0.4 °C. This process produced layers 4–5 μm thick.

6.2.2 Surface Morphology and Composition of LPE InSbBi Layers

Gao and Yamaguchi [33] obtained very smooth surfaces for layers grown from either In or Bi melt. On the other hand, layers grown from Sn solutions showed strong dependence on melt cooling rates. Higher cooling rates of 3 °C/min resulted in rough pyramidal structure instead of a smooth surface which the authors attributed to segregation of Bi and Sn particles. Mirror-smooth surfaces were also reported by Raczynska et al. [34] for layers grown from Bi solutions. InSbBi layers grown by us [38] had smooth surfaces as shown in the scanning electron micrograph (SEM) of Fig. 6.2. A few growth inclusions were, however, observed on the surface which is usual for LPE growth.

Chemical compositions of LPE InSbBi layers were obtained through EPMA [33] and energy dispersive X-ray (EDX) [36, 38] measurements or indirectly by using high resolution X-ray diffraction (HRXRD) technique, assuming Vegard's law is applicable over the range of interest. A typical EDX spectrum for a layer grown in our laboratory using 2.2 at.% Bi in the melt is shown in Fig. 6.3 from which a Bi

Fig. 6.2 SEM photograph of LPE InSbBi

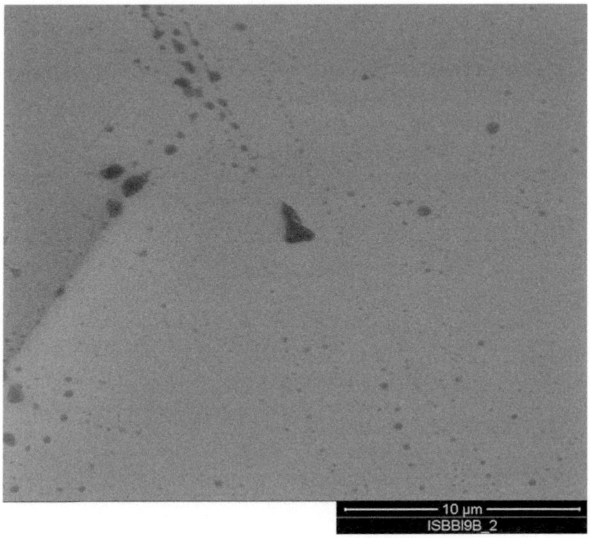

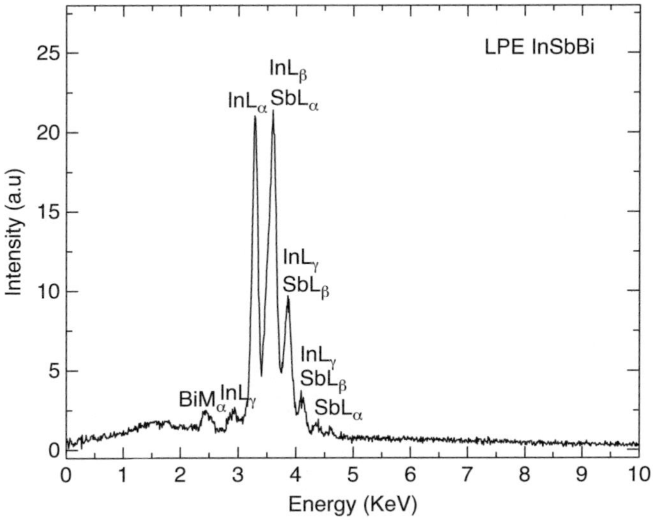

Fig. 6.3 EDX spectrum for an InSbBi layer grown from melt containing 2.2 at.% Bi. Reprinted from Das et al. Infrared Phys. Technol. **55**, 306 (2012) with permission from Elsevier

content of 0.4 at.% in the layer is calculated out. Linear EDX mapping for the same sample showed a uniform distribution (Fig. 6.4). This is further supported by a two-dimensional X-ray mapping of In, Sb, and Bi distribution over the layer surface as is seen from Fig. 6.5 [39].

X-ray diffraction pattern of an InSbBi layer grown over InSb substrate, reported by Dixit et al. [36], is reproduced here in Fig. 6.6 which indicates a lattice dilation due to Bi incorporation. This is in contrast with the results obtained by us [38] on our samples (Fig. 6.7) where the XRD peak due to the layer appeared at a higher

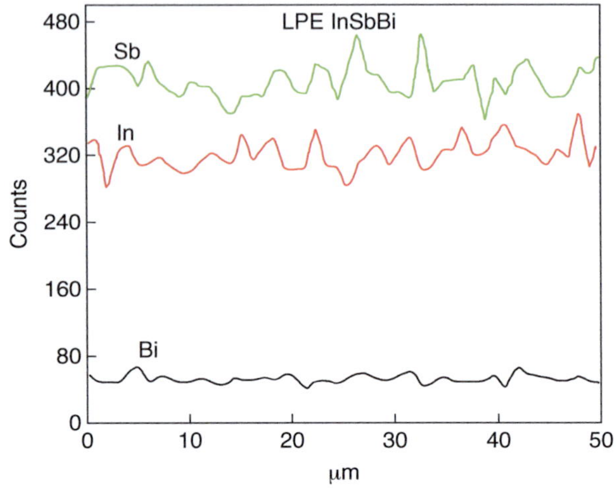

Fig. 6.4 Distribution of Bi on the surface of LPE-grown InSbBi. Reprinted from Das et al. Infrared Phys. Technol. **55**, 306 (2012) with permission from Elsevier

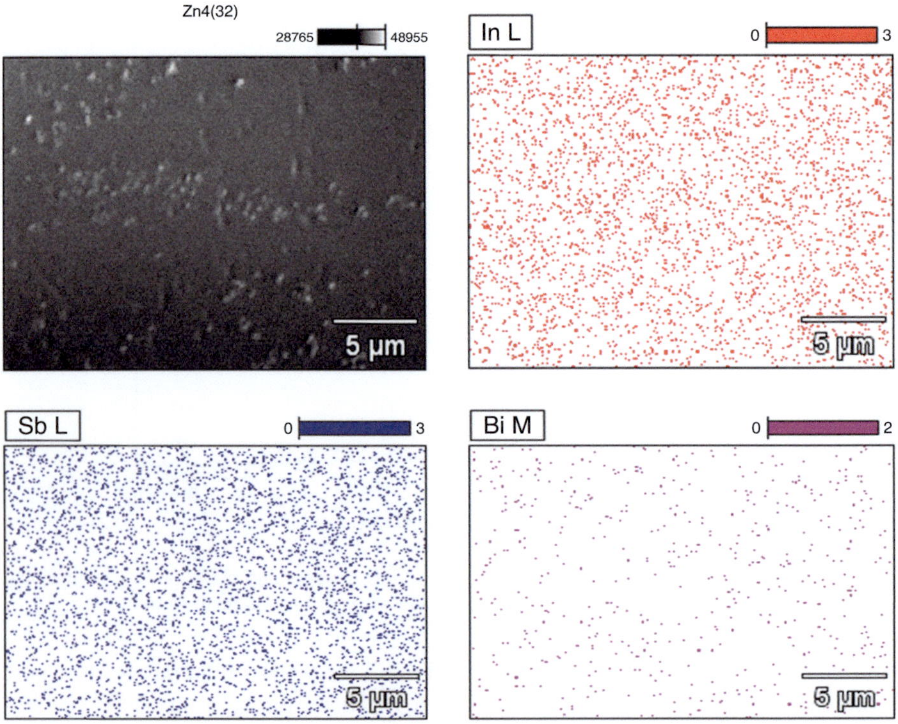

Fig. 6.5 X-ray mapping of InSbBi epitaxial layer surface showing distribution of In, Sb, and Bi

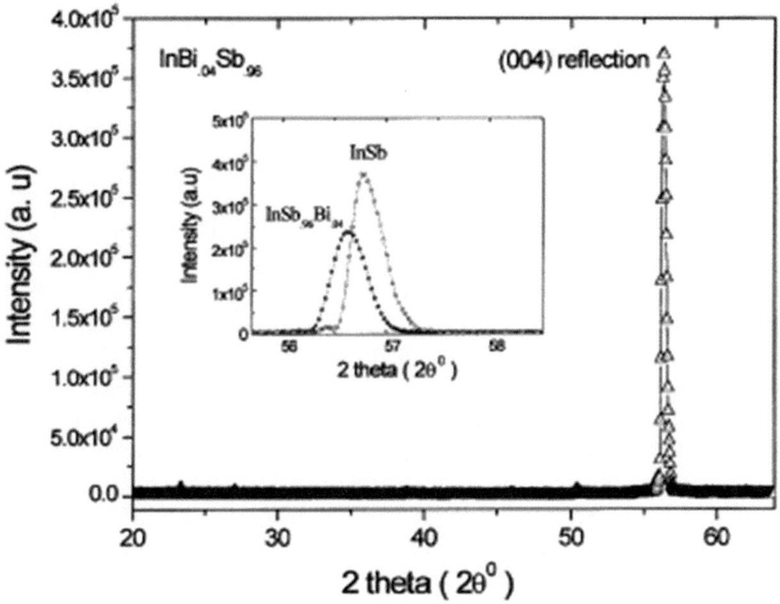

Fig. 6.6 X-ray diffraction pattern of $InSb_{0.96}Bi_{0.4}$ epitaxial layer. Reprinted from Dixit et al. J. Cryst. Growth **241**, 171 (2002) with permission from Elsevier Ltd.

Fig. 6.7 HRXRD pattern of LPE InSbBi grown with 2.2 at.% Bi in the growth melt. Reprinted from Das et al. Infrared Phys. Technol. **55**, 306 (2012) with permission from Elsevier

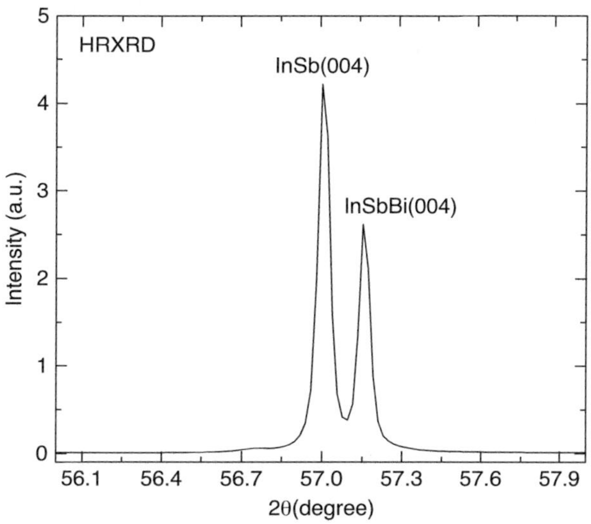

angle indicating a lattice contraction upon Bi incorporation. Similar results were observed in MBE-grown InSbBi and GaSbBi [5, 40]. Quite narrow XRD linewidth of the layer in Fig. 6.7, which is almost equal to that for the substrate, indicates high crystalline quality of the LPE-grown layer.

6.2.3 Optical Absorption/Transmission Measurements on LPE InSbBi

Optical transmission/absorption measurements have been used to determine the band gap of InSbBi. Results of optical transmission measurements by Gao and Yamaguchi [32], presented in Figs. 6.8 and 6.9, show that whereas the substrate InSb has a clear absorption edge at 1,300 cm^{-1}, that for InSbBi exhibits a second absorption edge at 983 cm^{-1} which is evidently due to the layer containing Bi and corresponds to an optical wavelength of 10.17 μm. Similar reduction in band gap absorption energy was observed by Dixit et al. [36] as shown in Fig. 6.10.

We have done room temperature optical absorption measurements on InSbBi in a Fourier transform infrared (FTIR) spectrometer. Figure 6.11 shows a plot of $(\alpha E)^2$ vs. photon energy E where α is the absorption coefficient. Data for LPE InSb and InSbBi layers grown with 0.55, 1.1, and 2.2 at.% Bi in the melt are shown. The band gap of the material, obtained from the intercept of the linear part of the absorption curve on the E axis, shows a continuous decline with increased Bi content in melt indicating increased incorporation of Bi in the solid. The resultant band gap reduction with respect to the band gap of InSb is shown in Fig. 6.12 which indicates a tendency for saturation at higher values of Bi in the melt.

The Bi content in the layer was obtained by linearly interpolating the energy band gap between InSb and InBi, the band gap of the latter being assumed to be −1.5 eV at 300 K [36]. For the layer grown with 2.2 at.% Bi in melt, a Bi content of 0.4 at.% is obtained in agreement with that obtained from EDX.

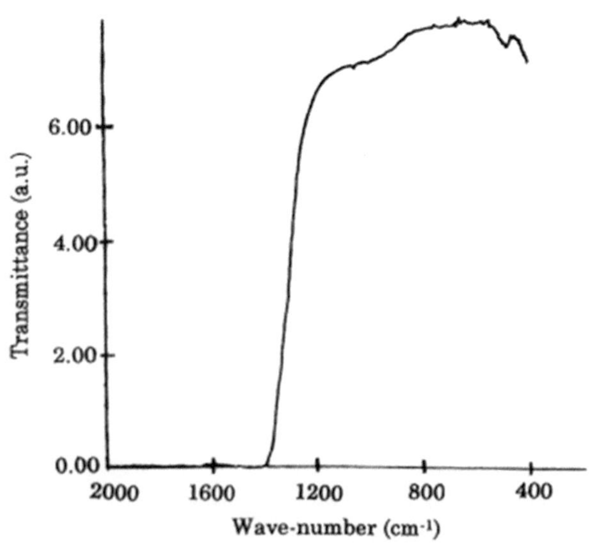

Fig. 6.8 Transmission spectrum for InSb substrate showing absorption edge at 1,300 cm^{-1} Reprinted from Gao and Yamaguchi, Cryst. Res. Technol. **34**, 285 (1999) with permission from John Wiley & Sons

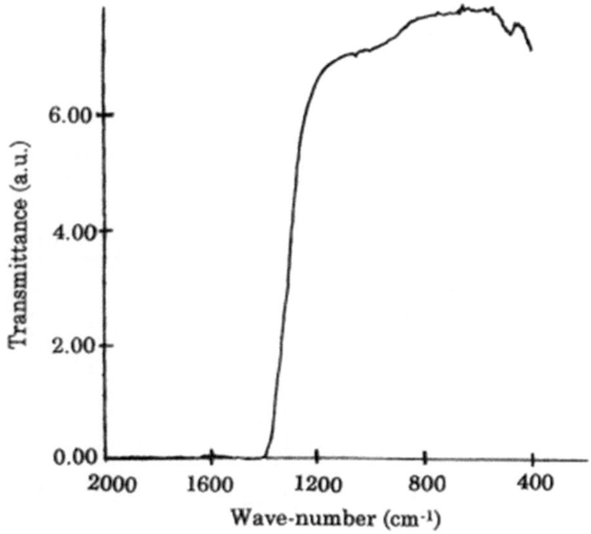

Fig. 6.9 Transmission spectrum for InSbBi. Reprinted from Gao and Yamaguchi, Cryst. Res. Technol. **34**, 285 (1999) with permission from John Wiley & Sons

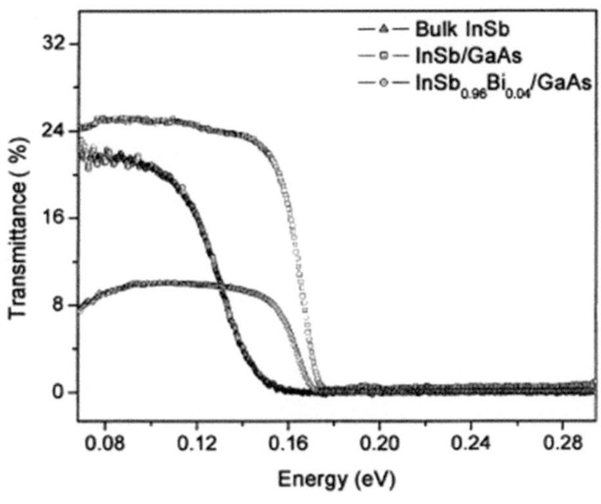

Fig. 6.10 Room temperature infrared transmission spectra of bulk InSb, InSb/GaAs, and $InSb_{0.96}Bi_{0.4}$/GaAs. Reprinted from Dixit et al. J. Cryst. Growth **241**, 171 (2002) with permission from Elsevier

6.2.4 Raman Spectroscopy of LPE InSbBi

Raman spectroscopy techniques can provide useful insights into the lattice dynamics, composition, and local bonding of elements in III–V compounds. The technique has been widely used for dilute III–V nitrides [41, 42] and for GaAsBi [43] and InAsBi [19]. We did room temperature Raman spectroscopy study of the InSbBi layers grown by us [38]. Figure 6.13 shows the Raman spectrum obtained on a sample grown with 2.2 at.% Bi in melt. Taking clue from earlier reports on Raman

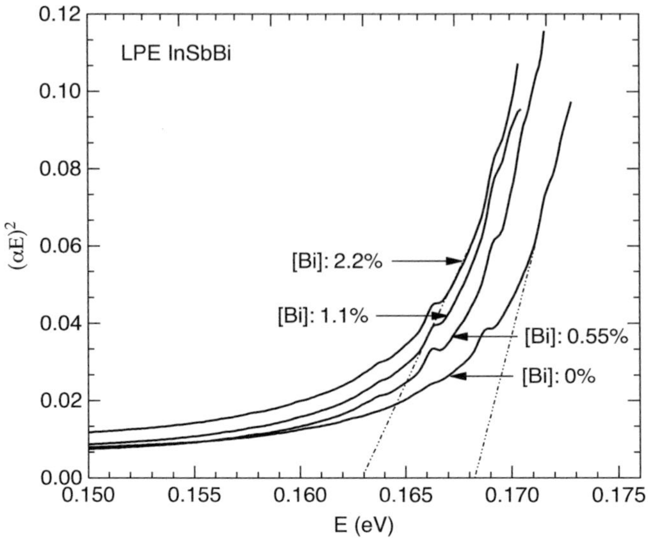

Fig. 6.11 Plot of $(\alpha E)^2$ against photon energy E for LPE-grown InSbBi layers grown from melts containing different amounts of Bi as indicated in the figure. Reprinted from Das et al. Infrared Phys. Technol. **55**, 306 (2012) with permission from Elsevier

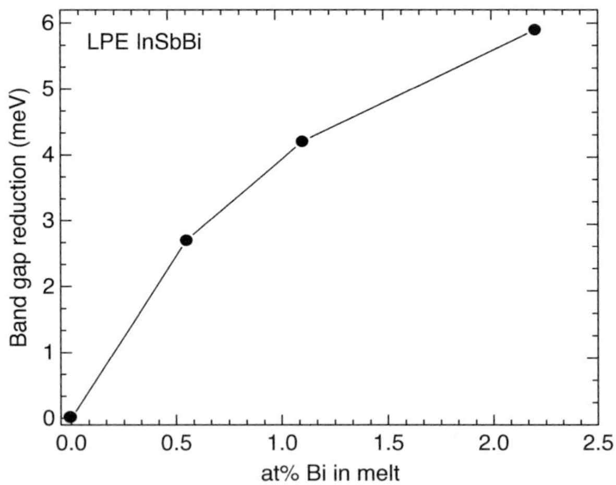

Fig. 6.12 Variation of the band gap of LPE InSbBi with Bi content in the growth melt. Reprinted from Das et al. Infrared Phys. Technol. **55**, 306 (2012) with permission from Elsevier

studies of InSb [44] and InAsBi [45] we associated the observed major peaks near 180 and 190 cm^{-1} with the InSb first order transverse optical (TO) and longitudinal optical (LO) modes and the peak at about 380 cm^{-1} to InSb second order LO mode. On the lower frequency side of the spectrum, three small peaks are found at approximately 130, 140, and 152 cm^{-1}. Verma et al. [19] compared Raman spectrum of pure Bi with that of GaAsBi and InAsBi and came to the conclusion that a

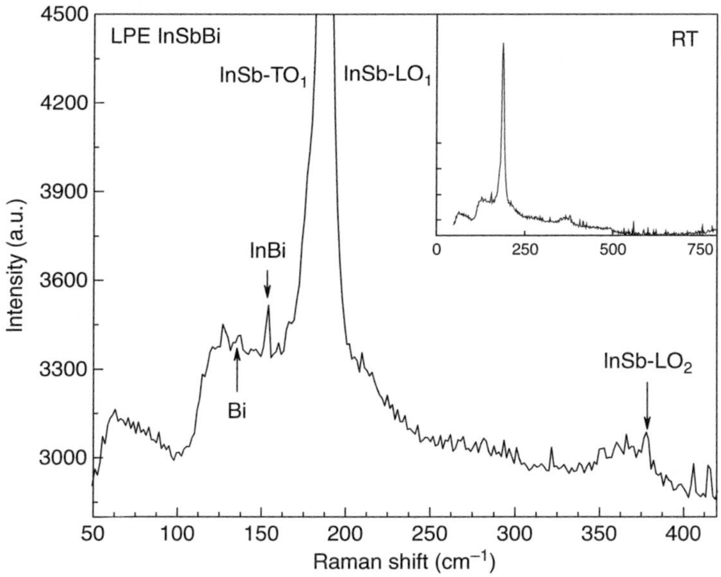

Fig. 6.13 Room temperature Raman spectrum for an LPE InSbBi layer grown from melt containing 2.2 at.% Bi. *Inset* shows the full spectrum. Reprinted from Das et al. Infrared Phys. Technol. **55**, 306 (2012) with permission from Elsevier

vibrational mode at 140 cm^{-1} observed in both the ternaries may be related to undissolved Bi. Further, from theoretical calculations, these authors estimated the TO and LO vibrational frequencies of InBi to be 161 and 155 cm^{-1}, respectively, and they also observed a weak structure at 160 cm^{-1} in InAsBi which they attributed to InBi TO mode. Since the Bi content in our material was only 0.4 %, it is possible that the small peak observed at 152 cm^{-1} is due to InBi LO phonons. The TO phonons are perhaps not observed for the fact that the Raman signal starts rising from this region due to the strong InSb TO mode. The possible existence of Raman vibrational modes due to both isolated Bi and Bi bonded to In atoms may thus indicate that Bi in the layer resided at the interstitial as well as at the substitutional sites.

6.3 LPE Growth of GaSbBi

GaSbBi has the potential to be used in mid-infrared optoelectronic devices though it did receive very little attention so far. We could locate only one very recent report on the MBE growth of GaSbBi [46]. LPE was used long back to grow GaSb layers from Bi + Ga solution in order to introduce Bi in the solid as an isovalent dopant [47, 48]. These studies showed that Bi could be incorporated in GaSb up to a maximum value of 0.3 at.% with a band gap lowering of 20 meV. This value was obtained in

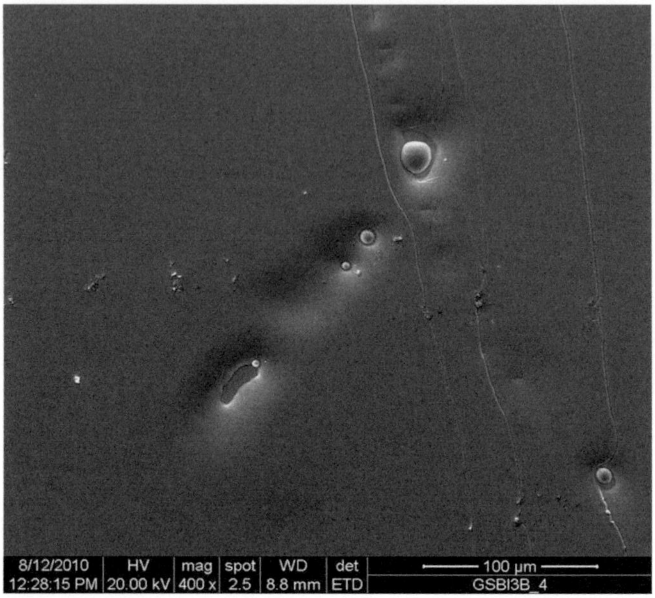

Fig. 6.14 SEM photograph of the surface of an LPE-grown GaSbBi layer

the range of 20–70 at.% Bi in the liquid and fell to lower values on either side of the range. Gladkov et al. [49] showed that only 0.15 at.% Bi incorporation resulted in GaSb layers grown from pure Bi melts with a very small band gap reduction of 0.8 meV. Similar results were obtained in GaSb bulk crystals grown from Bi solutions [50] and it was indicated that an excess of Bi in the melt gave rise to a very low Bi segregation coefficient of 1×10^{-5} in the solid. Our group, in collaboration with the mid-infrared optics group of Lancaster University, UK, first made a systematic study of the LPE growth and characterization of GaSbBi layers [51]. This work is being described below.

6.3.1 LPE Growth

The layers were grown from a Ga solution saturated with Sb at 500 °C. In separate growth runs, melts were prepared containing 0.2, 0.5, 0.8, and 1.6 at.% Bi. Growth was typically done for 8–10 min under a melt supercooling of 5–6 °C under a constant cooling rate of 0.2 °C/min. The thickness of the layers was in the range 5–6 µm as measured from the cleaved cross section under a *Nomarski* interference contrast microscope. The SEM of a typical GaSbBi layer in Fig. 6.14 shows a smooth surface with only a few growth inclusions. The morphology of the surface tended to degrade for Bi content in the melt >1.6 at.%.

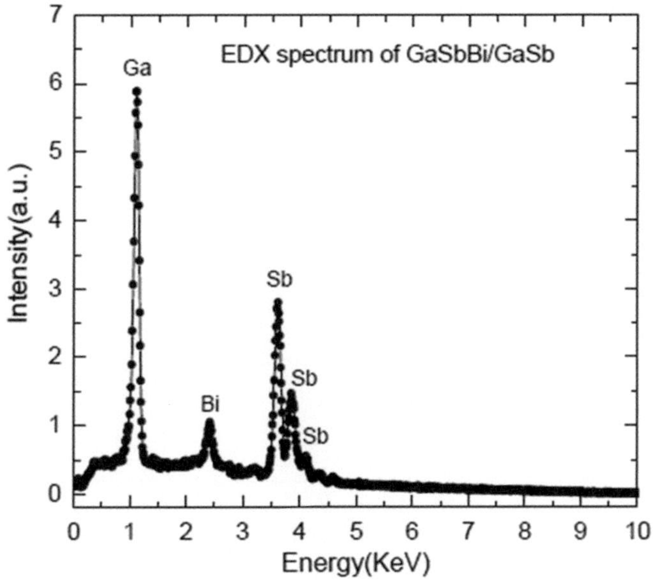

Fig. 6.15 EDX spectrum for LPE $GaSb_{0.996}Bi_{0.004}$. Reprinted from Das et al. Infrared Phys. Technol. **55**, 156 (2012) with permission from Elsevier

6.3.2 Chemical Composition and Structure of the Layers

The Bi content in the grown layer was estimated from EDX studies. EDX spectrum of a layer grown with 0.8 at.% Bi in the melt, shown in Fig. 6.15, shows distinct peak for Bi and, from the data, Bi content of about 0.4 at.% is measured in the solid. HRXRD pattern of the same sample in Fig. 6.16 shows two sharp peaks due to the layer and the substrate. In contrast to the results obtained by us on InSbBi (Fig. 6.7) and that observed in MBE GaSbBi [40], a lattice dilation is observed here due to Bi incorporation. The width of the GaSbBi peak is quite narrow and almost comparable to that of the substrate and this indicates very high crystalline quality of the grown material. It should be mentioned that MBE- and MOCVD-grown III–V bismides usually show much broader XRD peaks for the layer [17, 18, 20].

6.3.3 Photoluminescence Properties

Photoluminescence (PL) technique could be the most useful tool to characterize the details of electronic and optical transition properties in GaSbBi whose band gap falls within the easily accessible near infrared region. We have studied our samples at temperatures ranging from 4 to 330 K on layers grown from melts containing 0.2,

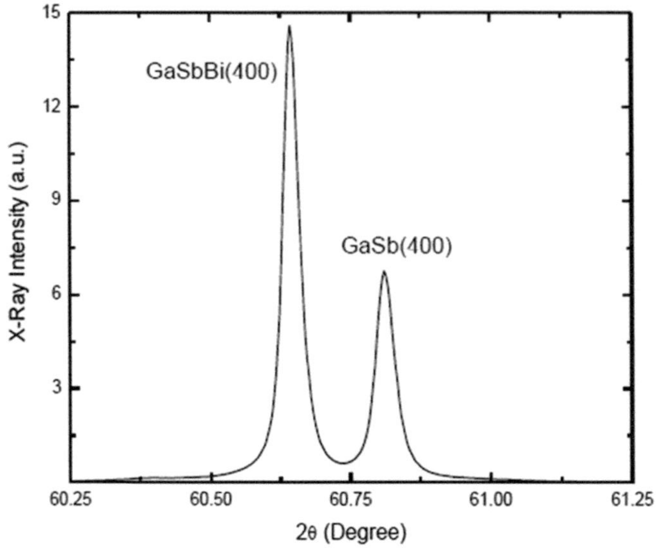

Fig. 6.16 HR Technol. **55**, 156 (2012) with permission from Elsevier HRXRD rocking curve for LPE $GaSb_{0.996}Bi_{0.004}$. Reprinted from Das et al. Infrared Phys. Technol. 55, 156 (2012)

0.5, 0.8, and 1.6 at.% Bi. Figure 6.17 shows the 4 K results where PL data for a layer grown without any Bi is also presented for comparison. GaSb layer grown without any Bi in the melt shows a single peak at 0.794 eV which we label as "A." PL emission at this energy is a characteristic of near band edge (bound exciton) recombination in GaSb [52]. On the other hand, all GaSbBi materials, except that with 0.4 at.% Bi, exhibit a lower energy peak "B." Peaks A and B are also identified separately after a Gaussian deconvolution of the broad PL peak observed for the sample with 0.4 at.% Bi. Peak A is found to shift to lower energies with increase in Bi content in the melt up to 0.8 at.%, above which again an increase in band gap is observed. The measured values of Bi introduced in the layer show that the maximum allowed Bi content in the layer is 0.4 at.% and, for higher Bi content of 1.6 at.% in the melt, Bi content in the solid is reduced to 0.3 at.%. This variation is being illustrated in Fig. 6.18. The change in energy position of peak A with respect to that for GaSb in Fig. 6.17 gives a measure of the corresponding change of band gap of the material as is being illustrated in Fig. 6.19. The result indicates a maximum decrease of 40 meV obtained for the layer grown with 0.8 at.% Bi in the melt. Further increase of Bi to 1.6 at.% in the melt did not produce anymore band gap reduction for the material. 0.4 at.% thus seems to be the limit of solubility of Bi in GaSb at the particular growth temperature and is more than the maximum Bi content obtained in bulk GaSbBi [47, 48]. It is also interesting to note that the segregation coefficient of Bi in GaSb appears to be unity for very low value of 0.2 at.% Bi in melt which tends to decrease for higher Bi concentrations in the melt. In contrast to the results obtained on bulk GaSb [47, 48], the saturation of Bi in the solid occurred at much

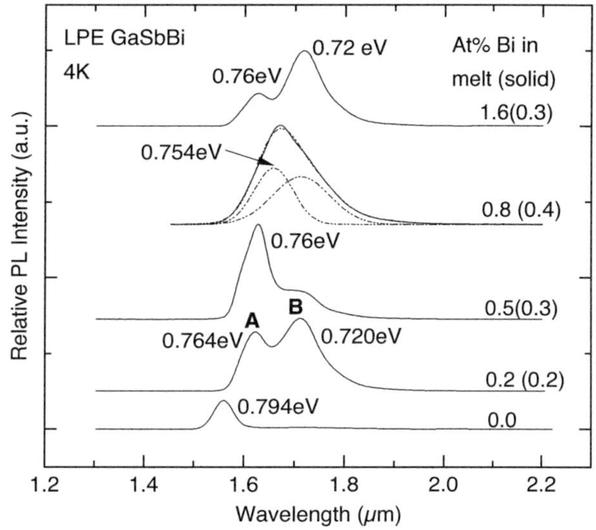

Fig. 6.17 4K PL spectra obtained on $GaSb_{1-x}Bi_x$ layers grown with 0, 0.2, 0.5, 0.8, and 1.6 at.% Bi in the growth melt. Bi contents in the solid, measured by EDX, are shown in the figure. The broad PL peak for $GaSb_{0.996}Bi_{0.004}$ is resolved into two component peaks using Gaussian deconvolution procedure

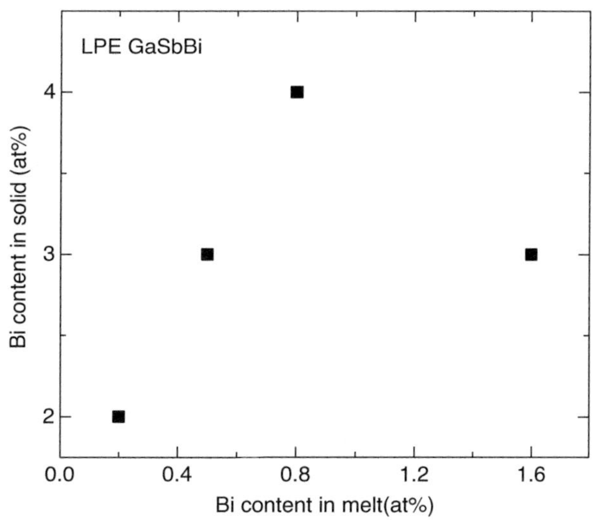

Fig. 6.18 Variation of Bi content in GaSbBi with respect to that in the growth melt

lower concentration of Bi here. We also observe an increase in the height of PL peak A with increase of Bi content in the layer which reached a maximum for 0.3 at.% Bi in the solid and decreases thereafter as shown in Fig. 6.20. We suppose that the initial rise in PL is due to Bi-gettering of shallow impurities in the melt. However, as the Bi concentration in the melt increases beyond 0.5 at.%, additional acceptors are created due to the formation of Bi-related complex defects, resulting in a decrease of the PL. Peak B is observed at a constant energy of 0.72 eV irrespective

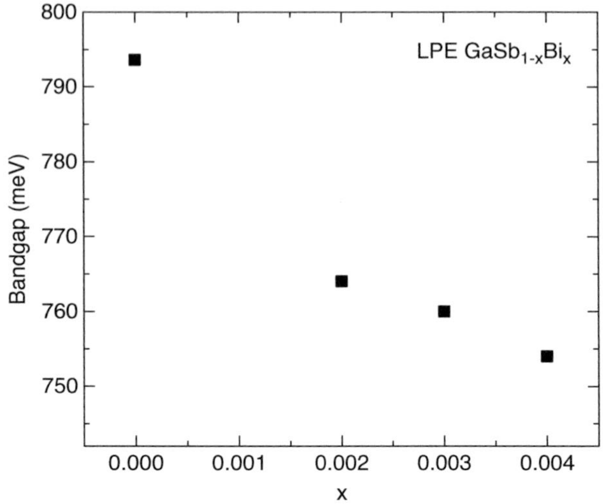

Fig. 6.19 Band gap of $GaSb_{1-x}Bi_x$ as a function of Bi mole fraction x

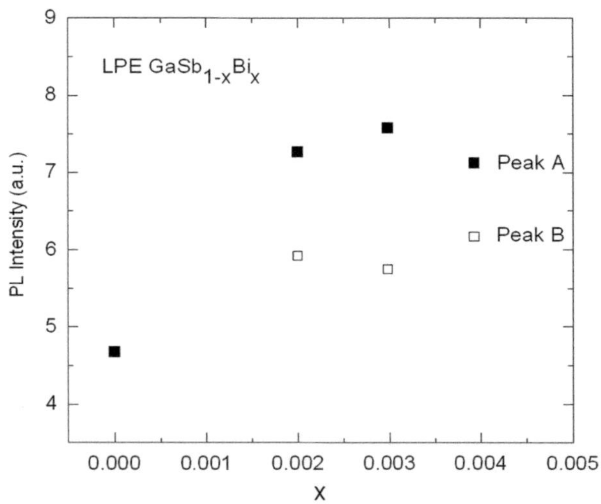

Fig. 6.20 Variation of PL intensity measured for peak A and peak B in Fig. 6.17. Reprinted from Das et al. Infrared Phys. Technol. 55, 136 (2012) with permission from Elsevier

of the Bi content in the material and hence appears to have its origin to a deep defect. Height of this peak tends to increase with Bi indicating its origin to Bi-related complexes. From the Arrhenius plot of the integrated PL intensity due to peak A, shown in Fig. 6.21, we note that the thermal quenching of PL is associated with activation energies of 55 and 57 meV in the low and mid temperature range. These values are much larger than the excitonic binding energy signifying that PL is quenched mostly due to non-radiative recombinations.

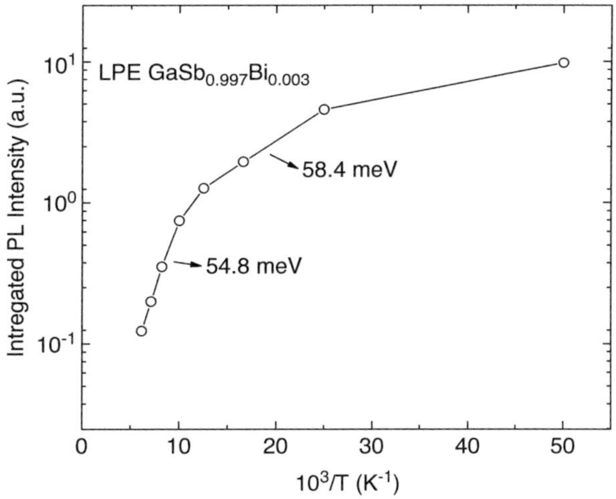

Fig. 6.21 Arrhenius plot of the integrated PL intensity associated with peak A in GaSbBi containing 0.3 at.% Bi. Reprinted from Das et al. Infrared Phys. Technol. 55, 136 (2012) with permission from Elsevier

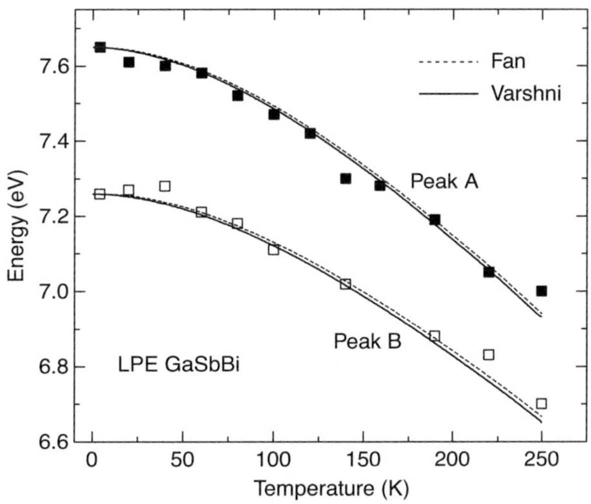

Fig. 6.22 Temperature dependence of the energy positions of peaks A and B of LPE $GaSb_{0.997}Bi_{0.003}$ (indicated by *squares*) fitted with empirical Varshni (*solid line*) and Fan (*dashed line*) equations

The temperature variation of the energy positions of peaks A and B was studied from 4 to 250 K. The points in Fig. 6.22 show the experimental data. The solid lines indicate fitting of the experimental data with empirical Varshni equation [53], given by

$$E_g(T) = E_g(0) - \frac{\alpha T^2}{T + \beta}$$

Table 6.1 Varshni and Fan parameters used in the fitting of data in Fig. 6.22

	Varshni parameters			Fan parameters	
Peak	$E_g(0)$ (eV)	α (10^{-4} eV/k)	β (K)	A (eV)	E_p (eV)
A	0.765	5.9	264	0.117	0.0229
B	0.726	4.97	264	0.099	0.0229

where $E_g(0)$ is the energy at 0 K and α and β are the Varshni parameters which are usually close to the temperature coefficient of band gap and the Debye temperature Θ_D, respectively. We have also fitted the same data with Fan equation [54], given by

$$E_g(T) = E_g(0) + \frac{A}{\exp(E_p / kT) - 1}$$

where A is the Fan parameter and E_p is the average phonon energy affecting the band gap variation. The dashed lines in Fig. 6.21 show the Fan fitting. Values of the Varshni and Fan parameters used in the fitting are given in Table 6.1. The values of α and β are in close agreement with the published reports [55] and also agree with the known values of the temperature coefficient of band gap and the Debye temperature for GaSb [56]. We also note that both the Varshni and Fan equations give almost similar agreement with experimental data for peaks A and B.

6.3.4 Raman Spectroscopy

We performed a room temperature Raman spectroscopy measurement on our LPE-grown GaSbBi to identify the local vibrational modes related to Ga, Sb, and Bi in the compound [51]. Figure 6.23 shows the results. Most prominent features of the Raman spectrum are a shoulder at ~227 cm^{-1} due to GaSb TO phonon and the peak at ~233 cm^{-1} is related to the longitudinal optical phonon (LO) mode of GaSb [57] with their second order modes being observed at 450–475 cm^{-1}. The structure spreading over 125–160 cm^{-1} might include the reported Raman mode at around 140 cm^{-1} due to free Bi atoms [19]. We also note a very weak shoulder around 213–215 cm^{-1} which is possibly due to GaBi phonon mode as suggested by Seong et al. [43]. The feature between 525 and 550 cm^{-1} is most likely due to carbon which is always present in LPE-grown materials and originates from the boat material [58].

6.4 Kinetics of the LPE Growth of InSbBi and GaSbBi

We have earlier used a one-dimensional diffusion model to investigate the transport of nitrogen atoms through the growth melt during LPE growth of InAsN and GaAsN [59]. The results obtained through this study were primarily used to optimize the

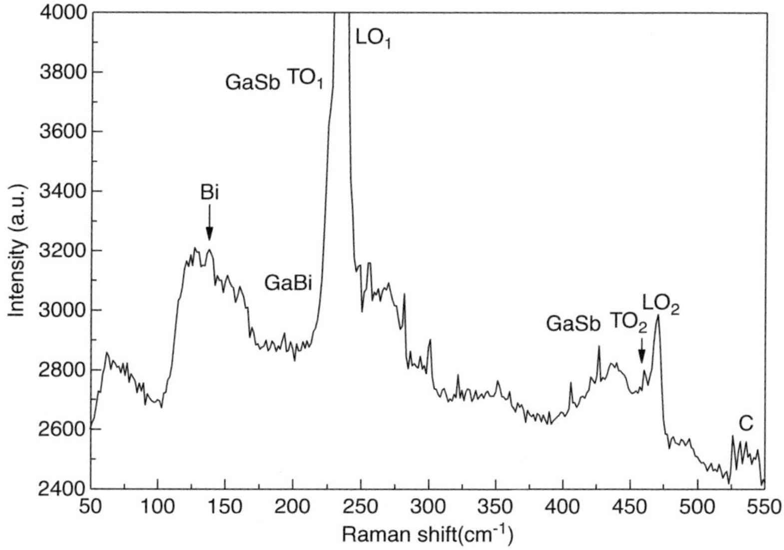

Fig. 6.23 Raman spectrum for LPE GaSbBi taken at room temperature

various LPE growth parameters. A similar kind of calculations has been made by us [60] to optimize the parameters associated with the LPE growth of InSbBi and GaSbBi.

One-dimensional diffusive transport equation for the diffusion of Bi atoms towards the growing interface can be written following Small and Barnes [61] as

$$\frac{\partial C}{\partial T} = D\frac{\partial^2 C}{\partial x^2} \tag{6.1}$$

where C is the concentration of the solute species and D is the diffusion coefficient of Bi atoms in the solution. The concentration in the jth segment is obtained using the finite difference approximation by solving (6.1) and can be written as [62]

$$C_{j,n+1} = C_{j,n} + \frac{D\tau}{\varepsilon^2}(C_{j-1,n} + 2C_{j,n} + C_{j+1,n}) \tag{6.2}$$

The factor $D\tau/\varepsilon^2$ is called modulus M. It is necessary to choose values of τ and ε for a given value of D in such a way so as to make $M \le 1/3$.

The diffusion coefficients of nitrogen in the solution have been calculated using Wilke and Chang correlation [63] for the liquid phase.

$$D_{(1,2)} = 7.4\times10^{-8}\frac{(\varphi M2)^{0.5}T}{\eta V1^{0.6}} \tag{6.3}$$

$D_{(1,2)}$ = the diffusion coefficient of solute 1 in solvent 2 (cm^2/s), T = temperature in K, η = viscosity of solvent 2 (centipoise), $M2$ = molecular weight of solvent 2, $V1$ = molar volume of solute (cm^3/mol), φ = association factor of the solvent which was taken as 1 for non-associated nonpolar solvent.

Value of the viscosity of Bi at equilibrium temperature was taken from the experimental value of Flinn et al. [64]. Viscosities of indium and gallium melt at the equilibrium temperature were calculated using

$$\eta_{\mathrm{In}} = \eta_0 + A_1 \exp[-(t - t_0)/B_1] \tag{6.4}$$

$$\eta_{\mathrm{Ga}} = 0.4359 \exp[481/T] \tag{6.5}$$

where $t_0 = 280$ °C, t is the temperature of the metal in °C, $A_1 = 0.7144$ and $B_1 = 242.2606$, and $\eta_0 = 1.0285$ (mPa s).

The concentration of Bi atoms in the solution was determined from the liquidus data of the corresponding growth melt

$$C_{\mathrm{Bi}} = \frac{C_{\mathrm{sol}} X_{\mathrm{Bi}}}{1 - X_{\mathrm{Bi}}} \tag{6.6}$$

where C_{sol} is the atomic concentration of solvent and X_{Bi} is the atomic fraction of bismuth atoms in the solution. Hence (6.6) reduces to

$$C_{\mathrm{Bi(In)}} = 6.398 \times 10^{22} \exp\left[8.4990 - \frac{8023}{T}\right] \tag{6.7}$$

$$C_{\mathrm{Bi(Ga)}} = 4.946 \times 10^{22} \exp\left[8.8732 - \frac{10723}{T}\right] \tag{6.8}$$

where $C_{\mathrm{Bi(In)}}$ and $C_{\mathrm{Bi(Ga)}}$ are the bismuth concentrations in In and Ga melt for the growth of InSbBi and GaSbBi, respectively. Densities of Bi and In at the corresponding equilibrium temperatures, required in the calculations, were obtained using standard relations. Density of Ga and In at the corresponding equilibrium temperatures was obtained using standard relations.

$$\rho(\mathrm{Ga}) = 6.32723 - 7.3743 \times 10^{-4} T + 1.37767 \times 10^{-7} T^2 \tag{6.9}$$

$$\rho(\mathrm{In}) = 7.1295 - 0.6798 \times 10^{-3} T \tag{6.10}$$

The melt equilibrium temperature and the growth temperature for InSbBi and GaSbBi were taken as 623 and 801 K, respectively. For the numerical calculations, we have considered a solution depth of 4 mm in front of the growing interface which is divided into 40 equally spaced segments, each having width h = 0.1 mm. Programs were developed to solve the diffusion equation taking into consideration the relevant boundary conditions. The computation was carried out at different cooling

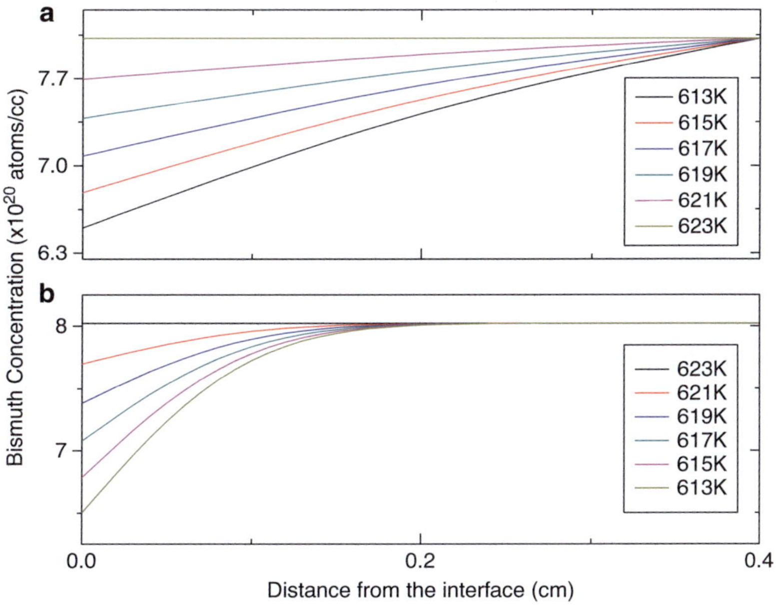

Fig. 6.24 Concentration profile of Bi in InSb + Bi solution at six different growth temperatures with respect to the melt equilibrium temperature of 623 K and for cooling rates of (**a**) 0.2 K and (**b**) 2 K per min

rates and for different melt supercooling rates which were selected in accordance with typical values used in actual growth. Figure 6.24 shows the calculated concentration profile of Bi near the melt–InSbBi interface for six different growth temperatures ranging from 613 to 623 K for two different cooling rates of 0.2 and 2 K. It is observed that for higher cooling rates of 2 K/min, Bi atoms rapidly depleted at the melt–solid interface whereas for a lower cooling rate of 0.2 K/min, there is a slow and almost linear variation of Bi for melt supercooling up to 10 K. This indicates that the solution remained homogeneous during growth which is required for the growth of layers with uniform concentration of Bi along the thickness. The results for Bi in GaSbBi, shown in Fig. 6.25, give us similar information.

6.5 LPE Growth of Other III–V Bismides

There is just one old report on the LPE growth of InP layers doped with Bi [65]. Here the growth was done in a horizontal tipping type LPE system at 600–700 °C using an In melt containing 0.07–7 wt% Bi and the authors estimated a Bi concentration in the solid in the range 5×10^{15} to 5×10^{17}. Photoluminescence did not, however, show any shift of the band edge luminescence peak and the work was primarily confined to the study of Bi as an isoelectronic dopant in InP. Akchurin et al. [66] reported successful LPE growth of InAsSbBi layers on (111)A InSb substrates at

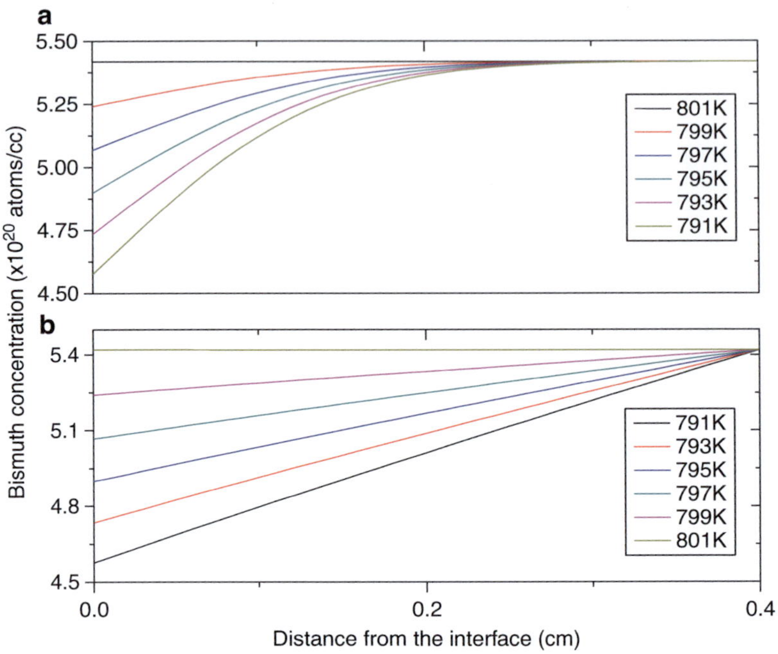

Fig. 6.25 Concentration profile of Bi in GaSb + Bi solution at six different growth temperatures with respect to the melt equilibrium temperature of 801 K and for cooling rates of (**a**) 0.2 K and (**b**) 2 K per min

temperatures in the range 380–450 °C. A Bi content of 0.16–0.36 at.% was introduced in the layer with a band gap lowering of 17–20 meV, measured by optical absorption measurements. LPE growth of InAsSb on GaAs substrates was reported by Dixit and co-workers [67]. The growth technique used was similar to that for the growth of InSbBi/GaAs reported by the same group [36] and a Bi incorporation of 2.6 at.% was obtained through this study.

Acknowledgements The author is grateful to his students S.K. Das, D.P. Samajdar, and T.D. Das for help in preparing this manuscript. Part of the work done by the author's group and described in this chapter was sponsored by the Department of Communication and Information Technology, Government of India and the "University with Potential for excellence" scheme of the University of Calcutta.

References

1. Tixier, S., Adamcyk, M., Tiedje, T., Francoeur, S., Mascarenhas, A., Wei, P., Schiettekatte, F.: Molecular beam epitaxy growth of $GaAs_{1-x}Bi_x$. Appl. Phys. Lett. **82**, 2245 (2003)
2. Fluegel, B., Francoeur, S., Mascarenhas, A., Tixier, S., Young, E.C., Tiedje, T.: Giant spin-orbit bowing in $GaAs_{1-x}$ Bi_x. Phys. Rev. Lett. **97**, 067205 (2006)

3. Alberi, K., Dubon, O.D., Walukiewicz, W., Yu, K.M., Bertulis, K., Kroktus, A.: Valence band anticrossing in $GaBi_xAs_{1-x}$. Appl. Phys. Lett. **91**, 051909 (2007)
4. Pettinari, G., Polimeni, A., Blokland, J.H., Trotta, R., Christiansen, P.C.M., Capizzi, M., Maan, J.C., Lu, X., Young, E.C., Tiedje, T.: Compositional dependence of the exciton reduced mass in $GaBi_xAs_{1-x}$. Phys. Rev. B **81**, 235211 (2010)
5. Lee, J.J., Kim, J.D., Razeghi, M.: Growth and characterization of InSbBi for long wavelength infrared photodetectors. Appl. Phys. Lett. **70**, 3266 (1997)
6. Lee, J.J., Kim, J.D., Razeghi, M.: Room temperature operation of 8–12 μm InSbBi infrared photodetectors on GaAs substrates. Appl. Phys. Lett. **73**, 602 (1998)
7. Wagner, M.C., Botha, J.R., Leitch, A.W.R.: Substitutional incorporation of arsenic from GaAs substrates into MOVPE grown InSbBi thin films. Physica B **308–310**, 866 (2001)
8. Henini, M. (ed.): Dilute Nitride Semiconductors. Elsevier, Amsterdam (2005)
9. Zhang, Y., Mascarenhas, A., Wang, L.-W.: Similar and dissimilar aspects of III-V semiconductors containing Bi versus N. Phys. Rev. B **71**, 155201 (2005)
10. Mascarenhas, A., Kini, R., Zhang, Y., France, R., Ptak, A.J.: Comparison of the dilute bismide and nitride alloys GaAsBi and GaAsN. Phys. Status Solidi B **246**, 504 (2009)
11. Zhang, Y., Mascarenhas, A., Xin, H.P., Tu, C.W.: Scaling of band gap reduction in heavily nitrogen doped GaAs. Phys. Rev. **B63**, 161303 (2001)
12. Kini, R.N., Mascarenhas, A., France, R., Ptak, A.J.: Low temperature photoluminescence from dilute nitrides. J. Appl. Phys. **104**, 113534 (2008)
13. Yoshida, J., Kita, T., Wada, O., Oe, K.: Temperature dependence of $GaAs_{1-x}Bi_x$ band gap studied by photoreflectance spectroscopy. Jpn J. Appl. Phys. **42**, 371 (2003)
14. Kini, R.N., Bhusal, L., Ptak, A.J., France, R., Mascarenhas, A.: Electron hall mobility in GaAsBi. J. Appl. Phys. **106**, 043705 (2009)
15. Cooke, D.G., Hegmann, F.A., Young, E.C., Tiedje, T.: Electron mobility in dilute GaAs bismide and nitride alloys measured by time-resolved tetrahertz spectroscopy. Appl. Phys. Lett. **89**, 122103 (2006)
16. Yoshimoto, M., Murata, S., Chayahara, A., Horino, Y., Saraie, J., Oe, K.: Metastable GaAsBi alloy grown by molecular beam epitaxy. Jpn. J. Appl. Phys. **42**, L1235 (2003)
17. Pacebutas, V., Bertulis, K., Dapkus, L., Aleksejenko, G., Krotkus, A., Yu, K.M., Walukiewicz, W.: Characterization of low-temperature molecular-beam-epitaxy GaBiAs layers. Semicond. Sci. Technol. **22**, 819 (2007)
18. Moussa, I., Fitouri, H., Chine, Z., Rebey, A., El Jani, B.: Effect of thermal annealing on structural and optical properties of the $GaAs_{0.963}Bi_{0.037}$ alloy. Semicond. Sci. Technol. **23**, 125034 (2008)
19. Verma, P., Oe, K., Yamada, M., Harima, H., Herms, M., Irmer, G.: Raman studies on $GaBi_xAs_{1-x}$ and $InAs_{1-x}Bi_x$. J. Appl. Phys. **89**, 1657 (2001)
20. Okamoto, H., Oe, K.: Structural and energy gap characterization of metal-organic-vapor-phase-epitaxy grown InAsBi. Jpn. J. Appl. Phys. **38**, 1022 (1999)
21. Feng, G., Yashimoto, M., Oe, K., Chayahara, A., Horino, Y.: New III-V semiconductor InGaAsBi alloy grown by molecular beam epitaxy. Jpn. J. Appl. Phys. **44**, L1161 (2005)
22. Lu, X., Beaton, J.A., Lewis, R.B., Tiedje, T., Zhang, Y.: Composition dependence of photoluminescence in $GaAs_{1-x}Bi_x$ alloys. Appl. Phys. Lett. **95**, 041903 (2009)
23. Ma, K.Y., Fang, Z.M., Cohen, R.M., Strigfellow, G.B.: Investigation of organometallic vapor phase epitaxy of InAs and InAsBi at temperatures as low as 275°C. J. Appl. Phys. **70**, 3940 (1991)
24. Devenson, J., Pacebutas, V., Butkute, R., Baranov, A., Krotkus, A.: Structure and optical properties of InGaAsBi with up to 7% Bi. Appl. Phys. Express **5**, 015503 (2012)
25. Capper, P., Mauk, M. (eds.): Liquid Phase Epitaxy of Electronic, Optical and Optoelectronic Materials. Wiley, Chichester (2007)
26. Kuznetsov, V.V., Kognovitskaya, E.A., Rubtsov, E.R.: Thermodynamic restrictions of LPE of bismuth-containing A^3B^5 solid solutions. J. Non Cryst. Solids **354**, 4375 (2008)
27. Kunzer, M., Jost, W., Kaufmann, U., Hobgood, H.M., Thomas, R.N.: Identification of the BiGa heteroantisite in GaAs:Bi. Phys. Rev. B **48**, 4437 (1993)

28. Joukoff, B., Jean-Louis, A.M.: Growth of $InSb_{1-x}Bi_x$ single crystals by Czochralski method. J. Cryst. Growth **12**, 169 (1972)
29. Jean-Louis, A.M., Hamon, C.: Phys. Status Solidi **34**, 329 (1969)
30. Jean-Louis, A.M., Ayrault, B., Vargas, J.: Phys. Status Solidi **34**, 341 (1969)
31. Mohan, P., Moorthy Babu, S., Santhanaraghavan, P., Ramasamy, P.: Vertical Bridgeman growth of $InSb_{1-x}Bi_x$ crystals for LWIR applications. J. Mater. Sci. Lett. **20**, 241 (2001)
32. Cui, Y., Ishihara, S., Liu, X.T., Ohnuma, I., Kainuma, R., Ohtani, H., Ishida, K.: Thermodynamic calculation of phase diagram in the Bi-In-Sb ternary system. Mater. Trans. **43**, 1879 (2002)
33. Gao, Y.Z., Yamaguchi, T.: Liquid phase epitaxial growth and properties of InSbBi films grown from In, Sn and Sn solutions. Cryst. Res. Technol. **34**, 285 (1999)
34. Raczynska, J., Rogalski, A., Rutkowski, J., Fronc, K.: Liquid phase epitaxial growth and characterization of In(Sb, Bi). Proc. SPIE **3725**, 66 (1999)
35. Ianowski, R.J., Heinonen, M.H., Raczynska, J., Fronc, K.: XPS analysis of surface compositional changes in $InSb_{1-x}Bi_x$ (111) due to low energy Ar^+ ion bombardment. Appl. Surf. Sci. **153**, 193 (2000)
36. Dixit, V.K., Keerth, K.S., Bera, P., Bhat, H.L.: Growth of $InBi_xSb_{(1-x)}$ films on GaAs (001) substrates using liquid phase epitaxy and their characterization. J. Cryst. Growth **241**, 171 (2002)
37. Ma, K.Y., Fang, Z.M., Cohen, R.M., Stringfellow, G.B.: Organometallic vapor phase epitaxy growth and characterization of Bi-containing III/V alloys. J. Appl. Phys. **68**, 4586 (1990)
38. Das, S.C., Das, T.D., Dhar, S.: Infrared absorption and Raman spectroscopy studies of InSbBi layers grown by liquid phase epitaxy. Infrared Phys. Technol. **55**, 306 (2012)
39. Das, T. D., Das, S. K., Dhar, S.: Structural and optical characterization of Bismuth containing III-V materials. Proceedings of the International Workshop on Nanomaterials (IWON): Engineering Photon and Phonon Transport, 181, Kolkata (2012)
40. Wang, S., Song, Y., Saha Roy, I.: Bismuth incorporation and lattice contraction in GaSbBi and InSbBi. Invited talk presented at the 13th International Conference on Transparent Optical Networks (ICTON 2011), Stockholm (2011)
41. Yu, G.Y., Shen, Z.X., Liu, L., Sun, W.X.: Raman scattering study of a GaAsN epitaxial layer. Mater. Sci. Semicond. Process. **4**, 581 (2001)
42. Talwar, D.N.: Assessment of microscopic lattice structures in dilute (AlGaIn)NAs laser materials by local mode spectroscopy. J. Appl. Phys. **99**, 123505 (2006)
43. Seong, M.J., Francoeur, S., Yoon, S., Mascarenhas, A., Tixier, S., Adamcyk, M., Tiedje, T.: Bi-induced vibrational modes in GaAsBi. Superlattices Microstruct. **37**, 394 (2005)
44. Gnezdilov, V.P., Lockwood, D.J., Webb, J.B.: Raman scattering in an $InSb/In_{1-x}Al_xAs$ strained-layer superlattice and in $In_{1-x}Al_xSb$ epilayers on InSb. Phys. Rev. B **48**, 11234 (1993)
45. Wagner, M.C., Botha, J.R., Leitch, A.W.R.: Substitutional incorporation of arsenic from GaAs substrates into MOVPE grown InSbBi thin films. Phys. B **308**, 866 (2001)
46. Song, Y.X., Wang, S.M., Roy, I.S., Shi, P.X., Hallen, A.: Growth of $GaSb_{1-x}Bi_x$ by molecular beam epitaxy. J. Vac. Sci. Technol. **B30**, 02B114 (2012)
47. Germogenov, V.P., Otman, Y.I., Chaldyshev, V.V., Shmartsev, Y.V., Epiktetova, L.E.: Appearance of 'natural' acceptors in GaSb due to isovalent doping with bismuth. Sov. Phys. Semicond. **24**, 689 (1990)
48. Chaldyshev, V.V., Germogenov, V.P., Shmartsev, Y.V.: Photoluminescence of bismuth doped GaSb. Key Eng. Mater. **65**, 109 (1992)
49. Gladkov, P., Monova, E., Weber, J.: Liquid phase epitaxy and photoluminescence characterization of p-type GaSb layers grown from Bi based melts. J. Cryst. Growth **146**, 319 (1995)
50. Danilewsky, A.N., Lauer, S., Meinhardt, J., Benz, K.W., Kauffmann, B., Hofmann, R., Dornen, A.: Growth and characterization of GaSb bulk crystals with low acceptor concentration. J. Electron. Mater. **25**, 1082 (1996)
51. Das, S.K., Das, T.D., Dhar, S., de la Mare, M., Krier, A.: Near infrared photoluminescence observed in dilute GaSbBi layers grown by liquid phase epitaxy. Infrared Phys. Technol. **55**, 156 (2012)

52. Ruhle, R., Jakowetz, W., Woelk, C., Linebach, R., Pilkuhn, M.: Optical studies of free and bound excitonic states in GaSb evidence for deep A^+-complexes. Phys. Status Solidi B **73**, 255 (1976)
53. Varshni, Y.P.: Temperature dependence of the energy gap in semiconductors. Physica (Amsterdam) **34**, 149 (1967)
54. Fan, H.Y.: Temperature dependence of the energy gap in semiconductors. Phys. Rev. **82**, 900 (1951)
55. Munez, M., Pollak, F.H., Zakia, M.B., Patel, N.B., Herrera-Perez, J.L.: Temperature dependence of the energy and broadening parameter of the fundamental band gap of GaSb and $Ga_{1-x}In_xAs_ySb_{1-y}$/GaSb ($0.07 < x < 0.22$, $0.05 < y < 0.19$) quaternary alloys using infrared photoreflectance. Phys. Rev. **B62**, 16600 (2000)
56. http://www.ioffe.rssi.ru/SVA/NSM/Semicond/GaSb/index.html
57. Feng, Z.C., Hou, F.C., Webb, J.B., Shen, Z.X., Rusli, E., Ferguson, I.T., Lu, W.: Optical investigation of GaSb thin films grown on GaAs by metalorganic magnetron sputtering. Thin Solid Films **516**, 5493 (2008)
58. Chen, X.K., Wiersma, R., Wang, C.X., Pitts, O.J., Dale, C., Bolognesi, C.R., Watkins, S.P.: Local vibrational modes of carbon in GaSb and GaAsSb. Appl. Phys. Lett. **80**, 1942 (2002)
59. Das, S.C., Dhar, S.: Transport of nitrogen atoms during liquid phase epitaxial growth of InAsN and GaAsN. Semicond. Sci. Technol. **26**, 085025 (2011)
60. Samajdar, D. P., Dhar, S.,: Transport of bismuth atoms during liquid phase Epitaxial growth of InSbBi and GaSbBi. Semicond. Sci. Technol. **28**, 065007 (2013)
61. Small, M.B., Barnes, J.F.: The distribution of solvent in an unstirred melt under the conditions of crystal growth by liquid epitaxy and its effect on the rate of growth. J. Cryst. Growth **5**, 9 (1967)
62. Jeganathan, K., Saravanan, S., Baskar, K.: Investigations on the concentration profiles of arsenic atoms during liquid phase epitaxial growth of GaAs from Ga-As-Bi solution. Mater. Chem. Phys. **49**, 141 (1997)
63. Wilke, C.R., Chang, P.: Correlations of diffusion coefficients in dilute solutions. AIChE J. **1**, 264 (1955)
64. Flinn, J.M., Jarzinsky, J., Litovitz, T.: Mechanism of volume viscosity in molten bismuth and lead. J. Chem. Phys. **54**, 4331 (1971)
65. Ruhle, W., Schmid, W., Meck, R., Stath, N., Fischbach, J.U., Strottner, I., Benz, K.W., Pilkuhn, M.: Isoelectronic impurity states in direct-gap III-V compounds: the case of InP:Bi. Phys. Rev. **B18**, 7022 (1978)
66. Akchurin, R.K., Zhegalin, V.A., Sakharova, T.V.: Epitaxial structures based on narrow bandgap $InAs_{1-x-y}Sb_xBi_y$ solid solutions. Proc. SPIE **3182**, 369 (1997)
67. Dixit, V.K., Keerthi, K.S., Bhat, H.L., Bera, P., Hedge, M.S.: Structural and compositional analysis of $InBi_xAs_ySb_{(1-x-y)}$ grown on GaAs (001) substrates by liquid phase epitaxy. Appl. Surf. Sci. **220**, 321 (2003)

Chapter 7
Spectroscopic Ellipsometry of AP-MOVPE-Grown $GaAs_{1-x}Bi_x$ Dilute Alloys

Nebiha Ben Sedrine

Abstract The complex dielectric function of dilute $GaAs_{1-x}Bi_x$ alloys ($0 \leq x \leq 3.7$ %), grown by atmospheric pressure metalorganic vapor-phase epitaxy (AP-MOVPE), is determined by using room temperature spectroscopic ellipsometry (SE). The interband transition energies in the energy range 1.4–5.4 eV are resolved using a line shape fitting on the numerically calculated dielectric function second derivatives. The bismuth effect on the critical point (CP) parameters is then determined. We have found that, similarly to previously reported $GaAs_{1-x}Bi_x$ alloys E_0 transition, the bismuth incorporation induces a shift in the E_1, $E_1+\Delta_1$, E_2, and E_0' transition energies but with a lower magnitude. We also observed a root-square-like increase of the E_1 broadening parameter (Γ_1) with respect to the bismuth composition.

7.1 Introduction

Bismuth containing III–V semiconductor alloys are attracting considerable attention due to their possible applications in electronic and optoelectronic devices [1, 2]. Being temperature-insensitive band gap materials [3, 4], of reduced GaAs bandgap [2–8], and successfully grown by MBE, low pressure metalorganic vapor-phase epitaxy (LP-MOVPE), and atmospheric pressure metalorganic vapor-phase epitaxy (AP-MOVPE) [9–14], they are the more practical and less expensive candidates in active layers for lasers used in the wavelength-division-multiplexing (WDM)

N. Ben Sedrine (✉)
Campus Tecnológico e Nuclear, Instituto Superior Técnico,
Universidade Técnica de Lisboa, Sacavèm 2686-953, Portugal
e-mail: bsnebiha@itn.pt

H. Li and Z.M. Wang (eds.), *Bismuth-Containing Compounds*, Springer Series in Materials Science 186, DOI 10.1007/978-1-4614-8121-8_7,

systems. In order to properly design semiconductor devices, it is important to know the optical properties of the alloy over a wide spectral range, which has been investigated by Yamashita et al. [15], and to understand the Bi alloying effect on the critical point (CP) parameters. For this purpose, spectroscopic ellipsometry (SE) is known to be a useful method to obtain accurate optical data [16–21]. Very recently, Tumėnas et al. [22] investigated MBE-grown $GaAs_{1-x}Bi_x$ alloys (x=3.5, 5.2, and 7.5 %) by spectroscopic ellipsometry in the energy range 0.8–2 eV.

In this chapter, we present spectroscopic ellipsometry study on $GaAs_{1-x}Bi_x$ (x=0.0, 0.2, 0.7, 1.3, 3.0, and 3.7 %) alloys, grown on GaAs substrate by AP-MOVPE. Spectroscopic ellipsometry overview, experimental technique, and data analysis are described in the next section in the next section, which is followed by a section dedicated to a detailed study of the effect of Bi incorporation on the dielectric function (DF) and critical points (CPs) in the $GaAs_{1-x}Bi_x$ alloy system, for the energy range 1.4–5.4 eV (results are adapted from ref. [23]).

7.2 Spectroscopic Ellipsometry

7.2.1 *Experimental Technique*

Spectroscopic ellipsometry (SE) is an optical, contactless, nondestructive, and highly sensitive technique that characterizes light reflection from samples. Spectroscopic ellipsometry measures the change of light polarization upon reflection on a sample. Indeed, after light reflection on the sample surface, the incident polarized light often becomes elliptical, from which the term ellipsometry comes. SE determines the complex reflectance ratio ρ defined in terms of the standard ellipsometric parameters ψ and Δ as:

$$\rho = \frac{r_\mathrm{p}}{r_\mathrm{s}} = (\tan\psi)\cdot \mathrm{e}^{i\Delta} \tag{7.1}$$

where r_p and r_s are the reflection coefficients for light polarized parallel (p) and perpendicular (s) to the sample's plane of incidence (Fig. 7.1), respectively.

The standard ellipsometric parameters ψ and Δ depend on the photon energy, the layer sequence within the samples, the dielectric function ε and the thickness d of each layer, the dielectric function of the substrate material, which supports the layer stack, and the angle of incidence Φ.

A common representation of the ellipsometric parameters ψ and Δ is the pseudo-dielectric function $\langle \varepsilon \rangle$. This transformation converts ψ and Δ into virtual bulk dielectric function values using the isotropic two-phase (ambient/substrate) model:

$$\langle\varepsilon\rangle = \langle\varepsilon_1\rangle + i\langle\varepsilon_2\rangle = \sin^2\phi + (\sin^2\phi)\cdot(\tan^2\phi)\cdot\frac{(1-\rho)^2}{(1+\rho)^2}$$

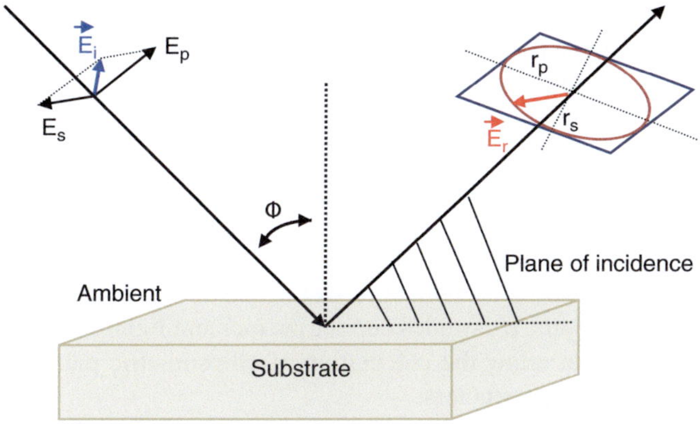

Fig. 7.1 Geometry of an ellipsometric experiment

Only in the case of an air ambient over a bare isotropic semi-infinite substrate with a perfectly smooth and uncovered surface, the measured pseudo-dielectric and the intrinsic dielectric functions are identical.

In this work, spectroscopic ellipsometry (SE) measurement was performed at room temperature using an automatic ellipsometer SOPRA GES5. Data were collected in the 1.4–5.4 eV photon energy region with the step of 10 meV, at incidence angles of $\Phi = 70$, 75, and 80°.

7.2.2 Data Analysis

Spectroscopic ellipsometry is well-known, thanks to its important and useful features, such as the high precision of the measurement, the very high sensitivity to layer thickness, in addition to the accurate determination of the optical constants and the film structures (thin film, quantum wells, superlattices [24], etc.). In addition, spectroscopic ellipsometry allows the direct access to the optical constants of the material, the refractive index n and extinction coefficient k, which is not the case of reflectance or transmittance measurements. However, it is an indirect technique, meaning that the data analysis requires an optical model defined by the optical constants and layer thicknesses of a sample. As a result, the data analysis provides the physical properties of the sample such as the optical constants and film thicknesses.

The ellipsometry data analysis is generally performed by using a linear regression analysis, where the sample properties, such as optical constants, film thickness, and composition, are determined by varying and adjusting fitting parameters until calculated and experimental data match perfectly. The use of a method known as mathematical inversion allows a rather straightforward extraction of the dielectric function of a sample. In this section, we will overview data analysis procedures and focus mostly on the mathematical inversion method used in this work (more details on the linear regression analysis can be found in ref. [20]).

Ellipsometry data analysis consists of three major parts: construction of an optical model, dielectric function modeling, and methodology used for fitting calculated to measured spectra.

7.2.2.1 Optical Model

In the data analysis, the optical model has to reflect the physical structure of the sample. The construction of an optical model is one of the most important steps in ellipsometry data analysis. If samples depolarize incident light, however, we have to use optical models that allow the calculation of ellipsometric parameters ψ and Δ spectra from the Stokes parameters.

In the case of our $GaAs_{1-x}Bi_x$ thin films, we have used the following four-phase layered model:

Ambient/GaAs native oxide overlayer/$GaAs_{1-x}Bi_x$ film/GaAs substrate.

The optical constants relative to GaAs substrate and GaAs native oxide were taken without change from previously published data [25].

7.2.2.2 Dielectric Function

Depending on the case, two methods can be used in order to determine an unknown dielectric function of a sample structure.

Modeling the Dielectric Function

There are different dielectric function models, and we need to select an appropriate model according to the optical properties of the sample. For dielectric function modeling in a transparent region, the Sellmeier or Cauchy models are commonly used. Data modeling using the Drude model is introduced if free-carrier absorption in the optical response occurs. In the visible–UV energy region, the optical response of a material can be modeled using different models, such as the Lorentz model, Tauc-Lorentz model [26], harmonic oscillator approximation (HOA), model dielectric function (MDF), and standard critical points model (CP).

Mathematical Inversion

As mentioned above, when the dielectric function of a material is not known, the dielectric function modeling is required. However, complete modeling of the dielectric function over a large energy range might be complex. When data analysis is performed in a selected energy region, the fitting quality is improved, providing accurate material's dielectric function only in such limited region. A good way to solve such complex case is the use of the mathematical inversion, where the dielectric function measured in the whole range can be extracted. This procedure, referred to as mathematical inversion or point-by-point fit, is only valid if one unknown parameter in the optical model structure is involved; which is the $GaAs_{1-x}Bi_x$ optical constants (real and imaginary parts of the dielectric function) in our case. In other words, mathematical inversion can always be employed to extract the optical constants of one layer if all the other parameters are known. The mathematical inversion is performed by using a linear regression analysis; for this reason, the mathematical inversion is also called optical constant fit. This data analysis procedure is an alternative and efficient way to determine the dielectric function of a sample structure, especially if the dielectric function modeling is complex in some energy regions. We can further parameterize the extracted dielectric functions using various dielectric function models (as enumerated in Section Modeling the Dielectric Function).

7.2.2.3 Methodology

When the dielectric function of a semiconductor is extracted by using the mathematical inversion, one can perform a standard critical points (CPs) analysis in order to characterize the structures of critical points also known as interband transitions. In fact, many semiconductor applications depend on the dielectric function ε (E) [16]. This function is related to the electronic band structure and can be used as a powerful source of experimental information on the material interband transitions. For GaAs semiconductor, for instance, the lowest direct bandgap E_0 is located at the Γ point of the Brillouin Zone [27]. The next higher interband transitions are labelled E_1 and $E_1+\Delta_1$. These closely spaced critical points (CPs) in GaAs occur along the $\langle 111\rangle$ directions or at the L point in the Brillouin Zone. In the energy region from 4 to 5.5 eV, several critical points have been resolved [28].

Since the semiconductor band structure varies according to alloy composition, temperature, strain, grain size, etc., the effect of these physical parameters can also be deduced from the critical points analysis. Theoretical expressions used for CP analysis can be derived from a dielectric function given by the standard critical point model following the equation [16, 29]:

$$\varepsilon(E)=\sum_{j=1}^{N}\left[C_j-A_je^{i\phi_j}(E-E_{cj}+i\Gamma_j)^n\right] \tag{7.2}$$

The critical point parameters, amplitude A, energy E_c, broadening Γ, and excitonic phase angle Φ, are determined by fitting numerically obtained second-derivative spectra $d^2\varepsilon(E)/dE^2$ of the experimental ε (E). The optical transitions in the spectral region around 3 eV are E_1 and $E_1+\Delta_1$. At room temperature and above, they are well represented by two-dimensional (2D) M_1 critical point (CP), so that the exponent n has the value of 0, i.e., logarithmic [30]. The standard critical point line shape and its second derivative are given by:

$$\varepsilon(E)=\sum_{j=1}^{N}\left[C_j-A_je^{i\phi_j}Ln(E-E_{cj}+i\Gamma_j)\right] \tag{7.3}$$

$$\frac{d^2\varepsilon}{dE^2} = \sum_{j=1}^{N} \left[A_j e^{i\phi_j} (E - E_{cj} + i\Gamma_j)^{-2} \right] \tag{7.4}$$

Equation (7.3) fulfills the Kramers–Kronig (K.K.) requirements of the dielectric function. The fitting of the second-derivative spectra $d^2\varepsilon(E)/dE^2$ enables us to accurately distinguish features attributed to interband critical points (CPs) in the measured spectra of our $GaAs_{1-x}Bi_x$ samples.

It has been known that the dielectric functions of semiconductors show significant changes by the following effects: alloy composition [31], temperature [16, 29], doping [32, 33], strain [34], grain size of polycrystalline semiconductors [35], quantum effects [36]. In these examples, the semiconductor band structure changes remarkably, thus, we can characterize these effects indirectly from critical point analyses.

The results of our study of the effect of Bi incorporation on the dielectric function (DF) and critical points (CPs) in the $GaAs_{1-x}Bi_x$ alloy system are presented in the next section.

7.3 Experimental Results

7.3.1 Samples

The study was performed on $GaAs_{1-x}Bi_x$ (x=0.0, 0.2, 0.7, 1.3, 3.0, and 3.7 %) samples grown by AP-MOVPE. These samples were grown in a horizontal geometry reactor on exactly (100) Zn-doped GaAs substrates. Trimethyl gallium (TMG), trimethyl bismuth (TMBi), and arsine (AsH_3) were used as precursors. $GaAs_{1-x}Bi_x$ layer nominal thicknesses are 120 nm for the $GaAs_{0.998}Bi_{0.002}$ sample, 22 nm for the $GaAs_{0.993}Bi_{0.007}$ and $GaAs_{0.987}Bi_{0.013}$ samples, and 50 nm for the others. Bi composition was determined by high resolution X-ray diffraction (HRXRD) measurements.

7.3.2 Results

Due to the multilayer nature of our samples, the $GaAs_{1-x}Bi_x$ optical constants cannot be obtained analytically from ψ and Δ. For this reason, we have used the mathematical inversion (based on the Newton–Raphson method) of the measured complex reflectance ratio ρ using the four-phase model: ambient/GaAs native oxide overlayer/$GaAs_{1-x}Bi_x$ film/GaAs substrate (see Sect. 7.2). This method was first applied to the 3.7 % Bi content $GaAs_{1-x}Bi_x$ sample measured at different incidence angles: 70, 75, and 80°. We have found that these incidence angles give identical results, which confirm that the extracted values can be interpreted as the true material data. For clarity of presentation, we only show the optical data recorded at 75° for all samples.

In Fig. 7.2, we present the extracted $GaAs_{1-x}Bi_x$ dielectric function real (a) ε_1 (E) and imaginary (b) ε_2 (E) parts in the 1.4–5.4 eV photon energy region, at an incidence angle of 75°. The feature shown by an arrow in Fig. 7.2b is the GaAs

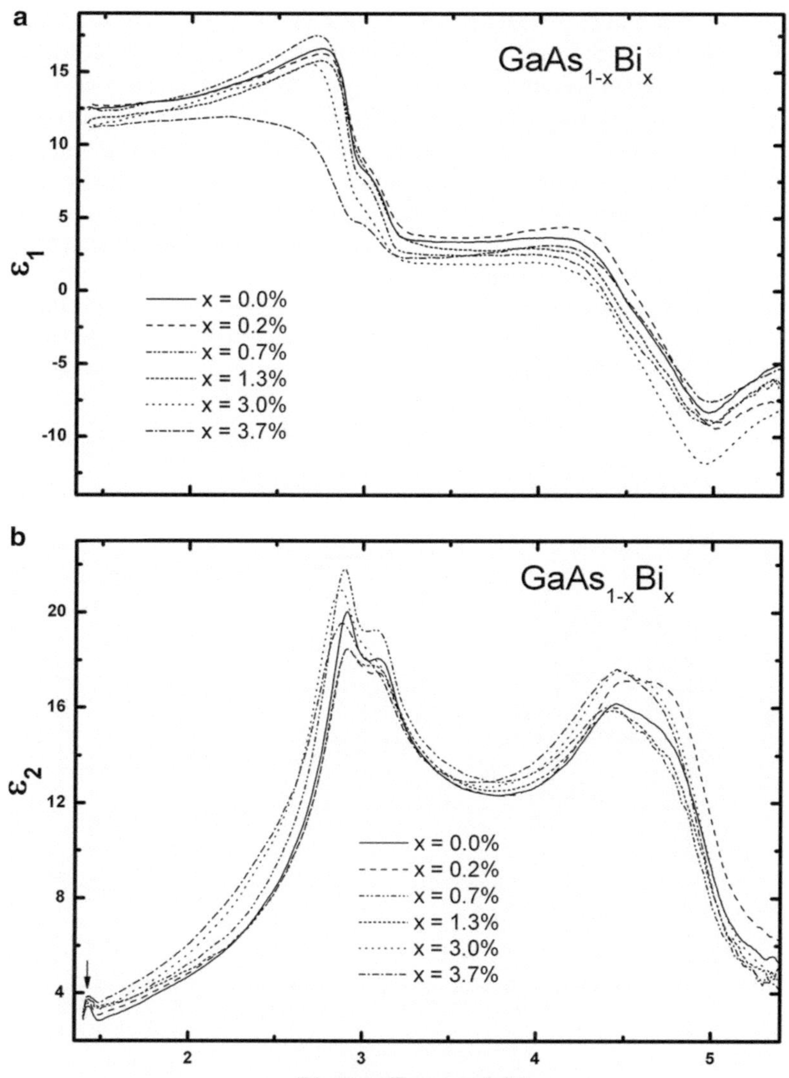

Fig. 7.2 Real (**a**) and imaginary (**b**) parts of the extracted $GaAs_{1-x}Bi_x$ (x=0.0, 0.2, 0.7, 1.3, 3.0, and 3.7 %) dielectric functions (modified from ref. [23])

absorption edge. In addition, $GaAs_{1-x}Bi_x$ dielectric function imaginary parts represented in Fig. 7.2b show different structures attributed to interband critical points (CPs).

The great similarity with GaAs spectra allows us to use the same nomenclature for the CP of the alloy as for bulk GaAs. These features result from interband transitions at high symmetry points in the Brillouin zone: E_0 (1.42 eV), E_1 (2.88 eV), $E_1+\Delta_1$ (3.11 eV) for the L point, E_2 (4.94 eV) for the X point, and E_0' (4.39 eV) above the fundamental bandgap at the Γ point [16]. The differences in the ε_2 (E) amplitudes in the 3 or 5 eV energy regions, for different Bi compositions, reflect typical differences in the quality surfaces of the samples and the presence of an oxide overlayer (ranging from 15 to 30 Å depending on the sample). Based on the work of Jung et al. [37], we assume that this is of minor effect on the E_1 critical point energies and broadenings. These structures can be analyzed in terms of standard analytic line shapes [16] to accurately access to their transition parameters.

At room temperature, in the case of GaAs [16, 38], the derivative spectra in the vicinity of the critical points (E_1, $E_1+\Delta_1$, E_2, and E_0') can be assumed as two-dimensional line shapes (see Sect. 7.2).

In Fig. 7.3, we show the dielectric function imaginary part second derivatives $d^2\varepsilon_2(E)/dE^2$ versus photon energy for the different Bi compositions. The best-fit calculated $d^2\varepsilon_2(E)/dE^2$ spectra (solid lines) from (7.4) are compared to the

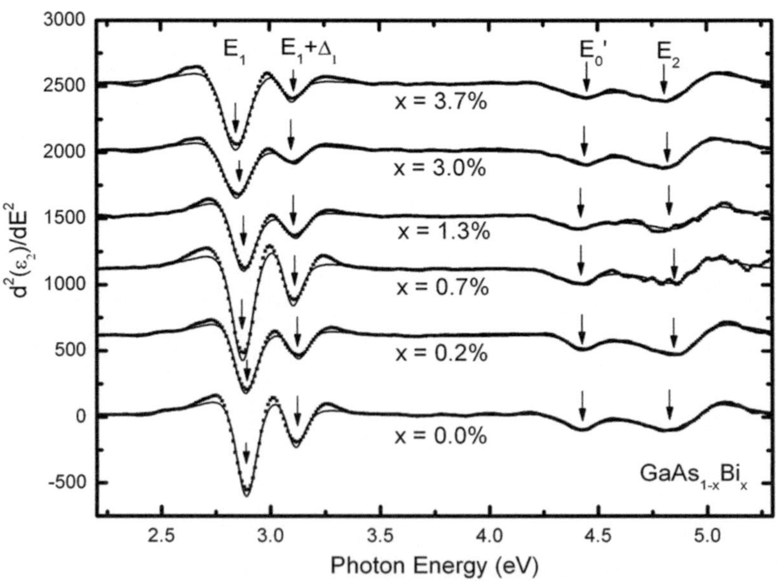

Fig. 7.3 Imaginary part second derivatives of the extracted (*scatters*) dielectric functions of $GaAs_{1-x}Bi_x$ layers (x=0.0, 0.2, 0.7, 1.3, 3.0, and 3.7 %), together with the best-fit calculated (*solid lines*) spectra using (7.4). The spectra with x=0.2, 0.7, 1.3, 3.0, and 3.7 % are shifted for clarity (used with permission from ref. [23])

numerically second derivatives-extracted results (scatters) using the Levenberg–Marquardt regression algorithm. In order to improve the quality of the fit, CP peaks (j) were fitted simultaneously by taking A_j, E_{cj}, Γ_j, and Φ_j as free parameters. As can be seen in Fig. 7.3, the ε_2 (E) second-derivative broadening of the $x = 0.0$ % containing $GaAs_{1-x}Bi_x$ sample is narrower than those of the other spectra.

Figure 7.4 represents the increase of the broadening parameter Γ_1, corresponding to the E_1 CP, versus bismuth content x. The broadening parameter Γ_1 shows a root-square-like dependence on x following $y = a + b\sqrt{x}$. The corresponding constant prefactors, a and b, obtained using the Levenberg–Marquardt regression algorithm are, respectively, 0.097 ± 0.002 and 0.130 ± 0.022 eV. Comparing to the GaAsN alloys ($\Gamma_1 = 0.091 + 0.509\sqrt{x}$ [39], $\Gamma_1 = 0.085 + 0.373\sqrt{x}$ [40]), the bismuth atom seems to less broaden the E_1 CP. The increase of the broadening parameter versus x content corresponds to potential fluctuations resulting from random Bi atomic placement in the alloy [41]. An increase of the broadening was similarly observed in the E_0 transition in $GaAs_{1-x}Bi_x$ [21].

The variation with composition of the transition energies is expected, as in other III–V semiconductors alloys [42] to be quadratic:

$$E_{cj}(x) = A + B \cdot x + C \cdot x \cdot (x - 1) \qquad (7.5)$$

where $A = E_{cj}(0)$, $B = E_{cj}(1) - E_{cj}(0)$, while C is a bowing parameter. The critical point energies E_{cj} determined by fitting the second derivatives are plotted in Fig. 7.5 (symbols).

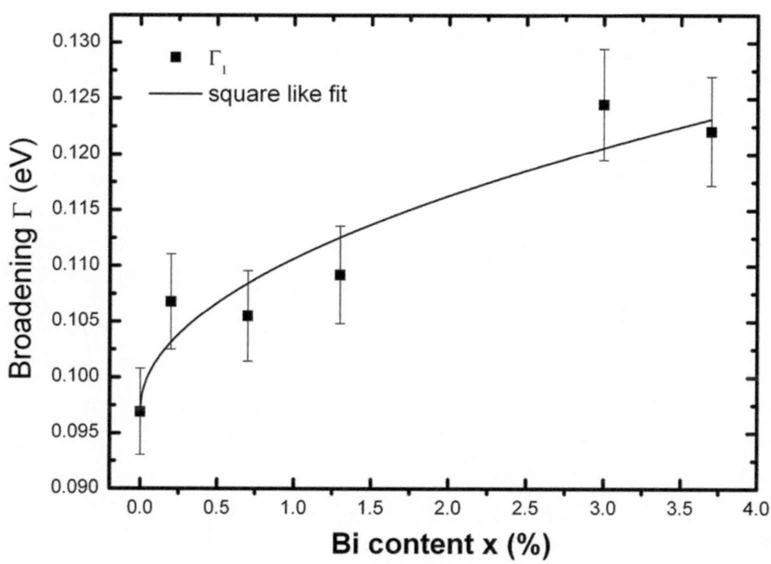

Fig. 7.4 $GaAs_{1-x}Bi_x$ broadening parameters Γ_1 for the E_1 critical point versus Bi content x (0.0, 0.2, 0.7, 1.3, 3.0, and 3.7 %). The *symbols* represent the results of fitting (7.4) to the second derivative of the experimental spectra. The *full line* represents the square-root-like ($y = a + b\sqrt{x}$) dependence of Γ_1 (modified from ref. [23])

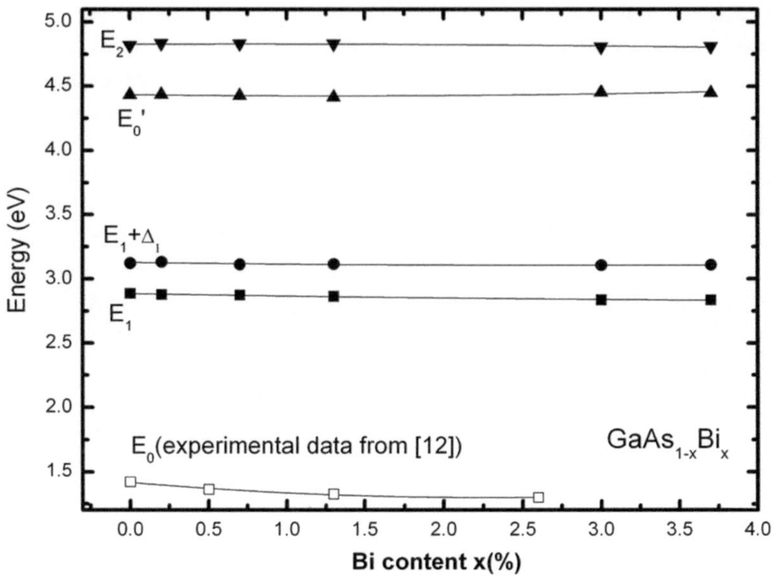

Fig. 7.5 $GaAs_{1-x}Bi_x$ transition energies (*symbols*): E_1, $E_1+\Delta_1$, E_0', and E_2 versus Bi content x. Experimental E_0 values are taken from [2]. *Solid lines* represent the best-fit to (7.5) (modified from ref. [23])

Table 7.1 Values of the A and B parameters obtained by fitting the compositional dependences of the E_0, E_1, $E_1+\Delta_1$, E_2, and E_0' CP energies to (7.4)

	A (eV)	B (eV)
E_0	1.415 (0.011)	−8.100 (0.500)
E_1	2.886 (0.003)	−1.858 (0.053)
$E_1+\Delta_1$	3.127 (0.005)	−1.226 (0.061)
E_0'	4.433 (0.007)	−0.888 (0.096)
E_2	4.827 (0.006)	0.134 (0.040)
Δ_1	0.231 (0.001)	0.690 (0.066)

The bowing parameters C are very low. Errors are given in parentheses

Solid lines in Fig. 7.5 represent the best-fitting to (7.5) using the Levenberg–Marquardt regression algorithm in the Bi composition range (0.0–3.7 %). We present, in Table 7.1, the best-fit parameters of the Bi energy dependence for the E_0, E_1, $E_1+\Delta_1$, E_2, and E_0' critical points. The values corresponding to the x=0.0 % $GaAs_{1-x}Bi_x$ content sample are found to be in good agreement with previous GaAs published data [16].

The experimental values relative to the fundamental band gap E_0 were taken from the work of Oe [2]. We have found that for E_0 the linearized dependence of the transition energy with respect to the band gap of GaAs is −81 meV/%Bi. Such a result is found to be in good agreement with previous ones [1]. For the bowing parameter $C(E_0)$, controversial values are found: –5.6 eV [1], 2.1 eV [43], and 1.74 eV [44], based on the band gap energy (−1.45 eV) of virtual compound GaBi [43]; the major source of uncertainty on the bowing coefficient originates from the uncertainty on the virtual zinc-blende GaBi band gap.

Madouri et al. [44] showed that among the three contributions, the different atomic electronegativities, the volume deformation, and the structural relaxation, the band gap bowing of $GaAs_{1-x}Bi_x$ alloys is mainly caused by both the latter structural effects.

Recently, the spectroscopic ellipsometry study of GaAsBi was analyzed in detail near the absorption edge [22]. The authors evidenced a decrease of the material's band gap with an increase of the Bi composition. In addition, at higher photon energies (≈1.7 eV) a weak optical feature was distinguished in the GaAsBi dielectric function imaginary part, which was attributed to the optical transitions from the spin-orbit split valence band *E*+, predicted by the GaAsBi valence band anticrossing model [6, 45]. The spin-orbit splitting Δ_0^+ was roughly estimated to be ≈0.7 eV.

We have found that all transition energies (E_1, $E_1+\Delta_1$, E_2, and E_0'), that occur above E_0, are characterized by very low bowing parameters ($C<0.1$ meV) leading to a linear Bi dependence. In addition, Fig. 7.5 shows that the Bi incorporation affects differently the E_0, E_1, $E_1+\Delta_1$, E_2, and E_0' transition energies. For instance, the E_1 energy follows a linear decrease with respect to Bi composition of −1.8 eV, which is rather small comparing to E_0 (−8.1 eV). This behavior is also observed for the $E_1+\Delta_1$ transition, with a decrease of −1.2 eV. It is interesting to note that the $GaAs_{1-x}Bi_x$ spin-orbit splitting Δ_1 between E_1 and $E_1+\Delta_1$ transition energies was found to increase linearly by 0.690 eV with Bi composition.

7.4 Conclusions

In summary, we have presented in this chapter a comprehensive study of the effect of Bi on the dielectric function of a set of $GaAs_{1-x}Bi_x$ layers grown by AP-MOVPE using room temperature spectroscopic ellipsometry. The optical constants were extracted in the energy region of 1.4–5.4 eV from the measured data for Bi content (x=0.0, 0.2, 0.7, 1.3, 3.0, and 3.7 %). By analyzing the $GaAs_{1-x}Bi_x$ transition energies (E_1, $E_1+\Delta_1$, E_2, and E_0') above the fundamental gap region using analytic line shapes, we have found an energy shift due to the Bi incorporation, a linear increase of the Δ_1 spin-orbit splitting, and an increase of the E_1 broadening parameter. The fitting results show that the dependences of the E_1, $E_1+\Delta_1$, E_2, and E_0' transition energies on Bi composition are weaker than that of E_0.

Acknowledgements The author gratefully acknowledge J.L. Stehle (SEMILAB-SOPRA, France), R. Chtourou, and V. Darakchieva for fruitful discussions. I. Moussa, H. Fitouri, A. Rebey, and B. El Jani are acknowledged for providing samples of this study. The author acknowledges financial support from FCT Portugal under grant REF: SFRH/BPD/66818/2009.

References

1. Francoeur, S., Seong, M.-J., Mascarenhas, A., Tixier, S., Adamcyk, M., Tiedje, T.: Band gap of $GaAs_{1-x}Bi_x$, 0<x<3.6%. Appl. Phys. Lett. **82**(22), 3874 (2003)
2. Oe, K.: Characteristics of semiconductor alloy $GaAs_{1-x}Bi_x$. Jpn. J. Appl. Phys. **41**, 2801–2806 (2002)
3. Yoshida, J., Kita, T., Wada, O., Oe, K.: Temperature dependence of $GaAs_{1-x}Bi_x$ band gap studied by photoreflectance spectroscopy. Jpn. J. Appl. Phys. **42**, 371 (2003)
4. Lu, X., Beaton, D.A., Lewis, R.A., Tiedje, T., Whitwick, M.B.: Effect of molecular beam epitaxy growth conditions on the Bi content of $GaAs_{1-x}Bi_x$. Appl. Phys. Lett. **92**, 192110 (2008)
5. Tixier, S., Adamcyk, M., Tiedje, T., Francoeur, S., Mascarenhas, A., Wei, P., Schiettekatte, F.: Molecular beam epitaxy growth of $GaAs_{1-x}Bi_x$. Appl. Phys. Lett. **82**, 2245 (2003)
6. Alberi, K., Dubon, O.D., Walukiewicz, W., Yu, K.M., Bertulis, K., Krotus, A.: Valence band anticrossing in $GaBi_xAs_{1-x}$. Appl. Phys. Lett. **91**, 051909 (2007)
7. Francoeur, S., Tixier, S., Young, E., Tiedje, T., Mascarenhas, A.: Bi isoelectronic impurities in GaAs. Phys. Rev. B **77**, 085209 (2008)
8. Bertulis, K., Krotkus, A., Alekseienko, G., Pacebutas, V., Adomavicius, R., Molis, G., Marcinkevicius, S.: GaBiAs: a material for optoelectronic terahertz devices. Appl. Phys. Lett. **88**, 201112 (2006)
9. Huang, W., Yoshimoto, M., Takehara, Y., Saraie, J., Oe, K.: $GaNyAs_{1-x-y}Bi_x$ alloy lattice matched to GaAs with 1.3 μm photoluminescence emission. Jpn. J. Appl. Phys. **43**, L1350 (2004)
10. Takehara, Y., Yoshimoto, M., Huang, W., Saraie, J., Oe, K., Chayahara, A., Horino, Y.: Lattice distortion of GaAsBi alloy grown on GaAs by molecular beam epitaxy. Jpn. J. Appl. Phys. **45**, 67 (2006)
11. Oe, K., Okamoto, H.: New semiconductor alloy $GaAs_{1-x}Bi_x$ grown by metal organic vapor phase epitaxy. Jpn. J. Appl. Phys. **37**, L1283 (1998)
12. Oe, K.: Metalorganic vapor phase epitaxial growth of metastable $GaAs_{1-x}Bi_x$ alloy. J. Cryst. Growth **237/239**, 1481 (2002)
13. Fitouri, H., Moussa, I., Rebey, A., Fouzri, A., El Jani, B.: AP-MOVPE of thin $GaAs_{1-x}Bi_x$ alloys. J. Cryst. Growth **295**, 114 (2006)
14. Moussa, I., Fitouri, H., Rebey, A., El Jani, B.: Atmospheric-pressure metalorganic vapour phase epitaxy optimization of GaAsBi alloy. Thin Solid Films **516**, 8372 (2008)
15. Yamashita, K., Yoshimoto, M., Oe, K.: Temperature-insensitive refractive index of GaAsBi alloy for laser diode in WDM optical communication. Phys. Status Solidi C **3**(3), 693 (2006)
16. Lautenschlager, P., Garriga, M., Logothetidis, S., Cardona, M.: Interband critical points of GaAs and their temperature dependence. Phys. Rev. B **35**, 9174 (1987)
17. Ihn, Y.S., Kim, T.J., Kim, Y.D., Aspnes, D.E., Kossut, J.: Optical properties of $Cd_{1-x}Mg_xTe$ ($x = 0.00$, 0.23, 0.31, and 0.43) alloy films. Appl. Phys. Lett. **84**, 693 (2004)
18. Logothetidis, S., Cardona, M., Garriga, M.: Temperature dependence of the dielectric function and the interband critical-point parameters of $AlxGa_{1-x}As$. Phys. Rev. B **43**, 11950 (1991)
19. Azzam, R.M.A., Bashara, N.M.: Ellipsometry and polarized light. North Holland, Amsterdam (1977)
20. Fujiwara, H.: Spectroscopic ellipsometry: principles and applications. Wiley, Chichester (2007)
21. Ben Sedrine, N., Bouhafs, C., Harmand, J.C., Chtourou, R., Darakchieva, V.: Effect of nitrogen on the $GaAs_{0.9-x}N_xSb_{0.1}$ dielectric function from the near-infrared to the ultraviolet. Appl. Phys. Lett. **97**, 201903 (2010)
22. Tumėnas, S., Karpus, V., Bertulis, K., Arwin, H.: Dielectric function and refractive index of $GaBi_xAs_{1-x}$ ($x = 0.035$, 0.052, 0.075). Phys. Status Solidi C **9**(7), 1633–1635 (2012)
23. Ben Sedrine, N., Moussa, I., Fitouri, H., Rebey, A., El Jani, B., Chtourou, R.: Spectroscopic ellipsometry study of $GaAs_{1-x}Bi_x$ material grown on GaAs substrate by atmospheric pressure metal-organic vapor-phase epitaxy. Appl. Phys. Lett. **95**, 011910 (2009)

24. Ben Sedrine, N., Darakchieva, V., Lindgren, D., Monemar, B., Che, S.B., Ishitani, Y., Yoshikawa, A.: Optical properties of InN/In0.73Ga0.27N multiple quantum wells studied by spectroscopic ellipsometry. Phys. Status Solidi C **8**(5), 1629 (2011)
25. Zollner, S.: Optical constants and critical-point parameters of GaAs from 0.73 to 6.60 eV. J. Appl. Phys. **90**, 515 (2001)
26. Jellison Jr., G.E., Modine, F.A.: Parameterization of the optical functions of amorphous materials in the interband region. Appl. Phys. Lett. **69**, 371 (1996)
27. Chelikowsky, J.R., Cohen, M.L.: Nonlocal pseudopotential calculations for the electronic structure of eleven diamond and zinc-blende semiconductors. Phys. Rev. B **14**, 556 (1976)
28. Aspnes, D.E., Studna, A.A.: Schottky-barrier electroreflectance: application to GaAs. Phys. Rev. B **7**, 4605 (1973)
29. Lautenschlager, P., Garriga, M., Vina, L., Cardona, M.: Temperature dependence of the dielectric function and interband critical points in silicon. Phys. Rev. B **36**, 4821 (1987)
30. Kim, C.C., Garland, J.W., Abad, H., Raccah, P.M.: Modeling the optical dielectric function of semiconductors: extension of the critical-point parabolic-band approximation. Phys. Rev. B **45**, 11749 (1992)
31. Aspnes, D.E., Kelso, S.M., Logan, R.A., Bhat, R.: Optical properties of AlxGa$_{1}$_x As. J. Appl. Phys. **60**, 754 (1986)
32. Vina, L., Cardona, M.: Optical constants of pure and heavily doped silicon and germanium: electronic interband transitions. Physica **117B/118B**, 356 (1983)
33. Aspnes, D.E., Studna, A.A., Kinsbron, K.: Dielectric properties of heavily doped crystalline and amorphous silicon from 1.5 to 6.0 eV. Phys. Rev. B **29**, 768 (1984)
34. Pickering, C., Carline, R.T., Emeny, M.T., Garawal, N.S., Howard, L.K.: Dielectric functions and critical points of strained InxGa$_{1-x}$As on GaAs. Appl. Phys. Lett. **60**, 2412 (1992)
35. Nguyen, H.V., Collins, R.W.: Finite-size effects on the optical functions of silicon microcrystallites: a real-time spectroscopic ellipsometry study. Phys. Rev. B **47**, 1911 (1993)
36. Herzinger, C.M., Snyder, P.G., Celii, F.G., Kao, Y.-C., Chow, D., Johs, B., Woollam, J.A.: Studies of thin strained InAs, AlAs, and AlSb layers by spectroscopic ellipsometry. J. Appl. Phys. **79**, 2663 (1996)
37. Jung, Y.W., Ghong, T.H., Kim, Y.D., Aspnes, D.E.: Effect of overlayers on critical-point parameters in the analysis of ellipsometric spectra. Appl. Phys. Lett. **91**, 121903 (2007)
38. Lukes, F., Gopalan, S., Cardona, M.: Effect of heavy doping on the optical properties and band structure of GaAs. Phys. Rev. B **47**, 7071 (1993)
39. Timoshevskii, V., Côté, M., Gilbert, G., Leonelli, R., Turcotte, S., Beaudry, J.-N., Desjardins, P., Larouche, S., Martinu, L., Masut, R.A.: Experimental and theoretical studies of the E+ optical transition in GaAsN alloys. Phys. Rev. B **74**, 165120 (2006)
40. Ben Sedrine, N., Rihani, J., Harmand, J.C., Chtourou, R.: Spectroscopic ellipsometry analys is of rapid thermal annealing effect on MBE grown $GaAs_{1-x}$–N_x. J. Telecommun. Inf. Technol. **1**, 51–56 (2009)
41. Kelso, S.M., Aspnes, D.E., Pollack, M.A., Nahory, R.E.: Optical properties of In_{1-x}GaxAsyP_{1-y} from 1.5 to 6.0 eV determined by spectroscopic ellipsometry. Phys. Rev. B **26**(12), 6669 (1982)
42. Thompson, G., Woolley, J.C.: Energy-gap variation in mixed III–V alloys. Can. J. Phys. **45**, 255 (1967)
43. Janotti, A., Wei, S.-H., Zhang, S.: Theoretical study of the effects of isovalent coalloying of Bi and N in GaAs. Phys. Rev. B **65**, 115203 (2002)
44. Madouri, D., Boukra, A., Zaoui, A., Ferhat, M.: Bismuth alloying in GaAs: a first-principles study. Comput. Mater. Sci. **43**, 818 (2008)
45. Alberi, K., Wu, J., Walukiewicz, W., Yu, K.M., Dubon, O.D., Watkins, S.P., Wang, C.X., Liu, X., Cho, Y.-J., Furdyna, J.: Valence-band anticrossing in mismatched III-V semiconductor alloys. Phys. Rev. B **75**, 045203 (2007)

Chapter 8
Effect of Bismuth Alloying on the Transport Properties of the Dilute Bismide Alloy, $GaAs_{1-x}Bi_x$

Rajeev N. Kini and Angelo Mascarenhas

Abstract We studied the effect of Bi incorporation on the carrier mobilities and electron effective mass in the dilute Bismide alloy, $GaAs_{1-x}Bi_x$, using electrical transport and photoluminescence techniques. We observed no significant degradation in the electron Hall mobility with Bi incorporation in GaAs, up to a concentration of 1.2 %. At higher Bi concentration (≥1.6 %) some degradation of the electron mobility was observed, although there is no apparent trend. Our measurements show that the hole Hall mobility decreases with increasing Bi concentration. Analysis of the temperature-dependent Hall transport data of p-type $GaAs_{1-x}Bi_x$ epilayers along with low-temperature PL measurements of p-doped and undoped epilayers suggests that Bi incorporation results in the formation of several trap levels above the valence band which we attribute to Bi–Bi pair states. The decrease in hole mobility with increasing Bi concentration can be explained as being caused by scattering at the isolated Bi and the Bi–Bi pair states. Magnetic field- and temperature-dependent resistivity measurements on n-type $GaAs_{1-x}Bi_x$ showed clear Shubnikov de Haas (SdH) oscillations up to a concentration of 0.88 %. The measured electron effective mass exhibits a gradual increase up to a concentration of 0.4 % Bi and then a gradual decrease. Accounting for the giant bandgap and spin-orbit bowing, the measured changes in the effective mass are in qualitative agreement with perturbation theory applied to these band changes, confirming that Bi mainly perturbs the valence band.

R.N. Kini (✉)
Indian Institute of Science Education and Research Thiruvananthapuram (IISER-TVM), CET Campus, Engineering College P.O., Kerala 695 016, India
e-mail: rajeevkini@iisertvm.ac.in

A. Mascarenhas
National Renewable Energy Laboratory (NREL), 1617 Cole Blvd., Golden, CO 80401, USA
e-mail: angelo.mascarenhas@nrel.gov

H. Li and Z.M. Wang (eds.), *Bismuth-Containing Compounds*, Springer Series in Materials Science 186, DOI 10.1007/978-1-4614-8121-8_8,

8.1 Introduction

The growing technological demand for new materials has led researchers to explore nonconventional semiconductor alloys. A great deal of effort has been devoted in exploring III–V alloys obtained by incorporating dilute amounts of either very light or very heavy isoelectronic elements into the binary compounds, for example, N in GaP and GaAs. Sometimes, not only is the growth of these materials challenging but understanding the electronic properties is also very complicated. Dilute III–V-N alloys have been studied extensively and various methods have been employed to understand the peculiar properties exhibited by these alloys, for instance, giant band gap bowing, various impurity states, and resonant optical transitions in GaAsN. The interest in the dilute GaAsN alloys was fuelled mainly by the possibility to tune the band gap by adding small amounts of N. Recently there has been tremendous interest in dilute III–V-Bi alloys like $GaAs_{1-x}Bi_x$, $In_pGa_{1-p}As_{1-q}Bi_q$, $GaSb_{1-y}Bi_y$, and $GaN_{1-z}Bi_z$ due to their potential technological applications [1–5]. From an application point of view there are certain advantages in using dilute Bismides, e.g., $GaAs_{1-x}Bi_x$ instead of $GaAs_{1-x}N_x$; incorporating N primarily perturbs the conduction band (CB), whereas Bi doping mainly distorts the valence band (VB). Since N perturbs the CB, the electron mobility would be affected and a drastic decrease in electron mobility was observed in GaAsN. The mobility degradation is detrimental for device application. Since Bi incorporation mainly perturbs the VB, it is expected that the electron mobility will be largely unaffected in dilute Bismides. Observation of some unusual properties like giant band gap bowing [1, 6] and spin-orbit bowing [2], in the dilute $GaAs_{1-x}Bi_x$ alloy and the potential to use this alloy in device applications, for example, in high-efficiency solar cells [7], heterojunction bipolar transistors (HBTs) [8], spintronics, and near-infrared devices, has sparked a lot of research activity on this alloy system. In this chapter we will present our studies to understand the effect of Bi incorporation on the carrier mobilities and electron effective mass in the dilute Bismide alloy, $GaAs_{1-x}Bi_x$, using electrical transport and photoluminescence (PL) techniques.

A comparison between the dilute III–V Nitride and Bismide systems can be found in ref. [9] and a detailed comparison between GaAsN and GaAsBi can be found in ref. [10]. Electronically N in GaAsN forms a resonant state in the conduction band and recombination of excitons bound to isoelectronic N impurities in GaAs under pressure has been reported [11]. Bound states due to N–N pairs and N-clusters have been studied theoretically and experimentally. In the case of GaAsBi, it was theoretically predicted that Bi forms a bound state in GaAs at 180 meV above the valence band maximum (VBM) [12]. The prediction of a Bi bound state in GaAs is rather surprising because the isolated Bi state in GaP lies only ~100 meV above the VBM [13] and, considering the valence band offset between GaAs and GaP (~500 meV) [14], one would expect the isolated Bi state to be resonant in the valence band in GaAs. A more recent band structure calculation predicted the Bi impurity state to be resonant at 80 meV below the VBM [9] which

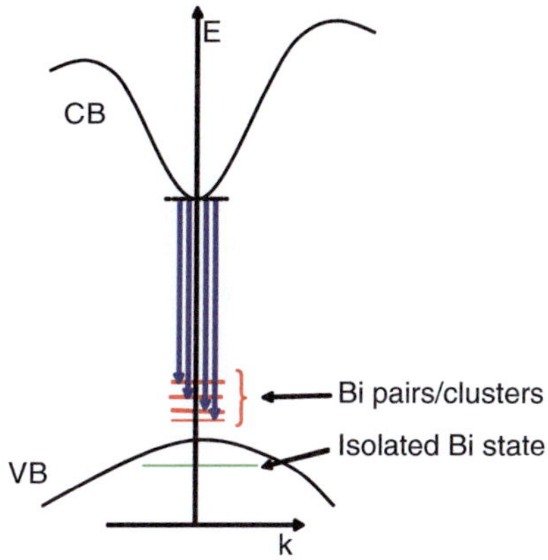

Fig. 8.1 Schematic representation of the isolated Bi and Bi–Bi pair states in GaAs

seems more reasonable. It was possible to study the isolated N-state which is resonant in the CB in GaAsN by applying hydrostatic pressure, which brings this state out of the CB [11]. However, in the case of Bi in GaAsBi, on applying hydrostatic pressure the isolated Bi state moves along with the VB and hence the isolated Bi state will not emerge out of the VB [9]. This makes the experimental study of this state and its interaction with the VB and/or CB extremely difficult. One possible way to study the interaction of this state is by using resonant tunneling experiments. In the case of isolated N-state in GaAsN, studies on an *n-i-n* resonant tunneling structure with an intrinsic GaAsN quantum well (QW) revealed the interaction of isolated N-state with the CB [15]. Since Bi interacts with the VB, similar studies on a *p-i-p* resonant tunneling diode with a GaAsBi QW will provide information about the interaction of the resonant Bi state with the bands. Apart from the isolated Bi state it is possible that Bi incorporation also introduces Bi–Bi pair states and cluster states as shown in Fig. 8.1. Experimental studies based on photoluminescence (PL) measurements showed some evidence for Bi–Bi pair/cluster states [16]. The isolated Bi and Bi–Bi pair/cluster states will affect the electronic and transport properties of this alloy. We will present our measurement using electrical transport and PL technique to study some of these effects.

8.2 The Experimental Details

All the measurements were done on $GaAs_{1-x}Bi_x$ epilayers of thickness 300–500 nm grown on GaAs substrates in an EPI930 molecular-beam epitaxy (MBE) system equipped with standard effusion cells for Ga and Bi, and an arsenic cracker that converted As_4 to As_2. The growth rate was around 2.0 μm/h and the substrate temperature was approximately 380 °C. The As/Ga ratio was kept as close to unity as possible without going to Ga-rich conditions. Beam equivalent pressures of Bi were in the approximate range of $(0.4–4)\times10^{-9}$ Torr. High resolution X-ray diffraction measurements (XRD) of asymmetric (224) and symmetric (004) reflections were used to determine the extent of relaxation and the concentration of Bi. XRD of all the samples showed clear Pendellosung fringes indicating smooth interfaces and no signs of relaxation, as expected. The narrow angular width of the diffracted alloy peak indicated a good crystalline uniformity over the 2 mm diffracted area. Interference fringes were fit with dynamical X-ray theory to extract the epilayer thicknesses. Bismuth composition was determined from the fits using the predicted GaBi lattice constant of 6.33 Å [17]. More details about the growth can be found in ref. [18].

To study the effect of Bi incorporation on the electron mobility and electron effective mass $GaAs_{1-x}Bi_x$ epilayers were grown on semi-insulating, undoped GaAs substrates and were n-doped with Si to concentrations of $\sim1\times10^{18}$ cm^{-3}. Before growing the epilayers, a 250 nm GaAs buffer layer was grown on the substrates and unintentional doping in the GaAs buffer layer is estimated to be $<1–2\times10^{14}$ cm^{-3}. Test structures with either eight-point Hall bar or van der Pauw geometries were fabricated from the epilayers and contacts were made using an annealed Au/Sn/Au alloy. p-Type GaAsBi epilayers were used to study the effect of Bi doping on the hole mobility and measurements were done on both carbon- and beryllium-doped epilayers. The p-doped epilayers were etched into a clover leaf pattern and electroplated Au was used as ohmic contacts. The carrier density and mobility data were obtained using the van der Pauw technique on as-grown samples.

Electrical transport measurements were carried out in the low electric field regime, i.e., in the "ohmic" regime that was identified from $I(V)$ conductivity measurements, from negative to positive bias, giving us a certainty about the linearity of the contact resistance in all the measured samples. For carrying out the Hall measurements, the samples were mounted on the cold finger of a closed cycle cryostat (Omniplex, Advanced Research Systems) placed between the poles of an electromagnet. Measurements were carried out at temperatures ranging from about 50 to 300 K. Electrical contacts between the cryostat and the Hall measurement system were made using triaxial cables and the measurements were performed in guarded-mode to avoid parasitic capacitance. Shubnikov de Haas (SdH) oscillations in the resistivity were measured using a superconducting magnet (Spectromag, Oxford Instruments) in magnetic fields up to 11 T, oriented either parallel or orthogonal to the current. Measurements were performed in He vapor at temperatures between 2 and 25 K.

For the temperature- and power-dependent photoluminescence (PL) measurements a frequency double Nd-YAG laser emitting at 532 nm was used as the excitation source. PL emission was dispersed by a Spex 270M spectrometer and detected using a cooled charge-coupled device (CCD). For time resolved PL (TRPL) measurements, frequency doubled pulses (~404 nm) from a Ti-Sapphire amplifier (Legend, Coherent) were used as the excitation source. The pulse duration was about 150 fs. PL was detected using an infrared photomultiplier tube (IRPMT).

8.3 Results and Discussion

8.3.1 Electron Mobility and Effective Mass

Measurement of the electron mobility and effective mass is important from a device application point of view. Figure 8.2 shows the room temperature (RT) electron Hall mobilities of the GaAsBi samples. For comparison, data for Se-doped $GaAs_{1-x}N_x$ films with carrier concentration of $5–7.63 \times 10^{18}$ cm^{-3} obtained from ref. [19] is also shown. For the n-type GaAsBi samples with Bi concentration ≤1.2 %, we obtained room temperature electron mobility in the range of ~2,100–2,300 cm^2/V s, which agrees reasonably well with values reported in literature for low-temperature-grown GaAs with carrier concentration of about 10^{18} cm^{-3} [20]. These direct transport measurements confirm that Bi does not degrade the electron mobility, in agreement with previous theoretical and experimental reports [9, 21]. In contrast, incorporation of N degrades the electron mobility even at very low concentrations (~0.1 %). N incorporation in GaAs causes a reduction in the electron mobility due to strong carrier scattering at the localized states due to isolated N and N-cluster states [22]. It has been theoretically shown that the isolated Bi impurity will form a resonant state in the valence band of GaAs because the 6*p* state of Bi is substantially higher than the 4*p* state of As, and Bi is likely to generate a potential trap for the holes. Thus, one might expect Bi alloying to affect the hole mobility rather than the electron mobility. Our observation confirms that the electron mobility is not affected substantially by the incorporation of Bi.

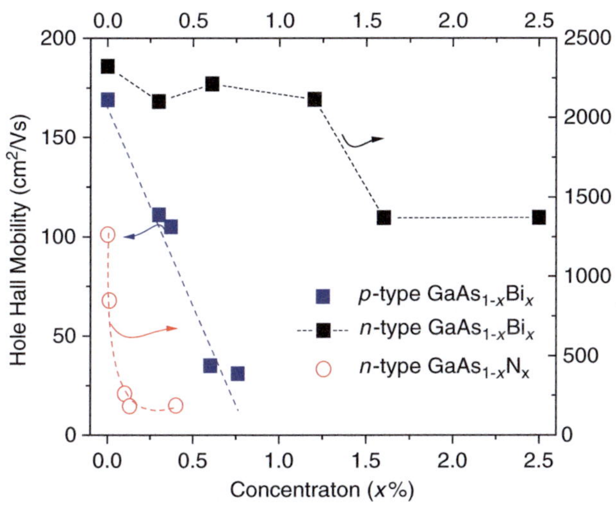

Fig. 8.2 Room temperature (RT) hole Hall mobility of n-type and p-type $GaAs_{1-x}Bi_x$ epilayers. For comparison electron Hall mobility data of n-type $GaAs_{1-x}N_x$ samples is also shown. The *dotted lines* are a guide to the eye. The $GaAs_{1-x}N_x$ data, obtained from ref. [19], is for Se-doped films with carrier concentration of $5–7.63 \times 10^{18}$ cm^{-3}

Samples with Bi concentration ≥1.6 % showed degradation in the electron mobility. One possible explanation for this is lattice relaxation occurring in the epilayers at higher Bi concentrations, probably resulting in higher dislocation densities. In the case of GaAsBi, there is a large mismatch between the isoelectronic impurity (Bi) and the host atom (As) it replaces, which can lead to global lattice expansion due to Bi incorporation. However, we have measured the asymmetric X-ray rocking curves of the epilayers containing Bi ≥ 1.6 % and we did not observe any lattice relaxation in the films. Hence, we believe that the degradation in the mobility is not due to lattice relaxation in the films but to some other effects such as Bi segregation or non-optimal growth. We were able to obtain samples with better RT electron Hall mobility (e.g., ~2,600 cm^2/V s at 2.5 % Bi incorporation) in the latter growth cycles. Hence the degradation is most likely due to non-optimal growth.

As discussed earlier, since Bi mainly perturbs the valence band, similar to the case of electron mobility, one would not expect the electron effective mass to be effected by the incorporation of Bi. For $GaAs_{1-x}Bi_x$, the sole reported carrier mass measurement is a recent result [23] in which the $GaAs_{1-x}Bi_x$ exciton reduced mass, μ_{exc}, was measured using the diamagnetic shift of bandedge PL over a wide range of bismuth concentrations, showing an unusual compositional behavior. The reduced mass μ_{exc} was observed to rise rapidly in the concentration range $x=0$–0.015, then oscillate around a value of $0.08\,m_0$ until $x=0.06$, followed by a drop that continued below the initial value of μ_{exc} in GaAs (m_0 is the free electron mass). This measurement probed the reduced effective mass μ_{exc}, rather than the electron or hole effective masses individually. Due to the relative size of those masses at $x=0$, the observed increase in μ_{exc} with $x>0$ could be attributed partially, but not entirely to changes in the hole effective mass. Thus from the diamagnetic shift results it was concluded that the electron effective mass increases up to a value $\geq 0.08\,m_0$ at low Bi concentrations, a result that would be surprising from $k{\cdot}p$ theory of the bands alone. The drop in the bandgap with increased Bi concentration should result in the electron effective mass decreasing, not increasing, due to the dominant term in the $k{\cdot}p$ perturbation of the electron mass at $k=0$. Incorporating the impurity state, predicted to lie near the valence band, into the calculation would be expected to modify the hole mass with less change to m_e^* [9, 12].

Addressing this discrepancy necessitates direct measurement of the electron effective mass in $GaAs_{1-x}Bi_x$. For the electron, m_e^* can be measured through magnetic field-dependent transport measurements in n-type samples wherein SdH oscillations appear in the resistivity when Landau levels in the density of states are swept past the Fermi energy. The technique is useful for materials with good mobility and therefore applicable to Bi concentrations below approximately 1 % [21, 24].

Figure 8.3 shows the results for the 0.88 % Bi sample. The oscillatory component of the resistivity, $(\rho-\rho_0)$, which was obtained from the resistivity data after subtracting a slowly varying non-oscillatory component, is plotted as function of the inverse magnetic field. It is fit very well by a single harmonic multiplied by the envelope $A(B,T)$ as shown. Following the approach of refs. [25, 26], we extracted the effective mass through the decay of this envelope with increasing temperature. Figure 8.4 shows the ratio of the oscillation envelopes for 2 and 20 K, fit to the function [27],

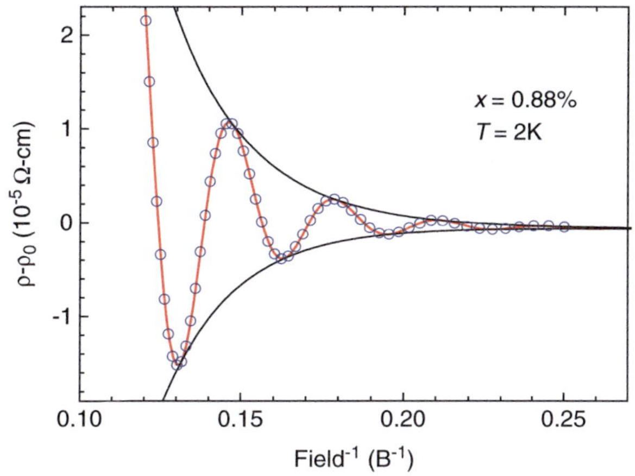

Fig. 8.3 Oscillatory component of the 2 K resistivity in $GaAs_{1-x}Bi_x$ (*open circles*) with $x = 0.88$ %, fit to a sinusoidal frequency with the envelope shown (*solid curves*). Resistivity was measured parallel to the current in a Hall bar and the magnetic field was normal to the sample surface

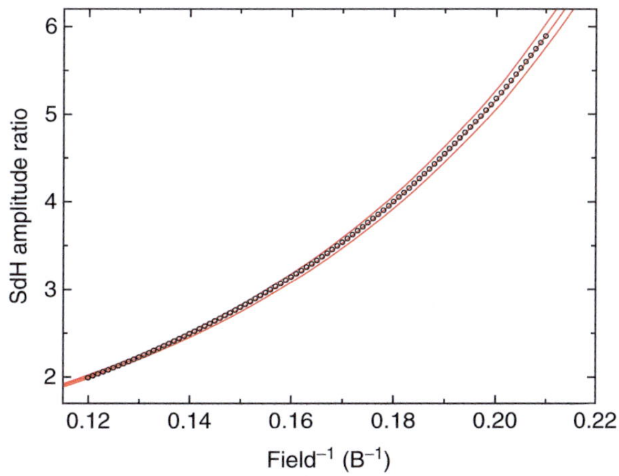

Fig. 8.4 Ratio of the resistivity oscillation amplitude from Fig. 8.3 at 2 K, and the same measurement at 20 K, showing its decay with temperature (points). The *solid curves* show the precision of the fit to the equation below. From upper to lower, they are evaluated for $m_e^* = 0.0627\, m_0$, $0.0622\, m_0$, and $0.0617\, m_0$

$$\frac{A(B,T_1)}{A(B,T_2)} = \frac{T_1 \sinh\left[\dfrac{2\ {}^2k_BT_2m_e^*}{\hbar eB}\right]}{T_2 \sinh\left[\dfrac{2\ {}^2k_BT_1m_e^*}{\hbar eB}\right]}$$

with m_e^* as an adjustable parameter. Results for m_e^* from samples corresponding to different Bi concentrations are plotted in Fig. 8.4 with error bars that were calculated from the repeated fitting at different fields and temperatures. Figure 8.5 includes a correction for the nonparabolicity of the conduction band [28] obtained using the measured carrier density and bandgaps estimated from refs. [1, 6] resulting in a reduction of all masses by an amount that is approximately constant, $0.0035\, m_0$. With this correction, the measured electron effective mass at 0 % Bi

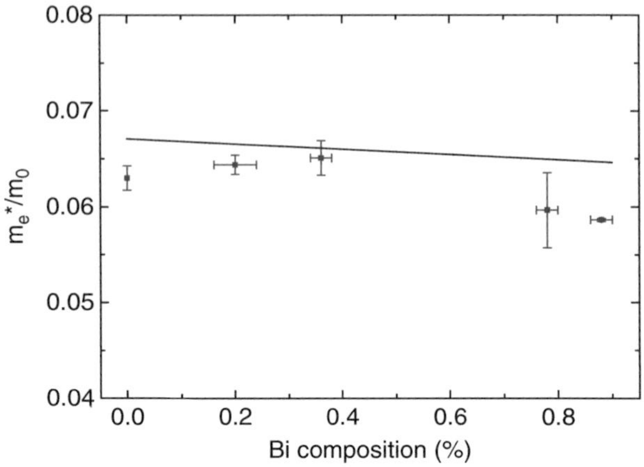

Fig. 8.5 Electron effective mass extracted from the temperature decay of the SdH oscillation envelope, for varying Bi concentration (points). *Solid curve* is a perturbation theory calculation that includes the upper valence band and spin-orbit band

differs slightly from the accepted value for GaAs [29]. We note, however, that all present samples were grown under low-temperature conditions optimized for $GaAs_{1-x}Bi_x$.

The solid curve in Fig. 8.5 shows a $k{\cdot}p$ calculation for m_e^* using only the perturbations from the valence and spin-orbit bands. Experimental values were used for the bandgaps [1, 2] and a constant P^2 was fit to the $x=0$ mass. As discussed above, neglecting the localized impurity states leads to a prediction for the concentration dependence of m_e^* that is much flatter than what is observed. The measured electron effective mass exhibits a gradual increase as a function of Bi concentration in the range of 0–0.4 % Bi and then a gradual decrease in the range 0.4–0.88 % Bi. The gradual rise in the electron effective mass can be attributed to the mixing of Bi impurity cluster states (as shown in Fig. 8.1) with the valence band and consequent coupling to the conduction band states via the valence band coupling parameter in the 12×12 Hamiltonian obtained using the $k{\cdot}p$ formalism [30]. This would correspond to the impurity band regime for $GaAs_{1-x}Bi_x$. A subsequent fall in the electron effective mass occurs from the rapid decrease in the bandgap with Bi composition due to coupling of the valence and conduction bands within the $k{\cdot}p$ formalism [31]. Parallel movements of the bandgap and electron effective mass are known in many ordinary III–V semiconductor alloys [32], and in the alloy regime, this effect predominates over the Bi clusters.

8.3.2 Hole Mobility and Concentration

Figure 8.2 also shows the RT hole Hall mobility of the samples. A decrease in the hole mobility with increasing x was observed for the p-type samples. As reported earlier and shown in Fig. 8.2, N incorporation in GaAs causes a reduction in the electron mobility due to strong carrier scattering at the localized states due to isolated N and N-cluster states [22]. Even though the degradation is not as drastic as in

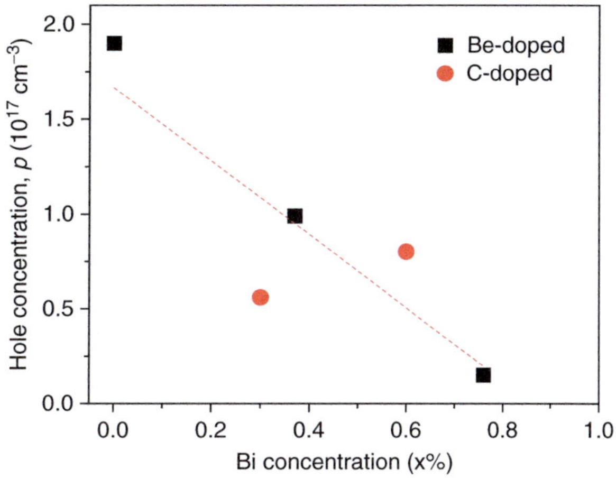

Fig. 8.6 The hole Hall concentration at RT for p-type $GaAs_{1-x}Bi_x$ epilayers. The *dotted line* is a guide to the eye

GaAsN, our observations show that Bi incorporation causes degradation of the hole mobility in GaAsBi. This can be understood by considering the fact that isolated Bi and pairs/clusters of Bi generate potential trap states for holes (schematically shown in Fig. 8.1) which can lead to a decrease in hole mobility due to scattering at these states. Our observation that Bi incorporation causes degradation of the hole mobility agrees well with another recent report [33]. Compared to our samples slightly better hole mobility over the entire range of Bi compositions was reported in ref. [33], but the same trend was observed; the hole mobility was found to decrease with increasing Bi content agreeing with our observations.

Figure 8.6 shows the variation of hole concentration at RT with Bi incorporation. A decrease in hole concentration with Bi incorporation was observed. The Be flux was maintained at the same level while growing the different samples and secondary ion mass spectroscopy (SIMS) measurements indicate that Be concentration is similar in all the samples ($\sim 2 \times 10^{17}$ cm^{-3}). This suggests that in the case of Be-doped samples the hole concentration decreases with increasing Bi concentration. SIMS measurements on C-doped samples indicate similar concentration of carbon in both the samples ($\sim 5 \times 10^{17}$ cm^{-3}). Comparing this with the observed hole concentration in Hall measurements (0.6–0.8×10^{17} cm^{-3}) of the C-doped samples, even though a monotonic decrease of the hole concentration with increasing Bi concentration cannot be unambiguously inferred, it is clear that Bi incorporation causes an overall decrease in the hole concentration. This suggests that the dopants are either compensated [34] or form complexes that passivate the dopants [35]. Similar behavior has been reported for GaInNAs alloys where the electron concentration was observed to decrease with increase in N concentration and it was suggested that N-related localized states or complexes were compensating the donors [34]. In the case of GaAsBi crystals grown using liquid encapsulation Czochralski method, it has been shown that a fraction (up to 10 %) of the total Bi content occupies the Ga site and acts as singly ionized double donors [36]. Using extended X-ray absorption fine-structure spectroscopy (EXAFS) measurements and density functional theory

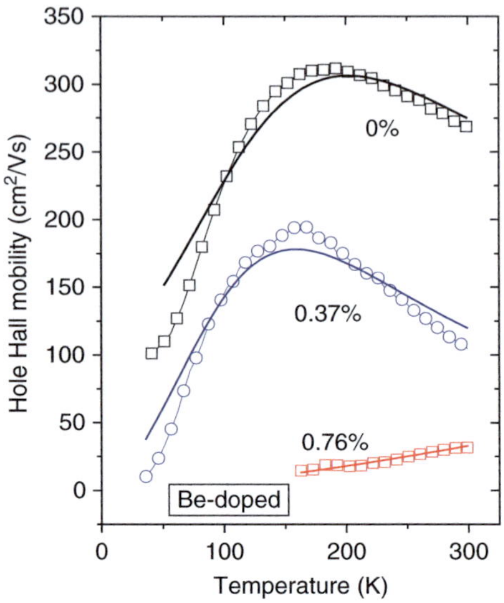

Fig. 8.7 The mobility for Be-doped $GaAs_{1-x}Bi_x$ samples as a function of temperature. The *solid lines* are fits to the data

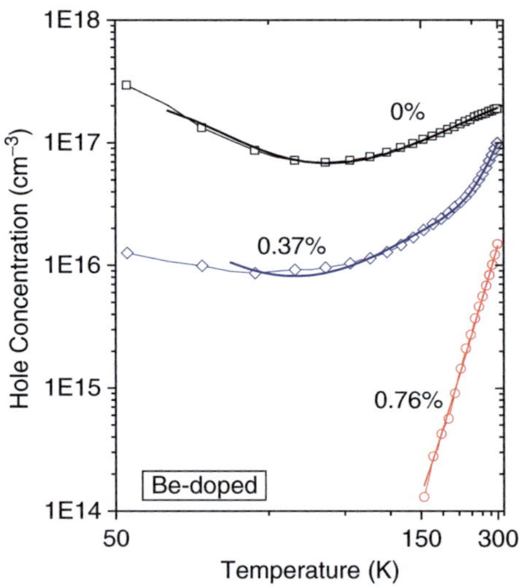

Fig. 8.8 The carrier concentration for Be-doped $GaAs_{1-x}Bi_x$ samples as a function of temperature. The *solid lines* are fits to the data

(DFT) calculations on MBE-grown GaAsBi, a maximum limit on the concentration of Bi_{Ga} heteroantisite was established [37]. It was shown that, for 1.2 % Bi concentration, the maximum possible Bi_{Ga} fraction was ~5 % (1.3×10^{19} cm^{-3}). Hence it is possible that these Bi_{Ga} heteroantisite defects result in the compensation of the acceptors thus reducing the effective hole concentration.

Figures 8.7 and 8.8 show the variable temperature Hall data obtained for Be-doped samples. Similar temperature dependence of the mobility and carrier concentration was observed for the C-doped samples also. The mobility of the sample with no Bismuth doping ($x=0$) shows a temperature dependence that is typical for p-type GaAs [38]. Due to the decrease in phonon scattering, the mobility increases as the temperature is decreased from 300 to 185 K. For $T<185$ K ionized impurity scattering becomes the dominant mobility-limiting scattering mechanism and mobility decreases as the temperature is decreased. The sample with 0.37 % Bi has a lower 300 K mobility compared to the sample with no Bismuth. The temperature dependence of the mobility is similar to the sample with no Bismuth—the mobility increases as the temperature is decreased up to 160 K, then decreases as the temperature is further decreased. The temperature dependence of the hole mobility can be analyzed assuming a typical temperature dependence for the various fundamental scattering mechanisms—for ionized impurity scattering, $\mu_{\mathrm{II}} \sim T^{3/2}$, and for lattice scattering, $\mu_{\mathrm{latt}} \sim T^{-3/2}$. If each scattering mechanism contributes simultaneously and independently, then the net mobility is given by the Matthiessen relation, $\mu=[1/\mu_{\mathrm{II}}+1/\mu_{\mathrm{latt}}]^{-1}$. The temperature dependence of the mobility for the sample with $x=0$ and 0.37 % can be estimated using ionized impurity and lattice scattering as shown in Figure 8.7. The temperature dependence of the hole mobility for the sample with 0.76 % Bi is quite different from the sample with 0.37 % Bi; the mobility decreases with decrease in temperature, which can be approximated by a $T^{3/2}$ dependence. This behavior can be explained using ionized impurity scattering-limited mobility ($\mu_{\mathrm{II}} \sim T^{3/2}$), which suggests that at higher Bi concentrations ionized impurity scattering is the dominant mobility-limiting mechanism.

Figure 8.8 shows the variable temperature carrier concentration data for Be-doped samples. For the sample with $x=0$, the hole concentration decreases as the temperature is lowered due to the freeze out of holes onto acceptor sites. The hole concentration reaches a minimum of $\sim 7\times 10^{16}$ cm^{-3} at ~92 K, and then appears to increase at lower temperatures indicating impurity band conduction. At this doping concentration ($\sim 10^{17}$ cm^{-3}) the dopants are close enough for their wave function to overlap and form an impurity band. This impurity band conduction is also evidenced in the variation of resistivity with temperature (not shown). Impurity band conduction is also seen in the sample with $x=0.37$ %. For the sample with $x=0.76$ % the carrier concentration decreases as the temperature is decreased and it does not show any impurity band formation. If an impurity band is formed, the effective hole concentration can be calculated as [39],

$$p_{\mathrm{h}} = \frac{(p_{\mathrm{v}} + \gamma p_{\mathrm{i}})^2}{(p_{\mathrm{v}} + \gamma^2 p_{\mathrm{i}})}$$

where p_{v} and p_{i} are the hole concentrations in the valence and impurity bands, respectively, and $\gamma=\mu_{\mathrm{i}}/\mu_{\mathrm{v}}$ is the ratio of the impurity and valence band mobility. The hole concentration in the valence band p_{v} can be calculated using the neutrality equation given by [40]

$$p_{\mathrm{v}} + N_{\mathrm{D}} = \sum_{i=1}^{k} \left[\frac{N_{\mathrm{Ai}}}{1+(p/\varnothing_{\mathrm{Ai}})} \right]$$

where $\varnothing_{Ai} = \frac{2[m_p^* k_B / 2\pi]^{3/2}}{\hbar^3} g_{Ai} T^{3/2} e^{-E_{Ai}/k_B T}$ and N_{Ai}, E_{Ai}, and g_{Ai} are the concentration, activation energy, and degeneracy factor, respectively, for the *i*th acceptor. N_{D} is the total donor concentration, m_p^* is the hole effective mass, and $\hbar$ is the reduced Planck's constant. The hole concentration in the impurity band is equal to

$$p_{\mathrm{i}} = \sum_{i=1}^{k} N_{\mathrm{Ai}} - N_{\mathrm{D}} - p_{\mathrm{v}}$$

The *p*-vs-*T* curve for the sample with $x=0$ was fitted using a single-acceptor model with $E_{\mathrm{A1}}=40$ meV. The data for the sample with $x=0.3$ and 0.37 % were fitted using a two-acceptor charge balance equation, and a single-acceptor model was sufficient for fitting the data for the samples with $x=0.6$ and 0.76 %. Table 8.1 gives the values of the parameters used for fitting the experimental data. Since there are a large number of free parameters while attempting to fit the experimental data to the above equation, it is difficult to find unique values for the activation energy and hole concentration. However, the fitting exercise gives a good estimate for the parameter values and shows that Bi incorporation introduces a deep trap level, E_{A2}, at about 167–224 meV above the valence band. The transport properties are mainly influenced by this level at higher Bi concentrations ($x \geq 0.6$ %). We believe that the E_{A1} level may be due to the acceptors. Localized states due to Bi incorporation have been reported by others using optical studies of dilute Bismides [41–43]. To correlate the trap levels observed in our transport measurements with those observed in optical studies we have done PL measurements which are described below.

Figure 8.9 shows the low-temperature (~10 K) PL data for Be-doped $GaAs_{1-x}Bi_x$ samples. For the sample with $x=0.37$ % it is possible to resolve two peaks at approximately 1.40 and 1.32 eV. For the sample with $x=0.76$ % the emission is dominated by the lowest energy peak at approximately 1.30 eV. The emission peaks show a red shift with increasing Bi concentration. If we assume that Bi causes a band gap reduction of 90 meV per % of Bi (see Fig. 8.12), and use the free excitonic bandgap of GaAs ($E_{\mathrm{FE}}^{\mathrm{GaAs}}$), we can calculate the binding energy corresponding to PL peaks, $E_{\mathrm{b}} = E_{\mathrm{FE}}^{\mathrm{GaAs}} - x \times 90\ \mathrm{meV} - E_{\mathrm{PL}}$. For the sample with $x=0.37$ % we obtained binding energies of about 80 and 160 meV and, for the sample with 0.76 % Bi, a binding energy of about 139 meV is obtained. Similar values were obtained for the C-doped samples. Hence PL data also suggest that Bi introduces a deep trap level (>100 meV) that qualitatively agrees with the transport data.

The concentration of the E_{A2} level obtained from fitting the carrier transport data is about two orders of magnitude smaller than the estimated concentration (0.6–1.5×10^{20} cm^{-3}) of isolated Bi atoms assuming a random distribution of Bi atoms in the lattice. This suggests that these deeper levels could be due to pairs or clusters of Bi, which are expected to be present in lower concentrations. The concentration of the clusters of Bi can be estimated as the product of the total Bi atoms ($2.2 \times 10^{22}\, x$ cm^{-3}) and the probability for each cluster [44]. The total concentration for different pair and triplet configurations is also given in Table 8.1. The total concentration was calculated assuming nearest neighbor (nn) and next nearest neighbor (nnn)

Table 8.1 Characteristics of the $GaAs_{1-x}Bi_x$ samples studied: parameters used for fitting the temperature-dependent transport data, the estimated concentration of Bi–Bi pairs and triplets, and the binding energy of the exciton corresponding to the PL peaks at ~10 K

	Analysis of transport data[a]					Calculated concentration of		Binding energy corresponding to the PL peaks, E_b (meV)	
Bi concentration (x%)	E_{A1} (meV)	E_{A2} (meV)	N_{A1} ($\times 10^{16}$ cm^{-3})	N_{A2} ($\times 10^{18}$ cm^{-3})	γ	Pairs ($\times 10^{18}$ cm^{-3})	Triplets ($\times 10^{18}$ cm^{-3})		
0	40	–	27	–	0.078	–	–	28	
0.3	56	182	1.3	1.25	0.0007	3.3	0.18	81	160
0.37	55	167	3.3	2	0.001	4.9	0.33	80	154
0.6	–	213	–	11	–	12	1.3	–	146
0.76	–	224	–	0.55	–	18	2.5	–	139

[a]$N_D=0$ was used for all the samples except for the sample with $x=0$ for which $N_D=1.5\times 10^{16}$ cm^{-3} gave a good fit to the experimental data

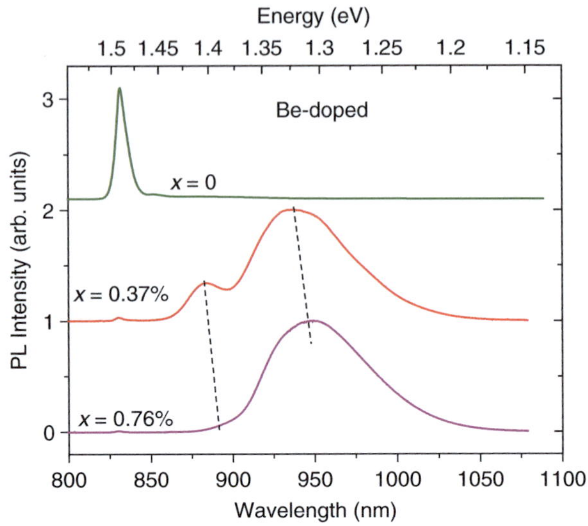

Fig. 8.9 Low-temperature (~10 K) PL data for Be-doped $GaAs_{1-x}Bi_x$ samples

interactions by following the method given in ref. [44]. There is reasonable agreement between the estimated pair concentration and the concentration of the E_{A2} level obtained from data fitting, except for the Be-doped sample with x=0.76 % sample. This shows that the E_{A2} level may be due to Bi–Bi pairs.

8.3.3 PL Measurements on Undoped Samples

To further understand the nature of the trap levels due to Bi, we performed PL measurements on a series of undoped epilayers in which the analysis of the data is slightly simplified since there are no acceptors. These measurements are described in this section. Figure 8.10 shows low-temperature (~10 K) PL for the undoped samples with different Bi concentrations, 0.05 $\% \leq x \leq 1.4$ %. For samples with $x \leq 0.25$ % several emission peaks labeled C1–C7 are seen and the spectrum is similar to that reported by Francoeur et al. [41]. In ref. [41] it was proposed that C1, C2 may be due to clusters of Bi and C3, C6, C7 were assigned as the phonon replicas of C1/C2. However, some of the peaks had to be excluded (C5 in ref. [41], N1 in our case) while assigning the peaks as phonon replicas and there is only an approximate agreement between the peak energy separation and the LO phonon energy. We believe that these emission peaks may be due to Bi–Bi pairs with different lattice spacing, similar to the case of N–N pairs in GaPN [45, 46], with the lower energy peaks corresponding to the pairs with the smallest pair separation. The following observations support this assignment:

1. As the Bi concentration is increased the lower energy peaks dominate the PL spectrum, because the probability of the closest pairs increases.

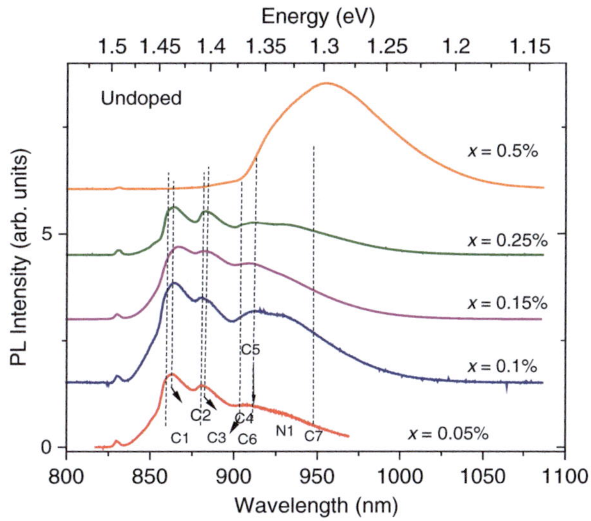

Fig. 8.10 Low-temperature (~10 K) PL data for undoped $GaAs_{1-x}Bi_x$ samples

2. As shown in Fig. 8.11a, when the temperature is increased, the intensity of the higher energy peaks reduces compared to the lower energy peaks. This is because as the temperature is increased the exciton which is bound to the center with a small binding energy is thermally released. Also, for the centers with small binding energy there will be opportunity for the exciton to tunnel to a lower energy state (closer pairs) and decay there [47]. This latter process also explains why the higher energy peaks are missing in the concentrated samples (x>0.25 %), since in the concentrated samples the probability of closer pairs and hence the tunneling probability to lower energy pairs is higher. As shown in Fig. 8.11a as the temperature is increased, the emission from the higher energy peaks quenches and eventually only a broad peak remains.
3. As shown in Fig. 8.11b, when the excitation intensity is increased the emission shifts to high energy peaks. This is because as the excitation intensity is increased the emission from the closest pairs saturates and recombination from the farthest pairs dominates.

These observations, the temperature and power dependence of the PL intensity of various peaks, cannot be explained by considering C3–C7 as phonon replicas of C1/C2. Comparing the PL spectra of undoped samples and the transport/PL measurements of p-doped samples discussed earlier suggests that each of the emission peaks is due to trap levels due to Bi–Bi pairs with different pair spacing. Hence it is possible to conclude that Bi incorporation in GaAs results in the formation of several trap states above the valence band as schematically shown in Fig. 8.1. These states act as scattering centers for holes and cause a reduction in the mobility.

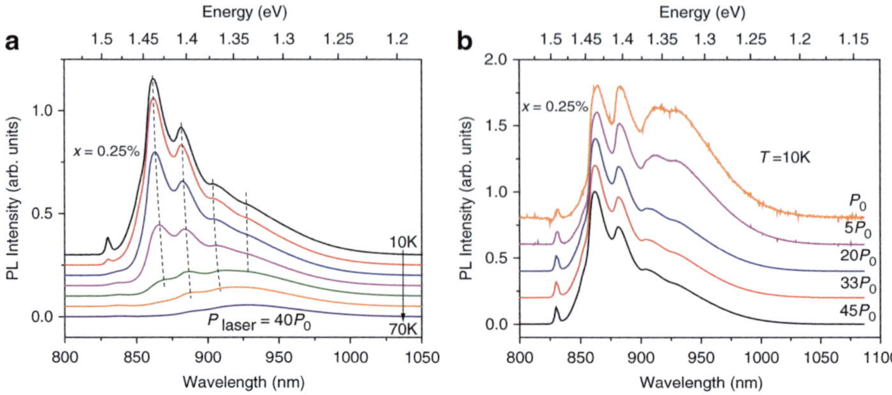

Fig. 8.11 (**a**) PL at different temperatures for $GaAs_{1-x}Bi_x$ sample with $x=0.25$ % at a laser excitation power of 1.2 mW. (**b**) PL curves for the sample with $x=0.25$ % at 10 K for different excitation powers. P_0 refers to a laser excitation power of 30 μW

Fig. 8.12 PL peak energy (*open squares*) as well as the bandgap, E_g, at 70 K (*closed squares*) as a function of Bi concentration, x. E_g was obtained using modulated reflectance measurements

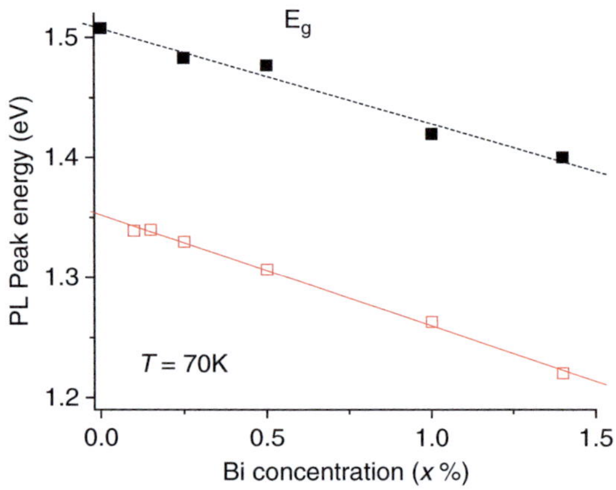

Figure 8.12 shows the energy of PL peak and band gap, E_g, as a function of the Bi concentration at 70 K for samples with 0.1 % $<x<$ 1.4 %. E_g was obtained using modulated reflectance measurements. From a linear fit to the PL peak data we obtain a red shift of 92 meV per % of Bi, which is close to the band gap reduction (90 meV per % of Bi) due to Bi incorporation. This shows that the emission from the trap levels closely follows the band edge.

We performed TRPL measurements to compare the lifetime (τ) and energy transfer processes in GaAsBi with GaPN [48]. Figure 8.13 shows the TRPL data obtained in a 10 nm window centered around 865, 884, 912, and 939 nm corresponding to C1/C2, C3/C4, N1, and C7 peaks, respectively. An exponential fit to the second half of the TRPL data was used to estimate the lifetime. Compared to the usual excitonic lifetime in GaAs ($\leq$1 ns), these emission peaks show long lifetime (~37–86 ns). Similar to the case of N–N lines in GaPN [48], as we move towards lower energy

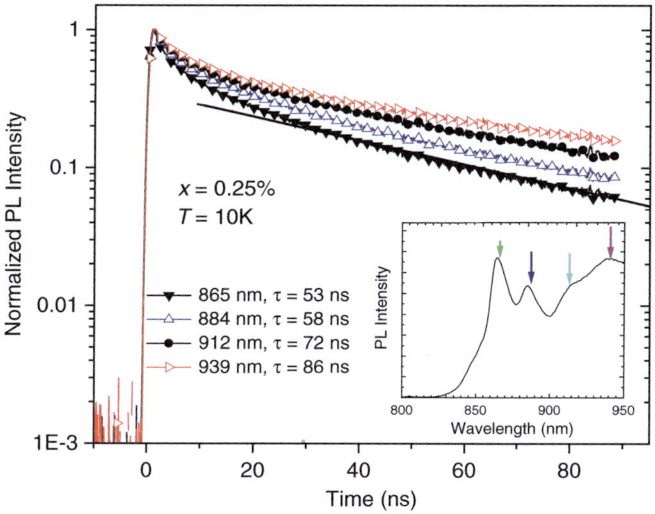

Fig. 8.13 TRPL data for the sample with $x=0.25$ % at 10 K obtained in a 10 nm window centered around 865, 884, 912, and 939 nm. The *inset* shows the PL obtained with 404 nm pulsed laser excitation

states, τ increases. However, in contrast to N–N pairs in GaPN, the rise time is similar at all the emission peaks (Bi–Bi pair states) and is very short (~100 ps), which indicates very efficient energy transfer between the levels.

8.4 Conclusions

In conclusion, we have studied the effect of Bi incorporation on the carrier mobilities and electron effective mass in the dilute Bismide alloy, $GaAs_{1-x}Bi_x$, using electrical transport and photoluminescence techniques. We observed no significant degradation in the electron Hall mobility with Bi incorporation in GaAs, up to a concentration of 1.2 %. At higher Bi concentration (≥1.6 %) some degradation of the electron mobility was observed, possibly due to non-optimal growth. Our measurements show that the hole Hall mobility decreases with increasing Bi concentration. Magnetic field- and temperature-dependent resistivity measurements on n-type $GaAs_{1-x}Bi_x$ showed clear SdH oscillations up to a concentration of 0.88 %. The measured electron effective mass exhibits a gradual increase up to a concentration of 0.4 % Bi and then a gradual decrease. Accounting for the giant bandgap and spin-orbit bowing, the measured changes in the effective mass are in qualitative agreement with perturbation theory applied to these band changes. Our observation can be understood by considering the fact that Bi mainly perturbs the VB due to the formation of Bi isolated state, which is possibly resonant in the VB and several Bi–Bi pair/cluster states. Electron mobility and effective mass is not affected much by the incorporation of Bi and hole mobility shows a decrease with increasing Bi concentration due to scattering at the isolated Bi and Bi–Bi pair/cluster states. Analysis of the temperature-dependent Hall transport data of p-type $GaAs_{1-x}Bi_x$ epilayers along with low-temperature PL measurements of p-doped and undoped epilayers suggests that Bi incorporation results in the formation of several trap levels above the valence band which we attribute to Bi–Bi pair states.

Acknowledgements This work was supported by the DOE/SC/BES/DMS under contract no. DE-AC36-08GO28308 with the National Renewable Energy Laboratory.

References

1. Francoeur, S., Seong, M.-J., Mascarenhas, A., Tixier, S., Adamcyk, M., Tiedje, T.: Band gap of $GaAs_{1-x}Bi_x$, 0<x<3.6%. Appl. Phys. Lett. **82**, 3874 (2003)
2. Fluegel, B., Francoeur, S., Mascarenhas, A., Tixier, S., Young, E.C., Tiedje, T.: Giant Spin-Orbit Bowing in $GaAs_{1-x}Bi_x$. Phys. Rev. Lett. **97**, 067205 (2006)
3. Levander et al.: GaN1−xBix: Extremely mismatched semiconductor alloys. Appl. Phys. Lett. **97**, 141919 (2010)
4. Petropoulos, J.P., et al.: Optical and electrical characterization of InGaBiAs for use as a mid-infrared optoelectronic material. Appl. Phys. Lett. **99**, 031110 (2011)
5. Das, S.K., et al.: Near infrared photoluminescence observed in dilute GaSbBi alloys grown by liquid phase epitaxy. Infrared Phys. Technol. **55**, 156 (2012)
6. Yoshida, J., Kita, T., Wada, O., Oe, K.: Temperature Dependence of GaAs1-xBix Band Gap Studied by Photoreflectance Spectroscopy. Jpn. J. Appl. Phys. **42**, 371 (2003)
7. Dimroth, F.: High-efficiency solar cells from III–V compound semiconductors. Phys. Status Solidi C **3**, 373 (2006)
8. Asbeck, P.M., Welty, R.J., Tu, C.W., Xin, H.P., Welser, R.E.: Heterojunction bipolar transistors implemented with GaInNAs materials. Semicond. Sci. Techonol. **17**, 898 (2002)
9. Zhang, Y., Mascarenhas, A., Wang, L.W.: Similar and dissimilar aspects of III–V semiconductors containing Bi versus N. Phys. Rev. B **71**, 155201 (2005)
10. Mascarenhas, A., Kini, R.N., Zhang, Y., France, R., Ptak, A.J.: Comparison of the dilute Bismide and Nitride alloys GaAsBi and GaAsN. Phys. Status Solidi B **246**, 504 (2008)
11. Wolford, D.J., Bradley, J.A., Fry, K., Thompson, J., King, H.E.: In: The nitrogen isoelectronic trap in GaAs. Chadi, J.D., Harrison, W.A. (eds.) Proceedings of the 17th International Conference on the Physics of Semiconductors, p. 627. Springer, New York (1984)
12. Janotti, A., Wei, S.-H., Zhang, S.B.: Theoretical study of the effects of isovalent coalloying of Bi and N in GaAs. Phys. Rev. B **65**, 115203 (2002)
13. Trumbore, F.A., Gershenzon, M., Thomas, D.G.: Luminescence due to the isoelectronic substitution of bismuth for phosphorus in gallium phosphide. Appl. Phys. Lett. **9**, 4 (1966)
14. Wei, S.-H., Zunger, A.: Calculated natural band offsets of all II–VI and III–V semiconductors: Chemical trends and the role of cation d orbitals, Appl. Phys. Lett. **72**, 2011 (1998)
15. Endicott, J., Patanè, A., Ibáñez, J., Eaves, L., Bissiri, M., Hopkinson, M., Airey, R., Hill, G.: Magnetotunneling spectroscopy of dilute Ga(AsN) quantum wells. Phys. Rev. Lett. **91**, 126802 (2003)
16. Imhof, S., et al.: Luminescence dynamics in Ga(AsBi). Appl. Phys. Lett. **98**, 161104 (2011)
17. Tixier, S., Adamcyk, M., Tiedje, T., Francoeur, S., Mascarenhas, A., Wei, P., Schiettekatte, F.: Molecular beam epitaxy growth of $GaAs_{1-x}Bi_x$. Appl. Phys. Lett. **82**, 2245 (2003)
18. France, R., Beaton, D.A., Alberi, K., Simon, J., Mascarenhas, A., Jiang, C.-S.: Kinetically limited growth of GaAsBi by molecular-beam epitaxy. J. Cryst. Growth **338**, 107 (2012)

19. Young, D.L., Geisz, J.F., Coutts, T.J.: Nitrogen-induced decrease of the electron effective mass in $GaAs_{1-x}N_x$ thin films measured by thermomagnetic transport phenomena. Appl. Phys. Lett. **82**, 1236 (2003)
20. Rhode, D.L.: Low field electron transport. In: Willardson, R.K., Beer, A.C. (eds.) Semiconductors and Semimetals, vol. 10. Academic, New York (1975)
21. Cooke, D.G., Hegmann, F.A., Young, E.C., Tiedje, T.: Electron mobility in dilute GaAs bismide and nitride alloys measured by time-resolved terahertz spectroscopy. Appl. Phys. Lett. **89**, 122103 (2006)
22. Fahy, S., Lindsay, A., Ouerdane, H., O'Reilly, E.P.: Alloy scattering of n-type carriers in GaN_xAs_{1-x}. Phys. Rev. B **74**, 035203 (2006)
23. Pettinari, G., Polimeni, A., Blokland, J.H., Trotta, R., Christianen, P.C.M., Capizzi, M., Maan, J.C., Lu, X., Young, E.C., Tiedje, T.: Compositional dependence of the exciton reduced mass in $GaAs_{1-x}Bi_x$ (x=0–10%). Phys. Rev. B **81**, 235211 (2010)
24. Kini, R.N., Bhusal, L., Ptak, A.J., France, R., Mascarenhas, A.: Electron Hall mobility in GaAsBi. J. Appl. Phys. **106**, 043705 (2009)
25. Shaw, R.W., Hill, D.E.: Shubnikov-de haas oscillations in n-type GaAs. Phys. Rev B **1**, 658 (1970)
26. Roth, L.M., Argyres, P.N.: Magnetic quantum effects. In: Willardson, R.K, Beer, A.C. (eds.) Semiconductors and Semimetals, vol. 1, pp. 159–202. Academic, New York (1966)
27. Hermann, C., Weisbuch, C.: k·p perturbation theory in III–V compounds and alloys: a reexamination. Phys. Rev. B **15**, 823 (1977)
28. Blakemore, J.S.: Semiconducting and other major properties of gallium arsenide. J. Appl. Phys. **53**, R159 (1982)
29. Adachi, S.: GaAs, AlAs, and $AlxGa_{1-x}As$: Material parameters for use in research and device applications. J. Appl. Phys. **58**, R1 (1985)
30. Alberi, K., Wu, J., Walukiewicz, W., Yu, K.M., Dubon, O.D., Watkins, S.P., Wang, C.X., Liu, X., Cho, Y.-J., Furdyna, J.: Valence-band anticrossing in mismatched III–V semiconductor alloys. Phys. Rev. B **75**, 045203 (2007)
31. Shokhovets, S., Ambacher, O., Gobsch, G.: Conduction-band dispersion relation and electron effective mass in III–V and II–VI zinc-blende semiconductors. Phys. Rev. B **76**, 125203 (2007)
32. Berolo, O., Woolley, J.C., Van Vechten, J.A.: Effect of disorder on the conduction-band effective mass, valence-band spin-orbit splitting, and the direct band gap in III–V alloys. Phys. Rev. B **8**, 3794 (1973)
33. Beaton, D.A., Lewis, R.B., Masnadi-Shirazi, M., Tiedje, T.: Temperature dependence of hole mobility in $GaAs_{1-x}Bi_x$ alloys. J. Appl. Phys. **108**, 083708 (2010)
34. Xin, H.P., Tu, C.W., Geva, M.: Effects of hydrogen on doping of GaInNAs grown by gas-source molecular beam epitaxy. J. Vac. Sci. Technol. B **18**, 1476 (2000)
35. Li, W., Pessa, M., Toivonen, J., Lipsanen, H.: Doping and carrier transport in $Ga1\text{-}3_xIn3_xN_xAs_{1-x}$ alloys. Phys. Rev. B **64**, 113308 (2001)
36. Kunzer, M., Jost, W., Kaufmann, U., Hobgood, H.M., Thomas, R.N.: Identification of the BiGa heteroantisite defect in GaAs:Bi. Phys. Rev. B **48**, 4437 (1993)
37. Ciatto, G., Alippi, P., Amore Bonapasta, A., Tiedje, T.: How much room for BiGa heteroantisites in $GaAs_{1-x}Bi_x$? Appl. Phys. Lett. **99**, 141912 (2011)
38. Stockman, S.A., Höfler, G.E., Baillargeon, J.N., Hsieh, K.C., Cheng, K.Y., Stillman, G.E.: Characterization of heavily carbon-doped GaAs grown by metalorganic chemical vapor deposition and metalorganic molecular beam epitaxy. J. Appl. Phys. **72**, 981 (1992)
39. Neumann, H.: Influence of impurity band conduction on the electrical characteristics of p-type GaAs. Cryst. Res. Technol. **23**, 1377 (1988)
40. Semiconductor Statistics, J.S.: Blakemore. Pergamon Press, New York (1960)
41. Francoeur, S., Tixier, S., Young, E., Tiedje, T., Mascarenhas, A.: Bi isoelectronic impurities in GaAs. Phys. Rev. B **77**, 085209 (2008)

42. Pettinari, G., Polimeni, A., Capizzi, M., Blokland, J.H., Christianen, P.C.M., Maan, J.C., Young, E.C., Tiedje, T.: Influence of bismuth incorporation on the valence and conduction band edges of $GaAs_{1-x}Bi_x$. Appl. Phys. Lett. **92**, 262105 (2008)
43. Kudrawiec, R., et al.: Carrier localization in GaBiAs probed by photomodulated transmittance and photoluminescence. J. Appl. Phys. **106**, 023518 (2009)
44. Kreitman, M.M., Barnett, D.L.: Probability tables for clusters of foreign atoms in simple lattices assuming next-nearest-neighbor interactions. J. Chem. Phys. **43**, 364 (1965)
45. Thomas, D.G., Hopfield, J.J., Frosch, C.J.: Isoelectronic traps due to nitrogen in gallium phosphide. Phys. Rev. Lett. **15**, 857 (1965)
46. Thomas, D.G., Hopfield, J.J.: Isoelectronic traps due to nitrogen in gallium phosphide. Phys. Rev. **150**, 680 (1966)
47. Wiesner, P.J., Street, R.A., Wolf, H.D.: Exciton energy transfer in GaP: N. Phys. Rev. Lett. **35**, 1366 (1975)
48. Mariette, H.: Picosecond spectroscopy in III–V compounds and alloy semiconductors. Phys. B **146B**, 286 (1987)

Chapter 9
Localized States in GaAsBi and GaAs/GaAsBi Heterostructures

Masahiro Yoshimoto and Takuma Fuyuki

Abstract Deep- and shallow-level defects in device-quality $GaAs_{1-x}Bi_x$ ($x \leq 10.9\%$) are investigated. Despite low-temperature growth, $GaAs_{1-x}Bi_x$ emits intense band-edge photoluminescence, and $GaAs_{0.975}Bi_{0.025}$ shows lasing operation by optical pumping. The deep-level trap density is suppressed on the order of 10^{15} cm^{-3} because of a surfactant effect of the Bi atoms. The Bi-induced localized states generated by the interaction between spatially localized Bi states and the valence band of GaAs are continuously located up to ~90 meV from the valence band with a density of $\sim 1 \times 10^{17}$ cm^{-3}. Despite concerns regarding the degradation of the hole mobility due to scattering at these Bi-induced localized states, the p-type doping masks the contribution of the Bi-induced states to the hole mobility, and a high hole mobility of 200 cm^2 V^{-1} s^{-1} is achieved. By characterizing the superlattices, (Al)GaAs/$GaAs_{1-x}Bi_x$ heterointerfaces have been proven to be smooth without distinct segregation and stable up to 700 °C. While the interface state density of $\sim 9 \times 10^{11}$ cm^{-2} eV^{-1} in a GaAs/$GaAs_{1-x}Bi_x$ heterointerface cannot be reduced by annealing, it can be reduced by half by the insertion of a Bi graded layer into the heterointerface, presumably due to the mitigation of the differences in the metallic $GaAs_{1-x}Bi_x$ and nonmetallic GaAs surfaces.

9.1 Introduction

Bismuth-containing III–V semiconductors such as $GaAs_{1-x}Bi_x$ [1–3], $GaN_yAs_{1-x-y}Bi_x$ [4, 5], and $Ga_{1-y}In_yAs_{1-x}Bi_x$ [6, 7] exhibit unusual properties owing to the large bowing of the band gap caused by Bi incorporation [3, 4, 7–9], a reduction in the

M. Yoshimoto (✉) • T. Fuyuki
Department of Electronics, Kyoto Institute of Technology,
Matsugasaki, Sakyo, Kyoto 606-8585, Japan
e-mail: yoshimot@kit.ac.jp; d1821006@edu.kit.ac.jp

H. Li and Z.M. Wang (eds.), *Bismuth-Containing Compounds*, Springer Series in Materials Science 186, DOI 10.1007/978-1-4614-8121-8_9,

temperature coefficient of the band gap [1, 4, 8, 9], and large spin-orbit splitting [10]. These dilute bismuthide III–V alloys are promising for laser diodes (LDs) with temperature-insensitive oscillation wavelengths [1], LDs with a reduction of the Auger recombination [11], solar cells [12], and long-wavelength photodetectors [7]. The recent achievement of laser oscillation for $GaAs_{0.975}Bi_{0.025}$ containing a temperature-insensitive oscillation wavelength [13] proves that metastable $GaAs_{1-x}Bi_x$ has sufficient functional properties and can be used in optoelectronic devices.

Similar to $GaAs_{1-x}Bi_x$, $GaN_{1-x}As_x$ and related dilute nitride III–V alloys show large band-gap bowing because of N incorporation; however, N incorporation is usually accompanied by the deterioration of the optical and electrical properties. The degradation of the electron mobility [14] is explained by the strong alloy scattering related to its electronic structure, in which spatially localized N states interact with the conduction band of GaAs [15]. The broad photoluminescence (PL) spectra emitted from $GaN_{1-x}As_x$ is, in fact, explained by the disorder of the conduction band of $GaN_{1-x}As_x$ [16]. In contrast, the Bi levels in $GaAs_{1-x}Bi_x$ cause the perturbation of the valence band of GaAs [17]. The broad PL spectra [18–22] and the degradation of hole mobility for $GaAs_{1-x}Bi_x$ with an increase in the GaBi molar fraction [23, 24] are explained in a manner similar to that for the degradation of the electron mobility and broad PL spectra in GaN_xAs_{1-x}.

It should be noted that the growth mode of $GaAs_{1-x}Bi_x$ is distinguished from that of GaN_xAs_{1-x} by a surfactant-like effect of the Bi atoms. It has been proposed that the Bi atoms work as a surfactant to enhance the migration of atoms during growth, as in other III–Vs grown at high temperatures [25, 26]. Although a portion of the Bi atoms incorporated into $GaAs_{1-x}Bi_x$ generate localized states, $GaAs_{1-x}Bi_x$ grown under the surfactant effect has the potential to show non-degraded optical and electrical properties.

In this chapter, after reviewing the growth conditions and optical properties of $GaAs_{1-x}Bi_x$, the Bi-induced localized states and deep-level states in $GaAs_{1-x}Bi_x$ grown layers are described. In the following section, the fabrication of heterostructures containing dilute bismuthide III–V alloys imperative for the realization of optoelectronic devices with $GaAs_{1-x}Bi_x$ is described. Then, the interface states of $GaAs/GaAs_{1-x}Bi_x$ heterostructures and the reduction of the interface states are discussed.

9.2 Structural and Optical Properties of $GaAs_{1-x}Bi_x$

$GaAs_{1-x}Bi_x$ was grown on GaAs (001) using molecular beam epitaxy (MBE) with solid sources of Ga, As, and Bi [2, 27]. Compared to the conventional III–V semiconductors such as GaAs, InP, and their alloys, the growth window is very narrow. The following are key factors for obtaining $GaAs_{1-x}Bi_x$ and related alloys, as summarized in ref. [9]: (1) low-temperature growth (<400 °C) to avoid immiscibility and the desorption of Bi; (2) the precise control of the As flux, the counterpart of the group-V atoms, in the limited range on the brink of As shortage on the growth

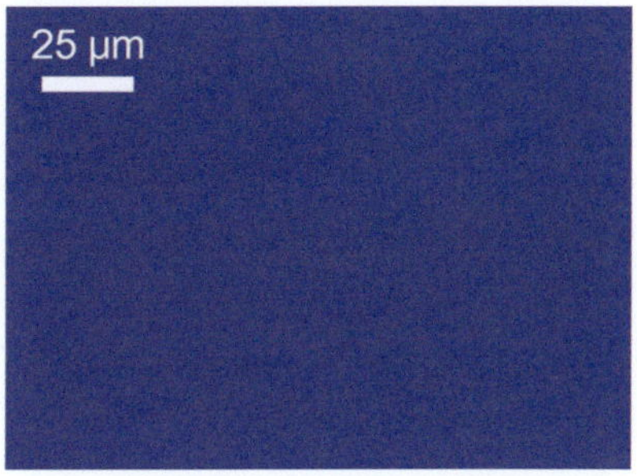

Fig. 9.1 Nomarski interference microscopic images of the surface of $GaAs_{1-x}Bi_x$

surface; and (3) scrupulous attention to the control of the Bi supply in response to the source supply of Ga and As and the growth temperature in order to avoid the formation of Bi droplets.

Under growth conditions in which Bi atoms are not incorporated into the grown layer, i.e., at a growth temperature greater than 400 °C, Bi atoms behave as a surfactant during the growth of III–V semiconductor alloys [25, 26]. The Bi surfactant provides a smoother surface/interface and higher PL intensity. The three extreme growth conditions for $GaAs_{1-x}Bi_x$ listed above are also mitigated by the surfactant-like effect of Bi atoms at growth temperatures below 400 °C [9]. On the basis of the results of ab initio calculations, photoemission, and scanning tunneling microscopy, the surfactant-like effects at low temperatures are proposed to be originated from the unusual metallic nature of the Bi-induced (2×1) reconstruction on a (100)-oriented GaAs surface [28]. The (2×1) reconstruction is usually observed on the $GaAs_{1-x}Bi_x$ growing surface, and it is not explained by the conventional electron counting model [29] that does explain the surface reconstruction of III–V semiconductors such as (2×4) reconstruction. The metallic nature of the Bi-induced (2×1) reconstruction possibly enhances the surface diffusion of adatoms [30, 31], as reported in a study of III–V nitrides [31]. The Bi-induced (2×4)α reconstruction observed at 300–400 °C during a surface analysis has also been proposed as the potential origin of the enhancement in the surface diffusion of adatoms [28, 32]. The smoother surface of (2×4)α reconstruction compared to (2×4)β possibly causes the enhancement in the surface diffusion [30]. The growth mode of $GaAs_{1-x}Bi_x$ is distinguished from that of GaN_xAs_{1-x} by this surfactant-like effect of the Bi atoms. While localized states such as Bi and N clusters have been reported for $GaAs_{1-x}Bi_x$ [21] and GaN_xAs_{1-x} [33], respectively, $GaAs_{1-x}Bi_x$ grown under the surfactant effect has the potential to show non-degraded optical and electrical properties, even for $GaAs_{1-x}Bi_x$ with a large GaBi molar fraction. The surfactant-like effect of the Bi atoms contributes to the improved quality of $GaAs_{1-x}Bi_x$ under extreme growth conditions that are different from the conventional conditions used for high-quality III–V semiconductors.

A mirror surface of $GaAs_{1-x}Bi_x$ can be obtained without Bi droplets, as shown in Fig. 9.1, by controlling the Bi supply with respect to the other source supplies and the growth temperature [9]. Even though $GaAs_{1-x}Bi_x$ was grown at a growth temperature below 400 °C, the $GaAs_{1-x}Bi_x$ grown layers with x up to 10.9 % emitted

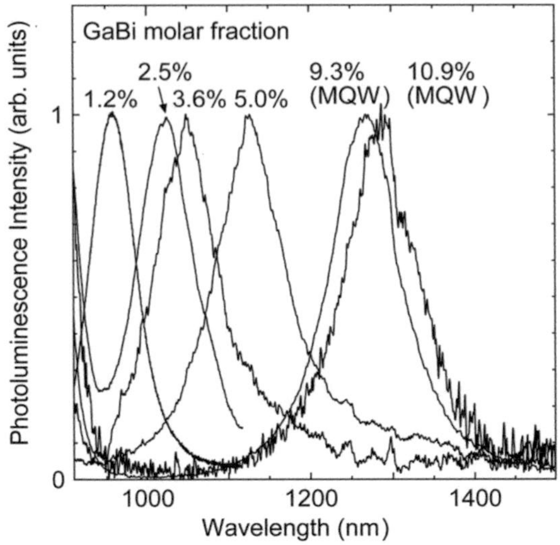

Fig. 9.2 Room temperature PL peaks of $GaAs_{1-x}Bi_x$ (1.2 % ≤ x ≤ 5.0 %), together with $GaAs_{1-x}Bi_x$/GaAs multiquantum wells (x = 9.3 and 10.9 %)

intense band-edge PL [4, 34]. As shown in Fig. 9.2, the PL peak energy decreases by increasing the GaBi molar fraction. The reduction of the PL peak energy as a function of the GaBi molar fraction can be approximated as −62 meV/%Bi for $GaAs_{1-x}Bi_x$ (x < 5 %) [35]. In addition, a larger constant of −88 meV/%Bi has been reported [3]. It should be noted that the nonexistence of GaBi binary alloys causes an inaccuracy in the estimation of the Bi molar fraction x in $GaAs_{1-x}Bi_x$ using Vegard's law. To date, a discrepancy exists in the estimated GaBi lattice constant. The lattice constant of 6.23 Å used in this study was derived from calibration experiments using X-ray diffraction (XRD) and Rutherford backscattering [36]. The lattice constant of GaBi (6.33 Å) estimated by other studies [18, 27] underestimates the GaBi molar fraction by a factor of ~17 % compared to the above-mentioned relationship. The discrepancy in the coefficients of −62 and −88 meV/%Bi is primarily caused by the discrepancy in the estimation of the GaBi molar fraction. Recently, the relationship between the band gap and the GaBi molar fraction (x ≤ 10.6 %) with a bowing parameter has been reported [18]. In addition, by adjusting the growth conditions, the lasing operation of $GaAs_{1-x}Bi_x$ with a reduced temperature coefficient of the lasing wavelength was achieved by optical pumping (Fig. 9.3) [13].

A reduction in the temperature coefficient of the PL emission energy was also observed in $GaAs_{1-x}Bi_x$ grown using metalorganic vapor phase epitaxy [1] and MBE [13, 19]. The temperature coefficient was reported to be between −0.15 meV/K (x = 2.5 %) [13] and −0.23 meV/K (x = 1.3 %) [19]. In terms of the temperature coefficient α of the band gap E_g in the Varshni equation $E_g(T) = E_0 - \alpha T^2/(T+\beta)$, α values of −0.39 and −0.45 meV/K (x = 2.2–6.4 %) [20] and −0.274 meV/K (x = 4–5 %) [21] have been reported, while the value of α for the GaAs band gap is −0.54 meV/K [37]. Here E_0 is the band gap at 0 K, and β is a fitting parameter.

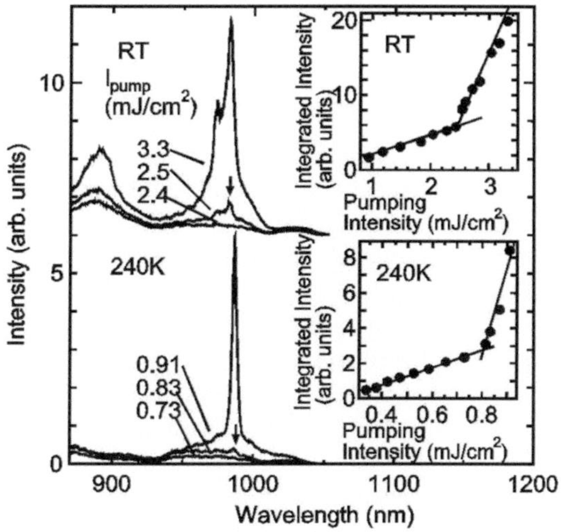

Fig. 9.3 Emission spectra from a $GaAs_{0.975}Bi_{0.025}$/GaAs laser chip at RT and 240 K. *Inset*: Integrated emission intensity from the laser chip as a function of the pumping density at RT and 240 K. © 2010 The Japan Society of Applied Physics [13]

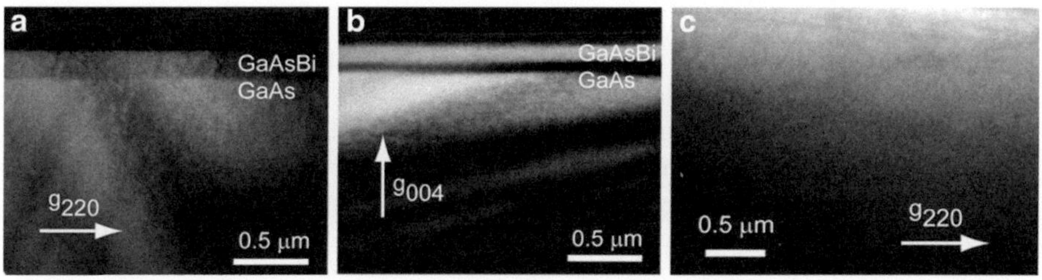

Fig. 9.4 (**a**) (110) bright-field cross-sectional TEM image taken under the 220 reflection, (**b**) (110) bright-field cross-sectional TEM image taken under the 004 reflection, and (**c**) bright-field plan-view TEM image taken under the 220 reflection. The observed sample: 0.2-μm-thick $GaAs_{0.941}Bi_{0.059}$ film. © 2011 IEEE [39]

Transmission electron microscopy (TEM) observation revealed that composition-modulated structures (CMSs) were observed in GaAsSb [38], while ordered structures were not observed in $GaAs_{1-x}Bi_x$ ($x \leq 10.9$ %) with a thickness less than the critical thickness [39]. Figure 9.4a, b shows the bright-field (110) cross-sectional TEM images of a 0.2-μm-thick $GaAs_{0.941}Bi_{0.059}$ film taken under the 220 and the 004 reflections, respectively. While a broad band-like region due to the bending of the thin specimen can be seen, grown-in defects such as dislocations, dislocation loops, precipitates, CMS, and ordered structures are not observed. In addition, in the plan-view TEM image (Fig. 9.4c), grown-in defects and CMSs are not observed. The satellite spots and streaks peculiar to CMSs are also not observed in the electron diffraction pattern (not shown here) [39]. Therefore, it was confirmed that neither grown-in defects nor CMSs generated in $GaAs_{0.891}Bi_{0.109}$ with a thickness of 60 nm less than the critical thickness [39].

9.3 Localized States in $GaAs_{1-x}Bi_x$

9.3.1 Bi-Induced Localized States

It is widely accepted that the N levels in GaN_xAs_{1-x} perturb the conduction band of GaAs, as mentioned above [15]. This interaction is usually explained by the band anti-crossing model [15]. In contrast, the spatially localized Bi states interact with the valence band of GaAs, resulting in the formation of localized states in $GaAs_{1-x}Bi_x$ [17–22]. Consequently, an S-shaped temperature dependence (redshift–blueshift–redshift) of the PL peak energy is observed in the PL spectra of $GaAs_{1-x}Bi_x$ (Fig. 9.5) similar to that of the peak energy observed in the PL spectra of dilute nitride materials such as GaN_xAs_{1-x} [16]. The undoped $GaAs_{1-x}Bi_x$ (Fig. 9.5a) and lightly doped p-$GaAs_{1-x}Bi_x$ with a hole density (p) of 1×10^{17} cm^{-3} (Fig. 9.5b) show S-shaped behavior in their temperature-dependent PL spectra. As summarized in Fig. 9.6, the PL peaks of the undoped and lightly doped $GaAs_{1-x}Bi_x$ (1.8 % $\leq x \leq$ 5.4 %) show a redshift, blueshift, and redshift with increasing temperature. For instance, the energy difference between the emission of localized states and the curve derived from the Varshni empirical equation indicate that the localized states reside ~50–90 meV above the valence band. Similar energy levels of 45 and 11 meV for the Bi-induced states were reported in a previous study [21].

The PL intensity (I_{PL}) as a function of temperature for the undoped $GaAs_{0.974}Bi_{0.026}$ shows a constant emission intensity (I_{max}) below 50 K, followed by temperature quenching, as depicted in Fig. 9.7. Considering the conventional relationship for

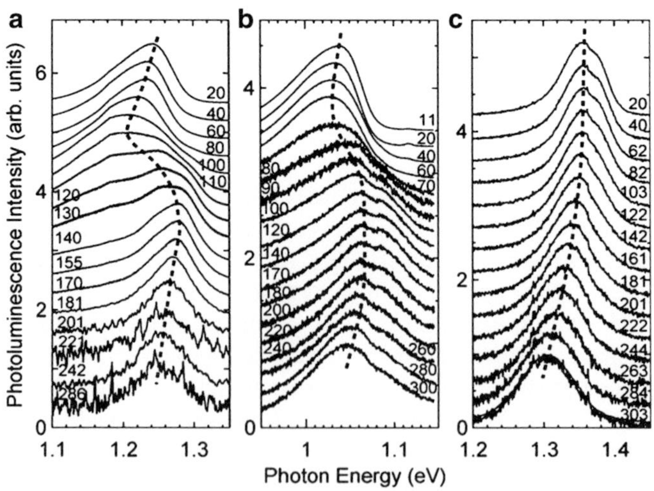

Fig. 9.5 Temperature-dependent PL of $GaAs_{1-x}Bi_x$. (**a**) x=2.6 %, undoped; (**b**) x=5.4 %, p=1×10^{17} cm^{-3}; (**c**) x=1.6 %, p=2×10^{18} cm^{-3}. The numbers indicate the absolute temperature (K) measurement values. © 2013 Elsevier B.V. [41]

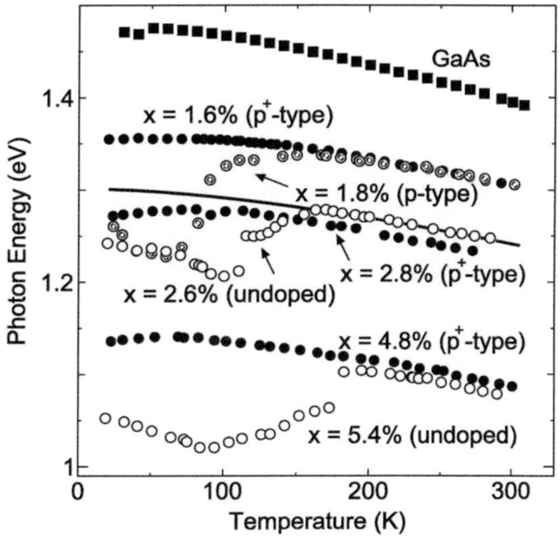

Fig. 9.6 Temperature dependence of the PL peak energy of GaAs and undoped p-type ($p = 1 \times 10^{17}$ cm^{-3}) and p$^+$-type ($p = 1 \times 10^{18}$ -2×10^{18} cm^{-3}) $GaAs_{1-x}Bi_x$. The curve fitted to the data for undoped $GaAs_{1-x}Bi_x$ ($x = 2.6$ %) is derived from the Varshni empirical equation. © 2013 Elsevier B.V. [41]

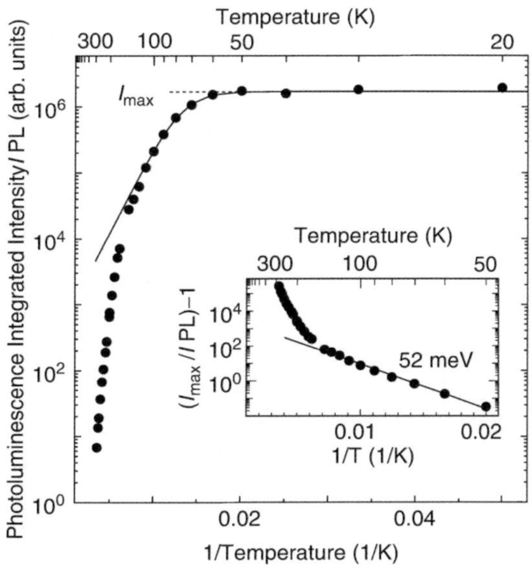

Fig. 9.7 Temperature dependence of the PL intensity (I_{PL}) for the undoped $GaAs_{1-x}Bi_x$ ($x = 2.6$ %). *Inset* shows the replot of the data as $(I_{max}/I_{PL}) - 1$ vs. temperature. I_{max}: the saturated I_{PL} below 50 K. © 2013 Elsevier B.V. [41]

thermal quenching, $I_{PL} = I_{max}\{1 - \exp(-\Delta E/kT)\}$ [40], the activation energy (ΔE) was estimated to be 52 meV [41].

The PL blueshift due to an increase in the excitation intensity for the undoped $GaAs_{1-x}Bi_x$ (Fig. 9.8) is another piece of evidence of the Bi-induced localized states. The increase in the excitation intensity induces a downward shift of the quasi-Fermi levels of the holes, which causes the hole occupation at a shallower level of the continuously distributed Bi-induced localized states. This resulted in the PL blue-shift for the undoped $GaAs_{1-x}Bi_x$. The difference in the energy of the PL peak

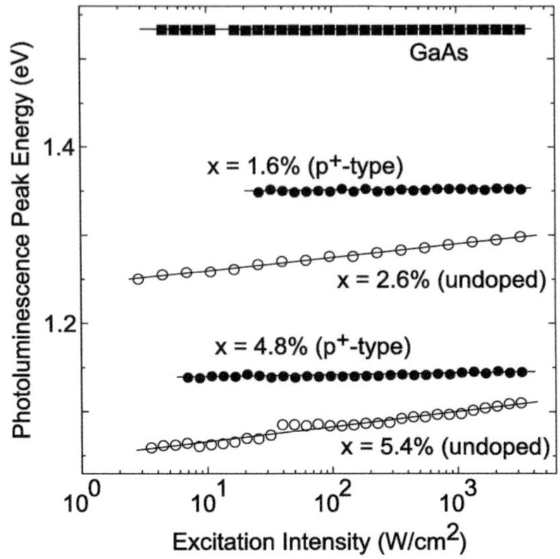

Fig. 9.8 PL peak energy as a function of the excitation intensity for undoped and p$^+$-GaAs$_{1-x}$Bi$_x$ ($p = 1 \times 10^{18}$–2×10^{18} cm^{-3}). © 2013 Elsevier B.V. [41]

between the excitation intensity of 3 and 3,000 W/cm^2 was measured to be 50 meV for the undoped GaAs$_{1-x}$Bi$_x$ with $x = 2.6$ % (Fig. 9.8). On the basis of the blueshift of 50 meV and ΔE of 52 meV derived from Fig. 9.7, the Bi-induced localized states are considered to be continuously located up to 52 meV from the valence band. Furthermore, the maximum PL redshift of ~90 meV for the undoped GaAs$_{0.974}$Bi$_{0.026}$ at 100 K (Fig. 9.6) implies the existence of the Bi-induced states that are continuously located up to ~90 meV from the valence band.

It should be noted that p$^+$-GaAs$_{1-x}$Bi$_x$ ($p = 1 \times 10^{18}$–2×10^{18} cm^{-3}) does not show S-shaped behavior. It exhibits a monotonic redshift with increasing temperature and shows normal behavior that can be described by the Varshni empirical equation, as shown in Figs. 9.5c and 9.6. Furthermore, the PL peak energy of p$^+$-type GaAs$_{1-x}$Bi$_x$ ($p = 2 \times 10^{18}$ cm^{-3}) is constant irrespective of the excitation intensity (Fig. 9.8). These results imply that a downward shift of the Fermi level induced by the p-type doping modifies the carrier occupation at the localized states and masks the S-shaped behavior of the PL peak energy as a function of temperature.

At measurement temperatures below 50 K, radiative recombination dominated in undoped, p$^-$, and p$^+$-GaAs$_{1-x}$Bi$_x$ compared to non-radiative recombination based on the relationship between the PL intensity and the excitation intensity [41]. Hence, assuming that the transition probability of a recombination involving the Bi-induced localized states is the same as that of band-to-acceptor recombination involving the acceptor as a localized state, it can be concluded that the energy shift of the PL peak is presumably associated with the carrier occupation at the Bi-induced level, Be-acceptor level, and valence band. Therefore, analyzing the PL spectra shown in Fig. 9.6 for the presence or absence of the S-shaped feature and with the assistance of the Shockley diagram, the density of the Bi-induced levels was estimated to be ~10^{17} cm^{-3} [41]. Accordingly, ~10^{-4} of the incorporated Bi atoms contribute to the formation of the localized states.

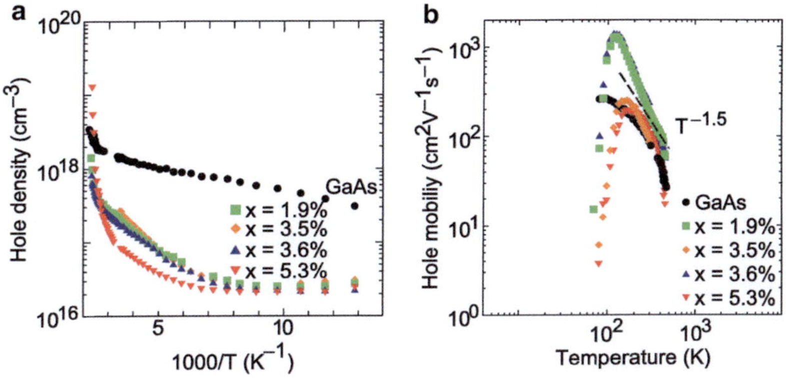

Fig. 9.9 (**a**) Temperature dependence of (**a**) the hole density and (**b**) the hole mobility of p-$GaAs_{1-x}Bi_x$ grown at T_{Be}=680 °C. © 2012 The Japan Society of Applied Physics [43]

9.3.2 Hole Conduction

Similar to $GaAs_{1-x}Bi_x$, $GaN_{1-x}As_x$ shows large band-gap bowing because of N incorporation. However, GaN_xAs_{1-x} alloys with x>~5 % become increasingly difficult to grow [42]. Even in GaN_xAs_{1-x} alloys with x<~5 %, N incorporation is usually accompanied by the deterioration of the electron mobility of GaN_xAs_{1-x} [14] because of the scattering of the N-induced localized states that are formed through the interaction of the N states with the conduction band of GaAs. Similarly, the degradation of hole mobility for $GaAs_{1-x}Bi_x$ with an increase in the GaBi molar fraction [23, 24] is explained by the scattering at the Bi-induced states.

The hole density and mobility of the lightly doped $GaAs_{1-x}Bi_x$ (1.9 % ≤ x ≤ 5.3 %) were measured using the van der Pauw method under a magnetic field of 300 mT, and the results are shown in Fig. 9.9a. The hole density decreases as the temperature decreases below room temperature because of the freeze-out of holes onto accepter sites. Below 110 K, the hole density appears to slightly increase at lower temperatures, indicating impurity band conduction. There is also an intrinsic range in which the hole density increases with temperatures higher than 400 K. The hole mobility is nearly inversely proportional to three-halves of the power of the temperature at 150–300 K (Fig. 9.9b), indicating the dominance of phonon scattering. A slight increase in the hole mobility with temperature has also been reported, and it was believed to result from the contribution of the localized states [23, 24]. In contrast to the previous report, typical conduction with a single carrier (hole) via the valence bands dominates in $GaAs_{1-x}Bi_x$ in Fig. 9.9b. The small density of the Bi-induced localized states (~10^{17} cm^{-3}) estimated in the previous section presumably results in the absence of a strong interaction between the holes and the Bi-induced localized states [43].

As can be seen in Fig. 9.10, the room temperature hole mobility of $GaAs_{1-x}Bi_x$ epilayers with strong scattering at the Bi-induced states (reported in refs. [23, 24])

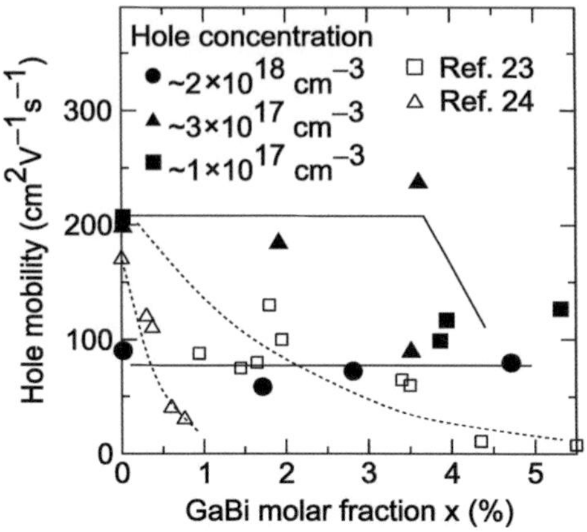

Fig. 9.10 Hole mobility in the Be-doped GaAs and $GaAs_{1-x}Bi_x$ at room temperature. © 2012 The Japan Society of Applied Physics [43]

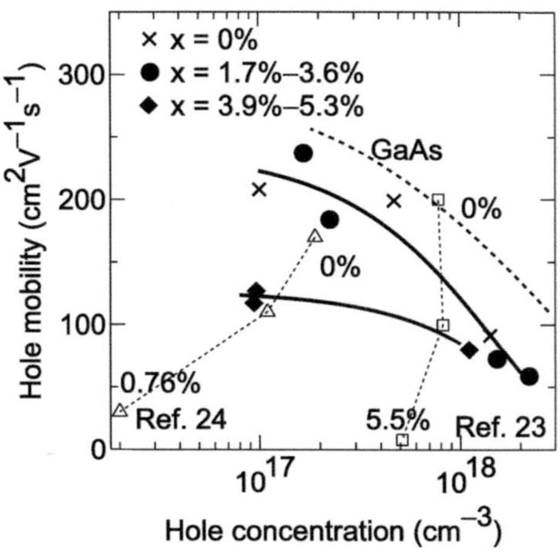

Fig. 9.11 Hole mobility of $GaAs_{1-x}Bi_x$ and GaAs as a function of the hole density. *Open squares* and *triangles* at a given GaBi molar fraction denote results taken from refs. [23, 24], respectively. The *dashed line* denotes conventional data for GaAs. © 2012 The Japan Society of Applied Physics [43]

decreases as the GaBi molar fraction increases. On the other hand, in the case of $GaAs_{1-x}Bi_x$ ($p = {\sim}2 \times 10^{18}$ cm^{-3}) with typical hole conduction via the valence band, the hole mobility is maintained at a nearly constant value that is equal to that of GaAs. By decreasing the hole density to $\sim 3 \times 10^{17}$ cm^{-3}, a constant hole mobility larger than that of the heavily doped sample ($p = {\sim}2 \times 10^{18}$ cm^{-3}) is initially maintained, followed by the degradation of the sample with a GaBi molar fraction of ~4 %. Note that the degraded mobility for samples with doping levels of $1\text{–}3 \times 10^{17}$ cm^{-3} is still larger than the previously reported values with unusual hole conduction. On the basis of the results of reciprocal lattice mapping, the degradation of the hole mobility for $GaAs_{1-x}Bi_x$ with $x = 4\text{–}5$ % was found to be due to the large GaBi molar fraction rather than lattice relaxation [43].

Figure 9.11 depicts the relationship between the hole mobility and hole density of $GaAs_{1-x}Bi_x$ and GaAs at room temperature. The hole mobility of $GaAs_{1-x}Bi_x$ ($x \leq 3.6$ %) is nearly the same as that of GaAs, and the degradation of hole mobility in samples with large GaBi molar fractions (x=3.9–5.3 %) was observed. In contrast, the hole mobility steeply degrades in the samples exhibiting unusual hole conduction due to the strong scattering at the Bi-induced states [23, 24]. In addition, the hole mobility with typical hole conduction is maintained at a larger value than previously reported. These results for the hole mobility suggest the existence of process windows for obtaining $GaAs_{1-x}Bi_x$ with a hole mobility equal to that of GaAs. A fluctuation in the incorporation scheme for Bi such as randomness or clustering, which cannot be observed by TEM, presumably affects the carrier mobility sensitivity. In fact, the density of the Bi-induced states (estimated to be ~10^{17} cm^{-3} in Sect. 3.1) cannot be observed by TEM. The p-type doping at a density greater than ~10^{17} cm^{-3} masks the contribution of the Bi-induced states (~10^{17} cm^{-3}) to the hole mobility.

9.3.3 Deep-Level Defects

Because low-temperature growth (<400 °C) is essential for realizing Bi incorporation in the epilayer [2, 9], point defects are of particular interest for Bi-containing materials. For instance, it has been reported that the low-temperature growth of GaAs using MBE is considered to enhance the drastic increase in the density of point defects because of poor migration [44]. In particular, it is well known that the MBE growth of GaAs at low temperature causes excess As incorporation into Ga antisites, denoted by As_{Ga}, and that the incorporation of excess As leads to lattice expansion [45]. In our growth apparatus, the lattice constant perpendicular to the surface ($a_\perp$) of GaAs epilayers grown at 200 °C was measured to expand up to 5.661 Å, while the lattice parallel to the surface ($a_\parallel$) had the same lattice constant as that of the GaAs substrate [36]. These lattice constants indicate the compressive distortion of the lattice that was probably caused by excess As incorporation. In the case of GaAs growth at 300 °C, the values for both $a_\perp$ and $a_\parallel$ of the epilayer were the same as those of the GaAs substrate, which corresponds with the previous report [45]. This result suggests that excess As incorporation in $GaAs_{1-x}Bi_x$ grown between 350 and 400 °C is suppressed below the measurable limit of the lattice constant. The relatively low-temperature growth, however, potentially generates antisite As measurable by deep-level transient spectroscopy (DLTS). Furthermore, the presence of bismuth antisites (Bi_{Ga}) in GaAs has been suggested by electron spin resonance analysis [46].

Next, DLTS measurements were performed with Be-doped p-$GaAs_{1-x}Bi_x$ Schottky diodes. The p-$GaAs_{1-x}Bi_x$ layers with x=1.2 and 3.4 % were grown at a substrate temperature (T_{sub}) of 370 °C. Figure 9.12a shows the DLTS spectra of p-GaAs grown at a T_{sub} of 560 °C and the p-$GaAs_{1-x}Bi_x$ samples with x=1.2 and 3.4 % with a rate window of 31 ms. Positive peaks corresponding to deep hole levels were observed at peak temperatures (T_p) of 320, 250, and 138 K for samples with x=0 %, 1.2 %, and 3.4 %, respectively. The activation energy of the deep hole levels was calculated to be 0.62 eV for p-GaAs and 0.43 and 0.23 eV for the p-$GaAs_{1-x}Bi_x$

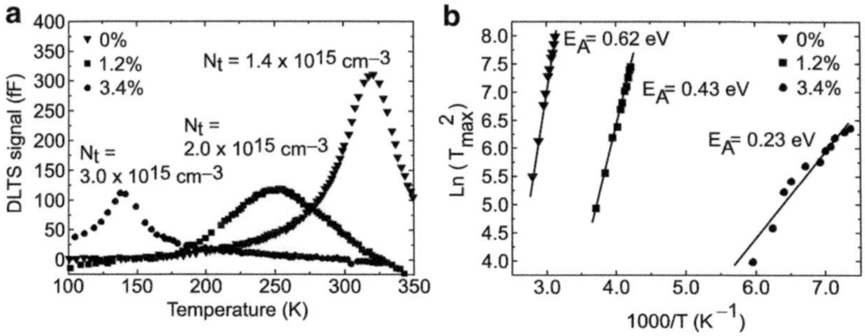

Fig. 9.12 (**a**) DLTS signals of p-$GaAs_{1-x}Bi_x$ Schottky diodes and (**b**) Arrhenius plots. © 2011 The Japan Society of Applied Physics [50]

Fig. 9.13 Schematic diagram of the energy level for a p-$GaAs_{0.966}Bi_{0.034}$ alloy depicting the energy levels of the Bi localized states detected by PL and deep-level defects observed by DLTS. The energy level of the Bi localized states is assumed to be the same as that of the $GaAs_{0.974}Bi_{0.026}$ alloy

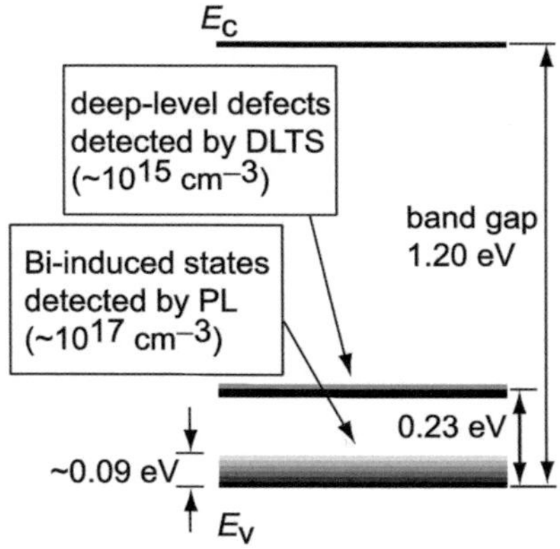

samples with $x = 1.2$ % and 3.4 %, respectively (Fig. 9.12b). A schematic diagram of the energy levels for a p-$GaAs_{0.966}Bi_{0.034}$ alloy is depicted in Fig. 9.13.

The ratio of the full width at half maximum (FWHM) to the peak temperature, FWHM/T_p, is close to 0.1 for the deep levels because of ideal point defects [47] such as As_{Ga}. Considering the reported activation energy of 0.62 eV and FWHM/T_p of 0.15, the deep hole levels observed in p-GaAs are likely related to arsenic antisites (As_{Ga}) [48, 49]. However, given the value of 0.32 for the FWHM/T_p and the GaBi molar fraction dependence of the activation energy, the distribution of isolated point defects such as Bi_{Ga} and As_{Ga} and/or relevant clusters of point defects may be the more likely sources for the deep hole traps observed in p-$GaAs_{1-x}Bi_x$ [50]. The deep hole trap density was calculated to be 1.4×10^{15}, 2.0×10^{15}, and 3.0×10^{15} cm^{-3} for samples with $x = 0$ %, 1.2 %, and 3.4 %, respectively. In addition, DLTS measurements have been performed on p-i-n diodes with i-$GaAs_{1-x}Bi_x$ layers grown using MBE [51]. The undoped $GaAs_{1-x}Bi_x$ layers were grown at a T_{sub} of 285–350 °C with a V/III ratio of 1–2. The sample with a 4.7 % GaBi molar fraction grown at a T_{sub} of 285 °C was found to have several traps in density of $\sim 5 \times 10^{15}$ cm^{-3}. In spite of the low-temperature growth, the deep trap density was suppressed on the order of 10^{15} cm^{-3}, indicating that the Bi atoms work as a surfactant.

9.4 Structural Properties of the Bi-Containing Superlattices

The heterostructure is critical for realizing devices consisting of III–V semiconductors. For instance, carrier confinement is one of the key requirements for realizing an LD. In this section, the fabrication of the Bi-containing superlattices is described. Because Bi is used as a surfactant in the growth of III–V semiconductors [52], it is segregated easily during growth. Therefore, in the fabrication of the Bi-containing heterointerfaces, there are concerns regarding incomplete formation and/or thermal stability. The Bi-containing superlattices $GaAs/GaAs_{1-x}Bi_x$ and $Al_yGa_{1-y}As/GaAs_{1-x}Bi_x$ were grown on a (001)-oriented GaAs substrate using MBE to carry out a structural investigation of the Bi-containing heterointerfaces. The $Al_yGa_{1-y}As/GaAs_{1-x}Bi_x$ heterointerface was studied with a view toward forming a barrier for electrons in the $GaAs_{1-x}Bi_x$ layer with a much larger band offset at the conduction band side using Al belonging to group III elements, similar to the case of the $Al_yGa_{1-y}As/GaAs_{1-x}Sb_x$ heterointerface [53].

The thermal treatment of the GaAs surface was performed at a T_{sub} of 560 °C. Prior to growth of the Bi-containing superlattices, an undoped 300-nm-thick GaAs buffer layer was grown at a T_{sub} of 560 °C under usual As-rich conditions. It was confirmed that the surface of the GaAs buffer layer showed a $(2\times4)\beta$ reconstruction by reflection high-energy electron diffraction (RHEED) analysis. During the growth of the $GaAs_{1-x}Bi_x$ layers, the RHEED pattern showed a streak (2×1) superstructure, indicating the existence of a metallic surface comprising Bi–Bi or Bi–As dimers [28, 54]. The $GaAs/GaAs_{1-x}Bi_x$ superlattices $(0.6\le x\le 10.9\ \%)$ were grown on the GaAs buffer layer at a T_{sub} between 350 and 400 °C. The thickness of the GaAs layer was between 13 and 16 nm, and that of the $GaAs_{1-x}Bi_x$ layer was between 3 and 14 nm [34]. For the $Al_yGa_{1-y}As/GaAs_{1-x}Bi_x$ superlattices $(3.5\le x\le 5\ \%, 0\le y\le 26\ \%)$ grown at a T_{sub} of 350 °C, the thickness of the $GaAs_{1-x}Bi_x$ and $Al_yGa_{1-y}As$ layers was 7 and 14 nm, respectively [55].

Figure 9.14 shows examples of the RHEED specular-beam intensity oscillations during the growth of the Bi-containing superlattices of (a) $GaAs/GaAs_{1-x}Bi_x$ and (b) $Al_yGa_{1-y}As/GaAs_{1-x}Bi_x$. In the case of the $Al_yGa_{1-y}As/GaAs_{1-x}Bi_x$ superlattice, growth interruption at 90 s was applied for the planarization of the growth surface. The persistent RHEED oscillation and streaked RHEED patterns during the growth of the $GaAs_{1-x}Bi_x$ layers can be observed in both the $GaAs/GaAs_{1-x}Bi_x$ and $Al_yGa_{1-y}As/GaAs_{1-x}Bi_x$ superlattices, indicating the layer-by-layer growth of the superlattices.

Figure 9.15 shows examples of the high-resolution XRD (HR-XRD) patterns of the (004) plane for the superlattices (a) $GaAs/GaAs_{0.952}Bi_{0.048}$ with a periodic number of 14 and (b) $Al_{0.19}Ga_{0.81}As/GaAs_{0.951}Bi_{0.049}$ with a periodic number of 10 obtained using θ–2θ scans. In both superlattices, clear satellite peaks attributed to the periodic structure can be observed. On the basis of the Laue function, the number of interference fringes in the area between the satellite peaks corresponds to the designed periodic number in both superlattices. The GaBi molar fraction of the superlattice samples was calculated from the position of the zeroth order peak, as in the case of ref. [34].

The structures of the $GaAs/GaAs_{0.952}Bi_{0.048}$ ($GaAs_{1-x}Bi_x$: 6 nm, GaAs: 15 nm) and $Al_{0.19}Ga_{0.81}As/GaAs_{0.951}Bi_{0.049}$ ($GaAs_{1-x}Bi_x$: 7 nm, GaAs: 14 nm) superlattices

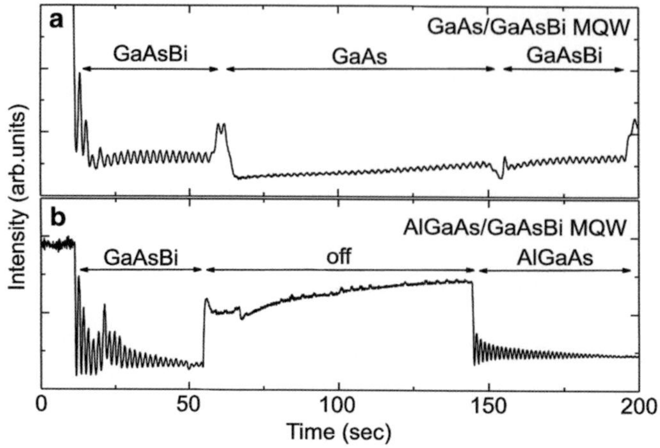

Fig. 9.14 RHEED specular-beam intensity oscillations during the growth of the Bi-containing superlattices of (**a**) GaAs/$GaAs_{1-x}Bi_x$ and (**b**) $Al_yGa_{1-y}As$/$GaAs_{1-x}Bi_x$. © 2010 The Japan Society of Applied Physics [55]

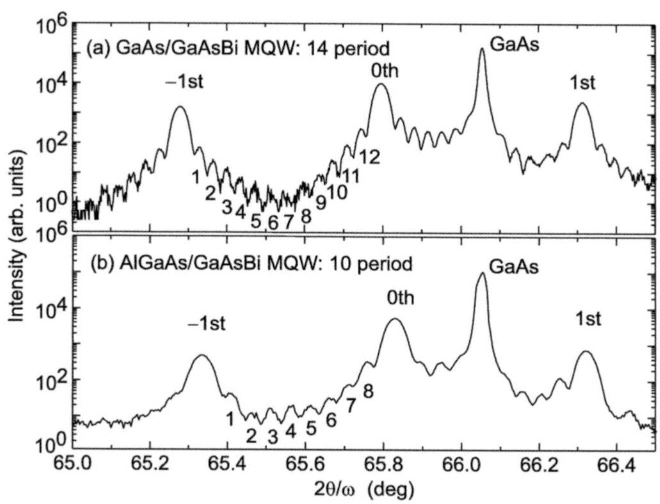

Fig. 9.15 HR-XRD patterns of the (004) plane for superlattices of (**a**) GaAs/$GaAs_{0.952}Bi_{0.048}$ with a periodic number of 14 and (**b**) $Al_{0.19}Ga_{0.81}As$/$GaAs_{0.951}Bi_{0.049}$ with a periodic number of 10. © 2010 The Japan Society of Applied Physics [55]

were analyzed by cross-sectional TEM. Figure 9.16a, b shows the low- and high-magnification images for the GaAs/$GaAs_{0.952}Bi_{0.048}$ superlattice, respectively. Figure 9.16c, d shows the low- and high-magnification images for the $Al_{0.19}Ga_{0.81}As$/$GaAs_{0.951}Bi_{0.049}$ superlattice, respectively. Figure 9.16a, c, the low-magnification TEM images, shows the successful fabrication of the GaAs/$GaAs_{0.952}Bi_{0.048}$ and $Al_{0.19}Ga_{0.81}As$/$GaAs_{0.951}Bi_{0.049}$ superlattices with smooth interfaces. The slight fluctuation in the

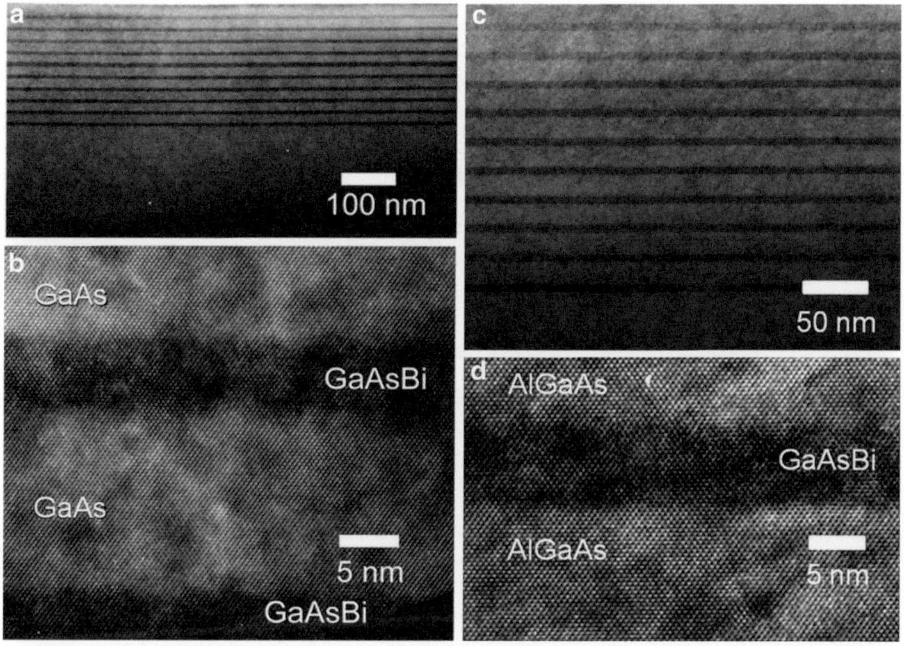

Fig. 9.16 Cross-sectional TEM images of $GaAs/GaAs_{0.952}Bi_{0.048}$ superlattices: (**a**) low magnification and (**b**) high magnification. Cross-sectional TEM images of $Al_{0.19}Ga_{0.81}As/GaAs_{0.951}Bi_{0.049}$ superlattices: (**c**) low magnification and (**d**) high magnification. © 2008 The Applied Physics Letters [34]. © 2010 The Japan Society of Applied Physics [55]

thickness of the layers is due to the manual switching of element flux. Clear lattice images of both superlattices are shown in the high-magnification TEM images in Fig. 9.16b, d. On the basis of the results of the HR-XRD and TEM analysis, it was concluded that the Bi-containing heterointerfaces with smooth interfaces can be fabricated without distinct Bi segregation. The microstructure of the $GaAs/GaAs_{1-x}Bi_x$ heterointerface was stable up to 700 °C, as determined by a thermal annealing system with halogen lamps under a N_2 flow [34]. It has been reported that a $GaAs_{1-x}Bi_x/GaAs/Al_yGa_{1-y}As$ separate confinement heterostructure (SCH) exhibits an enhanced PL intensity at room temperature that is more than 10 times that of a sample without SCH because of an $Al_yGa_{1-y}As$ barrier [56].

9.5 Localized States in $GaAs/GaAs_{1-x}Bi_x$ Heterostructures

9.5.1 Interface States in $GaAs/GaAs_{1-x}Bi_x$ Heterostructures

Interface states at the Bi-containing heterointerface can severely affect device performance; the interface states behave as non-radiative recombination centers and electrical carrier traps [57]. Thermal admittance spectroscopy (TAS) measurements

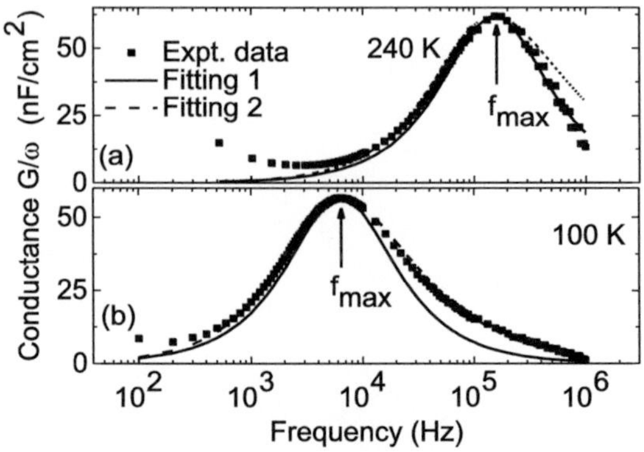

Fig. 9.17 Frequency dependence of the conductance under zero dc bias for samples of the GaAs/p-$GaAs_{0.982}Bi_{0.018}$ heterostructure: (**a**) at 240 K and (**b**) at 100 K. The *solid line* and *dashed line* show the simulation fitting using (9.1) and (9.2), respectively. The f_{max} indicates the peak frequency of the conductance spectra. © 2012 The Japan Society of Applied Physics [58]

were performed on a Schottky diode having GaAs/p-$GaAs_{0.982}Bi_{0.018}$ heterostructures to investigate the interface states at the heterointerface. The GaAs/p-$GaAs0_{.982}Bi_{0.018}$ heterostructures were grown on a p-type (100)-oriented GaAs substrate ($p = 1.0 \times 10^{19}$ cm^{-3}) using MBE at a T_{sub} of 370 °C [58]. When deep levels exist in the depletion layer, an equivalent parallel conductance (G_p) peak is observed because of the ac loss due to the capture and emission of carriers by the deep levels [59]. In the case of a single state level, the conductance follows the equation

$$\frac{G_p}{\omega} = \frac{qD_{it}}{(1+\omega^2\tau^2)}, \tag{9.1}$$

where $\tau = (2\pi f_{max})^{-1}$ is the time constant of the deep levels, f_{max} is the peak frequency of the conductance spectra, and D_{it} is interface state density. For a continuum state level, the conductance is

$$\frac{G_p}{\omega} = \frac{qD_{it}}{2\omega\tau} \ln(1+\omega^2\tau^2). \tag{9.2}$$

Figure 9.17a, b shows the frequency dependence of the conductance for the GaAs/p-$GaAs_{0.982}Bi_{0.018}$ heterostructure at 240 and 100 K, respectively. Well-defined conductance peaks due to deep hole levels were observed at both temperatures. Using (9.1) and (9.2), the experimental conductance data were fitted, and the corresponding simulation curves are depicted in Fig. 9.17a, b with the solid and dashed lines, respectively. As can be seen in Fig. 9.17, there is good agreement between the experimental data and the single state model for the heterostructure measured at 240 K, and the continuum state model for the heterostructure measured at 100 K. On the basis of these results and the Arrhenius plots of the TAS measurements for the

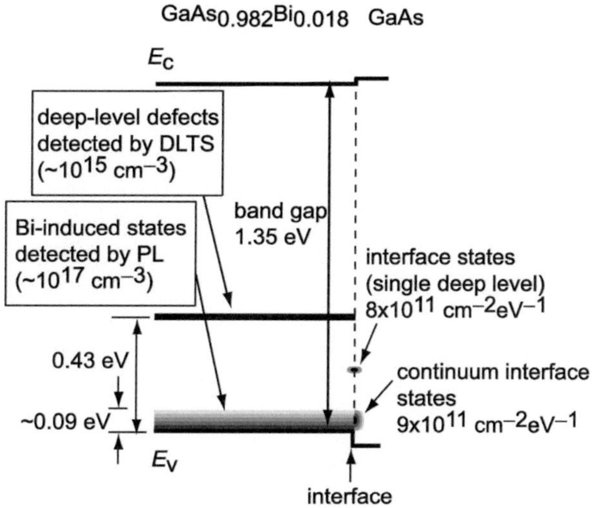

Fig. 9.18 Schematic diagram of the energy levels in the p-GaAs/p-$GaAs_{0.982}Bi_{0.018}$ heterointerface. The energy level of the deep-level defects is assumed be the same as that of the $GaAs_{0.988}Bi_{0.012}$ alloy

GaAs/p-$GaAs_{0.982}Bi_{0.018}$ heterostructure between 100 and 350 K [58], the difference in the activation energy between the high-temperature range and low-temperature range was observed. As shown in Fig. 9.17, the TAS spectrum in the high-temperature area is well fitted by a single state model, and, in the low-temperature area, it is well fitted by the continuum state model. This difference indicates that the nature of the interface states changes from continuum states to single states as the measurement temperature increases.

Assuming that the observed deep levels exist at the GaAs/p-$GaAs_{0.982}Bi_{0.018}$ interface, the deep-level density, D_{it}, was estimated to be 8.8×10^{11} cm^{-2} eV^{-1} for the continuum deep levels and 7.7×10^{11} cm^{-2} eV^{-1} for the single deep levels. It should be noted that TAS probes traps in the depletion layer; therefore, TAS signals potentially originate from traps in the GaAs and $GaAs_{0.982}Bi_{0.018}$ grown layers, as well as at the GaAs/$GaAs_{0.982}Bi_{0.018}$ interface. If the deep levels are located in the p-$GaAs_{1-x}Bi_x$ and/or GaAs grown layers, the deep-level density would be $\sim1\times10^{17}$ cm^{-3}. This value, however, is clearly greater than the deep-level density of $\sim10^{15}$ cm^{-3} that was measured by DLTS (see Sect. 3.3). Moreover, isothermal capacitance transient spectroscopy (ICTS) revealed that the trap density in p-$GaAs_{0.982}Bi_{0.018}$ was below the measurable limit of $\sim1\times10^{16}$ cm^{-3} [50]. Thus, it is concluded that the deep levels detected by TAS exist at the GaAs/p-$GaAs_{0.982}Bi_{0.018}$ heterointerface [58]. The schematic diagram of the energy levels of the GaAs/p-$GaAs_{0.982}Bi_{0.018}$ heterointerface is depicted in Fig. 9.18. The energy level of single deep level was evaluated to be 0.1–0.2 eV from the valance band [58].

For design of optoelectronic devices, it is imperative to clarify the band alignment of the Bi-containing heterointerfaces. The band alignment, however, remains unclear. To date, the band alignment of GaAs/$GaAs_{1-x}Bi_x$ has been proposed to be a type-I [10, 60], type-II [19, 61], and a nearly flat conduction band offset [13]. In fact, the interface state density on the order of 10^{11} cm^{-2} eV^{-1} can severely affect the

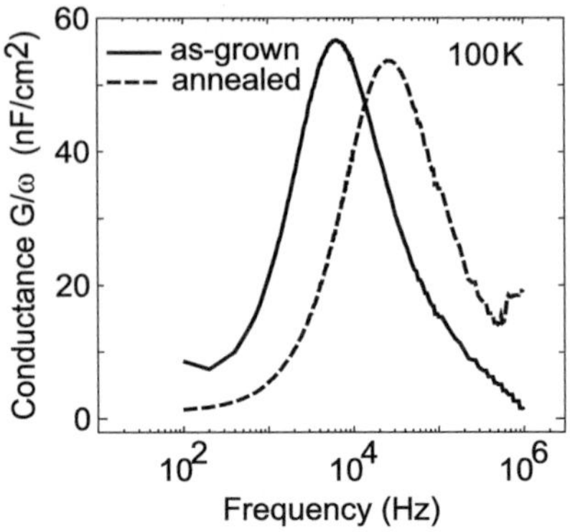

Fig. 9.19 Frequency dependence of the conductance at 100 K under zero dc bias for the as-grown and annealed samples of the GaAs/p-$GaAs_{0.982}Bi_{0.018}$ heterostructure. © 2012 The Japan Society of Applied Physics [58]

capacitance–voltage (*C*–*V*) methods used for the determination of the band alignment [57]. In other words, the electric field induced by the interface states at the GaAs/$GaAs_{1-x}Bi_x$ interface deteriorates the desired electronic structure of the GaAs/$GaAs_{1-x}Bi_x$ heterostructure, which is harmful for device performance.

9.5.2 Reduction of the Interface State Density

The reduction of the interface state density in GaAs/p-$GaAs_{1-x}Bi_x$ is necessary for the evaluation of the band alignment and improving device performance. Rapid thermal annealing (RTA), a powerful technique for reducing deep traps [62], was performed on the GaAs/p-$GaAs_{0.982}Bi_{0.018}$ heterostructure. The annealing temperature was set at 600 °C with a duration of 3 s to avoid the destruction of the GaAs/$GaAs_{1-x}Bi_x$ interface [63]. Figure 9.19 shows the frequency dependence of the conductance at 100 K under zero dc bias for the as-grown and annealed samples of the GaAs/p-$GaAs_{0.982}Bi_{0.018}$ heterostructure. The interface state density at the GaAs/p-GaAs0$_{.982}Bi_{0.018}$ heterointerface was not reduced by the RTA, as indicated by the similar peak intensity of both samples. However, the time constant of the interface states changed from 25 to 6 μs, which implied the transformation of deep levels upon receipt of the thermal energy during RTA.

As mentioned above, Bi atoms have been found to stabilize an unusual (2 × 1) surface reconstruction [28] because of the metallic nature of the $GaAs_{1-x}Bi_x$ surface [54]. On the other hand, the GaAs(100) surface reconstruction tends to be nonmetallic according to the famous electron counting model [29, 64, 65], which has been successfully used to explain various surface reconstructions of typical III–V alloys. Therefore, the surface nature of GaAs/p-$GaAs_{1-x}Bi_x$ was changed from metallic to nonmetallic during growth. For conventional growth at a heterointerface, an abrupt interface was formed by rapid closure of the Bi cell shutters. In this case, the drastic transformation of the surface nature from metallic to nonmetallic probably induces interface states at the GaAs/p-$GaAs_{1-x}Bi_x$ heterointerface. To avoid this situation, a 50-nm-thick Bi graded layer was inserted into the GaAs/p-$GaAs_{0.982}Bi_{0.018}$

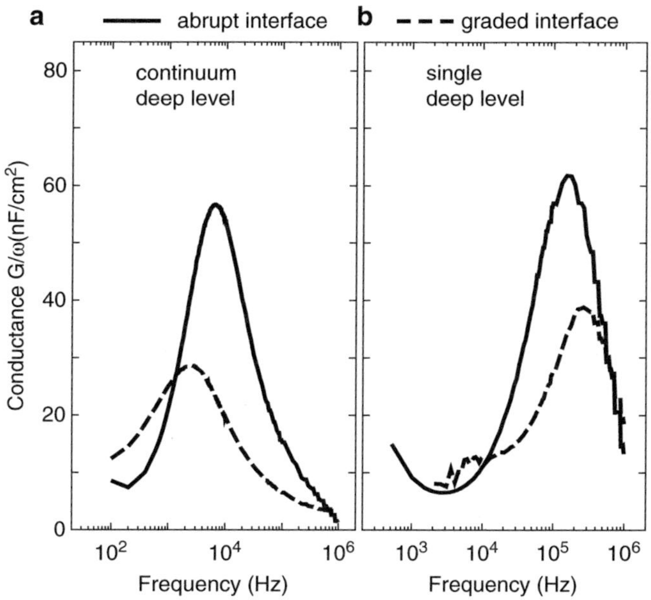

Fig. 9.20 Frequency dependence of the conductance under zero dc bias for GaAs/p-$GaAs_{1-x}Bi_x$ heterostructures with abrupt and graded interfaces: (**a**) for continuum deep levels measured at 100 K for both samples and (**b**) for a single deep level measured at 240 and 350 K for the abrupt interface and the graded interface, respectively. © 2012 The Japan Society of Applied Physics [58]

heterointerface to fabricate a graded interface by increasing both the growth temperature and the As flux.

Figure 9.20a, b shows the frequency dependence of the conductance fitted according to (a) the continuum deep level and (b) the single deep level under a zero dc bias for the GaAs/p-$GaAs_{0.982}Bi_{0.018}$ heterostructure with abrupt and graded interfaces. To compare the interface state density between the abrupt interface sample and the graded interface sample, the measurement temperature for the continuum deep levels was set at 100 K for both samples. The measurement temperature for the single deep levels was set at 240 and 350 K for the abrupt interface sample and the graded interface sample, respectively, on the basis of the results of the Arrhenius plots for GaAs/p-$GaAs_{0.982}Bi_{0.018}$ heterostructures (not shown here). By focusing on the peak intensity of the conductance spectra, the interface state density of the GaAs/ p-$GaAs_{0.982}Bi_{0.018}$ heterointerface was reduced by the insertion of the Bi graded layer into the heterointerface. The improved interface state density, D_{it}, was estimated to be ~4.4×10^{11} cm^{-2} eV^{-1} for the continuum deep levels and 4.8×10^{11} cm^{-2} eV^{-1} for the single deep levels. The interface state density of ~5×10^{11} cm^{-2} eV^{-1} at the GaAs/ p-$GaAs_{0.982}Bi_{0.018}$ heterointerface is on the same order as other III–V heterointerfaces, such as GaAs/$GaAs_{0.97}N_{0.03}$ and $In_{0.5}Ga_{0.5}P/Al_{0.25}Ga_{0.75}As$, grown at high temperature [57, 62]. Further reduction of the deep levels at the interface is expected by adjusting the growth sequence of the heterointerface.

9.6 Summary

$GaAs_{1-x}Bi_x$ ($x \leq 10.9$ %) with a mirror surface was grown on GaAs (001) using MBE. Compared to conventional III–V semiconductors, the growth window is very narrow for achieving Bi incorporation into the grown layer without the formation of Bi droplets on the surface. However, the surfactant-like effect of the Bi atoms contributes to the improved quality of $GaAs_{1-x}Bi_x$ under extreme growth conditions. Although $GaAs_{1-x}Bi_x$ was grown at a growth temperature below 400 °C, $GaAs_{1-x}Bi_x$ ($x \leq 10.9$ %) emits intense band-edge PL with a reduction of the temperature coefficient of the PL emission energy. Furthermore, lasing operation of $GaAs_{0.975}Bi_{0.025}$ by optical pumping was achieved with a reduced temperature coefficient of the lasing wavelength. Neither the CMSs that are observed in GaAsSb nor ordered structures were detected in the TEM analysis of $GaAs_{1-x}Bi_x$ ($x \leq 10.9$ %).

It is widely accepted that the spatially localized Bi states interact with the valence band of GaAs, resulting in the formation of localized states in $GaAs_{1-x}Bi_x$. On the basis of the experimentally determined temperature dependence of the PL peak energy and peak intensity as well as the PL peak shift caused by an increase in the excitation intensity, it can be concluded that the Bi-induced states are continuously located up to ~90 meV from the valence band. By analyzing the presence or absence of the S-shaped feature in the temperature-dependent PL of $GaAs_{1-x}Bi_x$ with different hole density and with the assistance of the Shockley diagram, the density of the Bi-induced levels was estimated to be ~10^{17} cm^{-3}. Accordingly, it can be concluded that ~10^{-4} of the incorporated Bi atoms contribute to the formation of localized states. Despite concerns regarding the degradation of the hole mobility in $GaAs_{1-x}Bi_x$ due to scattering at the Bi-related localized states near the valence band, a high hole mobility of 200 cm^2V^{-1} s^{-1} was achieved for Be-doped p-$GaAs_{1-x}Bi_x$ ($x=4$ %). The p-type doping with a density greater than ~10^{17} cm^{-3} masks the contribution of the Bi-induced states (~10^{17} cm^{-3}) to the hole mobility. With respect to deep-level traps measured by DLTS, the trap density was suppressed on the order of 10^{15} cm^{-3} in the p-type $GaAs_{1-x}Bi_x$ ($x=1.2$ and 3.4 %), in spite of the low-temperature growth. The low trap density suggests that the Bi atoms might contribute to an enhancement of migration to avoid the formation of point defects.

GaAs/$GaAs_{1-x}Bi_x$ ($0.6 \leq x \leq 10.9$ %) and $GaAs_{1-x}Bi_x/Al_yGa_{1-y}As$ ($0.02 \leq x \leq 0.05$, $0 \leq y \leq 0.26$) superlattices were successfully grown with smooth interfaces and can be fabricated without distinct segregation. The microstructure of the GaAs/$GaAs_{1-x}Bi_x$ heterointerface is stable up to 700 °C. The interface state density in a GaAs/$GaAs_{1-x}Bi_x$ heterointerface was evaluated to be ~9×10^{11} cm^{-2} eV^{-1} on the basis of a comparative study of capacitance dispersion, ICTS, DLTS, and TAS. The interface state density cannot be reduced by RTA, but can be reduced by half via the insertion of a Bi graded layer into the GaAs/p-$GaAs_{1-x}Bi_x$ heterointerface, resulting in an interface state density on the same order as other III–V heterointerfaces such as GaAs/GaAsN and InGaP/AlGaAs. The reduction of the interface states is presumably caused by the mitigation of the difference in the metallic $GaAs_{1-x}Bi_x$ and nonmetallic GaAs surfaces. Further reduction of the deep levels at the interface is expected by adjusting the growth sequence of the heterointerfaces with the goal of developing device applications for the GaAs/$GaAs_{1-x}Bi_x$ heterointerfaces.

Acknowledgements This work was partly supported by Grants-in-Aid for Scientific Research (A) and (B) from the Japan Society for the Promotion of Science. The authors would like to express their appreciation to Dr. W. Huang, Dr. G. Feng, Dr. Y. Takehara, Dr. Y. Tominaga, Mr. S. Murata, Mr. Y. Kinoshita, Mr. M. Yamakawa, Mr. K. Yamada, Mr. M. Itoh, Mr. S. Kashiyama, and Mr. M. Kado for their contribution to this work. The authors would also like to acknowledge Prof. O. Ueda, Kanazawa Institute of Technology, for his TEM observation. In addition, the authors would like to express their deep appreciation to Emeritus Prof. K. Oe, the pioneer of GaAsBi, for his continuous encouragement and stimulating discussions.

References

1. Oe, K., Okamoto, H.: New semiconductor alloy $GaAs_{1-x}Bi_x$ grown by metal organic vapor phase epitaxy. Jpn. J. Appl. Phys. **37**, L1283 (1998)
2. Yoshimoto, M., Murata, S., Chayahara, A., Horino, Y., Saraie, J., Oe, K.: Metastable GaAsBi alloy grown by molecular beam epitaxy. Jpn. J. Appl. Phys. **42**, L1235 (2003)
3. Francoeur, S., Seong, M.-J., Mascarenhas, A., Tixier, S., Adamcyk, M., Tiedje, T.: Band gap of $GaAs_{1-x}Bi_x$, $0<x<3.6\%$. Appl. Phys. Lett. **82**, 3874 (2003)
4. Huang, W., Oe, K., Feng, G., Yoshimoto, M.: Molecular-beam epitaxy and characteristics of $GaN_yAs_{1-x-y}Bi_x$. J. Appl. Phys. **98**, 053505 (2005)
5. Tixier, S., Webster, S.E., Young, E.C., Tiedje, T., Francoeur, S., Mascarenhas, A., Wei, P., Schiettekatte, F.: Band gaps of the dilute quaternary alloys $GaN_xAs_{1-x-y}Bi_y$ and $Ga_{1-y}In_yN_xAs_{1-x}$. Appl. Phys. Lett. **86**, 112113 (2005)
6. Feng, G., Yoshimoto, M., Oe, K., Chayahara, A., Horino, Y.: New III–V semiconductor InGaAsBi alloy grown by molecular beam epitaxy. Jpn. J. Appl. Phys. **44**, L1161 (2005)
7. Petropoulos, J.P., Zhong, Y., Zide, J.M.O.: Optical and electrical characterization of InGaBiAs for use as a mid-infrared optoelectronic material. Appl. Phys. Lett. **99**, 031110 (2011)
8. Yoshida, J., Kita, T., Wada, O., Oe, K.: Temperature dependence of $GaAs_{1-x}Bi_x$ band gap studied by photoreflectance spectroscopy. Jpn. J. Appl. Phys. **42**, 371 (2003)
9. Yoshimoto, M., Oe, K.: Molecular beam epitaxy of GaAsBi and related quaternary alloys. In: Henini, M. (ed.) Molecular Beam Epitaxy, pp. 159–170. Elsevier, Amsterdam (2013, in press)
10. Alberi, K., Dubon, O.D., Walukiewicz, W., Yu, K.M., Bertulis, K., Kroktus, A.: Valence band anticrossing in $GaBi_xAs_{1-x}$. Appl. Phys. Lett. **91**, 051909 (2007)
11. Broderick, C.A., Usman, M., Sweeney, S.J., O'Reilly, E.P.: Band engineering in dilute nitride and bismide semiconductor lasers. Semicond. Sci. Technol. **27**, 094011 (2012)
12. Ptak, A.J., France, R., Jiang, C.-S., Reedy, R.C.: Effects of bismuth on wide-depletion-width GaInNAs solar cells. J. Vac. Sci. Technol. B **26**, 1053 (2008)
13. Tominaga, Y., Oe, K., Yoshimoto, M.: Low temperature dependence of oscillation wavelength in $GaAs_{1-x}Bi_x$ laser by photo-pumping. Appl. Phys. Express **3**, 062201 (2010)
14. Young, D.L., Geisz, J.F., Coutts, T.J.: Nitrogen-induced decrease of the electron effective mass in $GaAs_{1-x}N_x$ thin films measured by thermomagnetic transport phenomena. Appl. Phys. Lett. **82**, 1236 (2003)
15. Shan, W., Walukiewicz, W., Yu, K.M., Ager III, J.W., Haller, E.E., Geisz, J.F., Friedman, D.J., Olson, J.M., Kurtz, S.R., Xin, H.P., Tu, C.W.: Band anticrossing in III-N-V alloys. Phys. Status Solidi B **223**, 75 (2001)
16. Harris Jr., J.S., Yuen, H., Bank, S., Wistey, M., Lordi, V., Gugov, T., Bae, H., Goddaard, L.: MBE growth and characterization of dilute nitride III–V alloys. In: Henini, M. (ed.) Dilute Nitride Semiconductors, pp. 1–92. Elsevier, Amsterdam (2005)
17. Alberi, K., Wu, J., Walukiewicz, W., Yu, K.M., Dubon, O.D., Watkins, S.P., Wang, C.X., Liu, X., Cho, Y.-J., Furdyna, J.: Valence-band anticrossing in mismatched III–V semiconductor alloys. Phys. Rev. B **75**, 045203 (2007)
18. Lu, X., Beaton, D.A., Lewis, R.B., Tiedje, T., Zhang, Y.: Composition dependence of photoluminescence of $GaAs_{1-x}Bi_x$ alloys. Appl. Phys. Lett. **95**, 041903 (2009)

19. Pettinari, G., Polimeni, A., Capizzi, M., Blokland, J.H., Christianen, P.C.M., Maan, J.C., Young, E.C., Tiedje, T.: Influence of bismuth incorporation on the valence and conduction band edges of $GaAs_{1-x}Bi_x$. Appl. Phys. Lett. **92**, 262105 (2008)
20. Mohmad, A.R., Bastiman, F., Hunter, C.J., Ng, J.S., Sweeney, S.J., David, J.P.R.: The effect of Bi composition to the optical quality of $GaAs_{1-x}Bi_x$. Appl. Phys. Lett. **99**, 042107 (2011)
21. Imhof, S., Thränhardt, A., Chernikov, A., Koch, M., Köster, N.S., Kolata, K., Chatterjee, S., Koch, S.W., Lu, X., Johnson, S.R., Beaton, D.A., Tiedje, T., Rubel, O.: Clustering effects in Ga(AsBi). Appl. Phys. Lett. **96**, 131115 (2010)
22. Kudrawiec, R., Syperek, M., Poloczek, P., Misiewicz, J., Mari, R.H., Shafi, M., Henini, M., Galvao Gobato, Y., Novikov, S.V., Ibanez, J., Schmidbauer, M., Molina, S.I.: Carrier localization in GaBiAs probed by photomodulated transmittance and photoluminescence. J. Appl. Phys. **106**, 023518 (2009)
23. Beaton, D.A., Lewis, R.B., Masnadi-Shirazi, M., Tiedje, T.: Temperature dependence of hole mobility in $GaAs_{1-x}Bi_x$ alloys. J. Appl. Phys. **108**, 083708 (2010)
24. Kini, R.N., Ptak, A.J., Fluegel, B., France, R., Reedy, R.C., Mascarenhas, A.: Effect of Bi alloying on the hole transport in the dilute bismide alloy $GaAs_{1-x}Bi_x$. Phys. Rev. B **83**, 075307 (2011)
25. Pillai, M.R., Kim, S.S., Ho, S.T., Barnett, S.A.: Growth of $In_xGa_{1-x}As$/GaAs heterostructures using Bi as a surfactant. J. Vac. Sci. Technol. B **18**, 1232 (2000)
26. Feng, G., Oe, K., Yoshimoto, M.: Temperature dependence of Bi behavior in MBE growth of InGaAs/InP. J. Cryst. Growth **121**, 301–302 (2007)
27. Tixier, S., Adamcyk, M., Tiedje, T., Francoeur, S., Mascarenhas, A., Wei, P., Schiettekatte, F.: Molecular beam epitaxy growth of $GaAs_{1-x}Bi_x$. Appl. Phys. Lett. **82**, 2245 (2003)
28. Punkkinen, M.P.J., Laukkanen, P., Komsa, H.-P., Ahola-Tuomi, M., Räsänen, N., Kokko, K., Kuzmin, M., Adell, J., Sadowski, J., Perälä, R.E., Ropo, M., Rantala, T.T., Väyrynen, I.J., Pessa, M., Vitos, L., Kollár, J., Mirbt, S., Johansson, B.: Bismuth-stabilized (2×1) and (2×4) reconstructions on GaAs(100) surfaces: Combined first-principles, photoemission, and scanning tunneling microscopy study. Phys. Rev. B **78**, 195304 (2008)
29. Pashley, M.D.: Electron counting model and its application to island structures on molecular-beam epitaxy grown GaAs(001) and ZnSe(001). Phys. Rev. B **40**, 10481 (1989)
30. Laukkanen, P.: Unusual Bi-induced surfaces of III–V semiconductors. 1st International Workshop on Bismuth Containing Semiconductor, Ann Arbor. http://www.bismides.net/ (2010)
31. Neugebauer, J., Zywietz, T.K., Scheffler, M., Northrup, J.E., Chen, H., Feenstra, R.M.: Adatom kinetics on and below the surface: The existence of a new diffusion channel. Phys. Rev. Lett. **90**, 056101 (2003)
32. Ahola-Tuomi, M., Laukkanen, P., Perälä, R.E., Kuzmin, M., Pakarinen, J., Väyrynen, I.J., Adell, M.: Structural properties of Bi-terminated GaAs(001) surface. Surf. Sci. **600**, 2349 (2006)
33. Kent, P.R.C.: Electronic structure evolution of dilute III–V nitride alloys. In: Henini, M. (ed.) Dilute Nitride Semiconductors, pp. 393–413. Elsevier, Amsterdam (2005)
34. Tominaga, Y., Kinoshita, Y., Oe, K., Yoshimoto, M.: Structural investigation of $GaAs_{1-x}Bi_x$/GaAs multiquantum wells. Appl. Phys. Lett. **93**, 131915 (2008)
35. Yoshimoto, M., Huang, W., Feng, G., Oe, K.: New semiconductor alloy GaNAsBi with temperature-insensitive bandgap. Phys. Status Solidi B **243**, 1421 (2006)
36. Takehara, Y., Yoshimoto, M., Huang, W., Saraie, J., Oe, K., Chayahara, A., Horino, Y.: Lattice distortion of GaAsBi alloy grown on GaAs by molecular beam epitaxy. Jpn. J. Appl. Phys. **45**, 67 (2006)
37. Madelung, O. (ed.): Semiconductor-Group IV Elements and III–V Compound, p. 101. Springer, Berlin (1991)
38. Ueda, O., Fujii, T., Nakata, Y.: TEM investigation of defects in GaAsSb crystal grown on (001) InP substrates by molecular beam epitaxy. In: Sumino, K. (ed.) Defect Control in Semiconductors, pp. 1091–1096. Elsevier, North Holland (1990)
39. Ueda, O., Tominaga, Y., Ikenaga, N., Yoshimoto, M., Oe, K.: Structural evaluation of$GaAs_{1-x}Bi_x$ mixed crystals by TEM. Proceedings of the 23rd International Conference on Indium

Phosphide and Related Materials (IPRM2011), p. 1, Berlin, INSPEC Accession Number: 12172587 (2011)
40. Yu, P.Y., Cardona, M.: Fundamental of Semiconductors, p. 354. Springer, Berlin (2005)
41. Itoh, M., Tominaga, Y., Oe, K., Yoshimoto, M.: Quantitative estimation of density of Bi-induced localized states in $GaAs_{1-x}Bi_x$ grown by molecular beam epitaxy. J. Cryst. Growth **378**, 73 (2013)
42. Vurgaftman, I., Meyer, J.R.: Band parameters for nitrogen-containing semiconductors. J. Appl. Phys. **94**, 3675 (2003)
43. Kado, K., Fuyuki, T., Yamada, K., Oe, K., Yoshimoto, M.: High hole mobility in $GaAs_{1-x}Bi_x$ alloys. Jpn. J. Appl. Phys. **51**, 040204 (2012)
44. Shiobara, S., Hashizume, T., Hasegawa, H.: Deep level and conduction mechanism in low-temperature GaAs grown by molecular beam epitaxy. Jpn. J. Appl. Phys. **35**, 1159 (1996)
45. Luysberg, M., Shon, H., Prasad, A., Specht, P., Liliental-Weber, Z., Weber, E.R., Gebauer, J., Krause-Rwhberg, R.: Effects of the growth temperature and As/Ga flux ratio on the incorporation of excess As into low temperature grown GaAs. J. Appl. Phys. **83**, 561 (1998)
46. Kunzer, M., Jost, W., Kaufmann, U., Hobgood, H.M., Thomas, R.N.: Identification of the BiGa heteroantisite defect in GaAs:Bi. Phys. Rev. B **48**, 4437 (1993)
47. Kimerling, L.C., Patel, J.R.: Defect states associated with dislocations in silicon. Appl. Phys. Lett. **34**, 73 (1979)
48. Weber, E.R., Enness, H., Kaufmann, U., Windscheif, J., Schneider, Y., Wosinski, T.: Identification of AsGa antisites in plastically deformed GaAs. J. Appl. Phys. **53**, 6140 (1982)
49. Lagowski, J., Lin, D.G., Chen, T.P., Skowronski, M., Gatos, H.C.: Native hole trap in bulk GaAs and its association with the double-charge state of the arsenic antisite defect. Appl. Phys. Lett. **47**, 929 (1985)
50. Fuyuki, T., Kashiyama, S., Tominaga, Y., Oe, K., Yoshimoto, M.: Deep-hole traps in p-type $GaAs_{1-x}Bi_x$ grown by molecular beam epitaxy. Jpn. J. Appl. Phys. **50**, 080203 (2011)
51. Jiang, Z., Beaton, D.A., Lewis, R.B., Basile, A.F., Tiedje, T., Mooney, P.M.: Deep level defects in $GaAs_{1-x}Bi_x$/GaAs heterostructures. Semicond. Sci. Technol. **26**, 055020 (2011)
52. Laukkanen, P., Ahola-Tuomi, M., Kuzmin, M., Perala, R.E., Vayrynen, I.J., Tukiainen, A., Pakarinen, J., Saarinen, M., Pessa, M.: Structural properties of Bi-stabilized reconstructions of GaInAs(100) surface. Appl. Phys. Lett. **90**, 082101 (2007). and references therein
53. Teissier, R., Sicault, D., Harmand, J.C., Ungaro, G., Le Roux, G., Largeau, L.: Temperature-dependent valence band offset and band-gap energies of pseudomorphic GaAsSb on GaAs. J. Appl. Phys. **89**, 5473 (2001)
54. Laukkanen, P., Punkkinen, M.P.J., Kosma, H.-P., Ahola-Tuomi, M., Kokko, K., Kuzumin, M., Adell, J., Sadowski, J., Perälä, R.E., Ropo, M., Rantala, T.T., Väyrynen, I.J., Pessa, M., Vitos, L., Kollár, J., Mirbt, S., Johansson, B.: Anomalous bismuth-stabilized (2×1) reconstructions on GaAs(100) and InP(100) surfaces. Phys. Rev. Lett. **100**, 086101 (2008)
55. Fuyuki, T., Tominaga, Y., Oe, K., Yoshimoto, M.: Growth of $GaAs_{1-x}Bi_x/Al_yGa_{1-y}As$ multi-quantum-well structures. Jpn. J. Appl. Phys. **49**, 070211 (2010)
56. Fan, D., Zeng, Z., Hu, X., Dorogan, V.G., Li, C., Benamara, M., Hawkridge, M.E., Mazur, Y.I., Yu, S.-Q., Johnson, S.R., Wang, Z.M., Salamo, G.J.: Molecular beam epitaxy growth of GaAsBi/GaAs/AlGaAs separate confinement heterostructures. Appl. Phys. Lett. **101**, 181103 (2012)
57. Cho, Y.-H., Choe, B.-D., Kim, Y., Lim, H.: Interface states in $In_{0.5}Ga_{0.5}P/Al_xGa_{1-x}As$ heterostructures grown by liquid phase epitaxy. J. Appl. Phys. **81**, 7362 (1997)
58. Fuyuki, T., Kashiyama, S., Oe, K., Yoshimoto, M.: Interface states in p-type GaAs/$GaAs_{1-x}Bi_x$ heterostructure. Jpn. J. Appl. Phys. **51**, 11PC02 (2012)
59. Nicollian, E.H., Brews, J.R.: MOS Physics and Technology, p. 183. Wiley, New York (2003)
60. Hossain, N., Marco, I.P., Jin, S.R., Hild, K., Sweeney, S.J., Lewis, R.B., Beaton, D.A., Tiedje, T.: Recombination mechanisms and band alignment of $GaAs_{1-x}Bi_x$/GaAs light emitting diodes. Appl. Phys. Lett. **100**, 051105 (2012)
61. Usman, M., Broderick, C.A., Lindsay, A., O'Reilly, E.P.: Tight-binding analysis of the electronic structure of dilute bismide alloys of GaP and GaAs. Phys. Rev. B **84**, 245202 (2011)

62. Krispin, P., Spuruytte, S.G., Harris, J.S., Ploog, K.H.: Electrical depth profile of p-type GaAs/Ga(As, N)/GaAs heterostructures determined by capacitance-voltage measurements. J. Appl. Phys. **88**, 4153 (2000)
63. Tominaga, Y., Oe, K., Yoshimoto, M.: Variations in the abruptness at $GaAs_{1-x}Bi_x$/GaAs hetero-interfaces caused by thermal annealing. In: Extract Abstract 38th International Symposium on Compound Semiconductor, p. 426 (2011)
64. Duke, C.B.: Semiconductor surface reconstruction: The structural chemistry of two-dimensional surface compounds. Chem. Rev. **96**, 1237 (1996)
65. Chadi, D.J.: Atomic structure of GaAs(100)-(2×1) and (2×4) reconstructed surfaces. J. Vac. Sci. Technol. A **5**, 834 (1987)

Chapter 10
Unusual Bi-Containing Surface Layers of III–V Compound Semiconductors

Pekka Laukkanen and Marko Punkkinen

Abstract In this chapter, it is first described how the surface science and engineering of the III–V compound semiconductors are relevant to developing the semiconductor-based materials, including bismuth (Bi) containing III–V films, for improved electronics and optoelectronics devices. After that the general properties of the III–V(100) surfaces, which usually undergo strong atomic rearrangements (reconstructions), are reviewed. Before focusing on the unusual Bi-induced III–V(100)(2 × 1) reconstructions and Bi-induced changes in the III–V(100)(2 × 4) surfaces, it is described the experimental and computational research methods used as well as the advantages obtained by combining the experimental and computational results in the analysis. Calculational results reveal an interesting correlation between the geometry and surface band structure of the (2 × 1) reconstruction, and the relative stability of the (2 × 1) reconstruction.

10.1 Role of Semiconductor Surfaces in Development of Device Materials

Monocrystalline III–V compound semiconductor (e.g., GaAs, InP, and GaSb) films are the established materials of devices like laser diodes (Fig. 10.1) and mobile-phone transistors, also in future very probably. Furthermore, the integration of III–V's films on silicon (Si) semiconductor substrates, the latter of which form the well-refined basis for most electronics components, has been intensively investigated for developing, for example, the transistors of radio-frequency and

P. Laukkanen (✉) • M. Punkkinen
Materials Research Laboratory, Department of Physics and Astronomy, University of Turku, Turku 20014, Finland
e-mail: pekka.laukkanen@utu.fi; marko.punkkinen@utu.fi

H. Li and Z.M. Wang (eds.), *Bismuth-Containing Compounds*, Springer Series in Materials Science 186, DOI 10.1007/978-1-4614-8121-8_10,

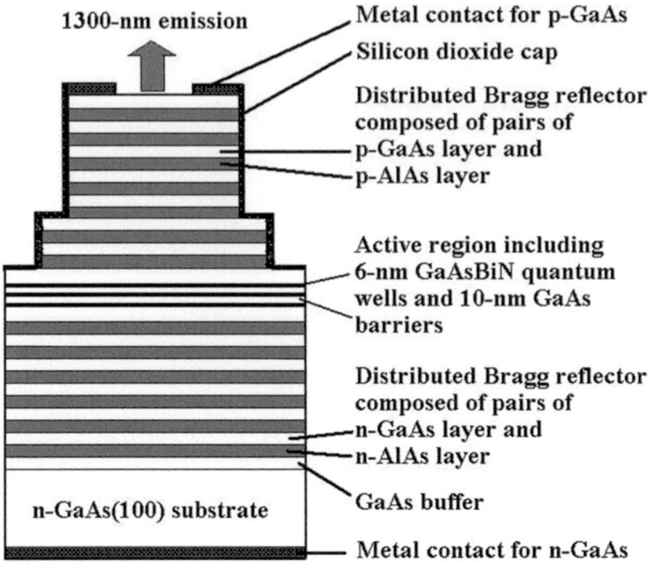

Fig. 10.1 Schematic laser diode structure which includes several monocrystalline, epitaxial III–V semiconductor films grown in the manner of atomic layer by layer

integrated-circuit applications (e.g., [1–5]). The common target in the research and development of semiconductor materials for the devices, that save energy and own extended lifetime, is to improve the crystal quality. Any improvement in the crystal order of the films and of their interfaces will decrease the density of harmful defects which impair the device performance.

In general, semiconductor interfaces have a crucial role in the operation of various devices like a laser diode of which schematic film structure (stack) is shown in Fig. 10.1. The interfaces affect, for instance, the flow, confinement, and recombination of electrons. The role of interfaces is expected to increase further in future because the physical size of many devices is continuously reduced.

Semiconductor film stacks for devices are usually produced by epitaxial growth techniques such as molecular beam epitaxy (MBE) and metal organic chemical vapor deposition (MOCVD). In these processes, a stack of semiconductor films grows in the manner of atomic layer by layer. The "epitaxy" means that a growing layer has the same crystal lattice as the substrate has. An MBE chamber provides the most clean growth environment to avoid the incorporation of harmful impurities (e.g., oxygen and carbon contamination) into the grown films, because the MBE is based on the ultrahigh vacuum (UHV) technique [6–8]. It should be also mentioned that the MBE is a relatively expensive method to produce the device materials. It is an advantage of the MBE technique that its growth chamber is usually equipped with the reflection high-energy electron diffraction (RHEED) method [6–9]. RHEED makes it possible to monitor the crystal structure and quality of the growing film in situ, in real time.

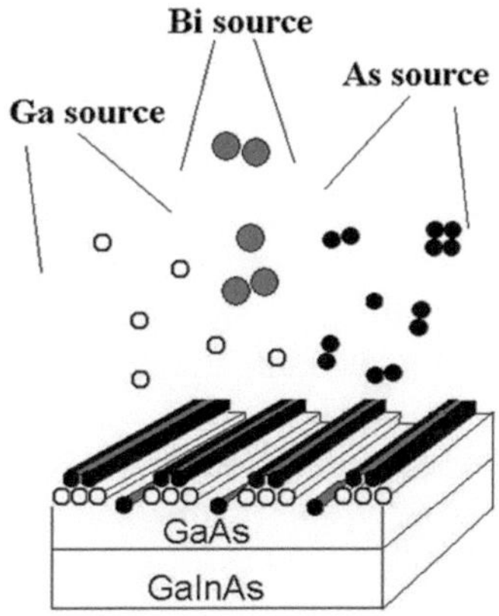

Fig. 10.2 Schematic MBE growth condition in which a surface layer of the growing film stack is reconstructed during the whole growth usually

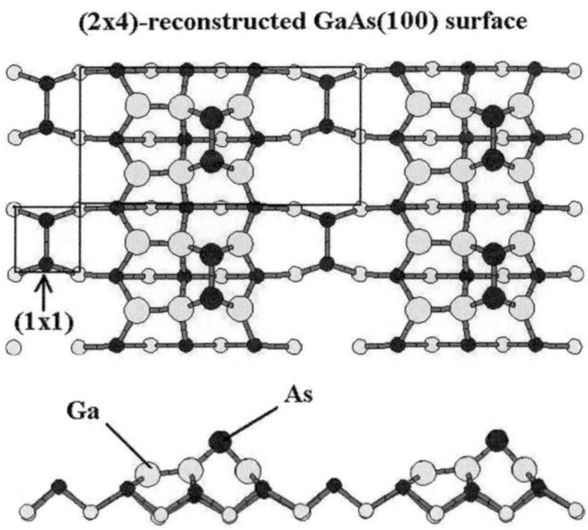

Fig. 10.3 Example of semiconductor surface reconstructions. A unit cell of the lattice of a surface layer is marked by a *rectangle*. A small *square* marks the unit cell of the corresponding bulk plane lattice

The epitaxial growth is essentially a surface process which is governed by the atomic and electronic structures of the semiconductor surface (i.e., the surface of a starting substrate or growing film). Usually a clean semiconductor surface is reconstructed; its structure is clearly different from that of the corresponding plane deeper in the bulk. (The main factors behind the reconstructions are described in the next chapter.) Thus, the growth front of a film or film stack is often reconstructed during the whole growth process (Fig. 10.2). A surface reconstruction can change during the growth, for example, when the composition of the layer is changed. These changes can be monitored (controlled) with RHEED.

A reconstruction example is shown in Fig. 10.3. It is a surface structure of the monocrystalline, cubic GaAs(100) substrate. The lattice unit cell of a surface layer is marked by a rectangle and the corresponding (1×1) unit cell of the (100) substrate plane is marked by a small square. The surface cell is 2 times larger than the

(1×1) cell in the direction of the vertical crystal vector (i.e., the [0–11] direction) and 4 times larger into the perpendicular direction (i.e., the [011] azimuth). Therefore, the surface reconstruction is called as (2×4). Its superlattice is simply related to the (100)(1×1) plane structure. All the surfaces considered in this chapter are such superstructures, and the surfaces are expressed in terms of the Wood notation [10–13]: $S(hkl)(n \times m)R\theta - A$, where $S(hkl)$ gives the crystal plane of the material with the chemical composition of S, n and m are the proportional lengths of the vectors defining the real-space surface lattice compared to the reference vectors of the (1×1) bulk plane, θ is the angle of possible rotation between the lattices defining the reconstruction, and A is the possible adsorbate inducing the reconstruction. The $R\theta - A$ term is not marked if the rotation angle is zero and the surface layer consists of only the substrate elements.

As can be seen from Fig. 10.3, the atomic structure of a semiconductor surface can be much more complex than that for the "ideal" bulk plane terminated (1×1) surface. Two common features of semiconductor reconstructions can be seen in Fig. 10.3: the formation of new bonding configurations and the corrugated geometry. First, the GaAs(100)(2×4) surface contains pairs (dimers) of two As atoms and also Ga–Ga pairs, which are not formed in a defect-free bulk GaAs. Second, the atomic-scale roughness of the GaAs(100)(2×4) arises from the third atomic layer that is exposed in addition to the first (i.e., topmost) and second layers. These surface properties affect the epitaxial growth processes (e.g., adsorption and surface diffusion of constituent atoms) as well as the characteristics of the grown interface (e.g., chemical composition).

The point is that the device material properties can be tailored by modifying the semiconductor reconstruction of the growth front. Indeed the literature includes several examples of the benefits of such surface engineering. Only three of them are shortly introduced here to motivate the rest part of this chapter. One of the earliest works, concerning the III–V surface engineering, is the research of Anan et al. [14]. They studied the InP/InGaAs quantum-well interface (i.e., infrared laser material) which is problematic because the As flux is changed to the P flux during the interface growth. The formation of Ga–P interfacial bonds has been found to be harmful to the material properties. Anan et al. found that it is useful to add a small amount of indium, about one monolayer (1 ML), on the InGaAs(100)(2×4) growth front before the As to P flux changes. That provides an In-stabilized $c(8 \times 2)$-reconstructed InGaAs(100) surface, as deduced by RHEED, to continue further the heteroepitaxial InP growth [14]. Anan et al. observed even 100-fold improvement in photoluminescence (PL) intensity for the InP/InGaAs quantum-well structure, in the growth of which the In-induced InGaAs(100)$c(8 \times 2)$ was prepared before the P_2 exposure, as compared to the structure grown without the In prelayer. It is worth noting that usually the PL-intensity increase is inversely proportional to the change in threshold current density of laser diodes.

The second example deals with the growth of III-Sb/InAs heterostructures for infrared devices. The group V flux change is problematic also here: an Sb exposure of the InAs(100)(2×4) growth front leads to a rough surface to continue the III-Sb layer growth. Nosho et al. [15] have found that this roughness, which is related to the above-described (Fig. 10.3) corrugated structure of the (2×4) reconstructions, can be eliminated by depositing a proper amount of In on the InAs(100)(2×4) before opening the Sb flux [15].

The growth of a rather new III–V compound, $GaAs_{1-x}Bi_x$ in combination with a surfactant effect of Bi, is the third example. It has been found that a low growth temperature, less than 400 °C, is needed to incorporate Bi into the crystal, which means that Bi atoms only float on the growth front at the normal growth temperatures (500–650 °C) of III–V films. This property makes Bi a potential surfactant. Indeed, it has been found that the Bi surfactant (a floating Bi layer) improves the III–V film quality [16], which is related to a Bi-induced change in the atomic structure of the growth front. In particular, it has been considered that the Bi surfactant is helpful to improve the crystal quality of III–V films which must be grown at low temperatures. Dilute nitrides, GaInAsN, are a good, notorious example. That material system has been intensely investigated from the early 1990s (e.g., [17–24]) because of various potential infrared applications of GaInAsN. It appears that many groups have given up the GaInAsN research and development due to the formation of high density of unknown N-induced defects. However, the recent results from the multijunction solar-cells that include GaInAsN films [25] are expected to generate a new boom, and here the Bi surfactant might be the key to further develop GaInAsN properties. Still, it is unclear which Bi-induced reconstruction(s) is helpful in the surfactant growth. In contrast, Masnadi-Shirazi et al. have recently [26] demonstrated that a Bi-induced (2×1) reconstruction, which is not normally seen during the III–V epitaxy, is a very useful growth front for $GaAs_{1-x}Bi_x$ films. As compared to other reconstructions observed during the $GaAs_{1-x}Bi_x$ growth, the unusual (2×1)-Bi surface provides a clear improvement in the Bi incorporation into the crystal and also in the photoluminescence intensity of the grown film. To understand and control the beneficial effects of the (2×1)-Bi surface as well as to improve the understanding of other Bi-induced reconstructions, potential for the surfactant-mediated epitaxy, it is helpful to determine the atomic and electronic structures of the Bi-containing III–V surfaces which appear in the experiments.

In this chapter, it is reviewed some results obtained from recent investigations of Bi-induced reconstructions of III–V surfaces. The focus is put on the Bi-induced surface structures of the GaAs(100) because they are relevant to the epitaxial growth of the $GaAs_{1-x}Bi$ and $GaAs_{1-x-y}Bi_xN_y$ films and also to the Bi surfactant-mediated growth of the GaAs-based compounds. Before that, reconstruction mechanisms and the used research methods are described.

10.2 General Principles Behind III–V Semiconductor Surface Reconstructions

Surface can be formed by cutting a solid sample along some direction. The cutting breaks the bonds in perpendicular to the opened surface. This leads, however, generally to a very high total energy of the formed system, i.e., to a large surface energy.

Therefore, the surface tends to minimize the number of unused broken bonds, so-called dangling bonds, by forming new bonds within the surface. Typically one finds that single surface-layer atoms pair by forming dimer bonds. Dimers can be formed by atoms of the same element (homodimers) or different elements (heterodimers). However, dimers can be formed also in the subsurface layer [27–29]. Thus, usually by dimer one means a pair of atoms which form a bond (in parallel to the surface) which is not characteristic for the underlying bulk. This description explains also why dimers are formed at semiconductor surfaces. The bonds are highly directional in semiconductors, and, therefore, the surface reconstruction is generally much more drastic on semiconductor surfaces compared to metal surfaces.

Perhaps the most famous general principle governing the formation of III–V semiconductor surfaces is the so-called electron counting model (ECM) or electron counting rule (ECR) [30]. This principle is rooted in the octet rule, which states that light atoms tend to combine in a way that they have eight electrons in their valence shell. This is satisfied in the bulk of III–V semiconductors as the atoms have four heteronuclear bonds, and every group III atom gives 0.75 electrons to a bond and every group V atom gives 1.25 atoms to a bond. (However, one should keep in mind that in reality the valence electrons are not confined to a particular bond and atom due to their delocalized nature, and that there is no such region in the crystal which we could unambiguously identify as a bond. This means that we should not use a simplifying model/rule to interpret the underlying physics. Instead, it should be used to identify those structures, which satisfy the rule.) Generally, at surfaces some of the atoms become three-coordinated, for example, the edge Ga atoms and As dimers in Fig. 10.3. This can mean a metallic state because the dangling bonds before the surface reconstruction are partially filled (0.75 electrons in group III dangling bonds and 1.25 electrons in group V dangling bonds). Let's consider a heterodimer; if the group III atom donates its dangling bond electrons to the group V atom, then the dangling bond of the group III atom becomes empty while that of the group V atom becomes fully occupied. The ECM states that a semiconductor surface is reconstructed in a way that the electron number in the surface region is equal to the number of bonds in the same region multiplied by 2. Here it is essential to include only the (full) group V dangling bonds in the total bond number. This principle is true, because the group III atomistic sp^3 hybrid orbitals lie at higher energies than the corresponding group V hybrid orbitals. The energy band associated with the dangling bond of the group III atom is within the conduction band, whereas the energy band associated with the dangling bond of the group V atom is within the valence band. Therefore, the semiconductor surface is semiconducting if it obeys the ECM. One has to identify the bonds to use the ECM. This is usually easy. If an atom has more than three bonds (coordination of five is also possible), then it does not have dangling bonds. Important is that within the ECM the semiconductor preserves its semiconducting nature also at the surface. Within the ECM, the octet rule is partially satisfied (for group V atoms). However, it turns out that the ECM, like the octet rule, has its own exceptions.

Mirbt et al. derived a more quantitative general principle which should be in general applied to those surface structures which obey the ECM. That takes into account the experimental surrounding conditions on the surface energy of the considered surface [31]. Repulsion between electronic charges (e.g., repulsion between filled dangling bonds [32, 33]) and reconstruction-induced surface strain/stress are other important factors affecting the surface energies and relative stabilities of different surface reconstructions.

Let's consider an example how the available electrons and the number of atomic bonds to be filled by electrons can be simply calculated to test whether an atomic model obeys the ECM. Let's examine the (2×4) unit cell in Fig. 10.3 and count first the valence electrons that As and Ga atoms contribute to the bonds of the (2×4) surface region which includes four atomic layers for clarity. (To apply the ECM to a surface, one has to consider, at least, all partially filled atomic layers and one full atomic layer below the partially filled layers.) *In the fourth atomic layer*, there are three Ga atoms inside of the (2×4)-cell area, eight edge atoms which give the half of their electrons to the neighboring cells, and four corner atoms each of which provides 0.25 electron to the (2×4) cell. It should be also noted that the fourth-layer atoms give the half of their electrons to the bonds below the considered slab toward the fifth layer. Thus, the total number of the electrons which the fourth-layer Ga atoms contribute to the (2×4) slab is $[3{\cdot}3+8{\cdot}3{\cdot}1/2+1/4{\cdot}4{\cdot}3]{\cdot}1/2=12$ electrons. *The third layer* includes five As atoms inside of the (2×4) cell area and six edge atoms, which makes $[5{\cdot}5+6{\cdot}5{\cdot}1/2]=40$ electrons. The second layer includes six inside Ga atoms [i.e., 18 electrons], and the first-layer As atoms give ten electrons. Thus there are 80 valence electrons available. Let's count next the number of the bonds that should be filled. *Between the fourth and third layers*, there are ten bonds inside of the (2×4) cell plus 12 edge bonds which get the half of the electrons from the neighboring cells (i.e., effectively six edge bonds). *In the third-layer As dimer* has one dimer bond and two dangling bonds to be filled. *Between the third and second layers*, there are 12 bonds. *The second layer Ga atoms* have 2 Ga–Ga dimer bonds (their dangling bonds should be empty). Furthermore, the slab includes *four bonds between the second and first layers* plus *one As–As dimer bond plus two dangling bonds*. Thus, the total number of the bonds per (2×4) slab is 40. Each bond is filled (saturated) by two electrons. So the number of the electrons needed to fill all the bonds (states below the valence-band maximum) is exactly the same as the number of the available valence electrons.

10.3 Research Methods

In this chapter, some experimental and computational techniques, which are useful in the investigations of atomic and electronic structures of the Bi-containing surfaces, are first introduced. The purpose is to describe shortly both strengths and limitations of each method, concerning the studies of the surface atomic and electronic structures. After that, it is discussed how the measured and calculated results can be combined to improve the analysis.

10.3.1 Experiments

In order to produce a clean well-ordered semiconductor surface for measurements, it is necessary to utilize UHV techniques because semiconductor surfaces are oxidized (and contaminated) quickly in air exposure and, in general, in any contact with an oxygen-containing medium. Therefore, the preparation and characterization of a semiconductor surface are usually performed in UHV chambers, of which base pressures are less than $1\cdot10^{-9}$ mbar. So, the standard surface-science facilities can be used (all the experiments described below have been performed in UHV systems). Here, it might be already addressed the question how well the surface-science experiments of such clean and well-ordered "model systems" correspond to reality. The correspondence is exceptionally good between the semiconductor surface science and semiconductor applications. For example, the epitaxial growth is carried out in very clean environment; in an MBE chamber, the base pressure is below $1\cdot10^{-9}$ mbar. Furthermore, as mentioned above, the heart of many devices is the monocrystalline semiconductor film stack, of which growth is carried out in the atomic layer-by-layer manner via reconstructed growth fronts.

Preparation of the clean starting surface of a semiconductor crystal for further processing (e.g., for the Bi deposition) is the first crucial step in experiments. A well-defined (ordered) semiconductor surface is a prerequisite for obtaining a justified interpretation of a measurement, and also for producing various high-quality device interfaces on the substrate. The cleaning might be a time-consuming, even frustrating, job before "the interesting experiments" on the substrate because a semiconductor surface layer is often the most defective part of the material. It can include badly oxidized areas for instance.

An MBE equipment which is in situ connected to a surface-science chamber provides perhaps the best method to prepare clean and long-range ordered III–V surfaces for investigations. The MBE equipped with RHEED is in particular helpful to finish a surface reconstruction before transferring the sample into the surface-science chamber without breaking the UHV environment. This method makes it possible to study the Bi behavior on the group V-stabilized (rich) III–V surfaces [e.g., GaAs(100)(2×4) and -c(4×4)]; the traditional surface-science system does not include an As or P source. Since Bi can be removed from the III–V surface simply by heating so that the substrate crystal is not decomposed, it is possible to perform several Bi deposition (and reevaporation) in situ tests in MBE on the same substrate before transferring an interesting and refined Bi-induced surface structure for further characterization, as it has been recently done [34]. The end of the surface preparation in MBE might however require a special trick(s) because the group V molecules (e.g., As_2, As_4, P_2, and P_4), which often become enriched in an MBE chamber due to the epitaxial growth, are easily adsorbed on the surface during its cooling.

The surface-science system equipped with the MBE-RHEED is still a rare tool. Therefore, a capping technique, which allows the sample transfer via air, is widely used in III–V surface studies. At least, As, P, and Sb cap layers protect the III–V substrate against the oxidation in air exposure, and these cap layers can be easily removed in a surface-science system by heating the substrate in UHV. For example, after the MBE growth of a GaAs film, the sample temperature is decreased to room temperature in the As_2 flux when an amorphous As cap layer grows on the GaAs film. After transferring the GaAs sample into a UHV chamber, the As cap layer including both As oxides and pure As can be removed by annealing around 350 °C. This approach has been previously utilized also in the investigations of Bi-induced reconstructions [35–38].

The third common cleaning procedure of III–V surfaces is an Ar-ion sputtering combined with the post-heating. This kind of Ar bombardment always destroys a substrate surface layer making it amorphous, but it is possible to recover the crystal structure at least to some extent by heating the substrate in UHV after the sputtering. Several cycles of the sputtering and post-heating are often required to obtain enough smooth and uniform substrate for surface studies. The Ar sputtering is the standard tool of surface-science chambers.

There are several *ways to evaporate Bi* in the UHV environment because a relative low temperature (less than 500 °C) is enough to remove Bi atoms and/or Bi dimers from a pure Bi metal piece. Therefore, no expensive cracker source is needed for the Bi deposition in MBE, but the common effusion cells (i.e., the group III sources) can be used for creating the Bi flux. In surface-science experiments, a small amount of Bi is often evaporated as compared to the MBE film growth. Thus, home-made evaporators can be utilized to obtain a proper Bi flux. Perhaps the most simple way is to wrap a Bi lump in a tungsten-wire coil which is heated by the direct current. This method was used recently [35–41]. To avoid dropping of the Bi piece from the coil into the chamber, a lump can be wrapped by a thin Ta sheet. The direct current can be driven through such a Ta envelope including Bi pieces.

Low-energy electron diffraction (*LEED*) is often the first method to characterize a sample in a surface-science chamber. The principles of LEED are found in many books (e.g., [10–12]). LEED provides quick responses to questions how well the surface cleaning procedure of a monocrystalline sample has succeeded and how the lattice of a surface layer has changed, as compared to the lattice of the (1 × 1) bulk plane. Although LEED is a very surface-sensitive probe, the LEED signal always includes some contribution from the atomic layers below the topmost one. The signal depth depends on the kinetic energy of elastic electrons [10–12], of which mean free path is energy-dependent. The minimum signal depth is around 1 nm (note that the signal depth is about 3 times the mean free path). On the other hand, this character makes it possible to probe the formation of the buried interfaces: it can be, for example, monitored whether a thin film obeys the planar lattice of the substrate, in the epitaxial manner, or starts to grow with a different crystal structure.

Figure 10.4 shows an example LEED pattern which visualizes the reciprocal lattice of a (2 × 1)-reconstructed III–V(100) surface with the square (1 × 1) planar cell. The integer diffraction spots, (0,0), (0,1), (1,0), (1,1), etc., set the reciprocal lattice of the bulk plane. In contrast, the presence of the fraction spots, (½, 0), (½, 1), etc., reveal the reconstruction (i.e., change in the lattice of a surface layer). The reciprocal unit cell of this kind of superstructure is shown by the white rectangle in

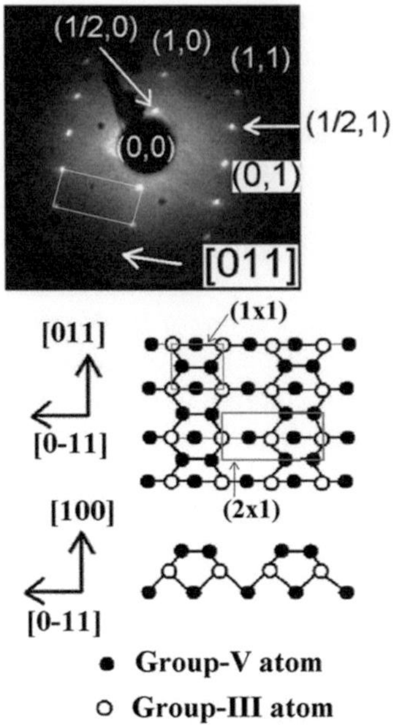

Fig. 10.4 LEED pattern showing a (2 × 1) reconstruction of the III–V(100) surface, and possible atomic arrangement of the III–V(100)(2 × 1) surface. The vectors, defined by the (00) and (10) spots and the (00) and (01) spots, represent the reciprocal lattice vectors for the bulk plane, as visualized by LEED. For a reconstructed III–V(100)(2 × 1) surface, the two vectors defined by the (00) and (1/2 0) spots and the (00) and (01) spots determine the reciprocal lattice of a surface layer. The *white rectangle* shows the (2 × 1) reciprocal unit cell. The *gray square* and *rectangle* represent the real-space bulk lattice cell and surface cell, respectively

Fig. 10.4. In the case of the III–V(100) surfaces, the 2× LEED periodicity often indicates the formation of dimers, but obviously much more data are needed to determine the atomic structure.

The substrate surface exhibiting a reasonable LEED pattern, like that in Fig. 10.4, could still have contaminated areas with poor crystalline quality. This is because the LEED electrons own a short coherence length (about 10–50 nm), which is a maximum width (diameter) of the surface area from which emitted electrons add coherently to the intensity. Thus LEED is not so sensitive to surface defects, and complementary measurements are needed to confirm the high quality of the surface.

LEED is a useful measurement for assessing the surface quality during the cleaning procedure. The absence of the diffraction spots indicates that a surface layer is amorphous (e.g., badly oxidized). If the bulk plane (1 × 1) pattern appears with a strong background intensity, the surface must be cleaned to obtain sharp spots. Furthermore, a LEED pattern from such a semiconductor surface should often include extra reconstructions spots, in addition to the (1 × 1) ones, because clean and well-ordered semiconductor surfaces are usually reconstructed, as discussed above.

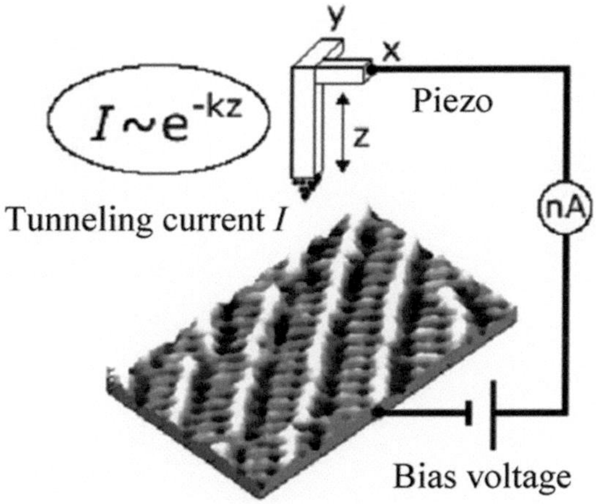

Fig. 10.5 Schematic operation of STM. Depending on the bias voltage, the electrons tunnel from the occupied surface states of the sample to the tip of the STM needle (i.e., filled-state mode) or from the tip to the empty states of the sample (i.e., empty state mode). In the constant-current mode, the tip is swept in the x and y directions over a chosen area. When the tunneling current locally increases (due to geometric and/or electronic structure) during the scanning, the piezo controller moves the tip away from the surface. This place is recorded as a white rise in this case. Here the white lines arise from the self-assembled Bi nanolines on the InAs(100)(2 × 1)-Bi template [40, 41]

Scanning tunneling microscopy/spectroscopy (*STM/STS*) is a complementary method to LEED. The physics of STM and STS has been described in detail in many books (e.g., [10–13, 42]) (Fig. 10.5). First of all, large-scale (e.g., 1 × 1 μm) STM images, which show smooth surface with two-dimensional islands (thickness of a few atomic layers), clearly support that a well-defined surface has been obtained for the measurements and further processing. STM reveals possible defective areas, of which presence cannot be excluded by LEED. Such microscopic info is irreplaceable in surface-science and engineering experiments, and can be obtained quickly. In contrast, the atomic-scale imaging that contributes to solving the atomic structure might be time-consuming.

An example STM image from a Bi-covered GaInAs(100) surface is shown in Fig. 10.6. The surface includes different areas: Bi-induced (2 × 1)- and (2 × 4)-reconstructed areas. The contour line in Fig. 10.6 shows that the atoms, which cause the white oval features marked by *A*, sit at higher geometric position than the gray features marked by *B*. However, it has been shown [38] that both the features reside in the same, topmost atomic layer, as can be seen from the schematic atomic model of Fig. 10.6. The contrast difference between the *A* and *B* features arises from an electronic structure rather than from the geometric structure. Sometimes it is difficult to conclude whether the geometric or electronic effect or the both cause STM features. In Fig. 10.6, the difference in the local electron density (filled-state image) is most likely related to the difference that the (2 × 4) phase is semiconducting and the (2 × 1) is metallic, as shown below.

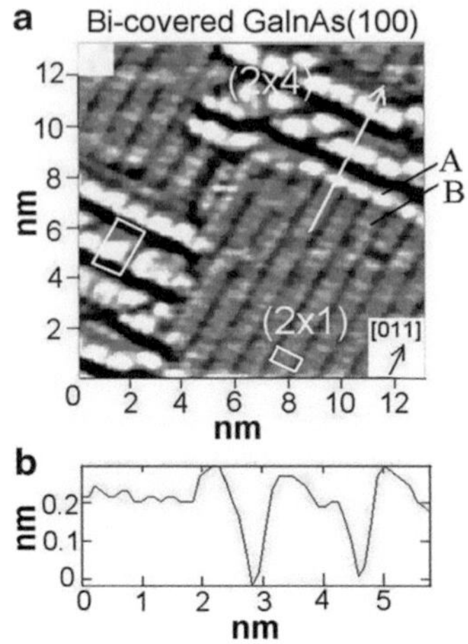

Fig. 10.6 (**a**) STM image from the GaInAs(100) surface, which includes areas with both the Bi-induced (2×1) and (2×4) reconstructions. (**b**) Contour line along the *white arrow*. The figure is adapted from Ref. [38] Appl. Phys. Lett. **90**, 082101 (2007)

The contour line in Fig. 10.6 shows that, in the (2×4) phase, the height of the topmost white feature measured from the black trench is about 0.3 nm. This indicates that also the third atomic layer is exposed because the distance between the neighboring group V and III layers is about 0.15 nm for $Ga_{0.5}In_{0.5}As$. From this kind of STM image, it can be determined the basic features which cause the reconstruction periodicities, and tentative atomic models can be proposed, taking into account also the previous knowledge on the common building blocks of the reconstructions. Obviously, more information is required for solving the detailed atomic structure.

To characterize the electronic properties of the Bi-induced III–V reconstruction, it has been previously [35, 37, 38] measured current–voltage (*I–V*) curves in the certain points on the surface by the STS mode. A chosen grid determines the points where *I–V* curves are measured. The benefit is that the topographic image is measured simultaneously, making it possible to choose a well-defined area inside of which the grid-point *I–V* curves are averaged and then numerically analyzed afterwards. Because the *I–V* curves are affected by the STM measurement parameters (e.g., the tip distance from the surface), it is important to measure *I–V* curves with various STM parameters. Here it is helpful to measure *I–V* curves from some reference sample of which surface band gap and the states around the gap are known for the comparison.

Core-level photoelectron spectroscopy (e.g., [12, 13, 43–46]) has been utilized to study the atomic and electronic structures of the Bi-covered *III–V* surfaces. In particular, the synchrotron-radiation facilities [43] at the MAX-lab in the Lund University (Sweden) have been used to obtain high-resolution photoelectron spectra. Figure 10.7 presents an example spectrum: a Bi 5*d* core-level spectrum measured from the Bi-induced GaAs(100)(2×4) reconstruction. It is worth noting that the Bi 5*d* emission is a double peak spectrum including both the $d_{5/2}$ and $d_{3/2}$ spin-orbit-split peaks. The separation of these peaks is about 3 eV, which makes it possible to analyze only one of the spin-orbit-split peaks, improving the justification

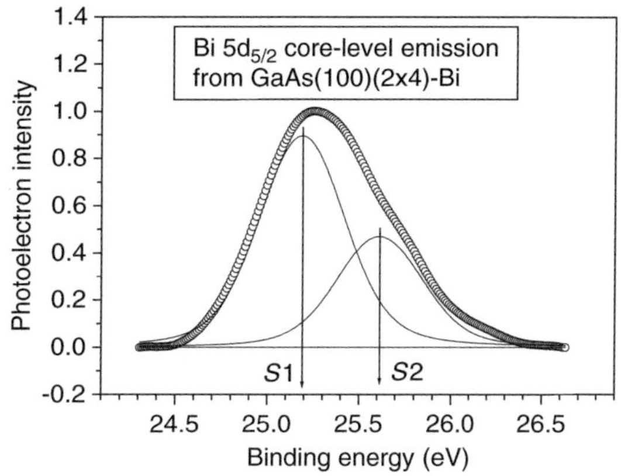

Fig. 10.7 Bi core-level photoelectron spectrum from a GaAs(100)(2×4)-Bi surface

of the data interpretation. The Bi spectrum in Fig. 10.7, first of all, confirms the presence of Bi on the surface (common LEED and STM do not provide such chemical identification). It can be also seen that the spectrum includes at least two components, *S*1 and *S*2, of which presence can be concluded from the pure line shape (i.e., from a clear shoulder). The presence of the two components means that Bi atoms have two clearly different bonding environments (i.e., two electronically different crystal sites) on the surface. Furthermore, there are three choices: Bi atoms occupy two different sites per the (2×4) cell and/or the (2×4) cells have only one Bi atom/cell, but the Bi site changes between the (2×4) cells. To make such conclusions about the atomic structure, it is essential that the sample surface included only the (2×4) reconstruction. The case would be obviously ambiguous if the sample surface included two or more different phases, as in Fig. 10.6. Because the photoelectron signal arises from a macroscopic surface area in the standard equipments, it is very crucial to prepare a uniform well-ordered surface for the measurements.

The intensity of the *S*1 component is about 2 times the *S*2 intensity, indicating that the number of the Bi atoms occupying the *S*1 sites is twice as large as the number of the Bi atoms in the *S*2 sites on the surface. Although the number of atoms directly contributes to the photoelectron intensity, the analysis is complicated by different attenuation effects for the atoms that possibly locate in different atomic layers. By changing the photoelectron emission angle or the photon energy (i.e., the photoelectron kinetic energy), it is possible to change the surface sensitivity of the signal and to study whether the intensity of some component(s) increases or decreases. Such changes indicate that different atomic sites reside in the different atomic layers.

By means of the pure experimental data (without ab initio calculations), it is challenging to suggest the atomic origins for the core-level shifts (CLSs) (i.e., the components like *S*1 and *S*2). In general, CLSs have been discussed with one or more of three common factors: changes in on-site atomic charges (i.e., ionic character), in Madelung potential (i.e., electrostatic potential caused by other atoms in crystal), and in final state screening (i.e., after the photoelectron removal), as compared to the bulk [47]. Previously, the origins for CLSs measured from III–V semiconductor surfaces have been under an intensive research (e.g., [48–60]). On the basis of this literature, it is now clear that the traditional reasoning of CLSs with the mere on-site atomic-charge picture is too simplified, unfortunately. For example, the recent results for the III–V(100)$c(8\times2)$ surfaces [60] support the conclusion according to which changes in the Madelung potential are the dominant factors behind the CLSs of the III–V surfaces. However, the results [58–60] indicate that the initial state electronic structure explains reasonably CLSs of a III–V surface (i.e., the final state screening has a minor role on the III–Vs).

Valence-band photoelectron spectroscopy readily provides information about the Fermi level position at a semiconductor surface, in relation to the valence-band maximum (possible surface band bending can be determined), and about the density of states (DOS) at the Fermi level (i.e., metallic character of a semiconductor surface). Usually a clean well-ordered III–V surface is semiconducting with a clear surface band gap, which is a bit smaller than the bulk band gap, and without any significant surface band bending. Thus, there should not be any significant DOS at the Fermi level on a clean III–V semiconductor surface. Then it can be asked how the Fermi level position can be determined if there is no electron at the Fermi level. That is obtained by measuring the Fermi level position (energy) from a metallic reference sample, such as a Ta plate, which is electrically connected to the backside of a semiconductor sample. That is, a Fermi edge valence-band spectrum is measured from the Ta plate using the same photon energy as for the semiconductor measurement. The determined Ta Fermi energy gives the Fermi level position also on the semiconductor surface because the Fermi levels in the Ta and semiconductor are aligned.

An angle-resolved spectrometer enables the mapping of surface bands, energy–crystal momentum (E–k) dispersions along the symmetry axes (e.g., [61]). Such band structures are sensitive to the surface atomic structures, which can be utilized in the determination of detailed atomic model.

10.3.2 Calculations

Ab initio thermodynamics (e.g., [62–64]) provides an efficient, reliable, and powerful method to determine the reconstruction with the lowest surface energy from a group of model surface structures. For a three-component system, like Bi/GaAs, the surface energy γ can be written in the form

$$\gamma A = E_{\text{tot,slab}} - \mu_{\text{GaAs}} N_{\text{As}} - (N_{\text{Ga}} - N_{\text{As}})\mu_{\text{Ga}} - N_{\text{Bi}}\mu_{\text{Bi}}, \tag{10.1}$$

where A, $E_{\text{tot,slab}}$, N_{Ga}, N_{As}, and N_{Bi} denote surface area, total energy of a slab unit cell, and the number of Ga, As, and Bi atoms, respectively, for the considered reconstruction. Here μ_{GaAs}, μ_{Ga}, and μ_{Bi} are the chemical potentials for bulk GaAs, Ga, and Bi, respectively. Thus, the reservoirs with which the surface is in equilibrium represent solid phases. The μ_{Ga} and μ_{Bi} are parameters reflecting experimental surface conditions. The upper limits for both chemical potentials are obtained from bulk values. (Ga [65] and Bi form in orthorhombic and rhombohedral unit cells, respectively.) The lower limit of μ_{Ga} (or μ_{As}) is obtained using the heat of formation ΔH_f of GaAs. Therefore, μ_{Ga} varies within the range

$$\mu_{\mathrm{Ga,bulk}} - \Delta H_{\mathrm{f}} < \mu_{\mathrm{Ga}} < \mu_{\mathrm{Ga,bulk}}. \tag{10.2}$$

The lower limit of μ_{Bi} is determined through the stabilization of reconstructions which do not include any Bi atoms.

Total energy calculations were performed using an ab initio density functional method within the local density approximation [66, 67]. The approach is based on plane-wave basis and projector-augmented wave method (PAW) [68, 69] (Vienna ab initio simulation package, VASP) [70–73]. Surfaces were modeled by slab unit cells having 11–12 atomic layers separated by vacuum of more than 20 Å wide. The optimization of the atomic structure was performed using conjugate-gradient minimization of the total energy with respect to the atomic coordinates. The dangling bonds of the bottom surface atoms were passivated by fractionally charged pseudohydrogen atoms [74], and two bottom layers were fixed to ideal bulk positions. Other atoms including pseudohydrogen atoms were relaxed until the remaining forces were less than 20 meV/Å. The number of k points in the surface Brillouin corresponded to 64 k points per (1×1) surface area. The k point sampling was performed by the Monkhorst–Pack scheme [75] with the origin shifted to the Γ point. The plane-wave cutoff energy was 350 eV. Theoretical bulk lattice constants were used in the surface calculations. These are 5.63, 5.88, and 6.47 Å for GaAs, InP, and InSb, respectively. The constant-current STM images were simulated within the Tersoff–Hamann approximation [76, 77]. The Ga 3d electrons were treated as core electrons in most of the calculations. Our tests showed that the inclusion of the Ga 3d electrons as valence electrons does not affect significantly the surface phase diagrams. Surface CLSs were calculated mostly within the initial state model, which has been shown to lead often to a quite good agreement between the theoretical and experimental values for semiconductor surfaces [58, 59, 78, 79]. Most notable exception is found in the CLSs of buckled Si (surface) dimer atoms for which the complete screening calculations provide more accurate CLSs [80]. Initial state CLSs were determined mainly by calculating the average electrostatic potential at the core of the atoms. Obtained values are quite close to those determined by the one-electron energies. Complete screening CLSs include both initial state and final state effects. However, the screening is complete, i.e., surrounding crystal screens the core-hole with one additional electron with respect to the ionized situation (surrounding crystal provides that electron). In reality, the screening might not be complete specially in semiconductors. (One has to also remember that the screening has to take place fast enough to have an effect on the energy of the photoemitted electron.) In the complete screening CLS calculation a single electron is excited from the core to the valence by generating the corresponding PAW potential [81].

10.3.3 Combining Calculations and Measurements

The energetically favored atomic models provide a strong basis to determine the detailed atomic structures of the surfaces appearing in experiments such as the MBE growth front. Because there might be several models, of which energies are close to each other, it is helpful to compare the measured properties (e.g., STM images and CLSs) with the corresponding calculated properties (e.g., calculated STM images and CLSs), which are simulated on the basis of the energetically favored ab initio models. Such comparison strengthens the analysis also because surface kinetics might affect the atomic structures formed in the experiments. That is, there might exist energy barriers for the surface reactions toward the reconstruction. These barriers might hinder the formation of the energetically most favored structure, and the metastable structure can be formed in practice. In particular, this might occur on the surfaces which are prepared at low substrate temperatures. The Bi-induced III–V surfaces are usually produced at low temperatures, as compared to many other III–V surfaces, because Bi is easily re-evaporated from the surfaces.

Figure 10.8 shows an example comparison of the measured and calculated STM images for an InSb(100)$c(2\times6)$-Bi surface [39]. It can be seen that the Model-2 reproduces best the measured filled-state image in Fig. 10.8c, among the Models-1–4. Also, the dimerization in the trench area of the Model-2 agrees with the observed contour line, which reveals the periodicity of about 0.9 nm corresponding to the doubling of the InSb(100)-plane lattice constant (about 0.45 nm). Moreover, the empty-state image, simulated on the basis of the Model-2, agrees best with the measurement in Fig. 10.8d, among the models considered. This comparison indicates that the measured surface includes, at least, the areas with the atomic structure described by the Model-2, in good agreement with the finding that the Model-2 is energetically stable.

It is interesting that the InSb(100)$c(2\times6)$-Bi surface described by the Model-2 should be metallic according to the calculated DOS data. Indeed, metallic areas have been observed on the InSb(100)$c(2\times6)$-Bi by the valence-band photoelectron spectroscopy and STS (Fig. 10.9), supporting the Model-2. In Fig. 10.9, the reference data have been measured from the pure InSb(100)$c(8\times2)$ surface which is semiconducting with a small band gap of about 0.2 eV. The photoemission intensity at the Fermi level is nonzero for the InSb(100)$c(2\times6)$-Bi, in contrast to the semiconducting reference, revealing that the InSb(100)$c(2\times6)$-Bi surface includes metallic areas. Note that this surface might include also semiconducting areas. Furthermore, differentiated I–V curves measured from very local areas, as compared to a macroscopic area of the photoelectron signal, support that the InSb(100)$c(2\times6)$-Bi includes metallic areas, because its d(I)/d(V) intensity is nonzero at the Fermi level. For the reference sample, the zero d(I)/d(V) intensity range is about 0.2 eV, as can be expected. Note that the substrate is n-type-doped and, therefore, Fermi level can be expected to locate near the conducting band minimum (a positive voltage range corresponds to the conduction band).

As mentioned above, the calculated CLS data clearly improve the analysis of the photoelectron spectra. Table 10.1 shows an example CLS comparison, also for the

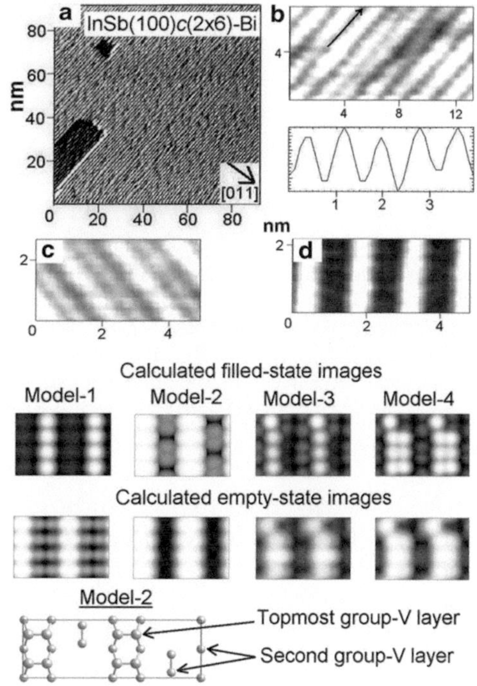

Fig. 10.8 Comparison of the measured STM data from an InSb(100)$c(2\times6)$-Bi surface and the calculated STM images for different atomic models. (**a**) Large-scale STM image. (**b**) Measured filled-state image with the contour line along the *black arrow* in the trench. (**c**) Measured filled-state image. (**d**) Measured empty-state image. The calculated images in the *bottom* for different models are simulated with the voltages similar to the measurements. The figure in Ref. [39] Phys. Rev. B **81**, 035310 (2010) has been modified

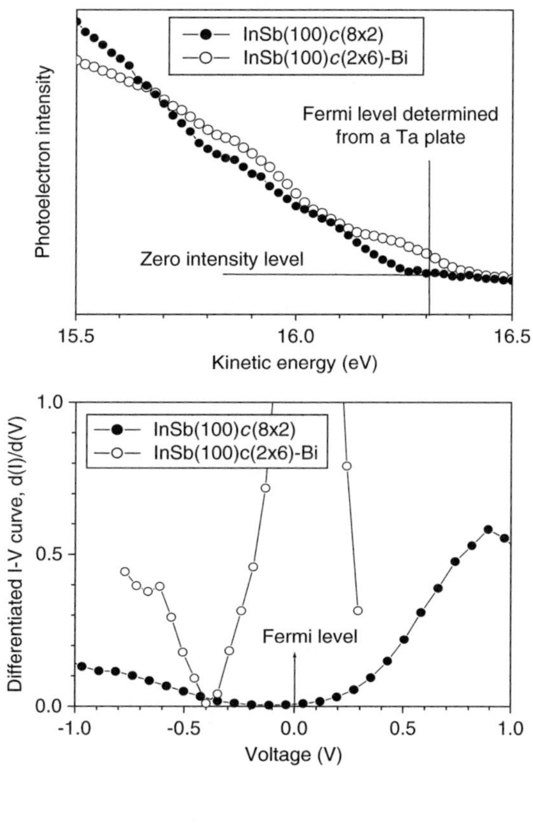

Fig. 10.9 *Top*: Valence-band photoemission from the pure InSb(100)$c(8\times2)$ surface and the Bi-induced $c(2\times6)$ one. *Bottom*: Differentiated STS *I–V* curves of the pure InSb(100)$c(8\times2)$ surface and the Bi-induced $c(2\times6)$ one. The figure in Ref. [39] Phys. Rev. B **81**, 035310 (2010) has been modified

Table 10.1 Measured surface core-level shifts (eV) and calculated ones (by VASP ab initio code) for the energetically stable structures of the InSb(100)-Bi surface within the initial state model

	Calculated Model-2	Calculated Model-3	Measured
In 4*d*	−0.09	−0.21, +0.23, +0.48	−0.33, +0.25
Sb 4*d*	+0.19	−0.37, −0.22, −0.12, +0.10, +0.38, +0.51	−0.35, +0.33
Bi 5*d*	−0.29, −0.06, +0.11	0, +0.17, +0.34	−0.39, 0, +0.24

Negative value indicates smaller binding energy (BE) as compared to the 0-eV reference, which is the bulk BE for the In 4*d* and Sb 4*d* levels. For the calculated Bi 5*d* shifts, the topmost dimer in the 4× kink site of the Model-3 phase is taken as the 0-eV reference. Because the 0-eV references for the measured and calculated Bi shifts are not aligned, the energy separation of different Bi shifts rather than individual shifts should be compared. Calculated shifts smaller than 0.05 eV are not listed. Data taken from Ref. [39] Phys. Rev. B **81**, 035310 (2010)

above-considered InSb(100)$c(2\times6)$-Bi surface. In the comparison of measured and calculated CLSs, generally, an important criterion is that the calculated maximum and minimum CLSs agree reasonably with the measured CLSs since the smaller shifts are most likely unresolved in the experiments. If the measured maximum CLS is significantly smaller than the calculated maximum CLS of an atomic model, then this model can be excluded (and vice versa for the minimum CLSs). If the measured maximum CLS is larger than the calculated one for an atomic model, then the exclusion of the corresponding model is not necessarily well justified, because the measured surface might include a significant amount of defects and other phases causing the maximum shift. Obviously, STM characterization is helpful here. On the other hand, the absence of the smaller CLSs in the measurements of course does not exclude the presence of the phases producing only the smaller shifts. For the presented InSb(100)$c(2\times6)$-Bi case, the CLS analysis is complicated by the presence of other reconstructed phases, different from the Model-2, as shown in [39]. However, the CLSs of the Model-2 do not disagree with the CLS measurements, in consistency with the above findings. Another clear benefit of the CLS calculations is that they provide the atomic origins for the measured CLS; the CLS of each atom in a model can be calculated. This advantage is exemplified later in Sect. 10.4.2.

10.4 Unusual Bi-Induced Reconstructions of III–V Semiconductor Surfaces

10.4.1 Bi-Induced (2 × 1) Reconstructions on III–V(100) Surfaces

The LEED results of Horng and Kahn [82] are most likely the first report on the formation of Bi-induced (2×1) reconstruction on III–V(100). This was found on the GaAs(100) substrate [82]. After that, the Bi-induced (2×1) reconstruction has been also observed, at least, on the $GaAs_{0.96}N_{0.04}(100)$ [37], $Ga_{0.47}In_{0.53}As(100)$ [38],

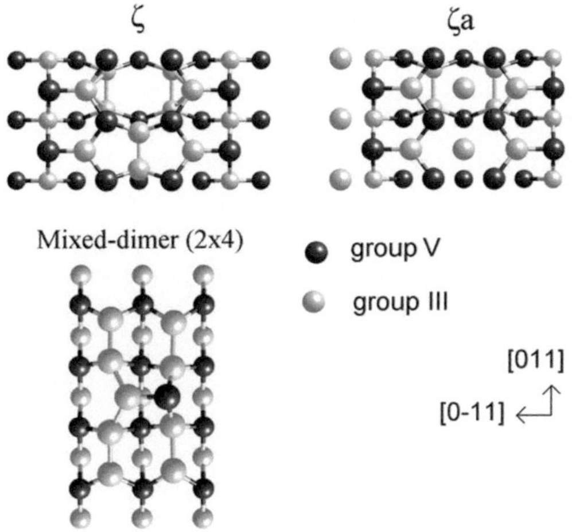

Fig. 10.10 Two most likely atomic models, ξ and ξa, which describe the structure of the starting III–V(100)c (8 × 2) surfaces, and also the mixed-dimer InP(100)(2 × 4), on which Bi is deposited

InAs(100) [40, 41], and InP(100) substrates [35, 37]. The tests on the InSb(100) substrate showed that no Bi-induced (2 × 1) forms on this surface. In contrast, the InSb(100)c(2 × 6)-Bi was found (Figs. 10.8 and 10.9) [39]. The Bi experiments on GaSb(100) and GaP(100) substrate have not yet been reported.

The (2 × 1) reconstruction is unusual because it does not obey the ECM. As described above, the ECM predicts that a (2 × 4) cell is the smallest unit cell on III–V(100) surfaces. In the defect-free (2 × 1) surface, all the dangling bonds of group V atoms cannot be filled completely; there remains a shortage of one electron per (2 × 1) cell. Therefore, a defect-free (2 × 1) surface can be expected to be metallic (i.e., to have a surface-related band which crosses the Fermi level). According to the literature, Bi is the only group V element that has been found to cause the (2 × 1) reconstruction in the experiments.

The starting III–V surface for preparing the Bi-induced (2 × 1) reconstructions has been usually a group V-poor substrate surface which has the c(8 × 2) reconstruction. The atomic structure of these III–V(100)c(8 × 2) surfaces has been widely accepted to be one of the two structures shown in Fig. 10.10 or their combination [27–29, 83]. The GaAs(100)c(8 × 2) surface has the ξ structure while the InAs(100)c(8 × 2) has the ξa structure rather than the ξ one. The InP(100) is an exception among the III–V(100) surfaces because it has no c(8 × 2) reconstruction. In contrast, the mixed-dimer (2 × 4) reconstruction is a P-poor phase of the InP(100) surface [84]. The point is that the starting surfaces, on which Bi is deposited in the previous surface-science experiments, have a rather complex atomic structure, which also changes drastically due to the presence of Bi and to post-heating.

Bi has been previously [35–41, 85] deposited on the above-described clean III–V(100) surfaces kept at room temperature. A quartz crystal flux meter was used to estimate the amount of the deposited Bi. Often several Bi depositions were necessary to obtain a enough uniform Bi-induced surface, in particular, for the photoelectron spectroscopy measurements, because it was observed that Bi was easily re-evaporated from the surface during the post-deposition heating. On the other hand, if too much Bi was deposited on the surface resulting in the formation of Bi clusters, the post-heating temperature, clearly higher than that needed for the (2×1) formation, was required to remove Bi clusters.

The post-heating around 250–300 °C produced the Bi-induced (2×1) reconstruction on the GaAs(100), GaAsN(100), GaInAs(100), and InAs(100) substrates systematically. It is worth noting that before any heating, LEED showed the Bi-covered surfaces to be amorphous. Increasing the post-heating temperature led to the formation of Bi-induced (2×4) structure, which usually coexisted together with the (2×1) areas, a shown in Fig. 10.6. The InP(100)(2×1)-Bi was exceptional among the III–V(100)(2×1)-Bi because it was found to stand up to the heating of 350–400 °C. Anyhow these temperatures used to prepare the III–V(100)(2×1)-Bi are low, as compared to the normal heatings (400–600 °C) used for III–V reconstructions.

Figure 10.11 shows STM and STS results for the GaAs(100)(2×1)-Bi surface. The STM images in Fig. 10.11a, b indicated that at least two different (2×1) phases exist. The modeled STM image based on the (2×1)-BiBi model in Fig. 10.11d agrees well with the measurement in Fig. 10.11a. Furthermore, the (2×1) structure in Fig. 10.11b is best reproduced by the (2×1)-BiV and (2×1)-sBiV models. A difference between these two models is that the (2×1)-sBiV model includes crystal defects, Bi_{Ga} antisites, which provide additional valence electrons so that all group V dangling bonds become completely filled. Therefore, if the (2×1)-sBiV structure appears, semiconducting (2×1) areas should be observed. Indeed, STS measurements show that the GaAs(100)(2×1)-Bi surface includes metallic and semiconducting areas, as shown in Fig. 10.11c, supporting the presence of the (2×1)-sBiV-like structure as well. Moreover, the measured and calculated core-level photoelectron properties clearly support the formation of (2×1)-sBiV phase in addition to the (2×1)-BiBi structure on the GaAs(100)(2×1)-Bi surface [35]. Importantly, the (2×1)-BiBi model has been found [35] to be energetically stable in the Bi-rich and As-poor conditions of the GaAs(100)-Bi surface. Also, the (2×1)-sBiV model becomes energetically favored among the different (2×1) models [35]. The above-presented findings for the GaAs(100)(2×1)-Bi largely apply also to the other III–V(100)(2×1)-Bi surfaces. The (2×1)-BiBi and (2×1)-sBiV models show that the III–V(100)(2×1)-Bi samples can be considered as ultrathin epitaxial III-Bi and III-$V0_{.5}Bi_{0.5}$ films, respectively.

The formation of Bi_{III} antisites can be supported by considering the amounts of group V and III atoms in the starting III–V(100)*c*(8×2) surfaces. For example, the GaAs(100)*c*(8×2) surface is well described [27] with the ξ model shown in Fig. 10.10. Considering the amounts of As and Ga atoms, which are needed to build up one layer of GaAs (i.e., one As layer plus the topmost Ga layer) during the Bi-induced rearrangement, it can be seen that there are exactly the correct number of As atoms for the full monolayer. However, the amount of Ga atoms is the ¾ monolayer (ML); there is a shortage of 2 Ga atoms per (4×2) cell. The (2×1)-sBiV model (Fig. 10.11) includes ¼ ML of Bi_{Ga} antisites. Interestingly, the other possible model for the III–V(100)*c*(8×2) surface, the ξa model, owns 1 ML of As atoms and 1 ML of Ga atoms. Thus, the ξa starting surface should provide a decrease in the

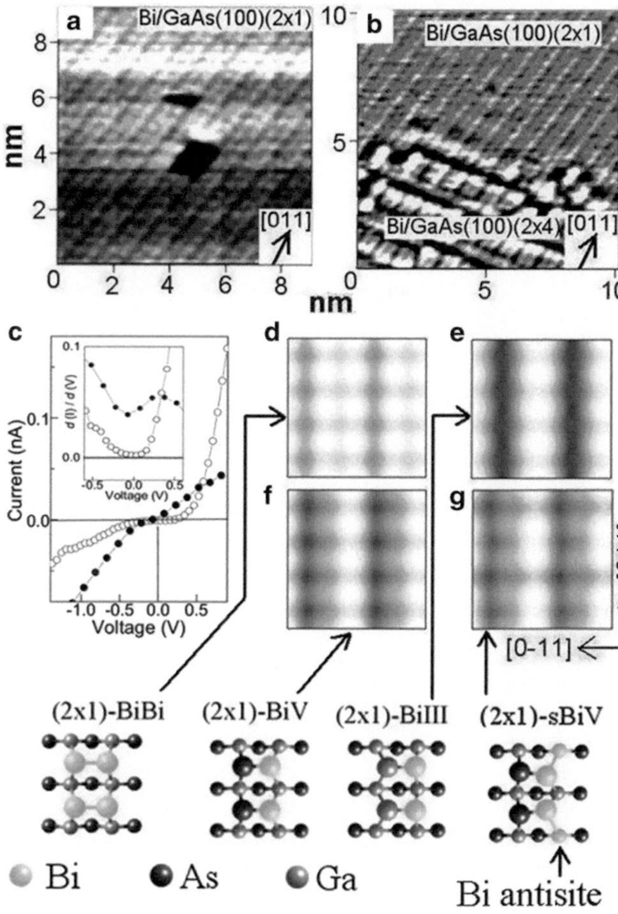

Fig. 10.11 STM and STS results from the GaAs(100)(2×1)-Bi surface. (**a**) Measured filled-state (0.09 nA, 1.70 V) image which agrees with the calculated image in (**d**). (**b**) Measured filled-state image (0.42 nA, 2.49 V) from the GaAs(100)(2×1)-Bi which agrees better with the calculated images in (**f**) and (**g**) than those in (**a**) and (**b**). STS curves in (**c**) show that the GaAs(100)(2×1)-Bi surface can be metallic or semiconducting. The figure in Ref. [35] Phys. Rev. Lett. **100**, 086101 (2008) has been modified

amount of the Bi antisites. Future experiments are required to confirm this. Tentatively it can be suggested that the amount of the Bi antisites is lower at the InAs(100)(2×1)-Bi than at the GaAs(100)(2×1)-Bi surface because the InAs(100)*c*(8×2) consists of the ξa building blocks rather than the ξ ones [85] (Fig. 10.12). Furthermore, it can be proposed that the mixed-dimer InP(100)(2×4) starting surface (Fig. 10.10) reduces the number of the Bi antisites but leads to an increase in the amount of the heterodimers, Bi–P and Bi–In. This kind of discussion

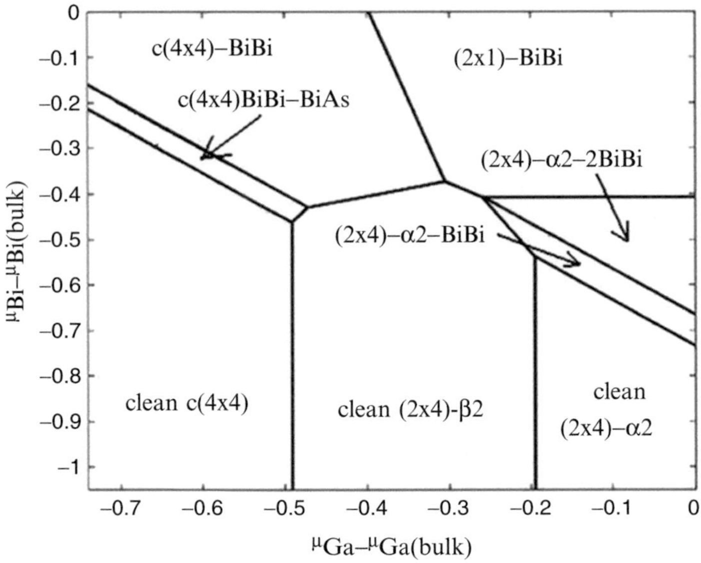

Fig. 10.12 Calculated phase diagram for the Bi-covered GaAs(100) surface showing the energetically most favored structures. The increase in the Bi (or Ga) chemical potential, μ_{Bi} (or μ_{Ga}), corresponds to an increase of the Bi (or Ga) amount on the surface. The figure in Ref. [35] Phys. Rev. Lett. **100**, 086101 (2008) has been modified

is of course speculative because strong and large-scale atomic rearrangements might occur during the Bi deposition and the post-heating. On the other hand, the Bi deposition is performed on the substrate kept at room temperature and the post-heating temperatures are relatively low as described above.

Angle-resolved valence spectra have been measured along the 2× direction (Figs. 10.13 and 10.14) from the GaAs(100)(2 × 1)-Bi surface. (No angle-resolved spectra have been reported so far for the perpendicular direction of [011].) Dispersion of the emission *b* is presented by the solid line in Fig. 10.14. The feature *b* is most likely the bulk-state. In the Γ spectrum (i.e., in the third spectrum from the bottom), its wave vector (*k*) along the surface normal appears to equal the center of the three dimensional Brillouin zone because the *b* peak disperses symmetrically around the Γ spectrum. When the angle is increased from the surface normal, the wave vector of the *b* state along the surface normal decreases. The spectra include at least two different emissions close to the valence-band maximum (VBM): emissions *s*1 and *s*2 of which angle dispersions are shown by open and solid dots, respectively. The *s*1 and *s*2 emissions can be surface states. However, they do not locate in the GaAs band gap, and, therefore, the origins of the *s*1 and *s*2 emissions remain unclear. It can be seen from Fig. 10.14 that the *s*2 dispersion obeys the 2× periodicity, which supports the surface-related origin of the state.

The right spectrum of Fig. 10.14 shows a valence emission around the Fermi level in more detail. Clearly, the photoelectron intensity at the Fermi level is

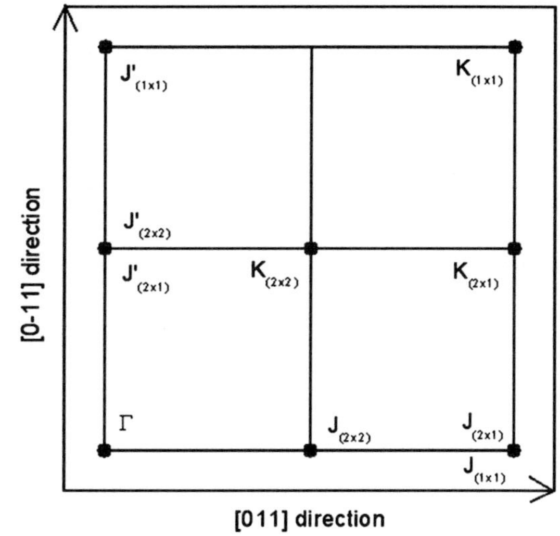

Fig. 10.13 The symmetry directions and *k* points for the square and rectangular irreducible parts of the surface Brillouin zones of the GaAs(100)(2 × 1)-Bi surface with the (2 × 1)-BiBi structure and of the GaAs(100) (2 × 2)-Bi surface with the (2 × 1)-sBiAs structure. The figure in Ref. [36] Phys. Rev. B **78**, 195304 (2008) has been modified

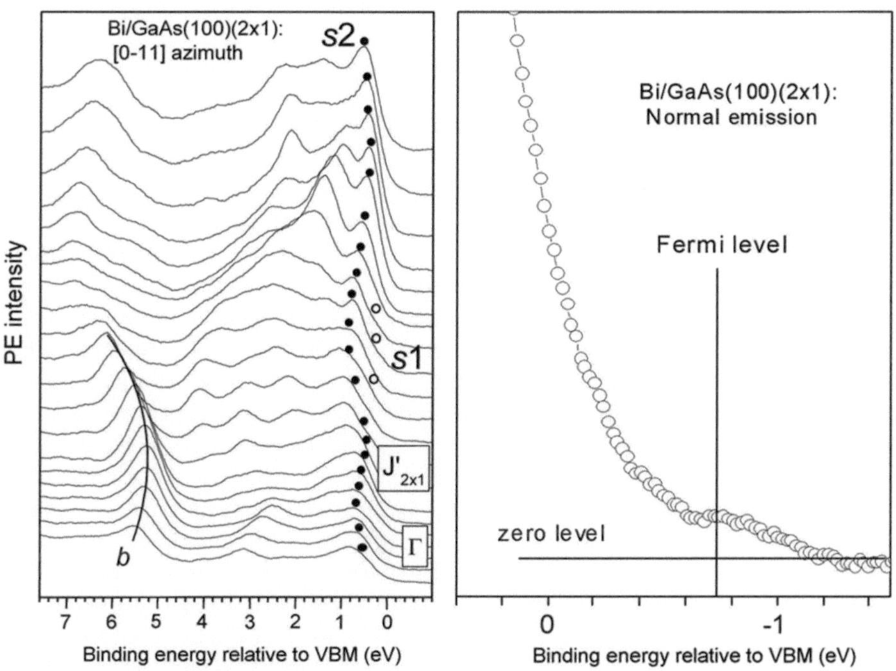

Fig. 10.14 *Left*: Measured angle-resolved valence spectra along the [0–11] direction (i.e., a spectrometer position has been changed along the 2x direction) around the normal emission, which corresponds to the Γ point. The angle change is 2° between each spectrum, and the photon energy is 29 eV. *Right*: Detailed valence emission around the Fermi level measured with the photon energy of 38 eV. The figure is adapted from Ref. [36] Phys. Rev. B **78**, 195304 (2008)

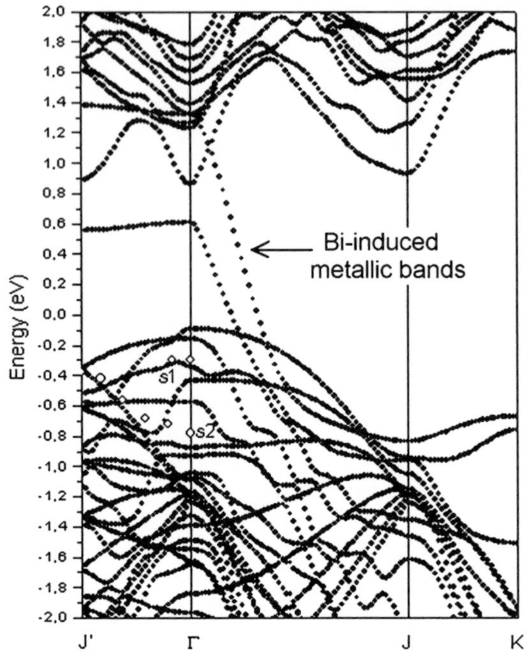

Fig. 10.15 Theoretical band structure along the main symmetry directions of the GaAs(100)(2 × 1)-Bi surface with the (2 × 1)-BiBi structure. *Open circles* denote the experimental results. The figure in Ref. [36] Phys. Rev. B **78**, 195304 (2008) has been modified

nonzero, which indicates that the GaAs(100)(2 × 1)-Bi surface includes metallic areas, in consistency with the STS results (Fig. 10.11). The intensity at the Fermi level is relatively low as compared to the other emission features in the left figure. Therefore, no clear Fermi level peak can be seen in the left spectra.

The calculated band structure in Fig. 10.15a for the GaAs(100)(2 × 1)-Bi surface with the (2 × 1)-BiBi structure agrees with the above measurements. First, two metallic bands which cross the Fermi level (i.e., 0 eV) are visible. These bands disperse strongly in the Γ–J direction (i.e., the [011] direction), as compared to the perpendicular Γ–J' direction. Indeed, stronger interaction between surface-related orbitals can be expected in the Γ–J direction, in which the distance between the neighboring dimers is 4 Å (and 8 Å in the perpendicular direction). The calculated band structure also supports the surface-related origin of the $s1$ and $s2$ emissions observed in the measurements, because the local orbital analysis shows the major contribution from Bi atoms to the corresponding theoretical bands.

Due to their metallic nature [86], the III–V(100)(2 × 1)-Bi surfaces were tentatively proposed to be beneficial growth fronts [36]. This idea was also supported by the atomic-scale smoothness of the III–V(100)(2 × 1)-Bi surfaces, as compared to the traditional (2 × 4)-reconstructed growth front (see Figs. 10.6 and 10.11). Indeed, the recent results from the epitaxial growth of GaAsBi films [26] confirm the previous prediction. Another evidence for the improved diffusion properties on the III–V(100)(2 × 1)-Bi surfaces is the finding that the deposition of extra Bi on the InAs(100)(2 × 1)-Bi template, kept at room temperature, produces a long-range

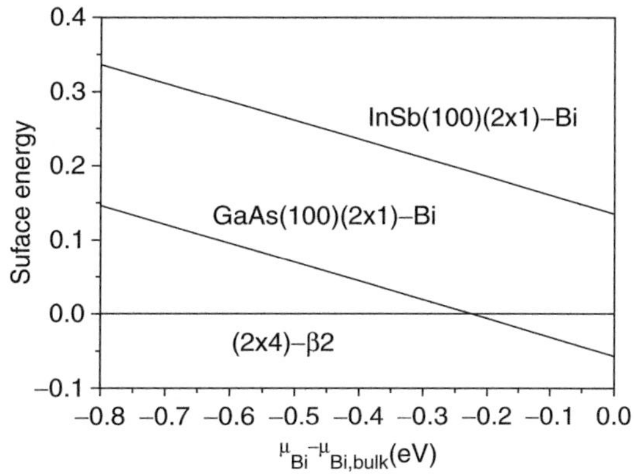

Fig. 10.16 Relative surface energies [eV per (1 × 1) surface area] of (2 × 1) and (2 × 4)-β2 reconstructions on GaAs(100)-Bi and InSb(100)-Bi surfaces as a function of the chemical potential of Bi in Ga-rich surface conditions

order of (2 × 6) without any post-heating [40]. This is unusual because the post-heating is usually necessary to arrange the surface atoms into an ordered structure.

Thus, Bi stabilizes the (2 × 1) reconstruction on several III–V(100) semiconductor surfaces. This is very uncommon, because the (2 × 1) reconstruction does not obey the ECM. Why the ECM is not obeyed?

First, it is noted that the (2 × 1) reconstruction is stabilized in relatively Ga-rich (or group III-rich) surface conditions, where the (2 × 4) reconstructions are also stable (Fig. 10.12). On the other hand, the *c*(4 × 4) reconstruction is stabilized in relatively As-rich (or group V-rich) surface conditions. Figure 10.16 shows relative surface energies of Bi-stabilized (2 × 1) and (2 × 4)-β2 reconstructions on GaAs(100) and InSb(100) surfaces in extremely group V-rich conditions as a function of the chemical potential of Bi. The Bi coverages for (2 × 1) and (2 × 4)-β2 surfaces are 1 and 0.75 monolayers (ML), respectively. It can be found that the (2 × 1) reconstruction is relatively much more stable on the GaAs(100) surface. The surface energy of the (2 × 1) reconstruction relative to that of the (2 × 4)-β2 is almost 0.2 eV per (1 × 1) surface area larger on the Bi/InSb(100) surface compared to the Bi/GaAs(100) surface, and this surface energy difference is very significant. This is in agreement with the experimental results as the (2 × 1) reconstruction is not seen on the InSb(100) surface. Instead, the (4 × 3) reconstruction is stabilized [39].

The symmetry of the (2 × 1) reconstruction depends strongly on the chosen adsorbant and the substrate. This is reflected on the angle between the backbonds of dimer atoms and second layer group III atoms. For example, this angle is 91.2° on the GaAs(100)(2 × 1)-Bi surface, whereas the corresponding angle is 105.7° on the hypothetical GaAs(100)(2 × 1) surface (Fig. 10.17) [GaAs(2 × 1) reconstruction is not found in experiments]. The surface bands (and consequently, the DOS) show considerable differences between these surfaces, which is supposed to reflect the relative stability of the GaAs(100)(2 × 1)-Bi surface [36]. It is very interesting that, if the adsorbants are interchanged, and the band structure is recalculated without relaxing the geometrical surface structure, the band structures are not changed

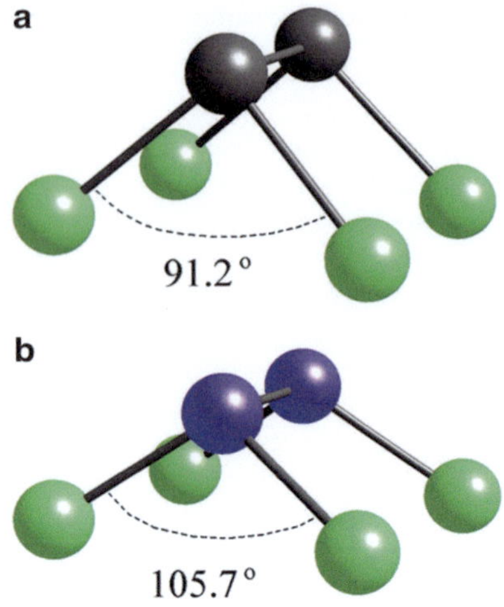

Fig. 10.17 Backbond angle on (**a**) GaAs(100)(2×1)-Bi and (**b**) GaAs(100)(2×1) surfaces. *Green*, *blue*, and *black* atoms denote Ga, As, and Bi atoms, respectively

significantly [36]. This means that, if we calculate the band structure for the GaAs(100)(2×1) using the geometrical structure of the GaAs(100)(2×1)-Bi, the resulting band structure is close to that obtained self-consistently for the GaAs(100)(2×1)-Bi (and vice versa). Important conclusion is that the atom size affects the band structure only indirectly through the geometrical structure.

It is interesting that the angle between the backbonds can be changed by varying the substrate even if we consider only the Bi-stabilized (2×1) surfaces. The backbond angle is 101.7 and 106.1° for hypothetical InSb(100)(2×1) and TlBi(100)(2×1) surfaces. We note that the calculated surface lattice parameter is 3.97, 4.55, and 4.79 Å for GaAs, InSb, and TlBi, respectively. Second, the surface band structure of the considered InSb(100)(2×1) and TlBi(100)(2×1) surfaces becomes more like of the GaAs(100)(2×1) surface as the backbond angle becomes similar to that of GaAs(100)(2×1). The band structures along the high-symmetry lines of the surface Brillouin zone are shown in Fig. 10.18 for the GaAs(100)(2×1)-Bi, InSb(100)(2×1)-Bi, and GaAs(100)(2×1) surfaces. Indeed, the energy band structure depends strongly on the backbond angle. Decreasing (increasing) backbond angle leads to increasing (decreasing) dispersion of surface energy bands in the direction parallel to dimer rows, which indicates a pronounced nanoline behavior within the dimer rows if the backbond angle is relatively small. On the other hand, decreasing (increasing) backbond angle leads to increasing (decreasing) localization of surface energy bands to dimer atoms. The charge associated with the quite dispersionless surface energy bands in the direction parallel to dimers might be called "dangling bond(s)". Increasing localization (decreasing covalency) suggests decreasing overlap with the bonding charge if the backbond angle is small. This finding is in agreement with the argument which states that the repulsion between the dangling bond and the backbonds is decreased when the backbond angle is decreased from the ideal one of the bulk zincblende structure (109.5°) [87]. These results show that

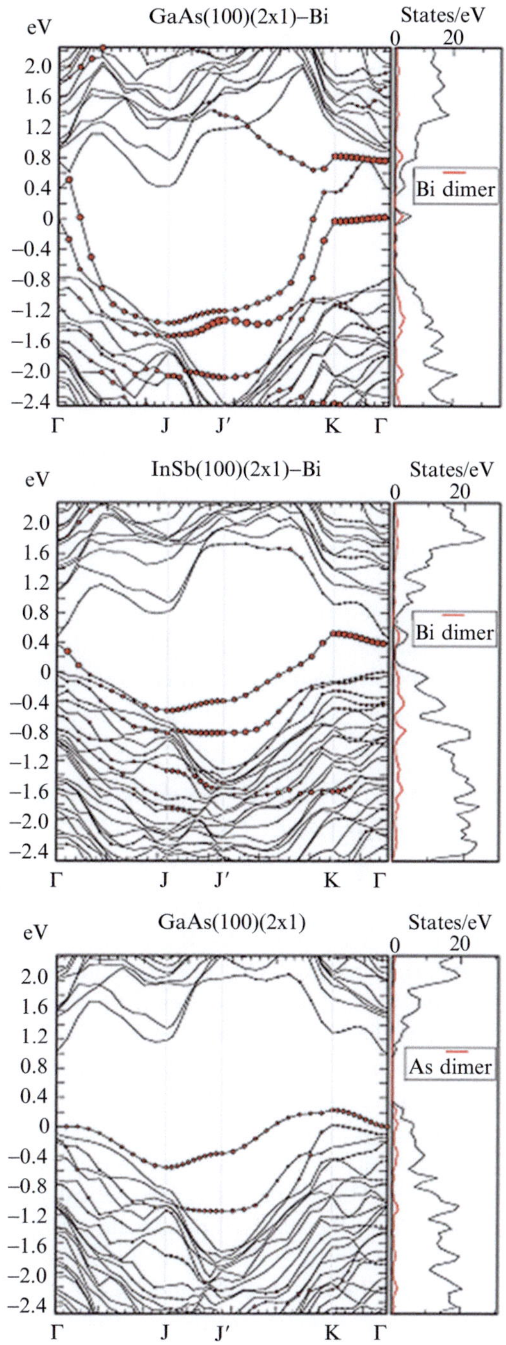

Fig. 10.18 Band structures of (**a**) GaAs(100)(2 × 1)-Bi, (**b**) InSb(100)(2 × 1)-Bi, and (**c**) GaAs(100)(2 × 1) surfaces along the high-symmetry lines of the surface Brillouin zone. The band dimer atom character is shown by *red spheres*. The larger the sphere, the more the charge density localized to dimer atomic spheres in the considered *k* point

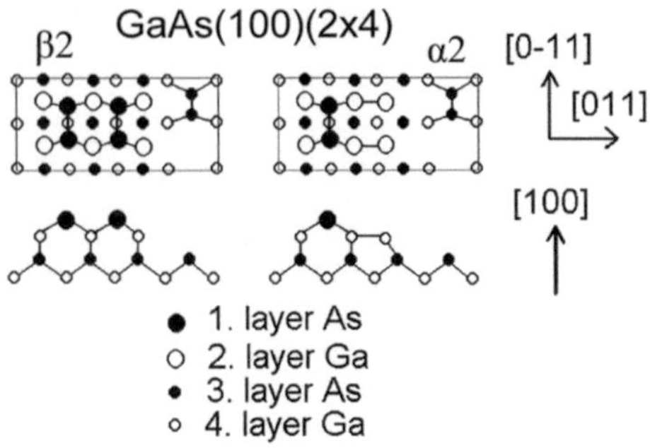

Fig. 10.19 The atomic models which describe the (2×4)-reconstructed GaAs(100) surface

there is a clear correlation between the backbond angle, surface band structure, and the relative stability of the surface reconstruction.

The backbond angle can be tuned by changing either the adsorbate or substrate. For example, the ratio of the lattice parameters of GaAs and GaBi determines the backbond angle in a reasonable accuracy. Based on the correlation explained above, it is suggested that when this ratio becomes smaller than 0.94–0.96, the (2×1) reconstruction is stabilized. This condition also reflects the fact that the (2×1) reconstruction is not found on two-component (pure) III–V surfaces.

10.4.2 Bi-Induced (2×4) Reconstructions on III–V(100) Surfaces

It is useful to summarize first the properties of the (2×4) reconstructions induced by other group V elements, As, P, and Sb on the III–V(100) surfaces, before describing Bi-induced changes. The (2×4) reconstruction is a very important III–V surface from the technological viewpoint because it usually appears during the epitaxial growth of the III-As and III-P materials. The most well-known (2×4) reconstruction is the pure GaAs(100)(2×4) surface, which is also one of the most intensely investigated solid surfaces in general. The information, accumulated via the research of about 20 years [32, 33, 88–99], on the properties of the GaAs(100)(2×4) clearly indicates that the clean well-ordered surface has the β2 type structure shown in Fig. 10.19. The β2 model is energetically stable and explains well the measurements. The other accepted structure for the GaAs(100)(2×4) is the α2 one (Fig. 10.19) which is also energetically stable. The α2 structure appears in the less As-rich conditions, as compared to the β2 conditions. However, a window of the preparation parameters for the α2 structure appears to be very narrow as compared to the preparation of the β2 structure on the GaAs(100) because usually the α2 structure coexists with the β2 one. It is also difficult to prepare a pure long-range ordered α2 structure on the GaAs(100), in contrast to the InAs(100)(2×4)α2 surface.

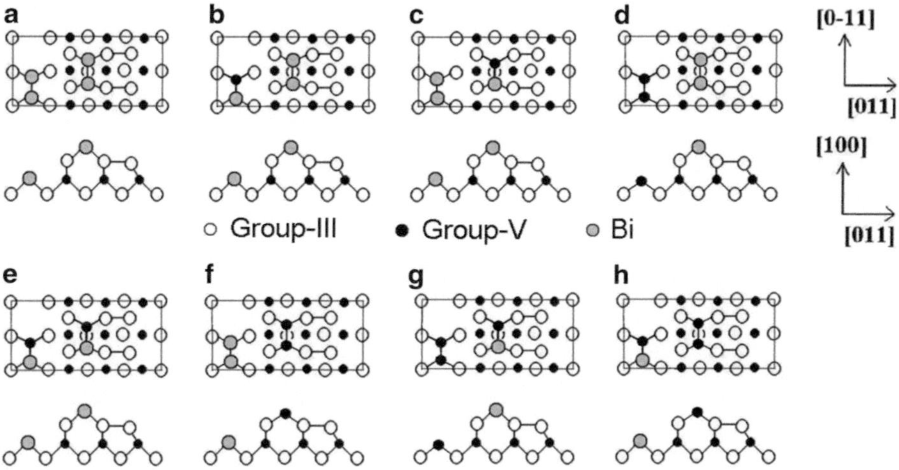

Fig. 10.20 Different configurations for Bi-containing (2×4) reconstructions of III–V(100) surfaces. The figure in Ref. [123] Surf. Sci. **601**, 3395 (2007) has been modified

The β2 and α2 type models have been found to describe also the surface structures of InAs(100)(2×4) [100–104]. For the InP(100)(2×4), the β2-like structure is more common than the α2 one [84, 105–112], similarly to the GaAs(100)(2×4) case. The InP(100)(2×4) has also a specific mixed-dimer (2×4) phase, of which model is shown in Fig. 10.10, and which also includes least P atoms among the (2×4) phases. Here it is worth noting that the III-Sb(100) alloy surfaces have not been observed to exhibit any (2×4) reconstruction.

In contrast, the Sb adsorbate has been found to induce a (2×4) reconstruction on the GaAs(100) surface [113–121]. The β2- and α2-like models describe also the GaAs(100)(2×4)-Sb surfaces well. However, it appears that Sb induces more ordered α2 phase than As does, and that a preparation window for the Sb-induced α2 is wider than that for the As-induced α2 on the GaAs(100).

The (2×4) reconstructions that include Bi atoms have been observed on many III–V(100) surfaces, at least, GaAs [36, 82], GaInAs [38], GaAsN [37], InAs [122], and InP [37, 123]. All these Bi-containing (2×4) surfaces have been found to have the α2-like structure; no β2-like structure has been reported. This is in agreement with the computational results for the GaAs(100)(2×4)-Bi [36] showing that the α2-like model is energetically more favored than the Bi-induced β2 structure. Therefore, the GaAs(100)(2×4)-Bi is different from the other III–V(100)(2×4) reconstructions.

There are still several possible unit-cell configurations for the Bi-containing α2-like structure, as shown in Fig. 10.20. For example, Bi atoms can occupy the first-layer or/and third-layer dimer site(s) and also form mixed V-Bi heterodimers. The previous studies of the GaAs(100)(2×4)-Bi surface [36, 85] have shown that it is difficult to determine the Bi sites with STM images. Although it might be possible

Table 10.2 Measured surface Bi core-level shifts (SCLS in eV) from the Bi/GaAs(100)(2×4) surface and calculated SCLS for the different Bi/GaAs(100)(2×4)-α2 structures with the different group V dimers in the first (topmost) and third atomic layers

GaAs(100)(2×4)-Bi	Bi 5*d*
Measured SCLS	−0.49 (*S*1), 0 (*S*2)
Calc.: (2×4)-$\alpha 2$ ((a) in Fig. 10.20) 1.layer:Bi–Bi, 3.layer:Bi–Bi	−0.44, 0
Calc.: (2×4)-$\alpha 2$ ((b) in Fig. 10.20) 1.layer:Bi–Bi, 3.layer:Bi–As	−0.36, −0.01
Calc.: (2×4)-$\alpha 2$ ((d) in Fig. 10.20) 1.layer:Bi–Bi, 3.layer:As–As	−0.03
Calc.: (2×4)-$\alpha 2$ ((c) in Fig. 10.20) 1.layer:Bi–As, 3.layer:Bi–Bi	−0.44, +0.09
Calc.: (2×4)-$\alpha 2$ ((e) in Fig. 10.20) 1.layer:Bi–As, 3.layer:Bi–As	−0.35, +0.08
Calc.: (2×4)-$\alpha 2$ ((g) in Fig. 10.20) 1.layer:Bi–As, 3.layer:As–As	+0.05

Negative values indicate smaller binding energy (BE) as compared to the reference, which is the topmost dimer of the Bi/GaAs(100)(2×4)-α2-2BiBi structure (i.e., first layer:Bi–Bi, third layer:Bi–Bi). Data taken from Ref. [36] Phys. Rev. B **78**, 195304 (2008)

to distinguish the STM feature of the first-layer Bi–Bi dimer from that of the Bi–As one [36], the identification of the third-layer dimers requires the complementary data. The high-resolution synchrotron-radiation photoelectron spectroscopy in combination with calculated CLSs has provided [36] a compelling evidence for the presence of Bi atoms in both the first- and third-layer dimer sites. Namely, the Bi 5*d* emission from the GaAs(100)(2×4)-Bi surface includes two distinguishable, main components, *S*1 and *S*2, as presented in Fig. 10.7. Comparison with the calculated CLSs (Table 10.2) confirms that the Bi atoms occupy both the first- and third-layer dimer sites on the GaAs(100)(2×4)-Bi surface. This result also supports the previous tentative suggestion that the two Bi core-level components, observed on the various III–V(100)(2×4)-Bi surfaces [122, 123], arise from the first and third-layer dimers. Furthermore, these calculations [36] justify the tentative assignment according to which the component at the higher binding energy (e.g., *S*2 in Fig. 10.7) is caused by the first-layer dimers and the other component at the lower binding energy (e.g., *S*1 in Fig. 10.7) is caused by the third-layer dimers. This assignment is consistent with the findings [36, 122, 123] showing that the higher binding-energy component is more surface-sensitive in the measurements (and therefore is originated from the atoms closer to vacuum) than the lower binding-energy component. The calculations [36] show that the same assignment is applicable also for the GaAs(100)(2×4)-Sb surface showing similar two Sb core-level components [120]. The first-layer Bi (or Sb) dimer atoms lose locally slightly more electrons to vacuum than the third-layer Bi dimer atoms, which contributes to the larger binding energy of the first-layer atoms. However, one should note that differences in the interatomic (Madelung) potentials of the considered atoms also contribute to the binding-energy differences [36, 60].

10.5 Summary and Outlook

The understanding and engineering of Bi-containing reconstructions of the III–V semiconductor surfaces are essential to develop the device materials, in particular, the properties of the III–V films and their interfaces. It has been found that Bi induces the unusual III–V surface structures as compared to the III–V surfaces induced by the other group V elements like As, P, and Sb. Among them, the most peculiar Bi-induced surface structure, found so far, is the (2×1)-Bi reconstruction. These III–V(100)(2×1)-Bi surfaces are metallic and very smooth at the atomic scale, as compared to many other III–V reconstructions. Correlation between the geometrical structure (and the surface band structure) and the relative stability of the (2×1) reconstruction is found. This suggests that the stabilization of the III–V(100)(2×1)-Bi surfaces is not necessarily a feature characteristic only for Bi-stabilized surfaces. It is clear, however, that by introducing a third component the variety of reconstructions becomes more rich. Because the III–V(100)(2×1)-Bi surfaces are smooth and metallic, they have been intuitively considered to be useful templates (growth fronts) to improve surface diffusion during the III–V epitaxy. This prediction is supported by the recent results [26] that the GaAs(100)(2×1)-Bi reconstruction improves the $GaAs_{1-x}Bi_x$ films. The (2×1)-Bi surface layers can be in fact considered as the thinnest possible epitaxial GaBi and InBi films. They might be also potential templates to improve the insulator/III–V interfaces, because the normal large III–V reconstructions such as (2×4) and $c(8\times2)$ can be removed with the (2×1)-Bi from the III–V starting surface on which an insulator film is deposited [124]. Future experiments will hopefully clarify that and the issue whether the metallic nature of (2×1)-Bi can be removed from the insulator interface.

The other Bi-induced change is found in the atomic structure of the (2×4) reconstruction, which is one of the most common III–V surface phases. When the III–V(100)(2×4) reconstruction contains Bi atoms, the formation of the $\beta2$ phase, which includes two adjacent dimers in the topmost layer, is hindered. In contrast, the $\alpha2$ phase of the (2×4) reconstruction, which includes only one dimer in the first layer per unit cell, becomes dominant due to the presence of Bi. The $\alpha2$ phase might also improve the surface diffusion along the dimer rows [125] because it decreases the formation of the kinks of (2×4) dimer rows. Thus, the III–V(100)(2×4)-Bi surface might be a useful growth front for the Bi surfactant-mediated epitaxy. However, it is often difficult to conclude the presence of Bi in the (2×4) growth fronts with RHEED because many III–V(100) surfaces exhibit the (2×4) reconstruction without Bi as well.

Acknowledgements We would like to thank MSc. Marja Ahola-Tuomi, Dr. Mikhail Kuzmin, Dr. Janusz Sadowski, Dr. Martin Adell, Dr. Johan Adell, Dr. Janne Pakarinen, MSc. Jouko Lång, and MSc. Niko Räsänen for their contributions to the experiments. We also are grateful to the MAX-lab staff and Mr. H. Ollila for their assistance. Furthermore, we thank Prof. Levente Vitos, Prof. Kalevi Kokko, Prof. Juhani Väyrynen, Prof. Mircea Guina, Prof. Markus Pessa, and Prof. Changsi Peng for the valuable discussions.

References

1. Ashley, T., Buckle, L., Datta, S., Emeny, M.T., Hayes, D.G., Hilton, K.P., Jefferies, R., Martin, T., Phillips, T.J., Wallis, D.J., Wilding, P.J., Chau, R.: Heterogeneous InSb quantum well transistors on silicon for ultra-high speed, low power logic applications. Electron. Lett. **43**, 777 (2007)
2. Stevenson, R.: Sematech builds III–V transistors on large Si wafers. Comp. Semicond. **17**(1), 13 (2011)
3. Oktyabrsky, S., Ye, P.D. (eds.): Fundamentals of III-V Semiconductor MOSFETs. Springer, New York (2010)
4. Ko, H., Takei, K., Kapadia, R., Chuang, S., Fang, H., Leu, P.W., Ganapathi, K., Plis, E., Kim, H.S., Chen, S.-Y., Madsen, M., Ford, A.C., Chueh, Y.-L., Krishna, S., Salahuddin, S., Javey, A.: Ultrathin compound semiconductor on insulator layers for high-performance nanoscale transistors. Nature **468**, 286 (2010)
5. Tomioka, K., Yoshimura, M., Fukui, T.: A III-V nanowire channel on silicon for high-performance vertical transistors. Nature **488**, 189 (2012)
6. Farrow, R.F.C. (ed.): Molecular Beam Epitaxy: Applications to Key Materials. Noyes, Park Ridge (1995)
7. Xue, Q.-K., Hashizume, T., Sakurai, T.: Scanning tunneling microscopy of III–V compound semiconductor (001) surfaces. Prog. Surf. Sci. **56**, 1 (1997)
8. Ohtake, A.: Surface reconstructions on GaAs(001). Surf. Sci. Rep. **63**, 295 (2008)
9. Patanè, A., Balkan, N. (eds.): Semiconductor Research: Experimental Techniques. Springer, Berlin (2012)
10. Zangwill, A.: Physics at Surfaces. Cambridge University Press, Cambridge (1988)
11. Unertl, W.N. (ed.): Handbook of Surface Science, vol. 1: Physical structure. Elsevier, Amsterdam (1996)
12. Lüth, H.: Solid surfaces, interfaces and thin films. Springer, Berlin (2001)
13. Bechstedt, F.: Principles of surface physics. Springer, Berlin (2003)
14. Anan, T., Sugou, S., Nishi, K.: Improvement of InP/InGaAs heterointerfaces grown by gas source molecular beam epitaxy. Appl. Phys. Lett. **63**, 1047 (1993)
15. Nosho, B.Z., Weinberg, W.H., Barvosa-Carter, W., Bennett, B.R., Shanabrook, B.V., Whitman, L.J.: Effects of surface reconstruction on III-V semiconductor interface formation: the role of III–V composition. Appl. Phys. Lett. **74**, 1704 (1999)
16. Tixier, S., Adamcyk, M., Young, E.C., Schmid, J.H., Tiedje, T.: Surfactant enhanced growth of GaNAs and InGaNAs using bismuth. J. Cryst. Growth **251**, 449 (2003)
17. Kondow, M., Uomi, K., Niwa, A., Kitatani, T., Watahiki, S., Yazawa, Y.: GaInNAs: a novel material for long-wavelength-range laser diodes with excellent high-temperature performance. Jpn. J. Appl. Phys. **35**, 1273 (1996)
18. Buyanova, I.A., Chen, W.M. (eds.): Physics and applications of dilute nitrides. Taylor & Francis, New York (2004)
19. Henini, M. (ed.): Dilute nitride semiconductors. Elsevier, Amsterdam (2005)
20. Guina, M., Leinonen, T., Härkönen, A., Pessa, M.: High-power disk lasers based on dilute nitride heterostructures. New J. Phys. **11**, 125019 (2009)
21. Wei, S.-H., Zunger, A.: Giant and composition-dependent optical bowing coefficient in GaAsN alloys. Phys. Rev. Lett. **76**, 664 (1996)
22. Perkins, J.D., Mascarenhas, A., Zhang, Y., Geisz, J.F., Friedman, D.J., Olson, J.M., Kurtz, S.R.: Nitrogen-activated transitions, level repulsion, and band gap reduction in $GaAs_{1-x}N_x$ with $x<0.03$. Phys. Rev. Lett. **82**, 3312 (1999)
23. Lindsay, A., O'Reilly, E.P.: Unification of the band anticrossing and cluster-state models of dilute nitride semiconductor alloys. Phys. Rev. Lett. **93**, 196402 (2004)
24. Karaiskaj, D., Mascarenhas, A., Adamcyk, M., Young, E.C., Tiedje, T.: Ultranarrow photoluminescence transitions of nitrogen cluster bound excitons in dilute GaAsN. Phys. Rev. B **74**, 035208 (2006)

25. Stevenson, R.: Triple junction CPV, dilute nitrides enhance infrared performance. Comp. Semicond. **17**(5), 18 (2011)
26. Masnadi-Shirazi, M., Beaton, D.A., Lewis, R.B., Lu, X., Tiedje, T.: Surface reconstructions during growth of GaAs1-xBix alloys by molecular beam epitaxy. J. Cryst. Growth **338**, 80 (2012)
27. Lee, S.-H., Moritz, W., Scheffler, M.: GaAs(001) surface under conditions of low As pressure: evidence for a novel surface geometry. Phys. Rev. Lett. **85**, 3890 (2000)
28. Kumpf, C., Marks, L.D., Ellis, D., Smilgies, D., Landemark, E., Nielsen, M., Feidenhans'l, R., Zegenhagen, J., Bunk, O., Zeysing, J.H., Su, Y., Johnson, R.L.: Subsurface dimerization in III–V semiconductor (001) surfaces. Phys. Rev. Lett. **86**, 3586 (2001)
29. Kumpf, C., Smilgies, D., Landemark, E., Nielsen, M., Feidenhans'l, R., Bunk, O., Zeysing, J.H., Su, Y., Johnson, R.L., Cao, L., Zegenhagen, J., Fimland, B.O., Marks, L.D., Ellis, D.: Structure of metal-rich (001) surfaces of III-V compound semiconductors. Phys. Rev. B **64**, 075307 (2001)
30. Pashley, M.D.: Electron counting model and its applications to island structures on molecular-beam epitaxy grown GaAs(001) and ZnSe(001). Phys. Rev. B **40**, 10481 (1989)
31. Mirbt, S., Moll, N., Kley, A., Joannopoulos, J.D.: A general rule for surface reconstructions of III–V semiconductors. Surf. Sci. **422**, L177 (1999)
32. Northrup, J.E., Froyen, S.: Structure of GaAs(001) surfaces: the role of electrostatic interactions. Phys. Rev. B **50**, 2015 (1994)
33. Schmidt, W.G., Mirbt, S., Bechstedt, F.: Surface phase diagram of (2×4) and (4×2) reconstructions of GaAs(001). Phys. Rev. B **62**, 8087 (2000)
34. Duzik, A., Thomas, J.C., Millunchick, J.M., Lång, J., Punkkinen, M.P.J., Laukkanen, P.: Surface structure of bismuth terminated GaAs surfaces grown with molecular beam epitaxy. Surf. Sci. **606**, 1203 (2012)
35. Laukkanen, P., Punkkinen, M.P.J., Komsa, H.-P., Ahola-Tuomi, M., Kokko, K., Kuzmin, M., Adell, J., Sadowski, J., Perälä, R.E., Ropo, M., Rantala, T.T., Väyrynen, I.J., Pessa, M., Vitos, L., Kollár, J., Mirbt, S., Johansson, B.: Anomalous bismuth-stabilized (2×1) reconstructions on GaAs(100) and InP(100) surfaces. Phys. Rev. Lett. **100**, 086101 (2008)
36. Punkkinen, M.P.J., Laukkanen, P., Komsa, H.-P., Ahola-Tuomi, M., Räsänen, N., Kokko, K., Kuzmin, M., Adell, J., Sadowski, J., Perälä, R.E., Ropo, M., Tukiainen, A., Rantala, T.T., Väyrynen, I.J., Pessa, M., Vitos, L., Kollár, J., Mirbt, S., Johansson, B.: Bismuth-stabilized (2×1) and (2×4) reconstructions on GaAs(100) surfaces: combined first-principles, photoemission, and scanning tunneling microscopy study. Phys. Rev. B **78**, 195304 (2008)
37. Laukkanen, P., Pakarinen, J., Ahola-Tuomi, M., Kuzmin, M., Perälä, R.E., Väyrynen, I.J., Tukiainen, A., Konttinen, J., Tuomisto, P., Pessa, M.: Structural and electronic properties of Bi-adsorbate-stabilized reconstructions on the InP(100) and $GaAs_xN_{1-x}$(100) surfaces. Phys. Rev. B **74**, 155302 (2006)
38. Laukkanen, P., Ahola-Tuomi, M., Kuzmin, M., Perälä, R.E., Väyrynen, I.J., Tukiainen, A., Pakarinen, J., Saarinen, M., Pessa, M.: Structural properties of Bi-stabilized reconstructions of GaInAs(100) surface. Appl. Phys. Lett. **90**, 082101 (2007)
39. Laukkanen, P., Punkkinen, M.P.J., Räsänen, N., Ahola-Tuomi, M., Sadowski, J., Adell, J., Kuzmin, M., Lång, J., Perälä, R.E., Ropo, M., Kokko, K., Vitos, L., Johansson, B., Pessa, M., Väyrynen, I.J.: Bismuth-stabilized *c*(2×6) reconstruction on a InSb(100) substrate: violation of the electron counting model. Phys. Rev. B **81**, 035310 (2010)
40. Ahola-Tuomi, M., Laukkanen, P., Punkkinen, M.P.J., Perälä, R.E., Väyrynen, I.J., Kuzmin, M., Schulte, K., Pessa, M.: Formation of an ordered pattern of Bi nanolines on InAs(100) by self-assembly. Appl. Phys. Lett. **92**, 011926 (2008)
41. Ahola-Tuomi, M., Punkkinen, M.P.J., Laukkanen, P., Kuzmin, M., Lång, J., Schulte, K., Räsänen, N., Perälä, R.E., Väyrynen, I.J.: Properties of self-assembled Bi nanolines on InAs(100) studied by core-level and valence-band photoemission, and first-principles calculations. Phys. Rev. B **83**, 245401 (2011)
42. Wiesendanger, R.: Scanning probe microscopy and spectroscopy: methods and applications. Cambridge University Press, Cambridge (1994)

43. Margaritondo, G.: Introduction to Synchrotron Radiation. Oxford University Press, Oxford (1988)
44. Briggs, D., Seah, M.P.: Practical Surface Analysis, vol. 1: Auger and X-ray Photoelectron Spectroscopy. Wiley, New York (1990)
45. Kevan, S.D.: Studies in Surface Science and Catalysis, vol. 74: Angle-Resolved Photoemission, Theory and Current Applications. Elsevier, Amsterdam (1992)
46. Horn, K., Scheffler, M. (eds.): Handbook of Surface Science, vol. 2: Electronic Structure. Elsevier, Amsterdam (2000)
47. Egelhoff, W.F.: Core-level binding-energy shifts at surfaces and in solids. Surf. Sci. Rep. **6**, 253 (1987)
48. Eastman, D.E., Chiang, T.C., Heimann, P., Himpsel, F.J.: Surface core-level binding-energy shifts for GaAs(110) and GaSb(110). Phys. Rev. Lett. **45**, 656 (1980)
49. Schmidt, W.G., Käckell, P., Bechstedt, F.: 3d core-level shifts at Se/GaAs(110). Surf. Sci. **357–358**, 545 (1996)
50. Davenport, J.W., Watson, R.E., Perlman, M.L., Sham, T.K.: Effect of the Madelung potential on surface core-level shifts in GaAs. Solid State Commun. **40**, 999 (1981)
51. Watson, R.E., Perlman, M.L., Davenport, J.W.: Ionic charge reduction at the surfaces of crystals. Surf. Sci. **115**, 117 (1982)
52. Watson, R.E., Davenport, J.W.: Ionic-charge modification at the surface of polar crystals. Phys. Rev. B **27**, 6418 (1983)
53. Mönch, W.: Charge-transfer from chemical-shifts at (110) surfaces of III–V compound semiconductors. Solid State Commun. **58**, 215 (1986)
54. McLean, A.B., Ludeke, R.: Surface core-level binding-energy shifts for the cleaved GaP(110) surface. Phys. Rev. B **39**, 6223 (1989)
55. Wilke, W.G., Hinkel, V., Theis, W., Horn, K.: Surface core-level shifts on InP(110)—Experiments and Madelung energy calculations. Phys. Rev. B **40**, 9824 (1989)
56. Mankefors, S., Nilsson, P.O.: Revised charge redistribution on semiconductor III–V(110) surfaces. J. Phys. Condens. Matter **13**, 823 (2001)
57. Sohn, S.H., Sohn, K.S.: Charge transfer in core-level shifts at (110) surfaces of compound semiconductors. Surf. Sci. **461**, 208 (2000)
58. Punkkinen, M.P.J., Laukkanen, P., Kokko, K., Ropo, M., Ahola-Tuomi, M., Väyrynen, I.J., Komsa, H.-P., Rantala, T.T., Pessa, M., Kuzmin, M., Vitos, L., Kollár, J., Johansson, B.: Surface core-level shifts of GaAs(100)(2×4) from first principles. Phys. Rev. B **76**, 115334 (2007)
59. Punkkinen, M.P.J., Kokko, K., Vitos, L., Laukkanen, P., Airiskallio, E., Ropo, M., Ahola-Tuomi, M., Kuzmin, M., Väyrynen, I.J., Johansson, B.: Calculation of surface core-level shifts within complete screening: Problems with pseudohydrogenated slabs. Phys. Rev. B. **77**, 245302 (2008)
60. Laukkanen, P., Punkkinen, M.P.J., Ahola-Tuomi, M., Lång, J., Schulte, K., Pietzsch, A., Kuzmin, M., Sadowski, J., Adell, J., Perälä, R.E., Ropo, M., Kokko, K., Vitos, L., Johansson, B., Pessa, M., Väyrynen, I.J.: Core-level shifts of the *c*(8×2)-reconstructed InAs(100) and InSb(100) surfaces. J. Electron Spectros. Relat. Phenomena **177**, 52 (2010)
61. Kevan, S.D. (ed.): Angle-Resolved Photoemission: Theory and Current Applications. Elsevier, Amsterdam (1992)
62. Qian, G.-X., Martin, R.M., Chadi, D.J.: First-principles study of the atomic reconstructions and energies of Ga-stabilized and As-stabilized GaAs(100) surfaces. Phys. Rev. B **38**, 7649 (1988)
63. Reuter, K., Scheffler, M.: Composition and structure of the RuO_2(110) surface in an O_2 and CO environment: implications for the catalytic formation of CO_2. Phys. Rev. B **68**, 045407 (2003)
64. Meyer, B.: First-principles study of the polar O-terminated ZnO surface in thermodynamic equilibrium with oxygen and hydrogen. Phys. Rev. B **69**, 045416 (2004)
65. Bernasconi, M., Chiarotti, G.L., Tosatti, E.: Ab initio calculations of structural and electronic properties of gallium solid-state phases. Phys. Rev. B **52**, 9988 (1995)

66. Ceperley, D.M., Alder, B.J.: Ground state of the electron gas by a stochastic method. Phys. Rev. Lett. **45**, 566 (1980)
67. Perdew, J.P., Zunger, A.: Self-interaction correction to density-functional approximations for many-electron systems. Phys. Rev. B **23**, 5048 (1981)
68. Blöchl, P.E.: Projector augmented-wave method. Phys. Rev. B **50**, 17953 (1994)
69. Kresse, G., Joubert, D.: From ultrasoft pseudopotentials to the projector augmented-wave method. Phys. Rev. B **59**, 1758 (1999)
70. Kresse, G., Hafner, J.: Ab initio molecular dynamics for liquid metals. Phys. Rev. B **47**, 558 (1993)
71. Kresse, G., Hafner, J.: Ab initio molecular dynamics simulation of the liquid metal amorphous semiconductor transition in germanium. Phys. Rev. B **49**, 14251 (1994)
72. Kresse, G., Furthmüller, J.: Efficiency of ab-initio total energy calculations for metals and semiconductors using a plane-wave basis set. Comput. Mater. Sci. **6**, 15 (1996)
73. Kresse, G., Furthmüller, J.: Efficient iterative schemes for ab initio total-energy calculations using a plane-wave basis set. Phys. Rev. B **54**, 11169 (1996)
74. Shiraishi, K.: A new slab model approach for electronic-structure calculation of polar semiconductor surface. J. Phys. Soc. Jpn. **59**, 3455 (1990)
75. Monkhorst, H.J., Pack, J.D.: Special points for Brillouin-zone integrations. Phys. Rev. B **13**, 5188 (1976)
76. Tersoff, J., Hamann, D.R.: Theory and application for the scanning tunneling microscope. Phys. Rev. Lett. **50**, 1998 (1983)
77. Tersoff, J., Hamann, D.R.: Theory of the scanning tunneling microscope. Phys. Rev. B **31**, 805 (1985)
78. Kuzmin, M., Punkkinen, M.P.J., Laukkanen, P., Lång, J.J.K., Dahl, J., Tuominen, V., Tuominen, M., Perälä, R.E., Balasubramanian, T., Adell, J., Johansson, B., Vitos, L., Kokko, K., Väyrynen, I.J.: Surface core-level shifts on Ge(111)*c*(2×8): Experiment and theory. Phys. Rev. B **83**, 245319 (2011)
79. Kuzmin, M., Punkkinen, M.P.J., Laukkanen, P., Lång, J.J.K., Dahl, J., Tuominen, M., Tuominen, V., Adell, J., Balasubramanian, T., Vitos, L., Kokko, K.: Dimer-T_3 reconstruction of the Sm/Si(100)(2×3) surface studied by high-resolution photoelectron spectroscopy and density functional theory calculations. Phys Rev. B **84**, 245322 (2011)
80. Pehlke, E., Scheffler, M.: Evidence for site-sensitive screening of core holes at the Si and Ge (001) surface. Phys. Rev. Lett. **71**, 2338 (1993)
81. Köhler, L., Kresse, G.: Density functional study of CO on Rh(111). Phys. Rev. B **70**, 165405 (2004)
82. Horng, S., Kahn, A.: Epitaxial growth of Bi on GaAs(100) surfaces. J. Vac. Sci. Technol. B **7**, 931 (1989)
83. Lång, J.J.K., Punkkinen, M.P.J., Laukkanen, P., Kuzmin, M., Tuominen, V., Pessa, M., Guina, M., Väyrynen, I.J., Kokko, K., Johansson, B., Vitos, L.: Structural properties of Bi-terminated GaAs(001) surface. Phys. Rev. B **81**, 245305 (2010)
84. Schmidt, W.G., Bechstedt, F., Esser, N., Pristovsek, M., Schultz, C., Richter, W.: Ab initio and scanning tunneling microscopy study of indium-terminated GaAs(100) surface: An indium-induced reconstruction change in the *c*(8×2) structure. Phys. Rev. B **57**, 14596 (1998)
85. Ahola-Tuomi, M., Laukkanen, P., Perälä, R.E., Kuzmin, M., Väyrynen, I.J., Pakarinen, J., Adell, M.: Atomic structure of InP(001)–(2×4): A dimer reconstruction. Surf. Sci. **600**, 2349 (2006)
86. Neugebauer, J., Zywietz, T.K., Scheffler, M., Northrup, J.E., Chen, H., Feenstra, R.M.: Adatom kinetics on and below the surface: The existence of a new diffusion channel. Phys. Rev. Lett. **90**, 056101 (2003)
87. Meade, R.D., Vanderbilt, D.: Origins of stress on elemental and chemisorbed semiconductor surfaces. Phys. Rev. Lett. **63**, 1404 (1989)
88. Pashley, M.D., Haberern, K.W., Friday, W., Woodall, J.M., Kirchner, P.D.: Structure of GaAs(001)(2×4)-*c*(2×8) determined by scanning tunneling microscopy. Phys. Rev. Lett. **60**, 2176 (1988)

89. Biegelsen, D.K., Bringans, R.D., Northrup, J.E., Swartz, L.-E.: Surface reconstructions of GaAs(100) observed by scanning tunneling microscopy. Phys. Rev. B **41**, 5701 (1990)
90. Skala, S.L., Hubacek, J.S., Tucker, J.R., Lyding, J.W., Chou, S.T., Cheng, K.-Y.: Structure of GaAs(100)-$c(8\times2)$ determined by scanning tunneling microscopy. Phys. Rev. B **48**, 9138 (1993)
91. Snyder, C.W., Sudijono, J., Lam, C.-H., Johnson, M.D., Orr, B.G.: Surface transformations on annealed GaAs (001). Phys. Rev. B **50**, 18194 (1994)
92. Hashizume, T., Xue, Q.-K., Zhou, J., Ichimiya, A., Sakurai, T.: Structures of As-rich GaAs(001)-(2×4) reconstructions. Phys. Rev. Lett. **73**, 2208 (1994)
93. Hashizume, T., Xue, Q.-K., Ichimiya, A., Sakurai, T.: Determination of the surface structures of the GaAs(001)-(2×4) As-rich phase. Phys. Rev. B **51**, 4200 (1995)
94. Yamaguchi, H., Horikoshi, Y.: Surface structure transitions on InAs(001) and GaAs(001) surfaces. Phys. Rev. B **51**, 9836 (1995)
95. Schmidt, W.G., Bechstedt, F.: Geometry and electronic structure of GaAs(001)(2×4) reconstructions. Phys. Rev. B **54**, 16742 (1996)
96. Garreau, Y., Sauvage-Simkin, M., Jedrecy, N., Pinchaux, R., Veron, M.B.: Atomic structure and faulted boundaries in the GaAs(001) $\beta(2\times4)$ surface as derived from x-ray diffraction and line-shape analysis. Phys. Rev. B **54**, 17638 (1996)
97. LaBella, V.P., Yang, H., Bullock, D.W., Thibado, P.M., Kratzer, P., Scheffler, M.: Atomic structure of the GaAs(001)-(2×4) surface resolved using scanning tunneling microscopy and first-principles theory. Phys. Rev. Lett. **83**, 2989 (1999)
98. Laukkanen, P., Kuzmin, M., Perälä, R.E., Ahola, M., Mattila, S., Väyrynen, I.J., Sadowski, J., Konttinen, J., Jouhti, T., Peng, C.S., Saarinen, M., Pessa, M.: Electronic and structural properties of GaAs(100)(2×4) and InAs(100)(2×4) surfaces studied by core-level photoemission and scanning tunneling microscopy. Phys. Rev. B **72**, 045321 (2005)
99. Ohtake, A.: Structures of the As-deficient phase on GaAs(001)-(2×4). Phys. Rev. B **74**, 165322 (2006)
100. Kendrick, C., LeLay, G., Kahn, A.: Bias-dependent imaging of the in-terminated InAs(001) $(4\times2)/c(8\times2)$ surface by STM: Reconstruction and transitional defect. Phys. Rev. B **54**, 17877 (1996)
101. Göthelid, M., Garreau, Y., Sauvage-Simkin, M., Pinchaux, R., Cricenti, A., Le Lay, G.: Atomic structure of the As-rich InAs(100) $\beta2(2\times4)$ surface. Phys. Rev. B **59**, 15285 (1999)
102. Ratsch, C., Barvosa-Carter, W., Grosse, F., Owen, J.H.G., Zinck, J.J.: Surface reconstructions for InAs(001) studied with density-functional theory and STM. Phys. Rev. B **62**, R7719 (2000)
103. Miwa, R.H., Srivastava, G.P.: Structure and electronic states of InAs(001)-(2×4) surfaces. Phys. Rev. B **62**, 15778 (2000)
104. Grosse, F., Barvosa-Carter, W., Zinck, J.J., Gyure, M.F.: Microscopic mechanisms of surface phase transitions on InAs(001). Phys. Rev. B **66**, 75321 (2002)
105. MacPherson, C.D., Wolkow, R.A., Mitchell, C.E.J., McLean, A.B.: Scanning tunneling microscopy study of InP(100)-(2×4): An exception to the dimer model. Phys. Rev. Lett. **77**, 691 (1996)
106. Schmidt, W.G., Bechstedt, F.: Atomic structure of InP(001)-(2×4): A dimer reconstruction. Surf. Sci. **409**, 474 (1998)
107. Mirbt, S., Moll, N., Cho, K., Joannopoulos, J.D.: Geometry and electronic structure of InP(001)(2×4) reconstructions. Phys. Rev. B **60**, 13283 (1999)
108. Esser, N., Schmidt, W.G., Bernholc, J., Frisch, A.M., Vogt, P., Zorn, M., Pristovsek, M., Richter, W., Bechstedt, F., Hannappel, T., Visbeck, S.: Cation-rich (100) surface reconstructions of InP and GaP. J. Vac. Sci. Technol. B **17**, 1691 (1999)
109. Li, L., Fu, Q., Li, C.H., Han, B.-K., Hicks, R.F.: GaP(001) and InP(001): Reflectance anisotropy and surface geometry. Phys. Rev. B **61**, 10223 (2000)
110. Lüdge, K., Vogt, P., Pulci, O., Esser, N., Bechstedt, F., Richter, W.: Determination of InP(001) surface reconstructions by STM and infrared spectroscopy of adsorbed hydrogen. Phys. Rev. B **62**, R11046 (2000)

111. Vogt, P., Ludge, K., Zorn, M., Pristovsek, M., Braun, W., Richter, W., Esser, N.: Clarification of the GaP(001)(2×4) Ga-rich reconstruction by scanning tunneling microscopy and ab initio theory. Phys. Rev. B **62**, R12601 (2000)
112. Laukkanen, P., Pakarinen, J., Ahola-Tuomi, M., Kuzmin, M., Perälä, R.E., Väyrynen, I.J., Tukiainen, A., Rimpiläinen, V., Pessa, M., Adell, M., Sadowski, J.: Surface structure of ordered InGaP(001): The (2×4) reconstruction. Surf. Sci. **600**, 3022 (2006)
113. Maeda, F., Watanabe, Y., Oshima, M.: Electronic and structural properties of the InP(100) (2×4) surface studied by core-level photoemission and scanning tunneling microscopy. Phys. Rev. B **48**, 14733 (1993)
114. Sugiyama, M., Maeyama, S., Maeda, F., Oshima, M.: Sb-induced surface reconstruction on GaAs(001). Phys. Rev. B **52**, 2678 (1995)
115. Lee, T.-L., Bedzyk, M.J.: X-ray standing wave study of an Sb-terminated GaAs(001)-(2×4) surface. Phys. Rev. B **57**, R15056 (1998)
116. Moriarty, P., Beton, P.H., Ma, Y.-R., Henini, M., Woolf, D.A.: High-resolution structural analysis of the Sb-terminated GaAs(001)-(2×4) surface. Phys. Rev. B **53**, R16148 (1996)
117. Whitman, L.J., Bennett, B.R., Kneedler, E.M., Jonker, B.T., Shanabrook, B.V.: Adsorbed and substituted Sb dimers on GaAs(001). Surf. Sci. **436**, L707 (1999)
118. Esser, N., Shkrebtii, A.I., Resch-Esser, U., Springer, C., Richter, W., Schmidt, W.G., Bechstedt, F., Del Sole, R.: The structure of Sb-terminated GaAs(001) surfaces. Phys. Rev. Lett. **77**, 4402 (1996)
119. Schmidt, W.G., Bechstedt, F.: Atomic structure of the Sb-stabilized GaAs(100)-(2×4) surface. Phys. Rev. B **55**, 13051 (1997)
120. Laukkanen, P., Perälä, R.E., Vaara, R.-L., Väyrynen, I.J., Kuzmin, M., Sadowski, J.: Antimony-stabilized GaAs(001)(2×4) reconstructions. Phys. Rev. B **69**, 205323 (2004)
121. Ohtake, A., Hirayama, M., Nakamura, J., Natori, A.: Electronic and structural analysis of Sb-induced GaAs(100)(2×4) and (2×8) surfaces. Phys. Rev. B **80**, 235329 (2009)
122. Laukkanen, P., Ahola, M., Kuzmin, M., Perälä, R.E., Väyrynen, I.J., Sadowski, J.: Variable stoichiometry in Sb-induced (2×4) reconstructions on GaAs(001). Surf. Sci. Lett. **598**, L361 (2005)
123. Laukkanen, P., Ahola-Tuomi, M., Adell, J., Adell, M., Schulte, K., Kuzmin, M., Punkkinen, M.P.J., Pakarinen, J., Tukiainen, A., Perälä, R.E., Väyrynen, I.J., Pessa, M.: Bi-induced (2×6), (2×8), and (2×4) reconstructions on the InAs(100) surface. Surf. Sci. **601**, 3395 (2007)
124. Lin, L., Robertson, J.: A comparative study of clean and Bi-stabilized InP(100)(2×4) surfaces by the core-level photoelectron spectroscopy. J. Vac. Sci. Technol. B **30**, 04E101 (2012)
125. Pakarinen, J., Polojärvi, V., Laukkanen, P., Tukiainen, A., Laakso, A., Peng, C.S., Tuomisto, P., Korpijärvi, V.-M., Puustinen, J., Pessa, M.: Passivation of interfacial defects at III–V oxide interfaces. Appl. Surf. Sci. **255**, 2985 (2008)

Chapter 11
MBE Growth of Thin Hexagonal Films Bi_2Te_3, Bi_2Se_3, and Their Alloys on Cubic GaAs (001) Substrates

Xinyu Liu, Y.P. Chen, D.J. Smith, Y.-H. Zhang, C. Liu, M.Z. Hasan, M. Dobrowolska, J.K. Furdyna, J. Fan, H. Cao, T.-L. Wu, and R.E. Pimpinella

Abstract Films of pseudo-hexagonal Bi_2Te_3, Bi_2Se_3, and their alloys were successfully grown by molecular beam epitaxy (MBE) on GaAs (001) substrates. The growth mechanism and structural properties of these films were investigated by reflection high-energy electron diffraction (RHEED), atomic force microscopy (AFM), X-ray diffraction (XRD), high-resolution transmission electron microscopy (HRTEM), and Raman spectroscopy and mapping. The results indicate that the epitaxial films are highly uniform and of high crystalline quality. The electronic band structure of these films was studied by angle-resolved photoemission spectroscopy (ARPES). The results confirm that these films are topological insulators, with a single Dirac cone for the topological surface states. Electric transport measurements were performed on the films, and weak anti-localization (WAL)

X. Liu (✉) • M. Dobrowolska • J.K. Furdyna • R.E. Pimpinella
Department of Physics, University of Notre Dame, Notre Dame, IN 46556, USA
e-mail: xliu2@nd.edu; mdobrowo@nd.edu; furdyna@nd.edu; rpimpine@nd.edu

Y.P. Chen • H. Cao • T.-L. Wu
Department of Physics, Purdue University, West Lafayette, IN 47907, USA
e-mail: yongchen@purdue.edu; cao2@purdue.edu; wu386@purdue.edu

D.J. Smith • J. Fan
Department of Physics, Arizona State University, Tempe, AZ 85287, USA
e-mail: david.smith@asu.edu; jin.fan@asu.edu

Y.-H. Zhang
Center for Photonics Innovation, Arizona State University, Tempe, AZ 85287, USA

School of Electrical, Computer and Energy Engineering, ASU, Tempe, AZ 85287, USA
e-mail: yhzhang@asu.edu

C. Liu • M.Z. Hasan
Department of Physics, Princeton University, Princeton, NJ 08544, USA
e-mail: cliu82@princeton.edu; mzhasan@princeton.edu

H. Li and Z.M. Wang (eds.), *Bismuth-Containing Compounds*, Springer Series in Materials Science 186, DOI 10.1007/978-1-4614-8121-8_11,

was observed in the magnetoresistance. Tuning of conductance by gating was demonstrated in top-gated field effect transistor (FET) devices fabricated from these films.

11.1 Introduction

In terms of conventional band theory, solids can be classified either as metals or as insulators. Traditional insulators include covalent insulators and atomic insulators, distinguished by the size of their band gaps, around 1.0 and 10 eV, respectively. Recently, a new type of quantum matter, the topological insulator (TI), has been brought to the attention of the science community, primarily by the ground-breaking theoretical work of Kane and Mele [1, 2]. Following this revolutionary theoretical milestone, quantum spin Hall effect (QSHE) was predicted and observed in two-dimensional (2D) HgTe quantum wells [3, 4], which possess a pair of crossing perfectly conducting edge states and exhibit a quantized resistance of $R = h/2e^2$ [5]. These advances have stimulated intensive search for three-dimensional (3D) TIs, a new state of matter with topologically nontrivial band structure originating from strong spin-orbit coupling [6, 7].

The 3D TIs possess fully gapped bulk states and gapless surface states, which can be described by the Rashba spin-orbit Hamiltonian [1, 8, 9]. These surface states are chiral, i.e., they possess a well-defined spin texture (referred to as spin-momentum locking), and they have an energy-momentum spectrum akin to the Dirac cone [10, 11]. Chiral Rashba surface states have a nontrivial band topology protected by time-reversal symmetry, and are thus robust against coherent backscattering and smooth perturbations invariant under time reversal [9]. Unlike 2D TIs, 3D TIs at present are experimentally established largely by surface-sensitive probes, such as scanning tunneling microscopy (STM) and angle-resolved photoemission spectroscopy (ARPES). These measurements have confirmed 3D topological insulating behavior in a number of materials—$Bi_{1-x}Sb_x$ [12], Bi_2Se_3 [13, 14], Bi_2Te_3 [15, 16], and Sb_2Te_3 [17]—all of which show an insulating energy gap in the bulk and gapless surface state(s) with a Dirac-like linear band dispersion. Further experimental studies of these systems have provided evidence of the existence of chiral surface states, and confirmed their protection by time-reversal symmetry [9]. For example, spin-resolved ARPES was used to measure spin polarization of the surface states, demonstrating the correlation between spin and momentum in these TI materials [17]. Moreover, a combination of STM and spin-resolved ARPES demonstrated that scattering between states with opposite momenta is strongly suppressed in the presence of time-reversal-invariant disorder in these materials [18]. The discovery of TIs has thus revolutionized our understanding of insulating behavior, and has stimulated intense investigation on the interaction between TIs and other forms of quantum matter. One should note that the rich new physics which occurs in TI-based structures, in addition to its fundamental interest, could also contribute to the development of a platform for many novel applications.

In order to study fundamental TI properties, high quality TI films need to be interfaced with superconductors, ferromagnets, or other complex materials [5, 9]. In this context, molecular beam epitaxy (MBE) is especially attractive, because of its capability for growing multilayer heterostructures under highly controlled conditions, so that defect formation is minimized during growth. Most efforts to fabricate TI films by MBE have so far been carried out using substrates with either hexagonal or threefold symmetric surface structures, such as Si (111) [19, 20], graphene [21], sapphire (0001) [21], $SrTiO_3$ (111) [22], and GaAs (111) [23]. Because representative spintronic materials, such as (Ga,Mn)As, are most commonly grown on GaAs (001) substrates [24], and Fe films of very high crystalline perfection can also be grown on GaAs (001) or (110) substrates [25], we are actively pursuing MBE growth of Bi_2Te_3, Bi_2Se_3, and their alloys on GaAs (001) substrates [26, 27], with the goal of combining these electronic materials into multifunctional device configurations in the future.

In this chapter, we describe MBE growth of hexagonal Bi_2Te_3, Bi_2Se_3, and their alloys on such symmetrically mismatched GaAs (001) substrates. This novel growth mode may enable one to combine almost any pair of layered materials together, thus allowing us to produce a variety of high quality hybrid heterostructures. Our work reveals the unique layer-by-layer growth of these materials in a pseudo-hexagonal layered structure—a crystalline form that involves sequences of five atomic layers [quintuple layers (QLs), e.g., Te-Bi-Te-Bi-Te or Se-Bi-Se-Bi-Se], each atomic Te/Se or Bi layer within the QL forming a 2D hexagonal lattice perpendicular to the growth axis (the *c*-axis). Our results thus suggest a powerful new possibility for combining the highly attractive properties of TI materials with traditional electronic materials that are compatible with the cubic structure, for the purpose of constructing new multifunctional device configurations.

11.2 MBE Growth of Bi_2Se_3 and Bi_2Te_3 Films

The TI films, including Bi_2Te_3, Bi_2Se_3, and their ternary alloys, were grown using a dual chamber Riber 32 solid-source MBE system. The Bi, Te_2, and Se_2 fluxes were generated by standard effusion cells installed in a modified II–VI MBE chamber. The structure and thickness of the films were monitored in situ by reflection high-energy electron diffraction (RHEED). The growth sequence was as follows. First, an epi-ready GaAs (001) substrate was heated to 600 °C for deoxidation in the III–V MBE chamber. This was followed by deposition of a 100-nm GaAs buffer layer. This modified substrate was then transferred to the II–VI MBE chamber through an ultrahigh vacuum connection. The growth of the TI film was initiated by deposition of a series of monolayers of Te-Bi-Te-Bi-Te or Se-Bi-Se-Bi-Se—a quintuple layer (QL)—in atomic layer epitaxy (ALE) fashion at room temperature. During this process, the (2×4) RHEED pattern of GaAs buffer (as shown in Fig. 11.1a, b) disappeared, indicating that an amorphous film had been deposited.

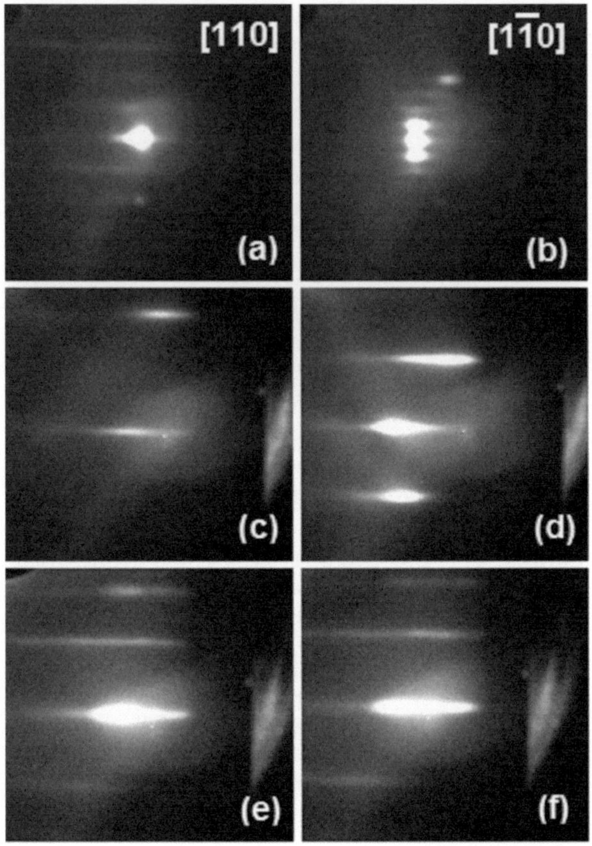

Fig. 11.1 RHEED patterns observed for two specific orientations of GaAs (001) substrate during MBE growth of Bi_2Te_3 and Bi_2Se_3. (**a**) and (**b**) correspond to GaAs (001) surface; (**c**) and (**d**) show single-phase growth mode (Bi_2Te_3); (**e**) and (**f**) indicate dual-phase growth mode (Bi_2Se_3). After ref. [30]

The substrate was then gradually heated to 300 °C, and a streaky unreconstructed RHEED pattern appeared (see Fig. 11.1c–f). The MBE growth of Bi_2Te_3, Bi_2Se_3, or their alloys was subsequently performed under Te_2- or Se_2-rich conditions with $T_{substrate}=300$ °C. The RHEED patterns shown in Fig. 11.1c–f were maintained throughout the entire growth process. It is important to note that the observed RHEED pattern showed six recurrences during each rotation of the substrate, which confirmed the *c*-axis growth of the TI films, with the *a*-axis lying along either the [110] or the [1$\bar{1}$0] direction of the GaAs (001) substrate. Two types of RHEED patterns were observed at this stage, depending on the annealing process. As shown in Fig. 11.1c, d typically observed in the case of Bi_2Te_3 (QLs of Bi_2Te_3 usually survive for a longer annealing time than for Bi_2Se_3), the annealing yielded an unreconstructed pattern with distinct features observed for GaAs [110] and [1$\bar{1}$0] directions, respectively. We attribute this to the hexagonal surface symmetry of the TI layer. Alternately, in the case of MBE growth of Bi_2Se_3, we typically observed the same RHEED patterns for both GaAs [110] and [1$\bar{1}$0] directions, as shown in Fig. 11.1e, f. Note that the RHEED pattern in Fig. 11.1e, f is actually a combination

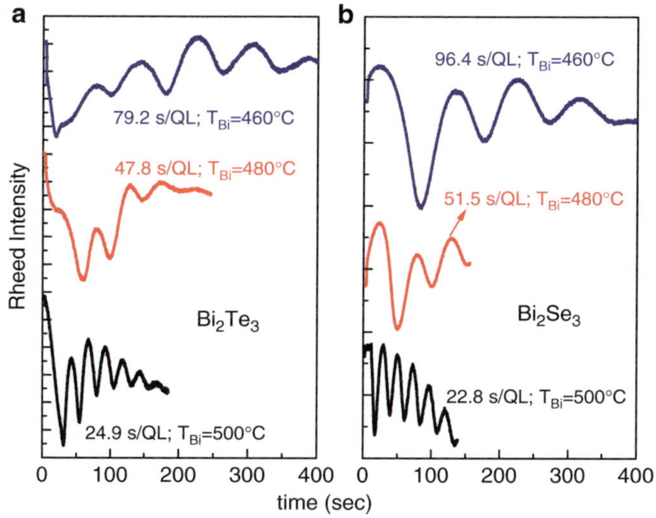

Fig. 11.2 RHEED intensity oscillations of the specular point versus growth time for different Bi cell temperatures T_{Bi}: (**a**) Bi_2Te_3; (**b**) Bi_2Se_3. After ref. [31]

of two distinct patterns, that for the [110] and the $[1\bar{1}0]$ directions in the Fig. 11.1c, d, respectively. We attribute this to the coexistence of two types of hexagonal surfaces perpendicular to each other.

The MBE growth of Bi_2Te_3 and Bi_2Se_3 was then performed under the condition of T_{Te} (or T_{Se}) < $T_{substrate}$ (300 °C) < T_{Bi} (500 °C) with a Se_2(or Te_2):Bi beam equivalent pressure (BEP) ratio ranging from 10:1 to 100:1. At the beginning of the growth clear RHEED oscillations of the specular spot were observed, each oscillation corresponding to the growth of one QL. Figure 11.2 shows RHEED oscillations observed with different temperatures of the Bi cell, T_{Bi}. As the Bi cell temperature was increased, the period of the oscillations decreased, indicating that the growth rate was directly controlled by the Bi flux, and that the growth of the TI films progressed in a QL-by-QL mode.

The same kind of growth is also observed on Ga-rich GaAs (001) surfaces, i.e., surfaces with a (4×6) RHEED pattern. We therefore attribute the successful growth of pseudo-hexagonal Bi_2Te_3 and Bi_2Se_3 on GaAs (001) substrates to the weak Van der Waals coupling between the substrate and the TI films, leading to immediate strain relaxation as the interface is forming. Note that the symmetry-mismatched interface between cubic (001) GaAs and pseudo-hexagonal (0001) Bi_2Te_3 or Bi_2Se_3 leads to the appearance of two hexagonal phases with the orientation relationships as either (110) || $(0\bar{1}10)$ or (110) || (1000), indicating an interface effect on the nucleation of the hexagonal phases.

Earlier studies of MBE growth of Bi_2Te_3 on cubic Si (001) substrates [22] appeared to suggest that a hexagonal structure of the substrate surface was essential for epitaxial growth of Bi_2Te_3 film to succeed. In contrast, our work shows that high quality Bi_2Te_3, Bi_2Se_3, and their alloys can form on GaAs (001) substrates with well-defined crystal orientations. This result suggests that the problems encountered in the MBE growth of Bi_2Te_3 films on Si (001) substrates could be caused by the reactivity of Te with Si [28], rather than being a result of mismatched symmetries at the substrate–TI interface. Our discovery shows that a self-correction process during growth of these layered honeycomb materials may play an important role in overcoming differences between crystal arrangements at interfaces during epitaxy.

11.3 Surface Morphology

Subsequent to MBE growth, the surface morphology of the TI samples was characterized ex situ by atomic force microscopy (AFM). Figure 11.3 shows AFM images of Bi_2Te_3 and Bi_2Se_3 films deposited at a growth rate of 2 nm/min, at the Te_2/Bi BEP

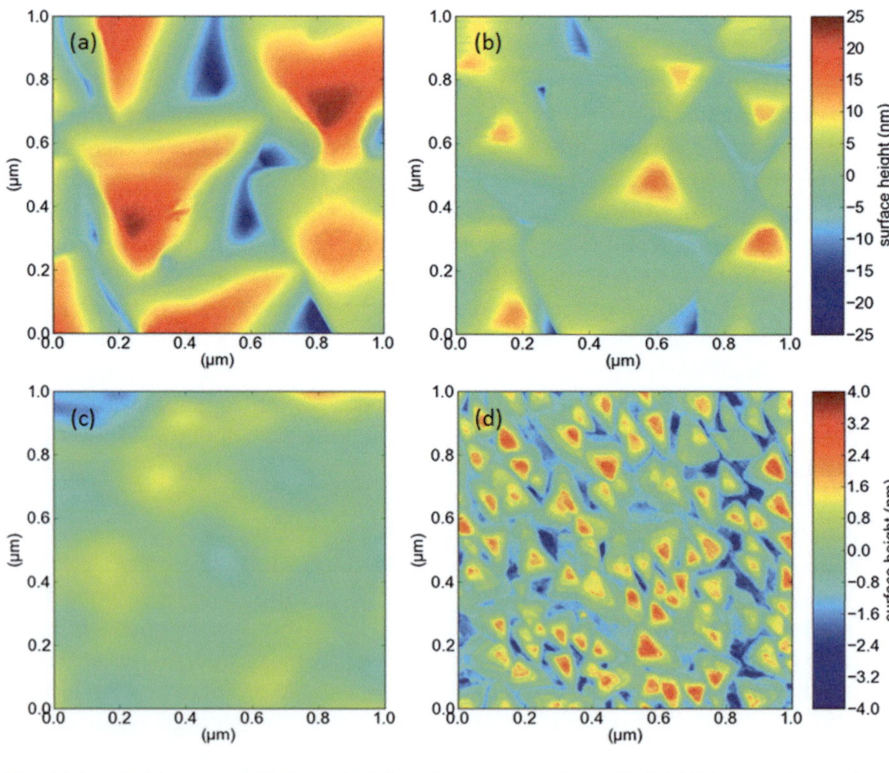

Fig. 11.3 AFM images of Bi_2Te_3 and Bi_2Se_3 films grown with the Te_2/Bi BEP ratio of 10 and Se_2/Bi BEP ratio of 20, respectively. (**a**) 210-nm-thick Bi_2Te_3; (**b**) 215-nm-thick Bi_2Se_3; (**c**) 15-nm-thick Bi_2Te_3; (**d**) 15-nm-thick Bi_2Se_3. After ref. [31]

ratio of 10 for Bi_2Te_3 and Se_2/Bi ratio of 20 for Bi_2Se_3. The thicknesses of the Bi_2Te_3 and Bi_2Se_3 layers shown in Fig. 11.3a, b are 210 and 215 nm, respectively, and the thicknesses of the films corresponding to Fig. 11.3c, d are 15 nm.

The images show many hills of triangular shape (spirals) aligned along specific orientations. Our results agree with other reports on Bi_2Te_3 [22] and Bi_2Se_3 [26] films, suggesting that the growth of TI films takes place by a spiral growth mode, with fast surface diffusion of Bi and Se/Te adatoms to existing Bi_2Se_3 or Bi_2Te_3 island edges due to weak interactions between these adatoms and the substrate [29, 30]. Such growth mode can be seen especially clearly in Fig. 11.3d where many small triangular terraces are observed for a 15-nm-thick Bi_2Se_3 layer, indicating initial 2D island formation for very thin films [27]. This result also suggests that the mobility of Bi adatoms is much slower on the Bi_2Se_3 surface than on Bi_2Te_3, most likely due to different chemical bond strengths of Bi–Te and Bi–Se. Note, however, that as the growth proceeds, the surface morphology of Bi_2Se_3 eventually becomes similar to that of Bi_2Te_3. It is already known in this connection that the surface morphology of TI films is dramatically affected by the group-VI/Bi BEP ratio and by the growth rate [22]. In the case of Bi_2Se_3 growth, as we decreased the Bi flux, the surface of thin Bi_2Se_3 layers became much smoother, becoming Bi_2Te_3-like.

The detailed growth mechanism of TI films is described in [34]: the growth of Bi_2Se_3 and Bi_2Te_3 on epitaxial GaAs (001) begins with 2D nucleation, and the spiral growth follows by pinning of the 2D growth fronts at domain boundaries that are created during the coalescence of the 2D islands. Winding of the growth fronts around these pinning centers leads to triangular shape domains (or spirals) that were already referred to. It should be noted that due to the symmetry mismatch between the hexagonal lattices of TI films and the cubic symmetry of the GaAs (001) surface, along with the weak interaction with the substrate surface, two types of rotation domains could form in the TIs films. This is displayed very clearly by the opposite orientation of the triangle islands in Fig. 11.3b.

11.4 Structure Investigations

The structures of TI samples grown by MBE were characterized by Raman spectroscopy, Raman mapping, transmission electron microscopy (TEM), and high-resolution X-ray diffraction (HR-XRD). Micro Raman spectroscopy was performed with laser excitation at 532-nm (power ~0.8 mW). The results, shown in Fig. 11.4a, b, reveal three of the characteristic peaks for Bi_2Se_3 [at ~71 cm^{-1} (A_{1g}^1), 131 cm^{-1} (E_g^2), and 174 cm^{-1} (A_{1g}^2)], and two of the characteristic peaks for Bi_2Te_3 [at ~102 cm^{-1} (E_g^2) and 134 cm^{-1} (A_{1g}^2)]. The peaks observed in these two TI materials are consistent with the lattice vibration modes reported earlier [31]. In Fig. 11.4c, d, representative Raman maps (showing Raman shift of the Raman peak E_g^2 as a function of position) are plotted for Bi_2Te_3 (Fig. 11.4c) and Bi_2Se_3 (Fig. 11.4d). These Raman maps show that the differences of the Raman shift are less than ~1 cm^{-1} within a scan area of 15×15 μm, indicating impressively a high uniformity of the films.

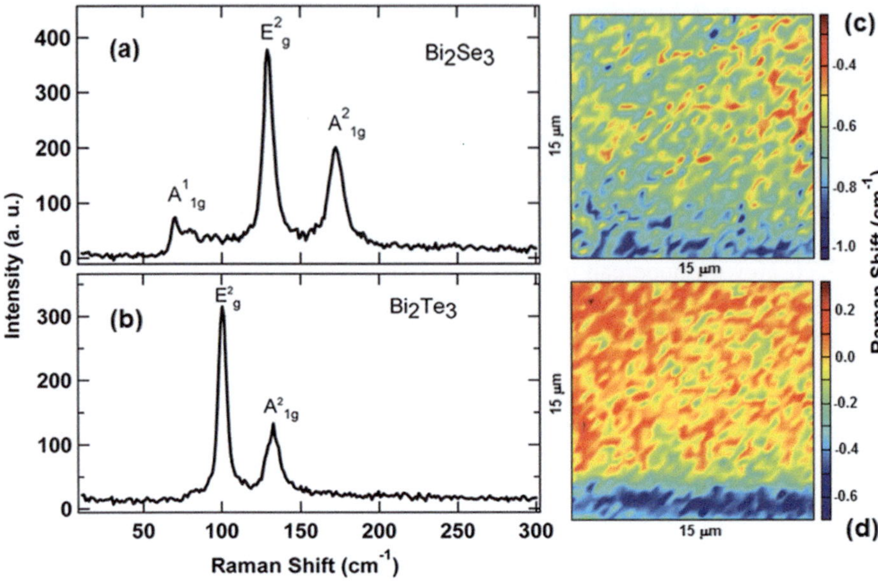

Fig. 11.4 (**a**, **b**) Raman spectra measured on thin Bi_2Se_3 and Bi_2Te_3 films. (**c**, **d**) Maps of Raman shift (relative to bulk) for Bi_2Se_3 and Bi_2Te_3 films, respectively, using the peak E_g^2 Raman peak. Adapted from ref. [30, 31]

The microstructure of the films was determined using cross section transmission electron microscopy (XTEM). Samples were prepared for TEM examination using standard mechanical polishing and argon-ion-milling, with the sample held at liquid-nitrogen temperature during the milling process in order to avoid unintentional ion-milling artifacts. In Fig. 11.5, XTEM images of Bi_2Te_3 and Bi_2Se_3 layers grown on GaAs (001) buffers show the lattice structure of both the TI films and the GaAs substrate near their interfaces. Clean interfaces without any amorphous phases are observed, similar to those reported for films grown on GaAs (111) substrates [27]. The highly parallel QLs—Te(Se)-Bi-Te(Se)-Bi-Te(Se)—are visible in both Bi_2Te_3 and Bi_2Se_3 films, marked by the symbol "**I**" in the figure. From Fig. 11.5 it appears that the highly parallel QLs in Bi_2Se_3 film extend over a significantly longer range than those in Bi_2Te_3 films, indicating that a particularly strong internal self-correction process is occurring in Bi_2Se_3 films as the growth proceeds. In addition, despite the symmetry mismatch between the hexagonal lattices of the TI films and the fourfold cubic symmetry of the GaAs (001) surface, the TEM images show that both TI films are highly uniform, and that their crystallinity is comparable to that of films grown on substrates with hexagonal surface structure.

In the case of Bi_2Se_3 films, domain boundaries can often be observed in XTEM images. In Fig. 11.6, the XTEM image of Bi_2Se_3 layer grown on GaAs (100) buffer clearly shows the boundaries between two rotation domains. Two interesting features can be seen in the images: first, the QLs crossing the domain boundaries are

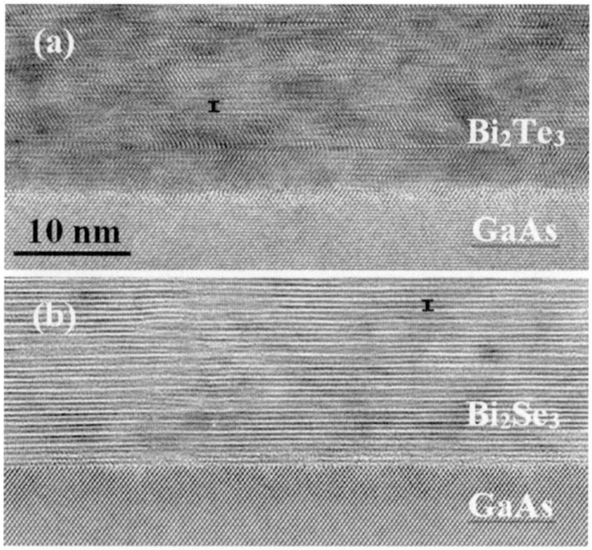

Fig. 11.5 HRTEM images showing cross sections of topological insulator (**a**) Bi_2Te_3 and (**b**) Bi_2Se_3 grown by MBE on a GaAs (001) substrate. The distances between QLs (~1 nm) are marked by "**I**." After ref. [31]

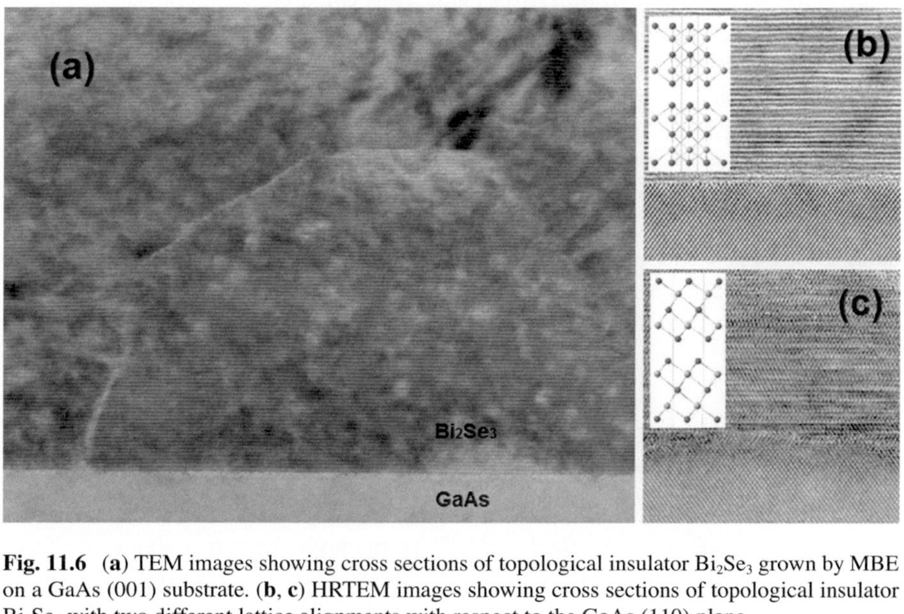

Fig. 11.6 (**a**) TEM images showing cross sections of topological insulator Bi_2Se_3 grown by MBE on a GaAs (001) substrate. (**b, c**) HRTEM images showing cross sections of topological insulator Bi_2Se_3 with two different lattice alignments with respect to the GaAs (110) plane

aligned on the same plane, suggesting that this boundary resulted from two types of rotation domains instead of antiphase boundaries or stacking faults; second, one domain is totally enclosed by the other, suggesting that some domains expand faster than other domains. In Fig. 11.6b, c, the high-resolution cross section images clearly

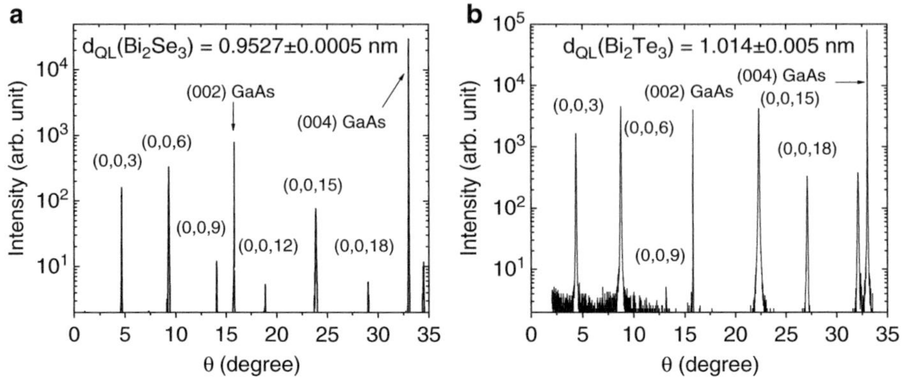

Fig. 11.7 XRD data obtained for (**a**) Bi_2Se_3 and (**b**) Bi_2Te_3 films grown on a GaAs (001) substrate

show two different lattice alignments with respect to the underlying GaAs lattice, as indicated schematically by the lattices shown in the insets. These TEM images thus confirm our observations from RHEED and AFM studies.

The overall crystallinity of the TI films was evaluated by HR-XRD using the Cu $K\alpha_1$ radiation line. Figure 11.7 shows XRD patterns that were obtained from (a) a 233-nm-thick Bi_2Te_3 film and (b) a 180-nm-thick Bi_2Se_3 film. Strong reflections from {003}-type lattice planes of the TI are clearly visible, indicative of the highly directional *c*-axis growth of the film. The full-width-at-half-maximum for the (0,0,6) plane indicates that the crystallinity of Bi_2Se_3 is considerably better than that of Bi_2Te_3 (108 versus 234 s). The QL thicknesses were calculated from the XRD data, giving $d_{QL}=1.014\pm0.005$ nm for Bi_2Te_3 and $d_{QL}=0.9527\pm0.0005$ nm for Bi_2Se_3, respectively (here the uncertainties correspond to 1σ). Both values are consistent with the values of 1.016 nm for bulk Bi_2Te_3 [32] and 0.9545 nm for bulk Bi_2Se_3 [33].

11.5 $Bi_2(TeSe)_3$ Ternary Alloy

In addition to exploration of the binary TI compounds described above, we have also grown a series of ternary Bi-chalcogenide alloys such as $Bi_2(Te_xSe_{1-x})_3$. Such alloys, including their ordered compounds such as $Bi_2Te_2Se_1$, are naturally expected to have a large bulk resistivity owing to several mechanisms [34, 35] that are not present in the binary TIs, thus making them of special interest for those TI studies where the high conductivity of the bulk acts to obscure the topological effects associated with surface states. The specific mechanisms which tend to increase the bulk resistivity of such alloys are primarily the following: (1) the formation of Se vacancies is expected to be suppressed in ternary alloys, because the Se ion trapped

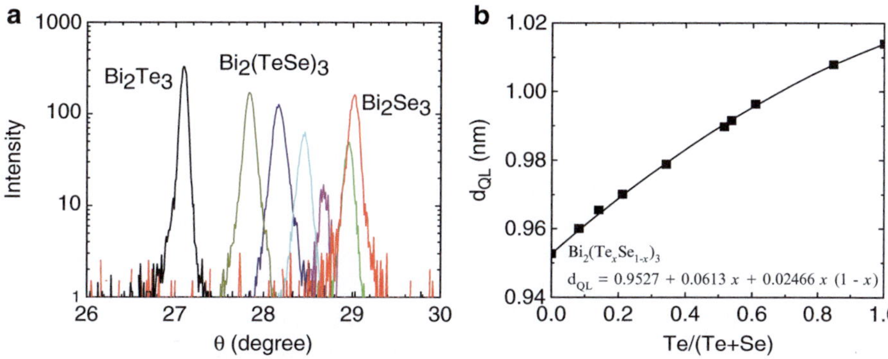

Fig. 11.8 (**a**) XRD spectra measured on a series of $Bi_2(TeSe)_3$ alloy films showing the (0,0,18) reflection peak. (**b**) QL thicknesses d_{QL} calculated from XRD data for a series of $Bi_2(TeSe)_3$ films plotted as a function of Te/(Te + Se) ratio

between two Bi atoms is less exposed to evaporation due to stronger chemical bonding with Bi that occurs in this position; (2) the formation of Te anti-site defects between Te and Se layers is also expected to be suppressed because of preferential bonding between Se and Bi, in contrast to the weak Se–Te bonding; (3) under proper conditions, the formation of ordered Te-Bi-Se-Bi-Te quintuple layers (QLs) can also be achieved in such ternary combinations of Bi, Te, and Se, minimizing the additional disorder that could be caused by Se/Te randomness, consecutively reducing the bulk carrier concentration.

With this in mind, we have also carried out the growth of a series of $Bi_2(TeSe)_3$ alloy films on GaAs (001) substrates using various $Te_2/(Te_2 + Se_2)$ BEP ratios. The ternary films were grown in a Te-rich regime by varying Se_2 flux, with a constant of Te_2/Bi BEP of around 10. As in the case of the binary TI films discussed above, the XRD spectra taken on the ternary $Bi_2(TeSe)_3$ alloy films reveal only reflections from the {003}-type lattice planes, again indicating highly directed *c*-axis growth of these films. X-ray rocking curves yielded a full-width-at-half maximum of 0.2–0.5°. Figure 11.8a shows the (0,0,18) reflection peaks of the XRD spectra obtained for a series of $Bi_2(TeSe)_3$ films, indicating that a wide spectrum of mixed $Bi_2(TeSe)_3$ alloys can be obtained, thus also representing a step toward realizing structures with ordered Te-Bi-Se-Bi-Te QLs. The QL thicknesses d_{QL} were calculated from the XRD data. As shown in Fig. 11.8b, the quintuple layer thickness d_{QL} is not fully linear with the Te/(Te + Se) ratio of the $Bi_2(TeSe)_3$ alloy composition, which was measured by Rutherford backscattering spectrometry. Such nonlinearity (bowing) suggests that the chemical stoichiometric ratios [i.e., Te/(Te + Se)] might be different between the center Te/Se plane and the top/bottom Te/Se planes of QL structure. We attribute this interesting nonlinearity to the fact that Bi favors bonding with Se rather than with Te. This result agrees with our AFM measurements.

11.6 ARPES and Transport

The ARPES measurements were performed at beamline 10.0.1 (HERS) of the Advanced Light Source, Berkeley, CA, equipped with a VG-Scienta R4000 electron analyzer. The 330-nm-thick film sample was cleaved in situ for full exposure of a pristine surface, and was measured at ~20 K in a vacuum better than 5×10^{-11} Torr.

Figure 11.9 shows ARPES data of a 330-nm-thick $Bi_{1.95}Te_{0.25}Se_{2.8}$ film, where the chemical stoichiometry of this alloy was measured by Rutherford backscattering spectrometry. Figure 11.9a presents typical ARPES constant energy maps at three different binding energies. The ARPES Fermi map (left panel of Fig. 11.9a) shows clearly the remnant photoemission intensity from the bulk conduction band (BCB) around Γ, and the topological surface state (Dirac cone) circumventing it. No sign

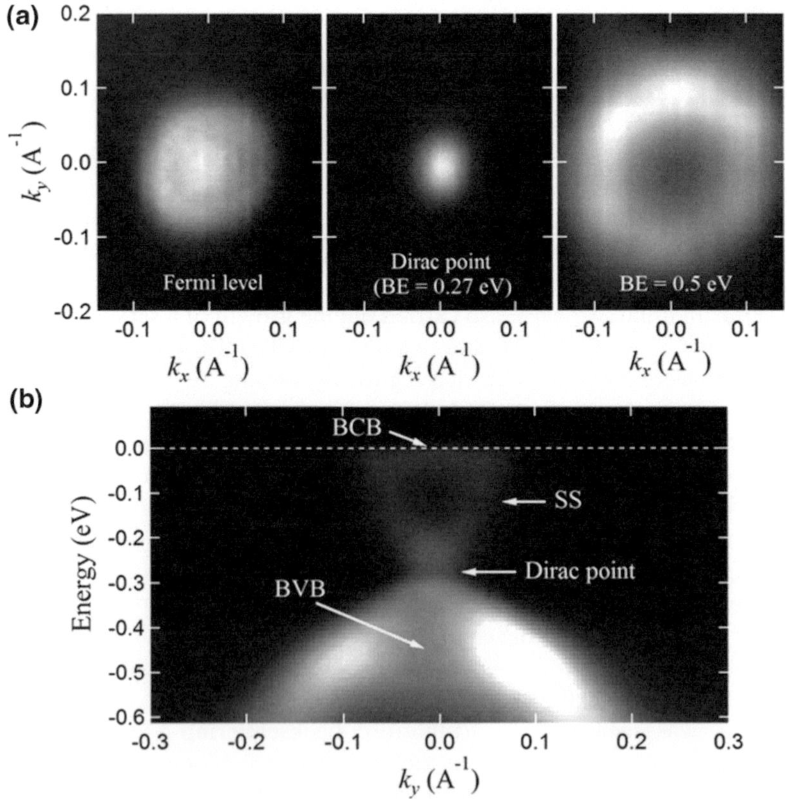

Fig. 11.9 ARPES data obtained on a 330-nm-thick $Bi_{1.95}Te_{0.25}Se_{2.8}$ film. (**a**) Constant energy maps at three selected binding energies (BEs). It is clear from the figure that the Dirac point is fully exposed for this film. (**b**) ARPES *k-E* map along the *Γ-M* direction (k_y). *BCB* bulk conduction band, *BVB* bulk valence band, *SS* surface state (*Dirac cone*)

of hexagonal warping is seen at the Dirac cone, since the Fermi level of this film is located at only 0.27 eV above the Dirac point, where the influence of the sixfold symmetry of the bulk bands on the surface state is weak. From the map taken at 0.27 eV binding energy (middle panel of Fig. 11.9b), it is seen that the Dirac point is fully exposed. The top of the bulk valence band (BVB) lies lower than the Dirac point. At higher binding energies, the influence of the bulk symmetry becomes more visible (right panel of Fig. 11.9a), as the band contour changes to a hexagonal shape. Figure 11.9b presents an ARPES *k–E* map taken along the *Γ-M* momentum space direction, where the BVB, BCB, and SS (surface state) are clearly marked. The presence of BCB below the Fermi level confirms the bulk conducting nature of the film. One noteworthy observation from this map is that the photoemission intensity experiences a slight suppression at the Dirac point, forming a gap-like feature. Several reasons might lead to this phenomenon. (1) The remaining film after the in situ cleaving may be so thin that the top and bottom layers of the film interact with each other, giving rise to a tunneling gap resolved by other studies. (2) Strong spatial fluctuations caused by surface impurities may prevent the momentum from being a good quantum number, which in turn gives rise to a gap-like feature in spatially averaged measurements such as ARPES [36].

We have carried out low-temperature (T) magnetotransport measurements on both Bi_2Se_3 and Bi_2Te_3 films. Our magnetotransport data show pronounced cusp-like behavior in both films, consistent with weak anti-localization (WAL) [37, 38]. As an example, we have performed low-T transport measurements on a Bi_2Se_3 film (6 nm). The magnetotransport data exhibit resistance suppression near zero B-field (Fig. 11.10c), consistent with WAL related to spin-momentum locking and spin-orbit coupling [9, 39, 40]. WAL has been recently extensively used in TI study to probe the coherent transport properties of both the bulk and the surface carriers, e.g., coherence channel number and length [42, 44].

In this work, all our Hall-bar devices are patterned into a 15-μm-width by standard lithography, with Cr/Au ohmic contacts deposited by an e-beam evaporator. To study electric field-modulated transport, the Bi_2Se_3 Hall-bars are capped by a 35-nm Al_2O_3 dielectric layer using atomic layer deposition. Figure 11.10a shows one of our typical top-gated Hall-bar field effect transistor (FET) devices with Al_2O_3 dielectric. The transport measurements are performed in cryostats using standard low-frequency lock-in techniques.

As shown in Fig. 11.10b, our 6-nm Bi_2Se_3 film exhibits a transition from a high-T metallic behavior to a low-T insulating behavior, indicating that the bulk conduction starts to be reduced at low-T. At the base T (1.5 K), the electron concentration (n_e) and mobility (μ) for this specific Bi_2Se_3 film extracted from the Hall data (Fig. 11.10c) are $n_e = 1.5–3.5 \times 10^{13}$ cm^{-2} (2.5–5.9 $\times 10^{19}$ cm^{-3}) and $\mu \sim 140$ cm^2/V s, respectively, indicating that there still remains a significant amount of bulk carriers. Moreover, near the zero B-field region, R_{xx} data shows a downward cusp-like behavior (WAL feature). Figure 11.10d shows the effects of the electric field on the sheet resistance (R_{xx}) at the base $T = 1.5$ K. In this sample with a substantial bulk carrier density, R_{xx} is only weakly modulated by the gate voltage (V_g). The observed WAL can be fit well by a two-component Hikami–Larkin–Nagaoka (HLN) equation in the strong

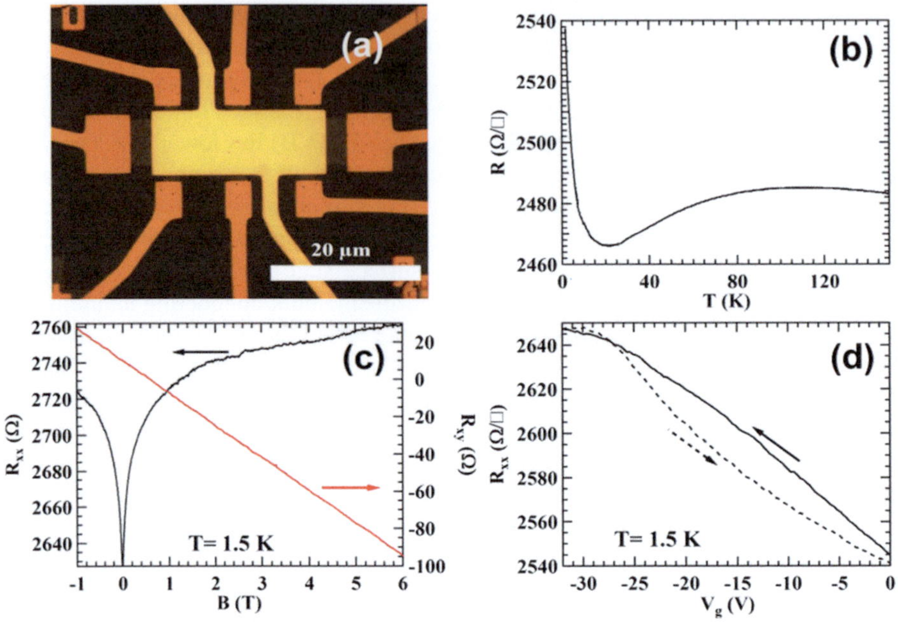

Fig. 11.10 (**a**) Optical image of a typical top-gated Bi_2Se_3 Hall-bar device with high-*k* dielectric (Al_2O_3). (**b**) Temperature dependence of the resistance for a typical 6 nm Bi_2Se_3 film. (**c**) Magnetoresistance (R_{xx} and R_{xy}) for the 6-nm Bi_2Se_3 device at $T=1.5$ K, showing a typical WAL behavior in the low *B*-field region. (**d**) Electric field (V_g) modulated sheet resistance at $T=1.5$ K

spin-orbit coupling limit [41]. The HLN analysis gives two important coherent transport parameters, phase coherence length ($\boldsymbol{l}_c$) and coherence channel number ($\boldsymbol{A}$), where $\boldsymbol{A}=-1$ corresponds to one single coherence channel. Our results for this sample show $\boldsymbol{A}=-1.2$ and $\boldsymbol{l}_c=42.6$ nm, both parameters being only weakly depending on V_g.

11.7 Concluding Remarks

In this exploratory study, we have successfully grown high quality epitaxial films of Bi_2Te_3, Bi_2Se_3, and their alloys on GaAs (001) substrates even though there is a mismatch between the hexagonal lattices of Bi_2Te_3 and Bi_2Se_3 TI films and the fourfold cubic symmetry of the GaAs (001) surface. The films are highly uniform, and their crystalline quality is comparable to that of films grown on substrates with hexagonal or threefold symmetric surfaces. Examination of the TI film surfaces by AFM reveals a spiral growth mode, the same as reported previously for TI films. Measurements of ARPES were also carried out, and confirmed the TI properties. Magnetotransport studies performed on these Bi_2Te_3 and Bi_2Se_3 films clearly show signatures of 2D transport, including effects of WAL. Future studies of TI films grown on GaAs (001) substrates should contribute toward a better understanding of the process of formation of TI-layered structures in MBE growth. Fabrication of TI films on substrates compatible with commonly used orientations also opens important options for future spin-based devices. Opportunities such as the realization of spin-momentum locking in the surface bands of interest for spintronic applications [42, 43, 45] and the realization of quantized magneto-electric effects [44, 45] suggest themselves in this context. As a more distant but highly exciting prospect, such studies would also lay the ground for further work for advancing magnetic-monopole-like image charges [46], as well as for obtaining structures that can host Majorana fermions at topological insulator/superconductor interfaces, that hold promise for applications in fault-tolerant topological quantum computing [47, 48].

Acknowledgements This work was supported by NSF Grant DMR10-05851 (ND); NSF grant ECCS10-02114 and an AFOSR Grant FA9550-10-1-0129 (ASU); and DARPA MESO program (Purdue). The authors acknowledge use of facilities in the John M. Cowley Center for High Resolution Electron Microscopy at Arizona State University. The authors also thank Dr. K.M. Yu at Lawrence Berkeley National Laboratory for providing the chemical stoichiometric ratios of $Bi_2(TeSe)_3$ alloys using Rutherford backscattering spectrometry.

References

1. Kane, C.L., Mele, E.J.: Quantum spin Hall effect in graphene. Phys. Rev. Lett. **95**, 226801 (2005)
2. Kane, C.L., Mele, E.J.: Z_2 topological order and the quantum spin Hall effect. Phys. Rev. Lett. **95**, 146802 (2005)
3. Bernevig, B.A., Hughes, T.L., Zhang, S.-C.: Quantum spin Hall effect and topological phase transition in HgTe quantum wells. Science **314**, 1757 (2006)
4. König, M., Wiedmann, S., Brüne, C., Roth, A., Buhmann, H., Molenkamp, L.W., Qi, X.-L., Zhang, S.-C.: Quantum spin Hall insulator state in HgTe quantum wells. Science **318**, 766 (2007)
5. Qi, X.-L., Zhang, S.-C.: The quantum spin Hall effect and topological insulators. Phys. Today **63**, 33 (2010)
6. Fu, L., Kane, C.L., Mele, E.J.: Topological insulators in three dimensions. Phys. Rev. Lett. **98**, 106803 (2007)
7. Moore, J.E., Balents, L.: Topological invariants of time-reversal-invariant band structures. Phys. Rev. B **75**, 121306 (2007)
8. Hasan, M.Z., Kane, C.L.: *Colloquium*: Topological insulators. Rev. Mod. Phys. **82**, 3045 (2010)
9. Bychkov, Y.A., Rashba, E.I.: Properties of a 2D electron gas with lifted spectral degeneracy. JETP Lett. **39**, 66 (1984)
10. Liu, C.-X., Qi, X.-L., Zhang, H., Dai, X., Fang, Z., Zhang, S.-C.: Model hamiltonian for topological insulators. Phys. Rev. B **82**, 045122 (2010)
11. Shan, W.-Y., Lu, H.-Z., Shen, S.-Q.: Effective continuous model for surface states and thin films of three-dimensional topological insulators. New J. Phys. **12**, 043048 (2010)
12. Hsieh, D., Qian, D., Wray, L., Xia, Y., Hor, Y.S., Cava, R.J., Hasan, M.Z.: A topological dirac insulator in a quantum spin Hall phase. Nature **452**, 970 (2008)

13. Xia, Y., Qian, D., Hsieh, D., Wray, L., Pal, A., Lin, H., Bansil, A., Grauer, D., Hor, Y.S., Cava, R.J., Hasan, M.Z.: Observation of a large-gap topological-insulator class with a single Dirac cone on the surface. Nat. Phys. **5**, 398 (2009)
14. Zhang, G., Qin, H., Teng, J., Guo, J., Guo, Q., Dai, X., Fang, Z., Wu, K.: Quintuple-layer epitaxy of thin films of topological insulator Bi_2Se_3. Appl. Phys. Lett. **95**, 053114 (2009)
15. Chen, Y.L., Analytis, J.G., Chu, J.H., Liu, Z.K., Mo, S.K., Qi, X.L., Zhang, H.J., Lu, D.H., Dai, X., Fang, Z., Zhang, S.-C., Fisher, I.R., Hussain, Z., Shen, Z.-X.: Experimental realization of a three-dimensional topological insulator, Bi_2Te_3. Science **325**, 178 (2009)
16. Hsieh, D., Xia, Y., Qian, D., Wray, L., Dil, J.H., Meier, F., Osterwalder, J., Patthey, L., Fedorov, A.V., Lin, H., Bansil, A., Grauer, D., Hor, Y.S., Cava, R.J., Hasan, M.Z.: Observation of time-reversal-protected single-dirac-cone topological-insulator states in Bi_2Te_3 and Sb_2Te_3. Phys. Rev. Lett. **103**, 146401 (2009)
17. Hsieh, D., Xia, Y., Wray, L., Qian, D., Pal, A., Dil, J.H., Osterwalder, J., Meier, F., Bihlmayer, G., Kane, C.L., Hor, Y.S., Cava, R.J., Hasan, M.Z.: Observation of unconventional quantum spin textures in topological insulators. Science **323**, 919 (2009)
18. Roushan, P., Seo, J., Parker, C.V., Hor, Y.S., Hsieh, D., Qian, D., Richardella, A., Hasan, M.Z., Cava, R.J., Yazdani, A.: Topological surface states protected from backscattering by chiral spin texture. Nature **460**, 1106 (2009)
19. Chen, X., Ma, X.-C., He, K., Jia, J.-F., Xue, Q.-K.: Molecular Beam Epitaxial Growth of Topological Insulators. Adv. Mater. **23**, 1162 (2011)
20. Li, H.D., Wang, Z.Y., Guo, X., Wong, T.L., Wang, N., Xie, M.H.: Growth of multilayers of Bi_2Se_3/ZnSe: Heteroepitaxial interface formation and strain. Appl. Phys. Lett. **98**, 043104 (2011)
21. Liu, M., Chang, C.-Z., Zhang, Z., Zhang, Y., Ruan, W., He, K., Wang, L.L., Chen, X., Jia, J.-F., Zhang, S.-C., Xue, Q.-K., Ma, X., Wang, Y.: Electron interaction-driven insulating ground state in Bi_2Se_3 topological insulators in the two-dimensional limit. Phys. Rev. B **83**, 165440 (2011)
22. Zhang, G.H., Qin, H.J., Chen, J., He, X.Y., Lu, L., Li, Y.Q., Wu, K.H.: Growth of topological insulator Bi_2Se_3 thin films on $SrTiO_3$ with large tunability in chemical potential. Adv. Funct. Mater. **21**, 2351 (2011)
23. Richardella, A., Zhang, D.M., Lee, J.S., Koser, A., Rench, D.W., Yeats, A.L., Buckley, B.B., Awschalom, D.D., Samarth, N.: Coherent heteroepitaxy of Bi_2Se_3 on GaAs (111)B. Appl. Phys. Lett. **97**, 262104 (2010)
24. Sadowski, J., Mathieu, R., Svedlindh, P., Domagała, J.Z., Bak-Misiuk, J., Światek, K., Karlsteen, M., Kanski, J., Ilver, L., Åsklund, H., Södervall, U.: Structural and magnetic properties of GaMnAs layers with high Mn-content grown by migration-enhanced epitaxy on GaAs(100) substrates. Appl. Phys. Lett. **78**, 3271 (2001)
25. Krebs, J.J., Jonker, B.T., Prinz, G.A.: Properties of Fe single-crystal films grown on (100) GaAs by molecular-beam epitaxy. J. Appl. Phys. **61**, 2596 (1987)
26. Liu, X., Smith, D.J., Fan, J., Zhang, Y.-H., Cao, H., Chen, Y.P., Leiner, J., Kirby, B.J., Dobrowolska, M., Furdyna, J.K.: Structural properties of Bi_2Te_3 and Bi_2Se_3 topological insulators grown by molecular beam epitaxy on GaAs(001) substrates. Appl. Phys. Lett. **99**, 171903 (2011)
27. Liu, X., Smith, D.J., Cao, H., Chen, Y.P., Fan, J., Zhang, Y.-H., Pimpinella, R.E., Dobrowolska, M., Furdyna, J.K.: Characterization of Bi_2Te_3 and Bi_2Se_3 topological insulators grown by MBE on (001) GaAs substrates. J. Vac. Sci. Technol. B **30**, 02B103 (2012)
28. Shufflebotham, P.K., Card, H.C., Kao, K.C., Thanailakis, A.: Amorphous silicon–tellurium alloys. J. Appl. Phys. **60**, 2036 (1986)
29. Li, H.D., Wang, Z.Y., Kan, X., Guo, X., He, H.T., Wang, Z., Wang, J.N., Wong, T.L., Wang, N., Xie, M.H.: The van der Waals epitaxy of Bi_2Se_3 on the vicinal Si(111) surface: an approach for preparing high-quality thin films of a topological insulator. New J. Phys. **12**, 103038 (2010)
30. Liu, Y., Weinert, M., Li, L.: Spiral growth without dislocations: Molecular beam epitaxy of the topological insulator Bi_2Se_3 on epitaxial Graphene/SiC(0001). Phys. Rev. Lett. **108**, 115501 (2012)

31. Richter, W., Kohler, H., Becker, C.R.: A Raman and far-infrared investigation of phonons in the rhombohedral V_2–VI_3 compounds Bi_2Te_3, Bi_2Se_3, Sb_2Te_3 and $Bi_2(Te_{1-x}Se_x)_3$ $(0 < x < 1)$, $(Bi_{1-y}Sb_y)_2Te_3$ $(0 < y < 1)$. Phys. Status Solidi B **84**, 619 (1977)
32. Jenkins, J.O., Rayne, J.A., Ure, J.R.W.: Elastic moduli and phonon properties of Bi_2Te_3. Phys. Rev. B **5**, 3171 (1972)
33. Nakajima, S.: The crystal structure of $Bi_2Te_{3-x}Se_x$. J. Phys. Chem. Solids **24**, 479 (1963)
34. Ren, Z., Taskin, A.A., Sasaki, S., Segawa, K., Ando, Y.: Large bulk resistivity and surface quantum oscillations in the topological insulator Bi_2Te_2Se. Phys. Rev. B **82**, 241306(R) (2010)
35. Sokolov, O.B., Skipidarov, S.Y., Duvankov, N.I., Shabunina, G.G.: Chemical reactions on the Bi_2Te_3–Bi_2Se_3 section in the process of crystal growth. J. Cryst. Growth **262**, 442 (2004)
36. Xu, S.-Y., Wray, L. A., Alidoust, N., Xia, Y., Neupane, M., Liu, C., Ji, H.-W., Jia, S., Cava, R. J., Hasan, M. Z.: Dirac point spectral weight suppression and surface "gaps" in nonmagnetic and magnetic topological insulators. arXiv:1206.0278 (2012)
37. He, H.T., Wang, G., Zhang, T., Sou, I.K., Wong, G.K.L., Wang, J.N., Lu, H.Z., Shen, S.Q., Zhang, F.C.: Impurity effect on weak antilocalization in the topological insulator Bi_2Te_3. Phys. Rev. Lett. **106**, 166805 (2011)
38. Steinberg, H., Laloe, J.-B., Fatemi, V., Moodera, J.S., Jarillo-Herrero, P.: Electrically tunable surface-to-bulk coherent coupling in topological insulator thin films. Phys. Rev. B **84**, 233101 (2011)
39. Moore, J.E.: The birth of topological insulators. Nature **464**, 194 (2010)
40. Qi, X.-L., Zhang, S.-C.: Topological insulators and superconductors. Rev. Mod. Phys. **83**, 1057 (2011)
41. Hikami, S., Larkin, A.I., Nagaoka, Y.: Spin-orbit interaction and magnetoresistance in the two dimensional random system. Prog. Theor. Phys. **63**, 707 (1980)
42. Ohno, H.: A window on the future of spintronics. Nat. Mater. **9**, 952 (2010)
43. Vobornik, I., Manju, U., Fujii, J., Borgatti, F., Torelli, P., Krizmancic, D., Hor, Y.S., Cava, R.J., Panaccione, G.: Magnetic proximity effect as a pathway to spintronic applications of topological insulators. Nano Lett. **11**, 4079 (2011)
44. Tse, W.-K., MacDonald, A.H.: Magneto-optical and magnetoelectric effects of topological insulators in quantizing magnetic fields. Phys. Rev. B **82**, 161104 (2010)
45. Essin, A.M., Moore, J.E., Vanderbilt, D.: Magnetoelectric polarizability and axion electrodynamics in crystalline insulators. Phys. Rev. Lett. **102**, 146805 (2009)
46. Qi, X.L., Li, R., Zang, J., Zhang, S.C.: Inducing a magnetic monopole with topological surface states. Science **323**, 1184 (2009)
47. Fu, L., Kane, C.: Superconducting proximity effect and majorana fermions at the surface of a topological insulator. Phys. Rev. Lett. **100**, 096407 (2008)
48. Nayak, C., Simon, S.H., Stern, A., Freedman, M., Das Sarma, S.: Non-Abelian anyons and topological quantum computation. Rev. Mod. Phys. **80**, 1083 (2008)

Chapter 12
Vapor Phase Deposition Synthesis of Bismuth-Based Topological Insulator Nanoplates and Their Electrostatic Properties

Guolin Hao, Xiang Qi, and Jianxin Zhong

Abstract Bismuth-based materials have been extensively investigated as the best thermoelectric materials. Recently they were confirmed to be topological insulators promising in future spintronics and topological quantum computation. Using vapor phase deposition method, we have synthesized few-layer topological insulator bismuth telluride, bismuth selenide, and ternary bismuth-based nanoplates with hexagonal, triangular, and truncated triangular nanostructures. The possible formation mechanism of these nanoplates with various nanostructures has been proposed. Electrostatic properties of the as-prepared nanoplates as well as charge injection have been systematically investigated by Kelvin probe force microscopy (KPFM) under ambient environment. We find bismuth-based nanoplates exhibit uniform surface potential and charge distributions. The work functions of topological insulator nanoplates grown on n-doped and p-doped silicon (111) substrates were determined by KPFM. Fermi level tuning of topological insulator nanoplates through substrates and doping were simultaneously investigated. These investigations may provide an effective route to investigate the electronic properties of topological insulators.

12.1 Introduction

Bismuth telluride (Bi_2Te_3), bismuth selenide (Bi_2Se_3), and their ternary compounds are typical layered narrow band gap semiconductors with rhombohedral structure, which have been employed as important materials for near room-temperature

G. Hao (✉) • X. Qi (✉) • J. Zhong (✉)
Hunan Key Laboratory for Micro-Nano Energy Materials and Devices, Xiangtan University, Xiangtan, Hunan 411105, P. R. China

Laboratory for Quantum Engineering and Micro-Nano Energy Technology, Faculty of Materials and Optoelectronic Physics, Xiangtan University, Xiangtan, Hunan 411105, P. R. China
e-mail: guolinhao@xtu.edu.cn; xqi@xtu.edu.cn; jxzhong@xtu.edu.cn

H. Li and Z.M. Wang (eds.), *Bismuth-Containing Compounds*, Springer Series in Materials Science 186, DOI 10.1007/978-1-4614-8121-8_12,

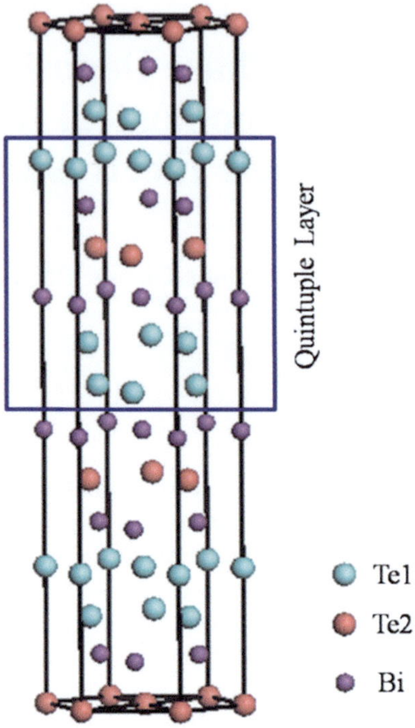

Fig. 12.1 Layer crystal structure of Bi_2Te_3, with five Bi and Te atomic sheets forming one quintuple layer

thermoelectric applications in thermoelectric industry [1–5]. Recently, theoretical studies predicted that Bi_2Te_3 and Bi_2Se_3 materials are three-dimensional topological insulators (TIs) consisting of single Dirac cone at the Γ point [6]. The feature of surface states is confirmed experimentally by angle-resolved photoemission spectroscopy [7–9], scanning tunneling microscopy [10], and corresponding transport measurements [11–13]. A TI is a gapped insulator in the bulk or gapless edge but possessing current-carrying massless Dirac surface states, which are protected by time-reversal symmetry on the surface. As a new state of quantum matter, a variety of intriguing phenomena have been predicted, such as Majorana fermions and magnetic monopoles, which have intrigued potential applications in spintronics and topological quantum computation [14–17]. Therefore, growth of high quality TIs becomes imperative for realistic applications [18]. Specially, growth of TIs on different substrates would further promote development in TI material engineering and integration.

Bi_2Te_3 and Bi_2Se_3 possess the same rhombohedral crystal structure belonging to the space group D_{3d}^5 (R-3m), consisting of five-atom layers arranged along the *z* axis known as quintuple layers. For example, as schematically shown in Fig. 12.1, along the *z* direction, the unit cell of Bi_2Te_3 consists of five atomic layers with a stacking sequence of Te (1)-Bi-Te (2)-Bi-Te (1) forming a quintuple layer. Covalently bonded quintuple layers are held together predominantly by weak van der Waals interactions. The strength of the van der Waals bonding is much weak compared to the ionic–covalent bonding inside the quintuple layer. Thus they can be easily cleaved parallel to the layers employing micromechanical exfoliation method [19]. Therefore, the growth of these TIs can be dominated by the anisotropic crystal structures via van der Waals forces.

Compared with bulk materials, low-dimensional TI materials exhibit extremely large surface-to-volume ratios, which are essential for the transport properties associated with topological surface states [20, 21]. For example, Aharonov–Bohm oscillation has been detected on Bi_2Te_3 and Bi_2Se_3 nanoribbons providing direct transport evidence of the robust, conducting surface states [20, 23]. And the Kondo effect has also been observed on magnetically doped Bi_2Se_3 nanoribbons [13]. In previous reports, TIs were also systematically characterized by angle-resolved photoemission spectroscopy as well as scanning tunneling microscopy [7, 23]. Although rapid progresses in measuring surface states, electrical transport, as well as optical properties have been achieved [24–27], little is known about the electrostatic properties on the TI surface. Particularly, with TI material thickness down to only several nanometers, the substrate effect on the TI electronic properties becomes non-negligible. Kelvin probe force microscopy (KPFM) has been considered to be one of the most powerful techniques to determine the electrostatic properties such as surface potential distribution and local work function on the nanoscale [14, 27–29].

Currently, various methods have been employed to prepare TI nanomaterials. TI thin films can be fabricated by molecular beam epitaxy [28, 29] and pulsed laser deposition methods [30]. However, the cost of molecular beam epitaxy and pulsed laser deposition is expensive and these methods are less accessible. Liquid and "graphene-like" exfoliation methods can also prepare TIs [31, 32], but with thick nanoflakes and irregular shapes. Vapor–liquid–solid (VLS) growth has been considered to be a useful route to synthesize high quality TI nanoribbons and nanowires, but only with moderate thickness range of 30–100 nm [33]. Herein, few-layer Bi_2Te_3, Bi_2Se_3, and ternary $Bi_2(Se_xTe_{1-x})_3$ nanoplates (NPs) with various morphologies were successfully synthesized on various substrates by vapor phase deposition method without any catalysts [22, 34–36]. The growth mechanism has been proposed to elucidate TI NPs with various morphologies [34]. Surface potentials on TI NPs have been systematically investigated by KPFM under ambient condition with 40 % relative humidity exhibiting uniform surface potential and charge distributions. Surface potential characteristics of TI NPs grown on different substrates indicate that electrostatic properties can be effectively modified [34]. Ambipolar charge injection and transport of few-layer TI NPs have also been investigated [35]. The local work function of Bi_2Te_3, Bi_2Se_3, and ternary $Bi_2(Se_xTe_{1-x})_3$ NPs is simultaneously obtained. These results pave a new route to explore the intriguing topological properties.

12.2 Kelvin Probe Force Microscopy

12.2.1 Introduction of Kelvin Probe Force Microscopy

KPFM was first reported in the 1991 [37], which is based on the atomic force microscopy (AFM) setup. It is also known as scanning Kelvin probe microscopy and is considered to be one of the most powerful techniques to investigate the surface potential as well as charge distributions of samples [37, 38]. The technique makes it possible to determine quantitative measurement of surface potential with high resolution, especially for nanostructures both under ambient and vacuum environment. The surface potential measurements of samples are characterized employing KPFM using a conductive probe. The contact potential difference (CPD) can be expressed as

$$\mathrm{CPD} = \frac{(\Phi_{\mathrm{tip}} - \Phi_{\mathrm{sample}})}{e} \tag{12.1}$$

where e is the elementary charge. Φ_{tip} and Φ_{sample} correspond to the work functions of metallic tip and sample, respectively.

12.2.2 Measurement Principle of KPFM

KPFM simultaneously performs sample topography as well as surface potential measurements. It is based on dynamic force microscopy by applying V_{AC}, V_{DC}, and $V_{\mathrm{DC}} + V_{\mathrm{AC}} \cdot \sin \omega t$ ($V_{\mathrm{DC}} = \mathrm{Vs} + V_{\mathrm{off}}$) voltages between a conductive cantilever and a sample with high spatial and electrical resolution, as schematically shown in Fig. 12.2. The output signal is separated from the cantilever into a frequency ingredient with lock-in amplifier. And the topography measurement is carried out by using the operation frequency ingredient (ω_{r}) of dynamic force microscopy, and the surface potential is determined simultaneously in the same area by using the frequency ingredient (ω) of the impressed V_{AC} voltage. Therefore, the voltage between the tip and the sample can be expressed by

$$V = V_{\mathrm{CPD}} - V_{\mathrm{DC}} + V_{\mathrm{AC}} \sin(\omega t) \tag{12.2}$$

The energy U in the parallel plate capacitor can be expressed as

$$U = \frac{1}{2} C V^2 \tag{12.3}$$

C represents the local capacitance between the tip and sample. Therefore, the electrostatic force, which is rate of change of energy with separation distance z between the tip and sample, can be expressed as

$$F_{\mathrm{c}} = -\frac{1}{2} \left(\frac{\partial C}{\partial z} \right) V^2 = F_{\mathrm{dc}} + F_{\omega} + F_{2\omega} \tag{12.4}$$

With component at DC

$$F_{\mathrm{dc}} = -\frac{1}{2} \left(\frac{\partial C}{\partial z} \right) \left[(V_{\mathrm{CPD}} - V_{\mathrm{DC}}) + \frac{1}{2} V_{\mathrm{AC}}{}^{2} \right] \tag{12.5}$$

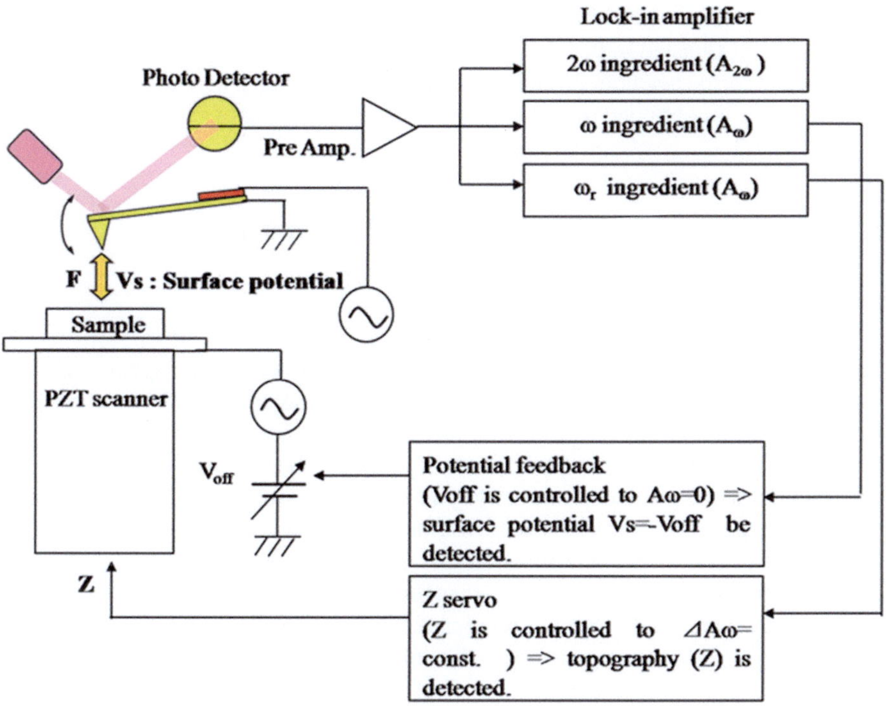

Fig. 12.2 Schematic diagram of KPFM principle

For frequencies ω and 2ω

$$F_{\omega} = -\left(\frac{\partial C}{\partial z}\right) \; [(V_{\mathrm{CPD}} - V_{\mathrm{DC}}) \;\; V_{\mathrm{AC}} \sin(\omega t)] \tag{12.6}$$

$$F_{2\omega} = -\frac{1}{4}\left(\frac{\partial C}{\partial z}\right) \; [V_{\mathrm{AC}}^{2} \cos(\omega t)] \tag{12.7}$$

Electrical properties of investigated samples can be determined through the systematic analysis of these three different force signals. In this case, the electrostatic force acts between the tip and sample. And a compulsive vibration of frequency ω_{r} and a vibration generating from the AC voltage of frequency ω and 2ω overlap with the cantilever. Herein, the static electricity capacity between the probe and sample is set to C and the compensation constant decided by the equipment is set to S. The frequency ω ingredient (A_{ω}) can be expressed by the following formula.

$$A \;\; = S \; \left(\frac{\partial C}{\partial z}\right) \;\; V_{\mathrm{DC}} \;\; V_{\mathrm{AC}} = S \; \left(\frac{\partial C}{\partial z}\right) \;\; (\mathrm{Vs} + V_{\mathrm{off}}) \;\; V_{\mathrm{AC}} \tag{12.8}$$

From this equation, the following result can be eliminated immediately. If the offset voltage, V_{off}, is adjusted so that the amplitude ingredient by static electricity force, A_ω, may become zero (potential feedback), it will be the surface potential of sample, Vs, Vs $= -V_{\text{off}}$ (V_{off} is controlled so that A_ω may always be set to zero) can be obtained.

12.2.3 *Parameters Adjustment in KPFM Mode*

12.2.3.1 Voltage of AC Electric Field, V_{AC}

The value of V_{AC} is critical for the detection of the electrostatic signal, which is dependent on the measured materials. If this voltage is set to zero, A_ω becomes zero. However, the feedback does not start at all and the electrostatic signal can't be detected. Although A_ω is zero finally, if the A_ω signal of basis can be detected greatly, the potential feedback and the sensitivity in KPFM measurement also become better. Therefore, it is better that V_{AC} is originally set to a comparatively large value (about 10 V). The case where resolution is thought as important, and the physical properties of sample surface may be changed in measurement, we must adjust the voltage to become small. On the other hand, when the feedback is not enough even if it is set to 10 V, we must adjust it to the enlarging value.

12.2.3.2 Frequency of AC Electric Field, ω

If the frequency ω approaches the resonance frequency (ω_0), the resolution of KPFM measurement and the flattery nature of surface potential can be improved. If ω and ω_0 are closer too much, the distinction of these two frequencies becomes difficult, and the measurement system is easy to oscillate. The frequency ω in the KPFM measurement condition is usually set to be 2 kHz smaller than the resonance frequency ω_0.

12.2.3.3 *I* Gain for Voltage Control

The monitor signal of voltage and the actual image is employed to direct the strength of voltage, which is controlled by the value of *I* gain parameter. If the flattery is bad, *I* gain will be enlarged. However, if it is likely to oscillate, *I* gain will be set as a small value. Although *I* gain value is set at least as 0.5 at first, when the signal magnification is changed, the optimal value of *I* gain is also changing considerably.

12.2.3.4 Signal Magnification

Signal magnification is the magnification of oscillating signal of cantilever amplified by the preamplifier. The oscillating ingredient, A_ω, by AC electric field is generally quite small comparing with the ingredient by the compulsive vibration of cantilever, A_{ω_r}.

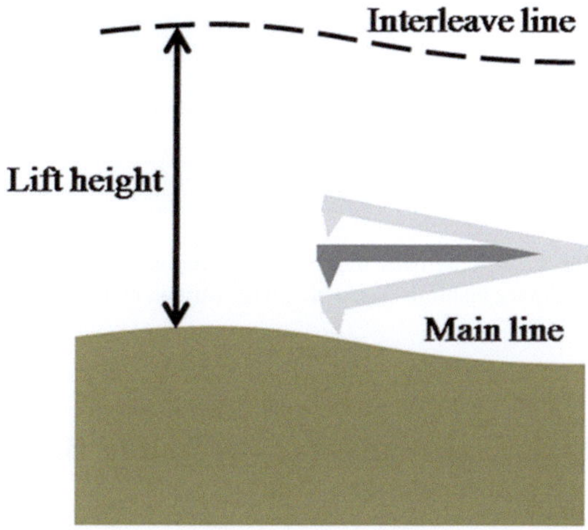

Fig. 12.3 Typical figure of active trace system KPFM

Therefore, it becomes the situation that potential feedback can't operate easily. In such a case, this signal magnification is enlarged. Although the signal magnification can be chosen from three kinds of 1 time, 10 times, and 50 times, a setup of experience of various samples to this value is good in starting with about 10 times.

12.2.4 Active Trace KPFM

The technique of the active trace KPFM has higher resolution and higher sensitivity employing lift height mode, which is schematically shown in Fig. 12.3. In the measurements, the tip scans twice on each scan line of the sample surface. During the first pass, the topography information is acquired in intermittent contact mode with the tip bias to zero. In the second scan, the tip is biased and retraced to a predetermined distance $z \sim 50$ nm above the average surface topography and scanned the same scan line again. The surface potential information is obtained simultaneously.

12.2.5 Applications of KPFM

KPFM is a versatile and spatially precise technique, which has been employed to investigate the electrostatic properties of various systems including organic thin films [38], few-layer graphene films [39, 40], and semiconductors as well as the dynamic properties of working devices [41]. KPFM could provide quantitative insight into other properties of the investigated architecture including work function [42], Fermi level tuning [34], band bending, and electrical polarization [43]. Herein, KPFM is used to systematically detect the electrostatic properties of bismuth-based TIs. The surface potential distributions, work functions, as well as Fermi level tuning of TI NPs have been studied.

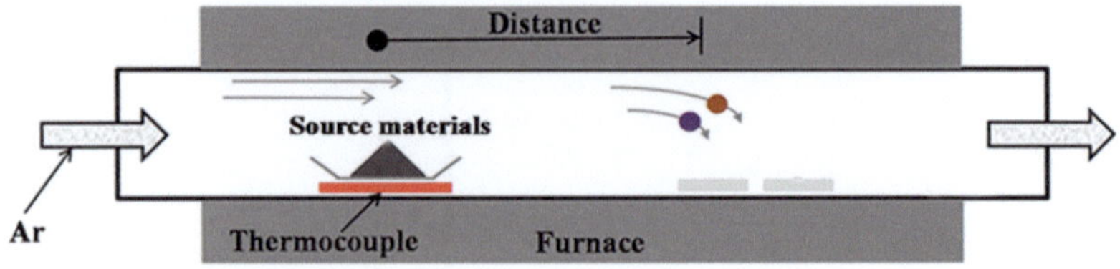

Fig. 12.4 Schematic diagram of the furnace employed for the synthesis of Bi_2Te_3, Bi_2Te_3, as well as ternary bismuth-based NPs with various nanostructures

12.3 Fabrication of Bi_2Te_3 and Bi_2Se_3 NPs

Vapor deposition method has been employed extensively to grow one-dimensional nanostructures through a VLS growth mechanism, in which Au nanoparticles function as universal one of the catalysts [33]. Recently, a catalyst-free vapor–solid synthesis was developed to grow two-dimensional nanomaterials [22, 25, 34, 35]. The reaction setup we used is shown in Fig. 12.4. Bi_2Te_3, Bi_2Se_3, and ternary bismuth-based NPs with various nanostructures were synthesized by vacuum vapor phase deposition method in a 12 in. horizontal tube furnace. In our experiments, the silicon wafers coated with 300 nm thermally grown silicon oxide are cleaned by ultrasonication, first in acetone during 10 min, afterwards for about 10 min in isopropanol, to remove organic residue and particles. Then the pre-treated substrates are rinsed with deionized water to ensure all remaining organic contaminants are removed. The source materials, Bi_2Te_3 powders (99.999 %), are employed as the precursors and placed in the hot center of the furnace. Wafers used for the growth of Bi_2Te_3 NPs are located in the expected locations 9–11 cm away from the hot center. Before experiment, the quartz tube is initially pumped down to 1 Torr and flushed several times with high purity Ar gas to remove oxygen residue. The furnace is heated to 460 °C at 66 °C per min, and then maintained for 5 min. After that, the furnace is cooled to room temperature naturally. During this period, the pressure is maintained at 33 Torr and Ar gas flow 20–100 sccm. In the controlled experiments, few-layer Bi_2Se_3 NPs have also been obtained employing the vapor phase deposition method on various substrates. Powers of source material (99.999 %, Bi_2Se_3) are located in the hot center of the horizontal tube furnace, followed by natural cooling down. The hot center is heated to 490 °C, which is found to be the optimal growth temperature in our experiments. The growth pressure is maintained at 200 Torr with 50–100 sccm Ar gas flow. The substrates are placed at the distance of 9–10 cm away from the hot center.

12.4 Characterization of Bi_2Te_3 and Bi_2Se_3 Nanostructures

12.4.1 *Morphology and Microstructure Characterization of Bi_2Te_3 NPs*

The crystal structures of the as-prepared Bi_2Te_3 NPs were characterized by X-ray diffraction (XRD) (Rigaku D/MAX-2500, Cu Kα radiation). The morphologies and microstructures of the synthesized samples were performed by employing optical microscopy, AFM (SEIKO, SPI3800N+300 HV), scanning electron microscopy (SEM, JEOL, JSM-6360), and high-resolution transmission electron microscopy (TEM, JEM2100) equipped with an energy dispersive X-ray spectrometer (EDS). Raman spectroscopy was excited under ambient environment by using Horiba-Jobin-Yvon LabRam HR confocal microscope.

The XRD pattern of as-synthesized samples is shown in Fig. 12.5a. All of the reflections of XRD pattern can be perfectly indexed to a rhombohedral Bi_2Te_3 phase, which are in good agreement with the parameters $a_H = 0.4384$ nm and $c_H = 3.045$ nm (JCPDS file 15-0863). This result demonstrates that pure phase Bi_2Te_3 was fabricated by the vapor phase deposition method. The dominant (006) and (0015) diffraction peaks suggest that the as-synthesized NPs have preferred orientation along (*00l*) direction indicating that the (001) facets have lower surface energies.

By the optical contrast in the optical microscope, Bi_2Te_3 NPs can be clearly identified on the 300 nm SiO_2/Si substrates, as displayed in Fig. 12.6a. The contrast provides a convenient way to identify different thickness of Bi_2Te_3 NPs. It is interesting to note that the Bi_2Te_3 NPs exhibit various morphologies including

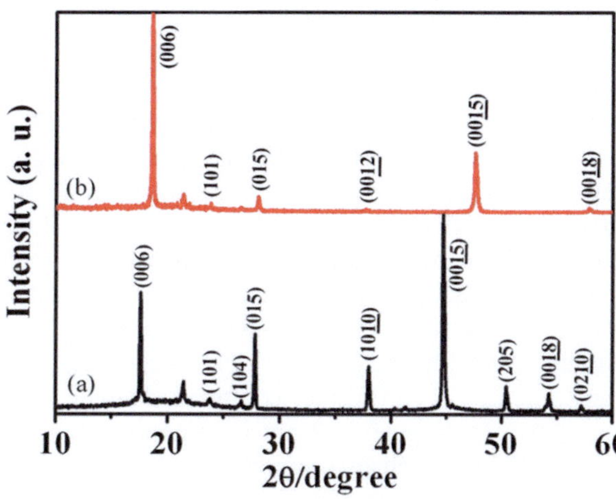

Fig. 12.5 XRD patterns of the as-prepared samples grown on SiO_2/Si substrate. (**a**) Bi_2Te_3. (**b**) Bi_2Se_3

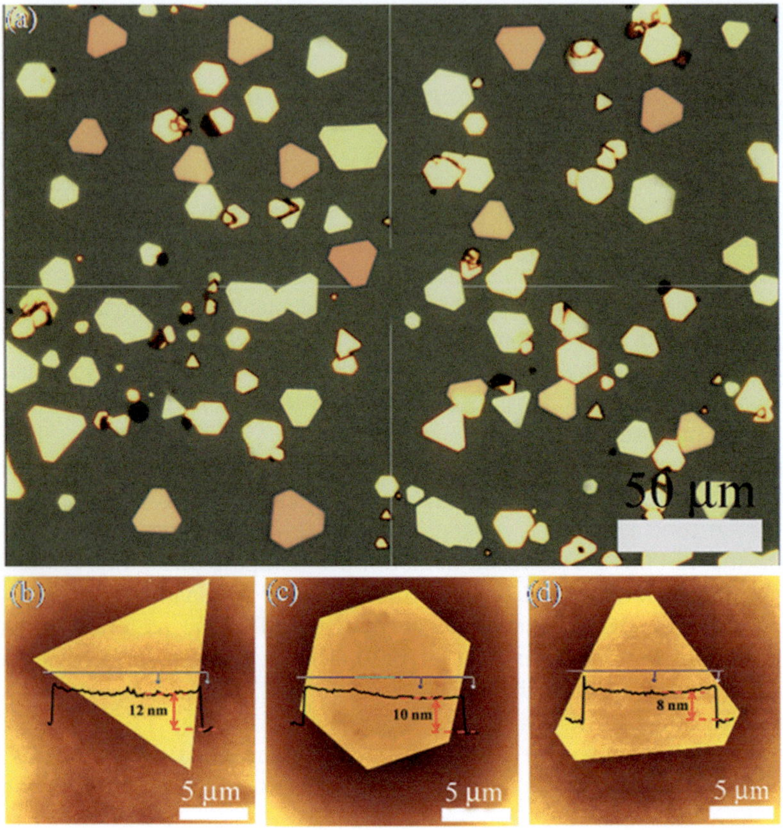

Fig. 12.6 (**a**) Optical image of Bi_2Te_3 NPs grown on SiO_2/Si substrate. (**b–d**) AFM topography images of triangular, hexagonal, and truncated triangular NPs with corresponding height profile, respectively

hexangular, triangular, and truncated triangular structures. Typical AFM topography images of Bi_2Te_3 NPs with various morphologies are shown in Fig. 12.6c, d. Few-layer Bi_2Te_3 NPs with thickness down to 10 nm can be successfully synthesized. Bi_2Te_3 NPs have atomically smooth surface morphologies with average lateral dimension ranging from several micrometers to 20 μm as shown in Fig. 12.6.

Further structural characterization of Bi_2Te_3 NPs was characterized by TEM. A typical TEM image of Bi_2Te_3 NPs is shown in Fig. 12.7a. In Fig. 12.7b, the high resolution of TEM images with clear lattice fringes and corresponding fast Fourier transformation (FFT) further confirm the single-crystalline nature of Bi_2Te_3 NPs. The lattice spacing of 0.22 nm is in good agreement with the (11–20) planes of layered Bi_2Te_3, which also demonstrates previous reports implying the growth along the [11–20] directions with facets as side surfaces and (0001) facets as top and

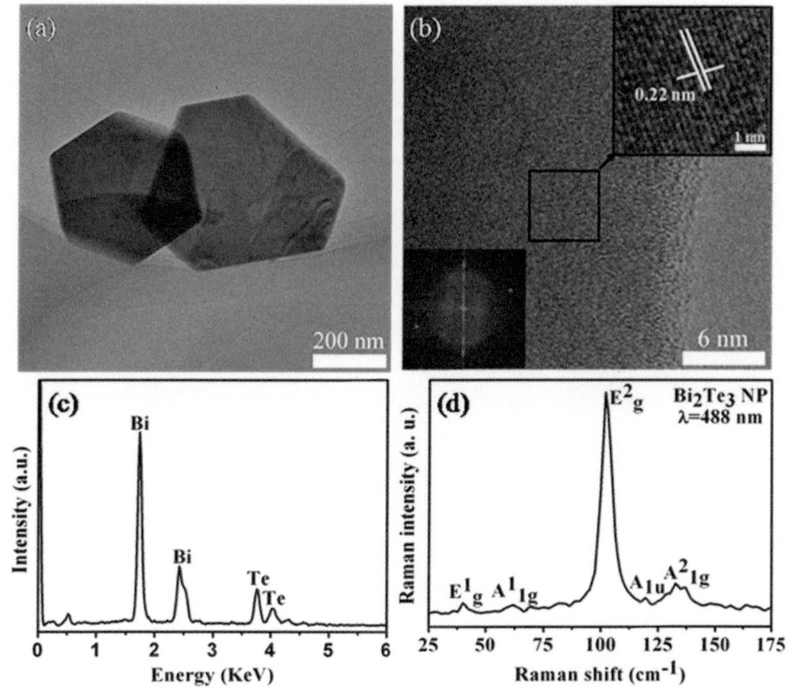

Fig. 12.7 (**a**) TEM image of Bi_2Te_3 NP. (**b**) HRTEM image of Bi_2Te_3 NPs. Corresponding magnified HRTEM and FFT pattern are shown in the *inset*. (**c**) Energy dispersive X-ray spectrometer spectrum of a single Bi_2Te_3 NP. (**d**) Raman spectra of Bi_2Te_3 NPs at 488 nm laser excitation [35]. Copyright belongs to AIP

bottom surfaces. EDS measurement, as shown in Fig. 12.7c, reveals that the atom contents of Bi and Te are 40.71 % and 59.29 %, respectively, for Bi_2Te_3 NP, which is very close to the formula of Bi_2Te_3. To obtain more accurate structure information of synthesized Bi_2Te_3 NPs, micro-Raman spectroscopy of Bi_2Te_3 NP was carried out ranging from 25 to 175 cm^{-1} with laser excitation light ($\lambda = 488$ nm). Due to Bi_2Te_3 NP having very low thermal conductivity, the maximum laser excitation power used in the experiment is 10 mW. Typical characteristic peaks of Bi_2Te_3 NP appear in the position of 39, 62, 101.7, 119.4, and 132.6 cm^{-1}, respectively, as shown in Fig. 12.7d, which are consistent with previous Raman peaks of Bi_2Te_3 crystals [44]. These peaks correspond to the E_g^1, A_{1g}^1, E_g^2, A_{1u}, and A_{1g}^2 modes, respectively. Among these peaks, A_{1g}^1, A_{1g}^2, and A_{1u} can be assigned to the out-of-plane vibration modes, and E_g^1, E_g^2 modes can be attributed to the in-plane vibration modes. However, the mode of A_{1u} is detected in few-layer Bi_2Te_3 NPs, which doesn't show up in Raman spectra of bulk samples. It is considered that the crystal symmetry in the third dimension is broken due to the limited thickness of few-layer Bi_2Te_3 NPs, which results in the appearance of A_{1u} mode.

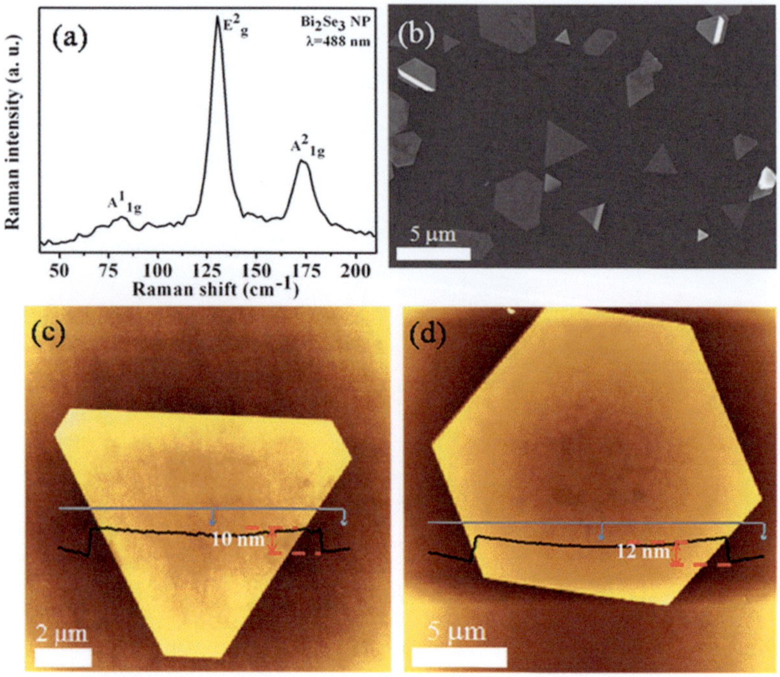

Fig. 12.8 (**a**) Raman spectrum of synthesized Bi_2Se_3 NPs in the 40–210 cm^{-1} region [35]. Copyright belongs to AIP. (**b**) An SEM image of Bi_2Se_3 NPs. (**c**, **d**) AFM topography image and height profile of one typical Bi_2Se_3 NP

12.4.2 Morphology and Microstructure Characterization of Bi_2Se_3 NPs

The morphologies and microstructures of as-grown samples using Bi_2Se_3 powders as source materials were also characterized. The XRD pattern of as-prepared samples is shown in Fig. 12.5b. The main peaks are all (0001) direction-oriented, which can be matched well with rhombohedral Bi_2Se_3 (JCPDS file 33-0214, $a_H = 0.414$ nm, $c_H = 2.8636$ nm). In our experimental results, the TEM image of Bi_2Se_3 NP, the HRTEM image, as well as corresponding FFT pattern prove that Bi_2Se_3 NP is also single crystalline similar to nanostructures of Bi_2Te_3. Corresponding EDS analysis confirms the correct stoichiometry [35]. Meanwhile, the three characteristic A^1_{1g}, E^2_g, and A^2_{1g} vibration modes of Bi_2Se_3 NP at the peak positions of 81.9, 130.5, and 172.4 cm^{-1} are characterized as expected (Fig. 12.8a). For Bi_2Se_3 NPs, the Raman peaks appear in the higher frequency range compared to Bi_2Te_3 Raman peaks, which is due to the stronger bonding forces than that of Bi_2Te_3. In addition, the Se atoms vibrate stronger producing Raman bands at a higher frequency. Raman measurements as well as other extensive characterizations indicate that the synthesized Bi_2Se_3 NPs have good crystallization nature. The prepared products grown on SiO_2/

Si substrates were characterized by SEM image (Fig. 12.8b). These Bi_2Se_3 NPs exhibit triangular or hexagonal morphologies with lateral dimensions of several micrometers. Typical AFM topography images of Bi_2Se_3 NPs with few-layer thickness are shown in Fig. 12.8c, d. Most of the NPs have atomically flat surface with uniform thickness across the lateral dimensions.

12.5 Growth Mechanism of Bi_2Te_3 and Bi_2Se_3 Nanostructures

The growth process of Bi_2Te_3 NPs has been investigated. The growth time, growth pressure, as well as substrate placement have been systematically conducted. Our experiments indicate subsamples which were obtained under the growth time of 2 min. EDS analysis confirms that Bi_2Te_3 nucleation is formed at first. With growth time of 5 min, large number of variable shape Bi_2Te_3 NPs occur, such as triangular, truncated triangular, and hexagonal of edge-length size ranging from 0.4 to 1.8 μm. When the growth time is prolonged to 30 min, the number of NPs is increased and most of the NPs grow together forming a film-like structure. However, when the growth pressure is increased to 200 Torr or much higher with other fixed conditions, there exist no Bi_2Te_3 NPs on the substrate demonstrating the effect of growth pressure on the formation of Bi_2Te_3 NPs. Interestingly, it is found that vertical placement of substrate can produce larger lateral dimension of Bi_2Te_3 NPs. When the substrate is put vertically on the quartz plate at 5 min. Triangular, truncated triangular, and hexagonal NPs are found to be the typical shape of NPs. The NPs with various morphologies, which grow on the substrate, have flat surfaces with thickness below 20 QLs. One can see that the lateral dimension of Bi_2Te_3 NPs grown on vertical substrate ranging from 4 to 16 μm with average size ~7 μm is larger than that grown on parallel substrate.

On the basis of the above experimental results, we propose that the growth mechanism for the formation of Bi_2Te_3 nanostructures follows two processes, as demonstrated in Fig. 12.9a: (1) the formation of Bi_2Te_3 nuclei; (2) the subsequent growth of these nuclei. At the initial stage, the dominant process is the formation of Bi_2Te_3 nuclei on the substrate. Due to its anisotropic bonding microstructure, the NPs subsequently grow along the [11–20] direction with (0001) facets as top and bottom surfaces and (01–10) facets as side surfaces. If the growth environment is uniform, the hexagonal plates would form as schematically illustrated in Fig. 12.9b. Nevertheless, the crystal growth would inevitably deviate from uniform growth along the directions (or to some extent) in the experiment system, leading to the poor symmetry of as-prepared NPs with triangular and truncated triangular Bi_2Te_3 plate formation. Figure 12.9c–e exhibits typical three-dimensional AFM images of corresponding Bi_2Te_3 nanostructures obtained in our experiments. Similar to the Bi_2Te_3, Bi_2Se_3 also belongs to the space group D_{3d}^5 (R-3m) with a stacking sequence of Se (1)-Bi-Se (2)-Bi-Se (1), forming a quintuple layer. Bi_2Se_3 NPs are synthesized due to anisotropic nature. Therefore, it is proposed that the growth mechanism for Bi_2Te_3 NPs is also applied to the Bi_2Te_3 NPs.

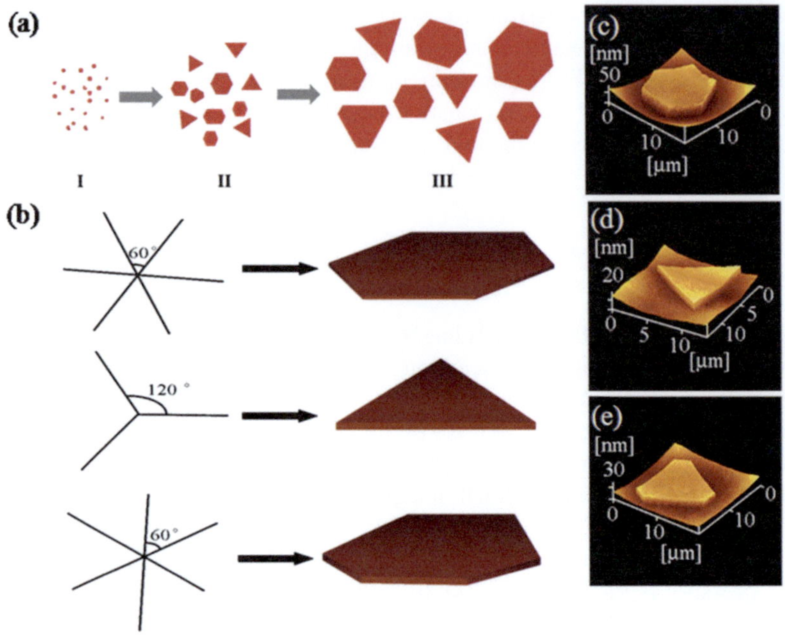

Fig. 12.9 (**a**) Growth processes for the formation of Bi_2Te_3 NPs. (**b**) Schematic illustrations of typical morphologies for Bi_2Te_3 NPs grown along different [11–20] directions. (**c–e**) Corresponding three-dimensional AFM images

12.6 Electrostatic Properties of Bi_2Te_3 and Bi_2Se_3 NPs

12.6.1 Electrostatic Properties of Bi_2Te_3 NPs

Electrostatic properties of Bi_2Te_3 NPs were performed by using KPFM (SEIKO, SPI3800N + 300 HV) under ambient environment with 40 % RH. A gold-coated Si probe (SI-DF3-A) was employed with force constant of 1.1 N/m and a typical tip radius about 15 nm. KPFM experiments were performed by a dual-pass mode in tapping AFM with tip lifted ~50 nm above the sample surface. The resonance frequency and the *Q*-factor of the cantilever are approximately 20 and 121 kHz, respectively. The cantilever is vibrated at a frequency slightly ~2 kHz lower than its resonance frequency. Ten volts AC at a frequency of 18 kHz is applied between the probe and sample. KPFM images of the sample surfaces are acquired at a probe scan rate of 0.25 Hz.

Figure 12.10a, b shows topography and surface potential of a truncated triangular Bi_2Te_3 NP that is typical of the majority of the sample structures grown on the SiO_2/Si substrate. The mapping of surface potential is acquired in the so-called "lift mode." In this case, the morphology of substrate does not influence the KPFM image. The surface potential mapping shows that potential of the Bi_2Te_3 NP is

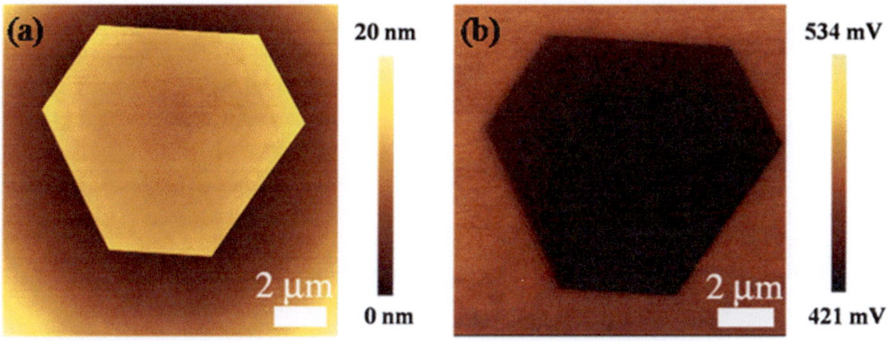

Fig. 12.10 (**a**) AFM image of Bi_2Te_3 NP. (**b**) Corresponding surface potential image

relatively homogeneous, indicating a uniform surface potential distribution. As shown in Fig. 12.10b, the surface potential of Bi_2Te_3 NP shows a dark contrast exhibiting 100 mV lower with respect to substrate, which is of paramount importance for understanding performance of Bi_2Te_3-based device performance.

We have performed surface potential measurements for Bi_2Te_3 NPs with various structures. These results demonstrate that the surface potential is uniform on the Bi_2Te_3 NP. Statistical measurement of the surface potential of amounts of Bi_2Te_3 NPs is conducted. The surface potential can be well fitted with Gaussian distribution and the mean surface potential is about 482 mV on SiO_2/Si substrates [34].

Although various nanostructures of Bi_2Te_3 NPs may have slight effect on the surface potential, it can be neglected compared to substrate modification. We propose that the charge exchange at interface between Bi_2Te_3 NPs and SiO_2 substrate plays an important role in the surface potential distribution of the Bi_2Te_3 NPs. Multiple-reflection attenuated-total-reflection FTIR spectrometry (ATR-FTIR) have found evidence for electrochemical trapping of electrons by SiOH silanol groups on the SiO_2 substrate [45, 46]. SiOH concentration on the SiO_2 surface has been determined to be ~$(3–7) \times 10^{13}$ cm^{-2} in a number of studies [47]. This concentration is nevertheless electronically significant, as it exceeds or is comparative to the typical carrier concentration of 10^{13} cm^{-2} found recently in single Bi_2Te_3 NP devices with a top gate and Bi_2Te_3 nanoribbon-based field-effect transistor (FET) [20, 22]. Such substrate surface condition leads to n-doping to the Bi_2Te_3 NPs with relatively large surface potential and broad distribution, which presents a practical challenge for probing TI surface states by transport measurements. Currently, worldwide efforts have been devoted to improve the quality of TIs with various approaches and tunable chemical potential in transport measurements via gating technique [11, 22]. However, many of the proposed applications of TI require a substrate and it is ongoing debate if and to what extent topological insulator's electronic properties are modified by the substrate.

In order to explore the influence of substrates on the surface potential distribution in the Bi_2Te_3 NP, the surface potential measurements of Bi_2Te_3 NPs grown on n-doped and p-doped Si (111) substrates have also been characterized [34, 37].

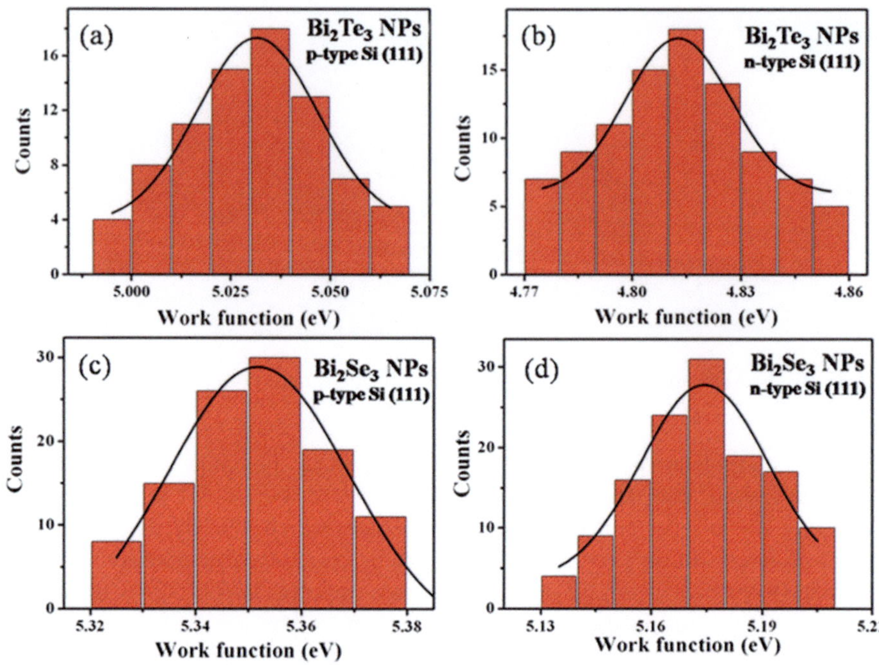

Fig. 12.11 (**a**, **b**) Statistics of work functions of a large number of Bi_2Te_3 NPs grown on p-type and n-type Si (111) substrates, respectively [52]. (**c**, **d**) Statistics of work functions of Bi_2Se_3 NPs grown on p-type and n-type Si (111) substrates, respectively [52]. Copyright belongs to Elsevier Ltd.

Our experiments also indicate that the topography of Bi_2Te_3 NPs grown on p-type silicon (111) substrate exhibits atomically flat surface with thickness of ~10 nm. In Fig. 12.11a, b, statistical measurements of the surface potentials of a large number of Bi_2Te_3 NPs on the two substrates have been conducted. The surface potentials were determined to be about 88 and −112 mV for Bi_2Te_3 NPs grown on n-doped and p-doped Si (111) substrates, respectively. The small surface potential distribution of ~0.1 V on the two substrates compared to SiO_2/Si substrate further confirms the reliability of KPFM technique. After calibration by highly ordered pyrolytic graphite (HOPG), the average work functions of a large number of Bi_2Te_3 NPs measurements were determined to be 5.03 and 4.80 eV on p-type and n-type Si (111) substrates, respectively.

For Bi_2Te_3 NPs grown on n-doped and p-doped substrates, we propose that the charge exchange exists at the interface due to different CPDs resulting in the Fermi level shift of Bi_2Te_3 NPs. KPFM measurement information of Bi_2Te_3 NP shows that the Fermi level of Bi_2Te_3 NPs grown on n-doped Si (111) substrate is ~0.2 meV lower compared to these Bi_2Te_3 NPs grown on p-doped Si (111) substrate. Our results demonstrate that the surface potential of Bi_2Te_3 NPs can be modified by choosing different type of substrates, which may be useful for tuning transport properties of TI-based devices on different substrates.

12.6.2 *Electrostatic Properties of Bi_2Se_3 NPs*

Among other such materials, Bi_2Se_3, a traditional thermoelectric material, is of particular importance due to its relatively large bulk band gap (~0.3 eV). More importantly, Bi_2Se_3 can be tuned into a superconductor by copper intercalation [48, 49], which may be employed to generate exotic Majorana fermions for promising applications in topological quantum computing. Recently, Bi_2Se_3 has been fabricated in the form of bulk crystals [38, 50, 51], thin films [7, 29], and NPs [22, 35]. The topological surface states of Bi_2Se_3 have been extensively investigated. However, little attention is paid on its electrostatic properties, such as surface potential and work functions of Bi_2Se_3. These electrostatic properties are important for us to understand the topological nature. Therefore, the electrostatic properties of Bi_2Se_3 NPs were also characterized by KPFM [52]. The Bi_2Se_3 NPs are prepared on n-type and p-type Si (111) substrates in order to investigate the substrate effects on the electronic properties of Bi_2Se_3 NPs. In order to reduce the error of measurements, a large number of Bi_2Se_3 NPs were performed. The work function of Bi_2Se_3 NPs on the substrate was determined to be ~5.35 eV from the statistics (Fig. 12.11c). To examine the doping effect from substrate, the work functions of Bi_2Se_3 NPs resting on the n-type Si (111) substrate have been simultaneously obtained corresponding to ~5.17 eV (Fig. 12.11d).

Our results demonstrate that the surface potential of Bi_2Te_3 and Bi_2Se_3 NPs can be modified by choosing different type of substrates, which may be useful for tuning transport properties of TI-based devices on different substrates. These results indicate the substrate doping effect on the TI NPs, which may provide a new route to design TI-based heterostructure devices in conventional semiconductor industry.

12.7 Ambipolar Charge Injection and Transport of Bi_2Te_3 and Bi_2Se_3 TI NPs

12.7.1 *Surface Potential and Charge Injection Investigations of Bi_2Te_3 NPs*

Charge injection combined with scanning probe microscopy technique provides a useful route to detect the electrostatic properties of materials in the nanoscale [25, 53, 54]. It allows us to measure the electrostatic forces in the vicinity of charged samples to gain insight into the charge and surface potential distributions within the objects. Surface potential images were characterized by KPFM using an AC voltage of 10 V at 19 kHz, which show uniform surface potential distribution on the Bi_2Te_3 NP surface [34]. This indicates all intrinsic charges uniformly distributed in the Bi_2Te_3 NP. Before any charge injection, the Bi_2Te_3 NP appears slightly dark in the KPFM image. This signal accounts for the effect of tip-substrate capacitance when the KPFM tip is moved over Bi_2Te_3 NP.

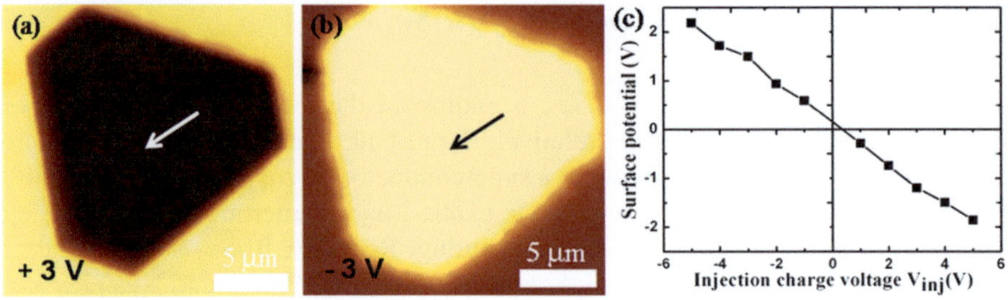

Fig. 12.12 (**a**) Surface potential image taken under the same conditions, after charge injection performed at the point indicated by an *arrow*, with $V_{inj}=+3$ V for 2 min. (**b**) Surface potential image corresponding to $V_{inj}=-3$ V for 2 min. (**c**) Surface potential after charge injection vs. voltage applied to the tip during charging process measured from KPFM taken after charge injection [35]. Copyright belongs to AIP

In our experiments, charge injection is achieved by pressing the conductive tip biased at V_{inj}, with respect to the substrate, with a typical 2 nN contact force on a given NP [35]. The charges are injected in the center of Bi_2Te_3 NP. Figure 12.12a, b exhibits the 2D distributions of charges injected in the Bi_2Te_3 NP characterized by KPFM after charge injection under +3 and −3 V. The KPFM images exhibit a dark (Fig. 12.12a) and bright (Fig. 12.12b) contrast corresponding to negative and positive charge injection, respectively.

It is shown that holes and electrons can be injected in the Bi_2Te_3 NP and delocalize over the whole NP upon injection. The delocalization of excess charges over microns is attributed to the charging of nanoplate–substrate capacitor. We determine the surface potential dependence on charge injection voltage and the surface potential measurements are shown as a function of V_{inj} after charge injection as shown in Fig. 12.12c. Therefore, to control quantitatively the charging process, the V_{inj} dependence of injection charges should be taken into account. The surface potential as a function of applied biases reveals an asymmetric bias dependence. It is noted that the surface potential values after negative biases are higher than those of corresponding biases, which indicates that the positive charge injection is easier than negative charge injection. We propose that this may be due to the n-type nature of Bi_2Te_3 NPs being ascribed to Te vacancies or antisite defects formed during the growth process. Previous studies had reported the ambipolar transport in the topological insulator-based FETs through various strategies [22, 54]. Our results provide the possibility to detect the charge and surface potential distributions directly in TI-based devices under different carrier type.

12.7.2 *Surface Potential and Charge Injection Investigations of Bi_2Se_3 NPs*

For the Bi_2Se_3 NP on SiO_2/Si substrate, electrons and holes can also be injected, and both carriers are delocalized over the whole NP as shown in Fig. 12.13a ($V_{inj}=+3$ V) and Fig. 12.13b ($V_{inj}=-3$ V). We find that the surface potentials measured as a

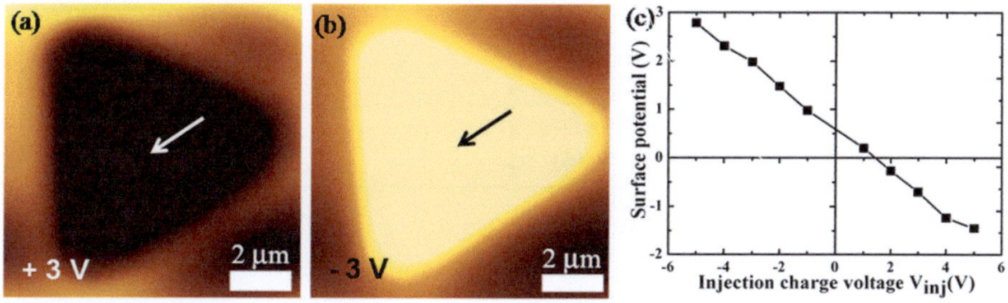

Fig. 12.13 (**a**, **b**) Surface potential images of Bi_2Se_3 NP corresponding to $V_{inj}=+3$ V and $V_{inj}=-3$ V, respectively. (**c**) The surface potential of Bi_2Se_3 NP as a function of V_{inj} after charge injection [35]. Copyright belongs to AIP

function of V_{inj} also exhibit a linear behavior. In the Bi_2Se_3 NP, the surface potential drops 963 mV by +3 V which is much lower than that bumps 1,329 mV by −3 V. Negative biasing produces a higher surface potential than positive biasing for the same absolute value of applied voltage as displayed in Fig. 12.13c. Since the potential changes are induced by the injected charges, bigger changes correspond to more charge injection. This indicates that it is also much easier to inject holes than electrons into the Bi_2Se_3 NP related to the n-type nature of Bi_2Se_3 NPs ascribed to the Se vacancies and antisite defects formed during the growth process [35].

By using scanning probe microscopy, we are able to inject electrons and holes in Bi_2Te_3 NPs and Bi_2Se_3 NPs. Surface potentials between the NP and the probe tip, characterized by KPFM, show an asymmetric feature with positive and negative tip biases. The n-type nature of Bi_2Te_3 and Bi_2Se_3 NPs can be illuminated from these results. These results indicate that KPFM measurements can be employed simultaneously to detect the modification of Fermi level using a gate voltage, doping, and other methods of TI nanostructures.

12.8 Fermi Level Tuning of $Bi_2(Se_xTe_{1-x})_3$ TI NPs

12.8.1 Fabrication and Characterization of $Bi_2(Se_xTe_{1-x})_3$ NPs

Bi_2Se_3 and Bi_2Te_3 are usually n-type in nature, which hinders the development of TI. In order to overcome this problem, various studies have been reported through extrinsic elemental doping to tune the Fermi level in Bi_2Te_3 and Bi_2Se_3 TIs, while extrinsic doping may introduce more disorders due to lattice mismatch, which will not benefit the novel surface transport properties. In previous studies, the surface state carrier mobility of ternary Bi_2Se_2Te is increased significantly compared to binary TIs [24]. Weak antilocalization was clearly observed in $Bi_2(Se_xTe_{1-x})_3$ nanoribbons [36]. Therefore, ternary TIs provide a new route for us to investigate the topological nature.

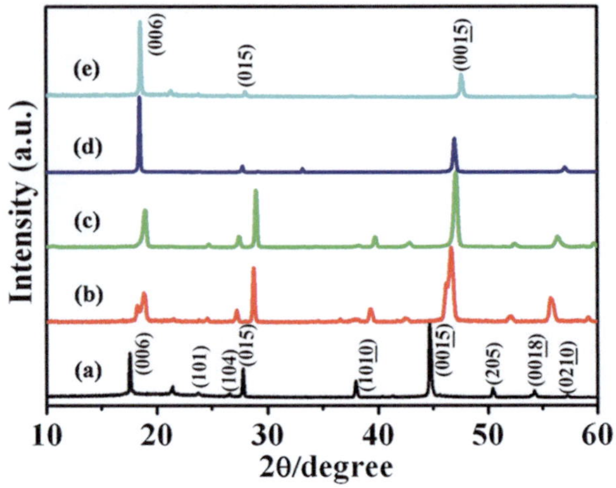

Fig. 12.14 XRD patterns of $Bi_2(Se_xTe_{1-x})_3$ NPs for (**a**) Bi_2Te_3, (**b**) $Bi_2(Se_{0.33}Te_{0.67})_3$, (**c**) $Bi_2(Se_{0.5}Te_{0.5})_3$, (**d**) $Bi_2(Se_{0.67}Te_{0.33})_3$, and (**e**) Bi_2Se_3. © AIP

By using vapor phase deposition method, we have synthesized ultrathin $Bi_2(Se_xTe_{1-x})_3$ NPs with various morphologies. A uniform mixture of Bi_2Te_3 (99.999 %) and Bi_2Se_3 (99.999 %) powders with different molecular ratio was employed as the precursor for evaporation. The source center was set to 480 °C for the growth of $Bi_2(Se_xTe_{1-x})_3$ NPs under low pressure about 30 Torr and Ar gas 50 sccm. Substrates were placed on the downstream for the growth of $Bi_2(Se_xTe_{1-x})_3$ NPs, which is 10–12 cm away from the hot center. The XRD patterns of the as-prepared are shown in Fig. 12.14. Compared with JCPDS data card no. 15-0863 for Bi_2Te_3 and 33-0214 for Bi_2Se_3, these ternary NPs can be indexed to a pure rhombohedral crystal geometry (space group: R-3m (166)) with no detectable impurities of other phases. The high quality of the NPs was confirmed by TEM, HRTEM, and EDS measurements.

12.8.2 Fermi Level Tuning of $Bi_2(Se_xTe_{1-x})_3$ NPs

The electrostatic properties and Fermi level tuning of different composition in ternary $Bi_2(Se_xTe_{1-x})_3$ NPs were quantitatively investigated employing KPFM. It is worthy to note that there are remarkable differences of work functions for $Bi_2(Se_xTe_{1-x})_3$ NPs with different ratios (x=0, 0.33, 0.5, 0.67, 1) grown on 300 nm SiO_2/Si substrate as shown in Fig. 12.15 ranging from 4.29 to 4.52 eV. The variation of work function is ascribed to the Fermi level shift of $Bi_2(Se_xTe_{1-x})_3$ NPs. Hence, the variations of x ranging from 0 to 1 lead to a significant shift of the Fermi level up to 0.23 eV (Fig. 12.15) with respect to reference value (Fermi level of Bi_2Se_3) measured by KPFM. For TIs with potential applications, the tuning of Fermi level is essential for the performance of surface states of $Bi_2(Se_xTe_{1-x})_3$-based devices. Our results indicate that the Fermi level can be effectively tuned by varying the composition of $Bi_2(Se_xTe_{1-x})_3$ NPs and provide a new route to study TIs.

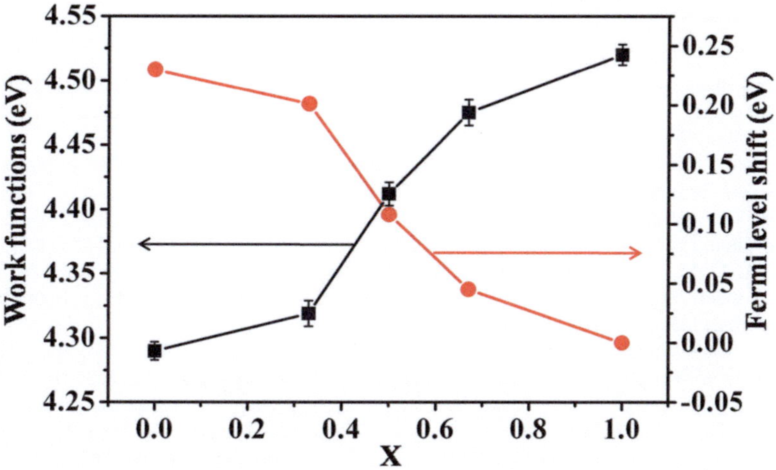

Fig. 12.15 The work functions of Bi_2Te_3, $Bi_2(Se_xTe_{1-x})_3$ (x=0, 0.33, 0.5, 0.67, 1), and Bi_2Se_3 NPs and Fermi level shifts as a function of x (selenium/tellurium composition ratio). © AIP

12.9 Conclusions

Single-crystalline Bi_2Te_3, Bi_2Se_3, and $Bi_2(Se_xTe_{1-x})_3$ NPs with various nanostructures have been synthesized on SiO_2/Si (300 nm) and n-doped and p-doped Si (111) substrates via vapor phase deposition method. The growth mechanism has been proposed to explain the formation of TI Bi_2Te_3 NP qualitatively. The surface potential of these NPs exhibits uniform surface potential and charge distributions. Simultaneously, the charge injection experiments indicate that both electrons and holes can be injected into the Bi_2Te_3 and Bi_2Se_3 NPs, which exhibits ambipolar charge injection phenomena. The surface potential as a function of applied biased voltages shows an asymmetric bias dependence resulting from n-type nature of Bi_2Te_3 and Bi_2Se_3 NPs. The work functions of TI NP–substrate systems were determined by KPFM technique to confirm the doping effect from the substrate, which may provide new opportunity to design TI-based heterostructure devices in conventional semiconductor industry. The electrostatic properties obtained by KPFM provide insight into understanding electronic properties of Bi_2Te_3 and Bi_2Se_3 as well as $Bi_2(Se_xTe_{1-x})_3$ bismuth-based compounds.

Acknowledgements This work was supported by the Grants from National Natural Science Foundation of China (Nos. 11074211, 51172191, and 51002129), National Basic Research Program of China (2012CB921303), and the Hunan Provincial Innovation Foundation for Graduate (No. CX2011B254).

References

1. Dresselhaus, M.S., Chen, G., Tang, M.Y., Yang, R.G., Lee, H., Wang, D.Z., Ren, Z.F., Fleurial, J.P., Gogna, P.: New Directions for Low-Dimensional Thermoelectric Materials. Adv. Mater. **19**, 1043 (2007)
2. Bejenari, I., Kantser, V., Balandin, A.A.: Thermoelectric Properties of Electrically Gated Bismuth Telluride Nanowires. Phys. Rev. B **81**, 075316, 14 (2010)
3. Qiu, X., Austin, L.N., Muscarella, P.A., Dyck, J.S., Burda, C.: Nanostructured Bi_2Se_3 Films and Their Thermoelectric Transport Properties. Angew. Chem. Int. Ed. **45**, 5656 (2006)
4. Dong, G.-H., Zhu, Y.-J., Chen, L.-D.: Microwave-Assisted Rapid Synthesis of Sb2Te3 Nanosheets and Thermoelectric Properties of Bulk Samples Prepared by Spark Plasma Sintering. J. Mater. Chem. **20**, 1976 (2010)
5. Venkatasubramanian, R., Siivola, E., Colpitts, T., O'Quinn, B.: Thin-Film Thermoelectric Devices with High Room-Temperature Figures of Merit. Nature **413**, 597 (2001)
6. Zhang, H.J., Liu, C.-X., Qi, X.-L., Dai, X., Fang, Z., Zhang, S.-C.: Topological Insulators in Bi_2Se_3, Bi_2Te_3 and Sb_2Te_3 with a Single Dirac Cone on the Surface. Nat. Phys. **5**, 438 (2009)
7. Zhang, Y., et al.: Crossover of the Three-Dimensional Topological Insulator Bi_2Se_3 to the Two-Dimensional Limit. Nat. Phys. **6**, 584 (2010)
8. Chen, Y.L., et al.: Experimental Realization of a Three-Dimensional Topological Insulator, Bi_2Te_3. Science **325**, 178 (2009)
9. Hsieh, D., et al.: Observation of Time-Reversal-Protected Single-Dirac-Cone Topological-Insulator States in Bi_2Te_3 and Sb_2Te_3. Phys. Rev. Lett. **103**, 146401 (2009)
10. Cheng, P., et al.: Landau Quantization of Topological Surface States in Bi_2Se_3 Phys. Rev. Lett. **105**, 076801 (2010)
11. Chen, J., et al.: Gate-Voltage Control of Chemical Potential and Weak Antilocalization in Bi_2Se_3. Phys. Rev. Lett. **105**, 176602 (2010)
12. Wang, Y., et al.: Gate-Controlled Surface Conduction in Na-Doped Bi_2Te_3 Topological Insulator Nanoplates. Nano Lett. **12**, 1170 (2012)
13. Cha, J.J., Williams, J.R., Kong, D., Meister, S., Peng, H., Bestwick, A.J., Gallagher, P., Goldhaber-Gordon, D., Cui, Y.: Magnetic Doping and Kondo Effect in Bi_2Se_3 Nanoribbons. Nano Lett. **10**, 1076 (2010)
14. Peng, H.L., Dang, W., Cao, J., Chen, Y., Di, W., Zheng, W., Li, H., Shen, Z., Liu, Z.: Topological Insulator Nanostructures for Nar-Infrared Transparent Flexible Electrodes. Nat. Chem. **4**, 281 (2012)
15. Hasan, M.Z., Kane, C.L.: *Colloquium*: Topological Insulators. Rev. Mod. Phys. **82**, 3045 (2010)
16. Qi, X.-L., Zhang, S.-C.: Topological Insulators and Superconductors. Rev. Mod. Phys. **83**, 1057 (2011)
17. Qi, X.-L., Li, R., Zang, J., Zhang, S.-C.: Inducing a Magnetic Monopole with Topological Surface States. Science **323**, 1184 (2009)
18. Kong, D.S., Cui, Y.: Opportunities in Chemistry and Materials Science for Topological Insulators and Their Nanostructures. Nat. Chem. **3**, 845 (2011)
19. Teweldebrhan, D., Goyal, V., Balandin, A.A.: Exfoliation and Characterization of Bismuth Telluride Atomic Quintuples and Quasi-Two-Dimensional Crystals. Nano Lett. **10**, 1209 (2010)
20. Xiu, F.X., et al.: Manipulating Surface States in Topological Insulator Nanoribbons. Nat. Nanotechnol. **6**, 216 (2011)
21. Hossain, M.Z., Rumyantsev, S.L., Shahil, K.M.F., Teweldebrhan, D., Shur, M., Balandin, A.A.: Low-Frequency Current Fluctuations in "Graphene-like" Exfoliated Thin-Films of Bismuth Selenide Topological Insulators. ACS Nano **5**, 2657 (2011)
22. Kong, D.S., Dang, W.H., Cha, J.J., Li, H., Meister, S., Peng, H.L., Liu, Z.F., Cui, Y.: Few-Layer Nanoplates of Bi_2Se_3 and Bi_2Te_3 with Highly Tunable Chemical Potential. Nano Lett. **10**, 2245 (2010)
23. Zhang, T., et al.: Experimental Demonstration of Topological Surface States Protected by Time-Reversal Symmetry. Phys. Rev. Lett. **103**, 266803 (2009)

24. Ren, Z., Taskin, A.A., Sasaki, S., Segawa, K., Ando, Y.: Large bulk Resistivity and Surface Quantum Oscillations in the Topological Insulator Bi_2Te_2Se. Phys. Rev. B **82**, 241306 (2010)
25. Hao, G.L., Qi, X., Wang, G., Peng, X.Y., Chang, S.L., Wei, X.L., Zhong, J.X.: Synthesis and Characterization of Few-Layer Sb_2Te_3 Nanoplates with Electrostatic Properties. RSC Adv. **2**, 10694 (2012)
26. Kim, D., Cho, S., Butch, N.P., Syers, P., Kirshenbaum, K., Adam, S., Paglione, J., Fuhrer, M.S.: Surface Conduction of Topological Dirac Electrons in Bulk Insulating Bi_2Se_3. Nat. Phys. **8**, 460 (2012)
27. Chen, J., He, X.Y., Wu, K.H., Ji, Z.Q., Lu, L., Shi, J.R., Smet, J.H., Li, Y.Q.: Tunable Surface Conductivity in Bi_2Se_3 Revealed in Diffusive Electron Transport. Phys. Rev. B **83**, 241304 (2011)
28. Wang, G., et al.: Atomically Smooth Ultrathin Films of Topological Insulator Sb_2Te_3. Nano Res. **3**, 874 (2010)
29. Li, H.D., et al.: The Van der Waals epitaxy of Bi_2Se_3 on the Vicinal Si(111) Surface: an Approach for Preparing High-Quality Thin Films of a Topological Insulator. New J. Phys. **12**, 103038 (2010)
30. Zhang, H.B., Yu, H.L., Bao, D.H., Li, S.W., Wang, C.X., Yang, G.W.: Magnetoresistance Switch Effect of a Sn-Doped Bi_2Te_3 Topological Insulator. Adv. Mater. **24**, 132 (2012)
31. Ren, L., Qi, X., Liu, Y.D., Hao, G.L., Huang, Z.Y., Zou, X.H., Yang, L.W., Li, J., Zhong, J.X.: Large-Scale Production of Ultrathin Topological Insulator Bismuth Telluride Nanosheets by a Hydrothermal Intercalation and Exfoliation Route. J. Mater. Chem. **22**, 4921 (2012)
32. Teweldebrhan, D., Goyal, V., Rahman, M., Balandin, A.A.: Atomically-Thin Crystalline Films and Ribbons of Bismuth Telluride. Appl. Phys. Lett. **96**, 053107, 3 (2010)
33. Kong, D.S., et al.: Topological Insulator Nanowires and Nanoribbons. Nano Lett. **10**, 329 (2009)
34. Hao, G.L., Qi, X., Yang, L.W., Liu, Y.D., Li, J., Ren, L., Sun, F., Zhong, J.X.: Growth and Surface Potential Characterization of Bi_2Te_3 Nanoplates. AIP Adv. **2**, 012114 (2012)
35. Hao, G.L., Qi, X., Liu, Y.D., Huang, Z.Y., Li, H.X., Huang, K., Li, J., Yang, L.W., Zhong, J.X.: Ambipolar Charge Injection and Transport of Few-Layer Topological Insulator Bi_2Te_3 and Bi_2Se_3 Nanoplates. J. Appl. Phys. **111**, 114312 (2012)
36. Cha, J.J., Kong, D., Hong, S.-S., Analytis, J.G., Lai, K., Cui, Y.: Weak antilocalization in $Bi_2Se_xTe_{1-x3}$ nanoribbons and nanoplates. Nano Lett. **12**, 1107 (2012)
37. Nonnenmacher, M., O'Boyle, M.P., Wickramasinghe, H.K.: Kelvin Probe Force Microscopy. Appl. Phys. Lett. **58**, 2921 (1991)
38. Palermo, V., Palma, M., Samorì, P.: Electronic Characterization of Organic Thin Films by Kelvin Probe Force Microscopy. Adv. Mater. **18**, 145 (2006)
39. Hao, G.L., Qi, X., Li, J., Yang, L.W., Yin, J.J., Lu, F., Zhong, J.X.: Surface Potentials of Few-Layer Graphene Films in High Vacuum and Ambient Conditions. Solid State Commun. **151**, 818 (2011)
40. Wang, R., Wang, S.N., Zhang, D.D., Li, Z.J., Fang, Y., Qiu, X.H.: Control of Carrier Type and Density in Exfoliated Graphene by Interface Engineering. ACS Nano **5**, 408 (2010)
41. Yu, Y.-J., Zhao, Y., Ryu, S., Brus, L.E., Kim, K.S., Kim, P.: Tuning the Graphene Work Function by Electric Field Effect. Nano Lett. **9**, 3430 (2009)
42. Shi, Y.M., Kim, K.K., Reina, A., Hofmann, M., Li, L.-J., Kong, J.: Work Function Engineering of Graphene Electrode via Chemical Doping. ACS Nano **4**, 2689 (2010)
43. Saraf, S., Rosenwaks, Y.: Local Measurement of Semiconductor Band Bending and Surface Charge Using Kelvin Probe Force Microscopy. Surf. Sci. **574**, L35 (2005)
44. Shahil, K.M.F., Hossain, M.Z., Goyal, V., Balandin, A.A.: Micro-Raman Spectroscopy of Mechanically Exfoliated Few-Quintuple Layers of Bi_2Te_3, Bi_2Se_3, and Sb_2Te_3 Materials. J. Appl. Phys. **111**, 054305 (2012)
45. Chua, L.-L., Zaumseil, J., Chang, J.-F., Ou, E.C.W., Ho, P.K.H., Sirringhaus, H., Friend, R.H.: General Observation of n-type Field-Effect Behaviour in Organic Semiconductors. Nature **434**, 194 (2005)

46. Nicollian, E.H., Berglund, C.N., Schmidt, P.F., Andrews, J.M.: Electrochemical Charging of Thermal SiO_2 Films by Injected Electron Currents. J. Appl. Phys. **42**, 5654 (1971)
47. Nagasawa, Y., Yoshii, I., Naruke, K., Yamamoto, K., Ishida, H., Ishitani, A.: The Study of the Thermal Oxide Films on Silicon Wafers by Fourier Transform Infrared Attenuated Total Reflection Spectroscopy. J. Appl. Phys. **68**, 1429 (1990)
48. Kriener, M., Segawa, K., Ren, Z., Sasaki, S., Ando, Y.: Bulk Superconducting Phase with a Full Energy Gap in the Doped Topological Insulator $Cu_xBi_2Se_3$. Phys. Rev. Lett. **106**, 127004 (2011)
49. Sasaki, S., Kriener, M., Segawa, K., Yada, K., Tanaka, Y., Sato, M., Ando, Y.: Topological Superconductivity in $Cu_xBi_2Se_3$. Phys. Rev. Lett. **107**, 217001 (2011)
50. Ren, Z., Taskin, A.A., Sasaki, S., Segawa, K., Ando, Y.: Observations of Two-Dimensional Quantum Oscillations and Ambipolar Transport in the Topological Insulator Bi_2Se_3 Achieved by Cd Doping. Phys. Rev. B **84**, 075316 (2011)
51. Checkelsky, J.G., Hor, Y.S., Cava, R.J., Ong, N.P.: Bulk Band Gap and Surface State Conduction Observed in Voltage-Tuned Crystals of the Topological Insulator Bi_2Se_3. Phys. Rev. Lett. **106**, 196801 (2011)
52. Hao, G.L., Qi, X., Yang, L.W., Huang, Z.Y., Cai, C.Y., Xue, L., Yu, S.X., Zhong, J.X.: Fermi Level Shift in Topological Insulator–Silicon Heterostructures. Solid State Commun. **152**, 2027 (2012)
53. Verdaguer, A., Cardellach, M., Segura, J.J., Sacha, G.M., Moser, J., Zdrojek, M., Bachtold, A., Fraxedas, J.: Charging and discharging of graphene in ambient conditions studied with scanning probe microscopy. Appl. Phys. Lett. **94**, 233105 (2009)
54. Kong, D.S., et al.: Ambipolar Field Effect in the Ternary Topological Insulator $(Bi_xSb_{1-x})_2Te_3$ by Composition Tuning. Nat. Nanotechnol. **6**, 705 (2011)

Chapter 13
Electronic and Optical Properties of Domain Walls and Phase Boundaries in Bismuth Ferrite

Jan Seidel

Abstract Physical phenomena involving domain walls in $BiFeO_3$ as nanoscale functional elements have recently received considerable attention. Their nanoscopic size and flexible arrangement using thin film growth engineering solutions and applied external electric fields offer unique possibilities for novel concepts in complex oxide nanoelectronics.

Transition metal oxides exhibit a rich physics that can be utilized in a wide variety of properties. Their properties are related to a delicate balance between lattice, charge, spin, and orbital degrees of freedom [1]. A large amount of work has been done on these materials, especially perovskites, as they offer the advantage of very high structural quality growth. Modern thin film synthesis methods such as pulsed laser deposition (PLD), molecular beam epitaxy (MBE), and metal-organic chemical vapor deposition (MOCVD), among others, enable the engineering of interfaces between complex transition metal oxides with atomic-scale precision. Interfaces introduce a variety of local changes, e.g., they break the symmetry of the material, they induce stress, and they lead to variation in the bonding between ions. This in turn gives rise to changes in electronic structure such as bandwidth, orbital interactions, and level degeneracy. Charge transfer in these correlated systems can induce changes in carrier densities at interfaces that result in physical properties which may be completely different from those of the parent bulk phase [2–6].

Nanoscale features of interfaces between strongly correlated oxides can be investigated using various scanning probe microscopy (SPM) approaches. Tuning and manipulating the physical properties of such interfaces between different complex oxide materials provides a new playground for research. In addition, it offers a novel nanoelectronics characterization platform for future nanotechnology [7].

J. Seidel (✉)
School of Materials Science and Engineering, University of New South Wales,
Sydney, NSW, Australia
e-mail: jan.seidel@unsw.edu.au

H. Li and Z.M. Wang (eds.), *Bismuth-Containing Compounds*, Springer Series in Materials Science 186, DOI 10.1007/978-1-4614-8121-8_13,

Domain walls are naturally occurring interfaces and topological boundaries in complex oxides as opposed to artificially created heterointerfaces between different materials [8]. They have been investigated with respect to the mechanical and leakage current behavior in various studies involving ceramics, thin films, and single crystalline material. However, the fact that their electronic conductivity can actually be utilized for nanoscale functional elements has sparked renewed interest and has led to the finding of unique properties associated with such domain walls. First findings in $BiFeO_3$ (BFO) have led to the investigation of other ferroelectrics and multiferroics, which also exhibit electronic conduction at domain walls, albeit in different forms. Similar considerations regarding electronic conductivity apply to structural phase boundaries [9], which might, e.g., occur in materials exhibiting morphotropic phase boundaries, which include BFO [10].

13.1 Nanoscale Electronic Transport

SPM with its many variations (e.g., conductive atomic force microscopy—c-AFM, piezoresponse force microscopy—PFM) is very well suited for the direct characterization of prototype oxide structures on the nanoscale, including domain walls [11]. With c-AFM one can probe local conductivity at interfaces and topological defects. PFM has been further developed and is now offering dynamic characterization of the switching process utilizing developments such as stroboscopic PFM and PFM spectroscopy [12–15]. Ferroelectric thin films typically contain various structural defect types such as cationic and anionic point defects, dislocations, and grain boundaries. The electric and stress fields around such defects in a ferroelectric thin film are inhomogeneous. It can therefore be expected that the switching behavior near such a structural defect will be different from the one found in the normal state. The role of single ferroelastic twin boundaries has been studied, e.g., in tetragonal lead zirconium titanate ($PbZr_{0.2}Ti_{0.8}O_3$) ferroelectric thin films [16]. In this work the potential required to nucleate a 180° domain was decreased near ferroelastic twin walls. The recent spike of interest in local conductivity measurements at domain walls arises from perspectives involving nonvolatile memory applications and a considerable potential for electroresistive memory devices [17]. Such functionality is also interesting for the characterization of oxygen vacancy movement and ionic battery materials on the nanoscale [18–20]. In many cases, the presence of extended defects and oxygen vacancies makes the identification of polarization-mediated transport mechanisms difficult. The combination of local electromechanical and conductivity measurements has been proven as an interesting approach and has revealed a connection between local current and pinning at bicrystal grain boundaries in bismuth ferrite [21]. More generally, electroresistance in ferroelectric structures has recently been reviewed by Watanabe [22]. The presence of extended defects and oxygen vacancy accumulation has also been shown to influence transport and nucleation mechanisms at domain walls [23–25].

As another local characterization method, scanning tunneling microscopy (STM) can be used to probe directly the electronic structure nanometer length scales. STM and scanning tunneling spectroscopy (STS) have been used, e.g., to investigate the electronic structure of ferroelastic twin walls in $YBa_2Cu_3O_{7-\delta}$ [26]. Twin boundaries in this material play an important role in pinning the vortices in this type II superconductor and thereby enhance the currents that the material can support while remaining superconducting. In the case of insulators, STM/STS is by definition considerably more difficult to use, primarily because tunneling currents can usually not be used for reliable SPM feedback. The emergence of semiconducting ferroelectrics with smaller band gaps and the possibility of conduction at domain walls, as described below, have stimulated renewed interest in exploring STM as a probe of local electronic structure in complex oxides. The emergence of combined AFM/STM or SEM/STM systems is especially interesting in terms of exploring the electronic properties of domain walls in such insulating materials. Investigations using such combined tools have been demonstrated and include measurements on BFO domain walls [27–30].

13.2 Conductivity at Domain Walls

Ferroic phases can arise in two or more distinct orientations; therefore, they can form domains, separated by domain walls. Domains are a representation of long-range order with respect to at least one macroscopic tensor property of the material (order parameter). Their orientation states can be changed, leading to movement of the domain walls. Many so-called smart materials and structures have at least one form of ferroic order. They can be designed to change this order in a preconceived manner through the application of external fields. Thus, given their intrinsic structure, ferroic materials are ideally suited for functionality on the nanoscale.

The changes in structure (and as a consequence electronic structure) that occur at ferroelectric (multiferroic) domain walls [31] can lead to changes in transport behavior. Indeed, domain wall conductivity has been shown in different ferroic materials, although with different transport behavior: the domain walls of $BiFeO_3$ were found to be more conductive than the domains [23, 32, 33] (Fig. 13.1), while those of $YMnO_3$ were found to be more insulating or conductive depending on their orientation [34, 35]. In multiferroic $YMnO_3$, a so-called improper ferroelectric, in which ferroelectricity is not the primary order parameter but rather induced by structural trimerization coexisting with magnetism, domain walls are found to be charged and stable and their conductivity depends on the specific orientation of the domain wall with respect to the polarization vector. The observed conduction suppression at domain walls at high voltages (still much less than the electric coercivity) is in striking contrast with what has been reported for BFO.

A useful clue for interpreting these results is perhaps the analysis of the paraphase. Whereas the high temperature, high symmetry phase of BFO is more conducting than the ferroelectric phase [36], the converse is true for $YMnO_3$ [30]. This

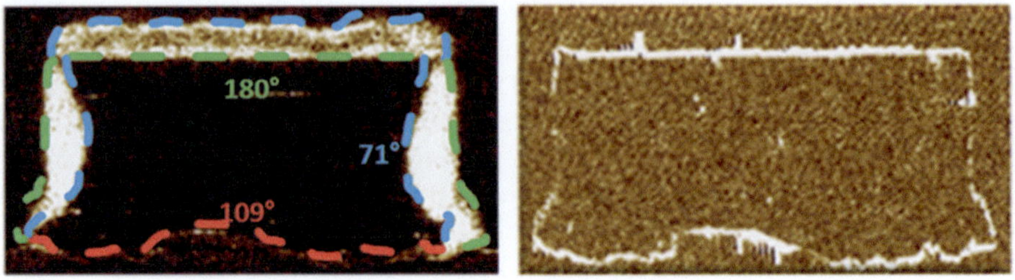

Fig. 13.1 The three different types of domain walls in rhombohedral bismuth ferrite as seen in an in-plane PFM image of a written domain pattern in a mono-domain BFO (110) thin film (*left*) and the corresponding c-AFM image showing conduction at both 109° and 180° domain walls (*right*). Adapted from Seidel et al. [23]

illustrates an important point: in some respects the internal structure of the walls can be considered to be in the paraelectric state; by way of trivial example, the 180° domain walls of a ferroic are nonpolar, just like its paraphase. The examples of BFO and $YMnO_3$ suggest that the paraphase-like behavior can be applied to domain wall properties other than just the polarization: the insulating paraelectric state of $YMnO_3$ is consistent with the insulating nature of its domain walls, and conversely the conducting state of the paraphase of BFO is consistent with its domain wall conductivity. Nevertheless, in the conductivity of the BFO walls at least there are several other considerations: octahedral rotations, electrostatic steps arising from rigid rotation of the polar vector, and increased carrier density at the wall are all thought to play a role in the domain wall conductivity of BFO and potentially also of other perovskites.

The electronic structure of domain walls in bismuth ferrite has been investigated by Lubk et al. [37]. Using density functional theory (DFT) the layer-by-layer densities of states were calculated showing that the domain walls have a significantly reduced band gap compared to the $R3c$ bulk structure. Structural changes at the walls lead to arrangements that approach the ideal cubic structure, in which the 180° Fe–O–Fe bond angle maximizes the Fe $3d$–O $2p$ hybridization and hence the bandwidth of the material. Figure 13.2 shows the local band gap extracted from the layer-by-layer densities of states across all three wall types in BFO. In all cases a reduction in the band gap in the wall can be seen, with the 180° wall showing the largest effect. Note that in no case does the band gap approach zero in the wall region. The same first principles calculations supporting experimental work of Seidel et al. [23] also give insight into the changes in Fe–O–Fe bond angle in BFO. In addition, the rotations of the oxygen octahedra are found to not change their phase across the walls, i.e., this scenario is significantly more favorable than that with rotation discontinuities. The calculations thus show that antiphase boundaries associated with octahedral rotations are energetically very costly.

The same DFT calculations also revealed steps in the electrostatic potential for all domain wall types, which is an important point, as they also contribute to the electronic conductivity. Steps in the electrostatic potential at domain walls are

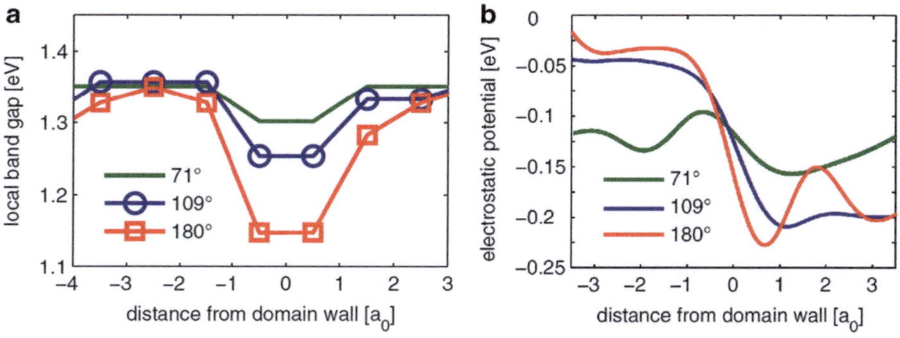

Fig. 13.2 (**a**) DFT calculations of local band gap at domain walls in BFO extracted from the layer-by-layer densities of states. (**b**) Macroscopically and planar-averaged electrostatic potential for the domain boundaries with continuous oxygen octahedral rotations. Adapted from Lubk et al. [37]

correlated with (and caused by) small changes in the component of the polarization normal to the wall [23]. These changes in normal polarization are a consequence of the specific rotation of the polar vector across the domain wall, and are also found in other materials. Tetragonal $PbTiO_3$, for example, shows a similar effect for a 90° wall [38]. Extended phase-field calculations for tetragonal $BaTiO_3$ also allow calculating the intrinsic electrostatic potential drop across the 90° domain wall, regardless of the consideration of the ferroelectric as n-type semiconductor or dielectric [39]. This potential change creates a large local electric field that promotes an asymmetric charge distribution around the walls, i.e., they foster the accumulation of carriers and oxygen vacancies close to the wall. The increased charge density plays a role in increased conductivity at such walls.

As mentioned before, the semirigid rotation of the polar vector across a ferroelectric domain wall leads to an electrostatic potential that is screened by free charges which enhance the local charge density and thus, presumably, the conductivity. Since this polar rotation [40] is not exclusive to BFO, other perovskite ferroelectrics should also be expected to display enhanced conductivity. This increased conductivity indeed has been found for lead zirconium titanate (PZT) [41], hexagonal manganites ($YMnO_3$, $ErMnO_3$) [30, 31], and lithium niobate under illumination [42]. For BFO, several other factors might be further helping the conductivity enhancement: Firstly, the magnetoelectric coupling between polarization and spin lattice is such that the magnetic sub-lattice rotates with the polarization [43, 44]. Since spins rotate rigidly, they might favor a more rigid rotation of the polarization and hence a bigger electrostatic step at the wall (and of course the polarization of BFO is itself bigger than that of other known perovskite ferroelectrics, which means that all other things being equal a rigid polar rotation in BFO will cause a bigger electrostatic step). But perhaps the most obvious consideration is the fact that $BiFeO_3$ has intrinsically smaller band gap than other prototype perovskite ferroelectrics (~2.7 eV instead of 3.5–4 eV). This means that the screening charges accumulated at the wall will be closer to the bottom of the conduction band and hence will more easily contribute to the conductivity.

As an additional factor, the role of defect accumulation at the domain walls also deserves close scrutiny. Localized states are known to appear in the spectrum of ferroelectric semiconductors, and states localized at the walls and inside the domain but close to the wall split off from the bulk continuum. These non-degenerate states typically have a high dispersion, in contrast with the "heavy-fermion" states at an isolated domain wall [45]. This behavior can lead to charged double layers that can be formed due to coupling between polarization and space charges at ferroelectric–ferroelastic domain walls [46]. Charged domain wall energies are about one order larger than the uncharged domain wall energies [47], and phenomenological calculations show decoration of walls by defects such as oxygen vacancies, which are always present in oxides. The presence of charge and defect layers at the walls means that such walls can cause electrical failure in ferroelectric capacitor structures by providing a high conductivity pathway from electrode to electrode [43].

Apart from the unwanted effects associated with defect accumulation at domain walls, the intentional manipulation of electronic structure at walls by doping and strain in ferroelectric and ferroelastic oxides provides a way to effectively engineer and control nanoscale functionality in such materials. For the case of $BiFeO_3$ A-site doping with Ca, and magnetic B-site substitution such as Co or Ni, might prove to be a viable way to achieve new domain wall properties by manipulating the electronic structure, spin structure, and dipolar moment in this material [14]. Of obvious future interest is the question of what sets the limits to the current transport behavior at walls: can one "design" the topological structure of the domain wall to controllably induce electronic phase transitions within the wall arising from the correlated electron nature? Is it possible to trigger an Anderson transition by doping of domain walls or straining them? Domain wall (super)conductivity has been shown before by Aird and Salje [48]. Exposing WO_3 to sodium vapor, they observed preferential doping along the ferroelastic domain walls. Transport measurements showed superconductivity with a critical temperature of 3 K, while magnetic measurements did not, suggesting that superconductivity was confined to the domain walls only, which provided a percolating superconductive path while occupying a very small volume fraction of the crystal.

Recently, the observation of tunable electronic conductivity at domain walls in La-doped BFO linked to oxygen vacancy concentration has been reported [24]. Specific growth conditions have been used to introduce varying amounts of vacancies in thin film samples with ordered stripe domain patterns [49]. The conductivity at 109° walls in such samples was shown to be thermally activated with activation energies of 0.24–0.5 eV. From a broader perspective, these results are a first step towards realizing the tantalizing possibility of inducing an insulator–metal transition locally within the confines of the domain wall through careful design of the electronic structure, the state of strain, and chemical effects at the domain wall. For actual device applications it would be desirable to increase the magnitude of the wall currents from currently achieved values. The choice of the right shallow-level dopant and host material might prove to be key factors in this respect.

Defect–domain wall interaction is a very important area of research that deserves increased attention [50]. Point defects, for example, can broaden domain walls [51, 52]. The width of twin walls in $PbTiO_3$, for example, can be strongly modified by the presence of point defects within the wall. The intrinsic wall width of $PbTiO_3$ is about 0.5 nm, but clusters of point defects can increase the size of the twin wall up to 15 nm [53]. Trapped defects at the domain boundary play a significant role in the spatial variation of the antiparallel polarization width in the $BaMgF_4$ single crystal as investigated by PFM [54]. Asymmetric charge distribution close to 90° domain walls in $BaTiO_3$ has also been reported, where electrons and oxygen vacancies concentrate on the opposite sides of the wall [46].

Interaction between the order parameter and the point defect concentration can cause them to accumulate within twin walls [52]. This effect in turn leads to defects contributing to wall kinetics and altered hysteresis behavior [25]. The underlying reason for this is that defects tend to clamp the walls. Oxygen vacancies in particular have been shown to have lower formation energies in the domain wall than in the bulk, confirming the tendency of these defects to migrate to and pin the domain walls [55]. This leads to a scenario where the domain wall can have a memory of its location during annealing cycles [43].

The general concept of doping with various ions has also been applied to bismuth ferrite in an attempt to modify the electronic and magnetic properties and to reduce leakage currents in capacitor devices [14, 56, 57]. B-site doping of BFO with Ti^{4+} has been applied to reduce leakage by over three orders of magnitude while doping with Ni^{2+} leads to a higher conductivity by over two orders of magnitude [58]. Likewise, doping with Cr or Mn has also been proven to greatly reduce leakage currents in BFO thin films [59]. These concepts can readily be extended to the domain wall properties and should be explored. As an example, Yang et al. investigated Ca doping of BFO with results showing strong similarity to phase diagrams of high-T_C superconductors and colossal magnetoresistive manganites where a competition between energetically similar ground states is introduced by hole doping. Control of the conductive properties by band-filling was observed in Ca-doped BFO. Application of an electric field enabled this control to the extent that a p–n junction can be formed, erased, and inverted in this material. A "dome-like" feature in the phase diagram is observed around a Ca concentration of 1/8, where a new pseudo-tetragonal phase appears and the electric modulation of conduction is found to be largest. c-AFM measurements reveal that the material exhibits resistive switching and that subsequent application of electric fields can reverse the effect. The observed reversible modulation of electric conduction accompanied by the modulation of the ferroelectric state is a consequence of the spatial movement of naturally produced oxygen vacancies under an electric field that act as donor impurities to compensate Ca acceptors and maintain a highly stable Fe^{3+} valence state. This observation might lead to new concepts for merging magnetoelectrics and magnetoelectronics at room temperature by combining electronic conduction with electric and magnetic degrees of freedom [60–62]. With this in mind, Ca doping of domain walls and phase boundaries [63] in BFO constitutes an interesting future aspect of domain wall research in BFO (Fig. 13.3).

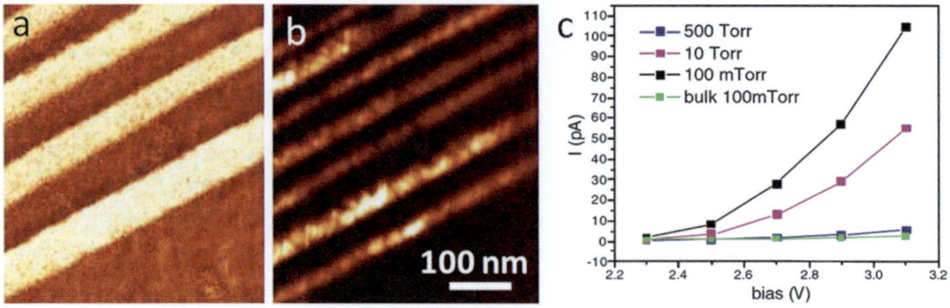

Fig. 13.3 (**a**) PFM image showing ordered 109° stripe domains in BFO. (**b**) Simultaneously acquired c-AFM image of the same area showing that each 109° domain wall is electrically conductive. (**c**) Current levels for samples with different oxygen cooling pressure and thus varying density of oxygen vacancies. Adapted from Seidel et al. [23]

13.3 Light Interaction with Domain Walls

BFO exhibits an anomalous photovoltaic effect in thin films arising from a unique mechanism—namely, structurally driven steps of the electrostatic potential at nanometer-scale domain walls [64–66]. In conventional solid-state photovoltaics, electron–hole pairs are created by light absorption in a semiconductor and separated by the electric field spanning a micrometer-thick depletion region. The maximum voltage these devices can produce is equal to the semiconductor electronic band gap. Interestingly, domain walls can give rise to a fundamentally different mechanism for photovoltaic charge separation, which operates over a distance of 1–2 nm and produces voltages that are significantly higher than the band gap. Recent investigations using c-AFM under light illumination reveal these high photovoltages at 71° and 109° domain walls in BFO (see Fig. 13.4) [66]. The charge separation happens at previously unobserved nanoscale steps of the electrostatic potential that naturally occur at ferroelectric domain walls in the complex oxide $BiFeO_3$. Electric-field control over domain structure allows the photovoltaic effect to be reversed in polarity or turned off.

Currently, the overall efficiency of this photovoltaic effect in devices is limited by the conductivity of the bulk bismuth ferrite material. Methods to increase the carrier mobility, as well as inducing the spatially periodic potential in an adjacent material with a lower gap than BFO, are possible routes to achieve larger current densities under white light illumination, and, more generally, they would demonstrate what the source of periodic potential and the PV current flow can be in different materials. Low band gap semiconductors with asymmetric electron and hole mobilities are possible candidates to show such an effect. In addition, photoelectrochemical effects at domain walls are a possible further interesting route, e.g., for applications in water splitting [67].

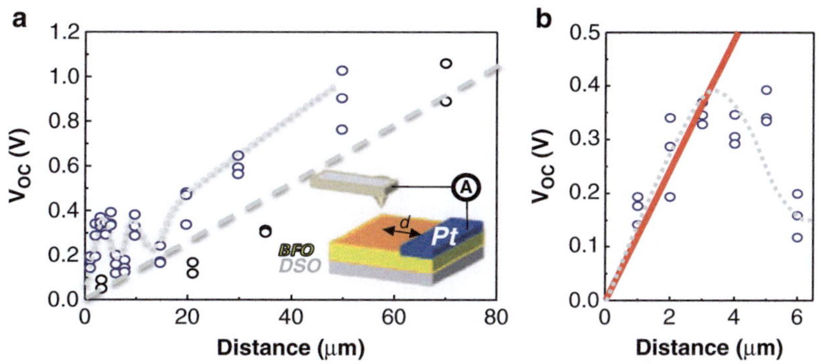

Fig. 13.4 Local measurements of open circuit voltage V_{OC} at domain walls in BFO. (**a**) 109° domain walls (*blue*) show different, oscillating behavior of V_{OC} with distance as compared to 71° domain walls (*black*). *Inset*: Schematic nanoscale PV measurement setup using a conductive AFM tip as a variable-distance counter electrode for *I–V* characterization. (**b**) Initial large slope (*red line*) indicating large PV effect at 109° walls. Adapted from Seidel et al. [66]

13.4 Scanning Tunneling Spectroscopy Investigation of Domain Walls

Over the last decade, many groups have explored the field of complex oxide interfaces. Many interesting phenomena have been observed in epitaxial heterostructures such as metallic and superconducting behavior at the interface between two insulators [3, 68], metals and insulators [69], or gated oxide surfaces [70], promoted by electronic reconstruction due to charge transfer or carrier redistribution effects. Charge transfer has been found to trigger novel magnetic phases at the manganite–ruthenate interfaces [71, 72]. Unusual phenomena such as inducing magnetism in a superconductor and the rearrangement of the magnetic domains in superconductor/ferromagnet heterostructures due to orbital rearrangement and strong hybridization [73] as well as other ways of combining superconductors with metallic ferromagnets [74] have been explored.

Another interesting aspect of nanoscale oxide materials is their potential usage in tunneling devices. Control of the growth of these materials on the unit cell level has led to research on the application of high quality ultrathin ferroelectric films as switchable tunnel barriers for various applications [75]. The polarization reversal of a ferroelectric barrier has been predicted to show large changes in tunneling resistance [76] due to the control of polarization charges at the interface. Using this effect may lead to the development of novel resistive memory devices.

STM and STS measurements in cross-sectional samples have been used to directly investigate the nature of the local electronic conductivity at ferroelectric domain walls in multiferroic BFO [30]. Samples with engineered stripe domain

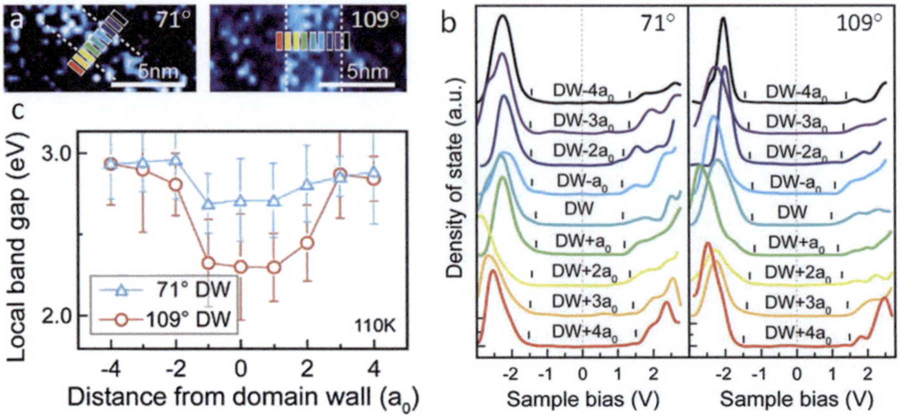

Fig. 13.5 Layer-by-layer d*I*/d*V* measurements across 71° and 109° domain walls in $BiFeO_3$ acquired at 110 K. *Bars* in (**a**) denote positions where the electronic spectra are probed, and (**b**) show the corresponding STS spectra. The band edges are indicated by *black tick marks* in (**b**). (**c**) Extracted local band gap across the domain walls. Adapted from Chiu et al. [30]

array were used in order to simplify the location of a single wall in the STM setup for local spectroscopy. It was found that in situ-cleaved samples with ordered stripe arrays show decreases of the band gap at the domain boundaries (Fig. 13.5). In addition, a shift towards the Fermi level in the band edges of 109° and 71° domain walls has been observed. The demonstrated approach in this work serves as a model technique to investigate and understand electronic structure at complex oxide interfaces.

13.5 Magnetic Properties of Domain Walls

An important aspect of the multiferroic domain walls in BFO concerns their true state of magnetism. Temperature-dependent transport measurements have been performed and are a possible route to follow to understand the actual spin structure and whether it exhibits a glasslike or ordered ferromagnetic state [83]. Of additional interest is the effect of extra carriers introduced into the system, e.g., by doping or electric gating, on magnetism. This includes doping with magnetic ions like Co or Ni. Is there a way to change the magnetic interaction from superexchange to double exchange in BFO, maybe only locally at the walls? The strength of the coupling between the ferroelectric and antiferromagnetic walls in BFO is an issue that still needs to be resolved from both a theoretical and an experimental perspective. The role of the dimensionality on electrical and magnetoelectrical transport needs to be elucidated and put into perspective with regard to known systems, such as manganites [77, 78]. The interaction between ferroelectric and antiferromagnetic domain walls has been studied in model multiferroics such as $YMnO_3$ [79] and BFO [80]. In both cases it has been shown that the antiferromagnetic domain walls are significantly wider (by ~1–2 orders of magnitude) compared to the ferroelectric walls. This is also in agreement with the phenomenological predictions of Daraktchiev et al. for coupling-mediated wall broadening [81].

13.6 Outlook

Nanoscale conduction phenomena at domain walls in complex oxides form an exciting and growing field of interest in functional materials. With the current developments surrounding their conductive properties there are many remaining questions and new areas to be explored. For example, the investigation of dynamic conductivity at domain walls is an exciting aspect [82]. This addresses important factors: a possible electric-field-induced distortion of the polarization structure at the domain wall; the dependence of conductivity on the degree of distortion; and weak-pinning scenarios of the distorted wall. The domain wall is very likely not a rigid electronic conductor, instead offering a quasi-continuous spectrum of voltage-tunable electronic states [79]. This aspect has been explored only rudimentary, while ferroelectric domains were shown to exhibit concrete conductance levels [17, 18]. The intrinsic dynamics of domain walls and other topological defects are expected not only to influence future theoretical and experimental interpretations of the electronic phenomena but also to pose a possibility to find unique properties of multiferroic domain walls, e.g., magnetization and magnetoresistance within an insulating antiferromagnetic matrix [83], influenced by order parameter coupling and localized secondary order parameters [52]. Of obvious future interest is the question of what sets the limits to the current transport behavior at walls. It might be possible to "design" the topological structure of the domain wall to controllably induce electronic phase transitions within the wall arising from the correlated electron nature. The observation of superconductivity in ferroelastic walls of WO_3 is an indication of various exciting and unexplored areas of domain boundary physics [48].

The investigation of topological defects beyond the classic domain walls is also a new exciting area of research. As an example, arrays of ferroelectric nanowires have been shown to contain switchable quadrupoles which make them potentially interesting in nano-device applications [84]. Exotic topological defects in nanostructures are currently a very active area of research. The vortex cores in BFO have been demonstrated to be dynamic conductors controlled by the coupled response of polarization and electron–mobile-vacancy subsystems with external bias [85].

Several applications have been proposed to make use of domain walls in ferroelectric materials based on their additional functionalities as well as their affects on existing devices [86]. Uses that have been mentioned are as a local strain sensor incorporated on an AFM probe, reconfigurable nano-circuits, or a multilevel resistance-state device that is written by an electrical current [87]. Other possibilities include nonvolatile memories, piezoelectric actuators, ultrasound transducers, surface acoustic wave devices, and optical applications [50]. For existing devices, the discovery of conducting domain walls indicates the possibility to engineer the right domain walls that could prevent leakage.

The ferromagnetic properties of ferroelectric walls in paramagnetic and antiferromagnetic materials suggest that domain walls in multiferroics are a new interesting field of research which includes the dynamics of domain walls in these materials [79, 81, 88]. Artificially engineered oxide interfaces, including domain walls, may pave the way to novel tailored states of matter with a wide range of electronic properties. Many ferroelectric relaxor-like systems nanoscale phase separated materials are interesting as well since they have intrinsic nanodomains. Domain wall electronics, particularly with ferroelectrics and multiferroics, may become interesting for nanotechnology applications [89].

References

1. Imada, M., Fujimori, A., Tokura, Y.: Metal-insulator transitions. Rev. Mod. Phys. **70**, 1039–1263 (1998)
2. Ogale, S.B.: Thin Films and Heterostructures for Oxide Electronics. Springer, New York (2005)
3. Ohtomo, A., Muller, D.A., Grazul, J.L., Wang, H.Y.: Artificial charge-modulationin atomic-scale perovskite titanate superlattices. Nature **419**, 378–380 (2002)
4. Dagotto, E.: When oxides meet face to face. Science **318**, 1076–1077 (2007)
5. Mannhart, J., Schlom, D.G.: Oxide interfaces—an opportunity for electronics. Science **327**, 1607–1611 (2010)
6. Yamada, H., et al.: Engineered interface of magnetic oxides. Science **395**, 646–648 (2004)
7. Zubko, P., Gariglio, S., Gabay, M., Ghosez, P., Triscone, J.-M.: Interface physics in complex oxide heterostructures. Annu. Rev. Condens. Matter Phys. **2**, 141–165 (2011)
8. Heber, J.: Enter the oxides. Nature **459**, 28–30 (2009)
9. Seidel et al.: (submitted)
10. Zhang, J.X., Xiang, B., He, Q.: Large field-induced strains in a lead-free piezoelectric material. Nat. Nanotechnol. **6**, 98 (2011)
11. Eng, L.M.: Nanoscale domain engineering and characterization of ferroelectric domains. Nanotechnology **10**, 405 (1999)
12. Kalinin, S.V., et al.: Local polarization dynamics in ferroelectric materials. Rep. Prog. Phys. **73**, 056502 (2010)
13. Gruverman, A., Rodriguez, B.J., Dehoff, C., Waldrep, J.D., Kingon, A.I., Nemanich, R.J., Cross, J.S.: Direct studies of domain switching dynamics in thin film ferroelectric capacitors. Appl. Phys. Lett. **87**, 082902 (2005)
14. Rodriguez, B.J., Jesse, S., Baddorf, A.P., Zhao, T., Chu, Y.H., Ramesh, R., Eliseev, E.A., Morozovska, A.N., Kalinin, S.V.: Spatially resolved mapping of polarization switching behavior in nanoscale ferroelectrics. Nanotechnology **18**, 405701 (2007)
15. Jungk, T., Hoffmann, A., Soergel, E.: Impact of elasticity on the piezoresponse of adjacent ferroelectric domains investigated by scanning force microscopy. J. Appl. Phys. **102**, 084102 (2007)
16. Choudhury, S., et al.: The influence of 180° ferroelectric domain wall width on the threshold field for wall motion. J. Appl. Phys. **10**(4), 084107 (2008)
17. Yang, C.-H., Seidel, J., Kim, S.Y., Rossen, P.B., Yu, P., Gajek, M., Chu, Y.-H., Martin, L.W., Holcomb, M.B., He, Q., Maksymovych, P., Balke, N., Kalinin, S.V., Baddorf, A.P., Basu, S.R.,

Scullin, M.L., Ramesh, R.: Electric modulation of conduction in multiferroic Ca-doped $BifeO_3$ films. Nat. Mater. **8**, 485 (2009)
18. Seidel, J., et al.: Prominent electrochromism through vacancy-order melting. Nat. Commun. **3**, 799 (2011)
19. Balke, N., Jesse, S., Morozovska, A., Eliseev, E., Chung, D., Kim, Y., Adamczyk, L., Garcia, R.: Nanoscale mapping of ion diffusion in a lithium-ion battery cathode. Nat. Nanotechnol. **5**, 749–754 (2010)
20. Kumar, A., Ciucci, F., Morozovska, A.N., Kalinin, S.V., Jesse, S.: Nanoscale electrochemistry: Feeling the strain. Nat. Chem. **3**, 707–713 (2011)
21. Rodriguez, B.J., Chu, Y.H., Ramesh, R., Kalinin, S.V.: Ferroelectric domain wall pinning at a bicrystal grain boundary in bismuth ferrite. Appl. Phys. Lett. **93**, 142901 (2008)
22. Watanabe, Y.: Review of Resistance Switching of Ferroelectrics and Oxides in Quest for Unconventional Electronic Mechanisms. Ferroelectrics **349**, 190 (2007)
23. Seidel, J., Martin, L.W., He, Q., Zhan, Q., Chu, Y.-H., Rother, A., Hawkridge, M.E., Maksymovych, P., Yu, P., Gajek, M., Balke, N., Kalinin, S.V., Gemming, S., Wang, F., Catalan, G., Scott, J.F., Spaldin, N.A., Orenstein, J., Ramesh, R.: Conduction at domain walls in oxide multiferroics. Nat. Mater. **8**, 229 (2009)
24. Seidel, J., Maksymovych, P., Katan, A.J., Batra, Y., He, Q., Baddorf, A.P., Kalinin, S.V., Yang, C.-H., Yang, J.-C., Chu, Y.-H., Salje, E.K.H., Wormeester, H., Salmeron, M., Ramesh, R.: Domain wall conductivity in La-doped $BiFeO_3$. Phys. Rev. Lett. **105**, 197603 (2010)
25. Fan, W., Cao, J., Seidel, J., Gu, Y., Yim, J.W., Barrett, C., Yu, K.M., Ji, J., Ramesh, R., Chen, L.Q., Wu, J.: Large kinetic asymmetry in the metal-insulator transition nucleated at localized and extended defects. Phys. Rev. B **83**, 235102 (2011)
26. Maggio-Aprile, I., Rennet, C., Erb, A., Walker, E., Fischer, O.: Critical currents approaching the depairing limit at a twin boundary in YBa2Cu3O(7-δ). Nature **390**, 487–490 (1997)
27. Wiessner, A., Kirschner, J., Schafer, G., Berghaus, T.H.: Design considerations and performance of a combined scanning tunneling and scanning electron microscope. Rev. Sci. Instrum. **68**, 3790 (1997)
28. Yang, B., Park, N.J., Seo, B.I., Oh, Y.H., Kim, S.J., Hong, S.K., Lee, S.S., Park, Y.J.: Nanoscale imaging of grain orientations and ferroelectric domains in $(Bi_{1-x}La_x)_4Ti_3O_{12}$ films for ferroelectric memories. Appl. Phys. Lett. **87**, 062902 (2005)
29. Garcia, R.E., Huey, B.D., Blendell, J.E.: Virtual piezoforce microscopy of polycrystalline ferroelectric films. J. Appl. Phys. **100**, 064105 (2006)
30. Chiu, Y.-P., Chen, Y.-T., Huang, B.-C., Shih, M.-C., Yang, J.-C., He, Q., Liang, C.-W., Seidel, J., Chen, Y.-C., Ramesh, R., Chu, Y.-H.: The evolution of local electronic structure across multiferroic domain walls. Adv. Mater. **23**, 1530 (2011)
31. Lubk, A., et al.: Evidence of sharp and diffuse domain walls in $BiFeO_3$ by means of unit-cell-wise strain and polarization maps obtained with high resolution scanning transmission electron microscopy. Phys. Rev. Lett. **109**, 047601 (2012)
32. Farokhipoor, S., Noheda, B.: Conduction through 71° domain walls in $BiFeO_3$ thin films. Phys. Rev. Lett. **107**, 127601 (2011)
33. Seidel, J., Singh-Bhalla, G., He, Q., Yang, S.-Y., Chu, Y.-H., Ramesh, R.: Domain wall functionality in $BiFeO_3$. Phase Transit. **86**, 53–66 (2013). doi:10.1080/01411594.2012.695371
34. Choi, T., et al.: Insulating interlocked ferroelectric and structural antiphase domain walls in multiferroic $YMnO_3$. Nat. Mater. **9**, 253–258 (2010)
35. Meier, D., Seidel, J., Cano, A., Delaney, K., Kumagai, Y., Mostovoy, M., Spaldin, N.A., Ramesh, R., Fiebig, M.: Anisotropic conductance at improper ferroelectric domain walls. Nat. Mater. **11**, 284 (2012)
36. Palai, R., Katiyar, R.S., Schmid, H., Tissot, P., Clark, S.J., Robertson, J., Redfern, S.A.T., Catalan, G., Scott, J.F.: β Phase and γ-β metal-insulator transition in multiferroic $BiFeO_3$. Phys. Rev. B **77**, 014110 (2008)
37. Lubk, A., Gemming, S., Spaldin, N.A.: First-principles study of ferroelectric domain walls in multiferroic bismuth ferrite. Phys. Rev. B **80**, 104110 (2009)

38. Meyer, B., Vanderbilt, D.: Ab initio study of ferroelectric domain walls in $PbTiO_3$. Phys. Rev. B **65**, 104111 (2002)
39. Hong, L., Soh, A.K., Du, Q.G., Li, J.Y.: Interaction of O vacancies and domain structures in single crystal BaTiO3: two-dimensional ferroelectric model. Phys. Rev. B **7**(7), 094104 (2008)
40. Borisevich, A.Y., et al.: Mapping octahedral tilts and polarization across a domain wall in $BiFeO_3$ from Z-contrast scanning transmission electron microscopy image atomic column shape analysis. ACS Nano **4**, 6071 (2010)
41. Guyonnet, J., Gaponenko, I., Gariglio, S., Paruch, P.: Conduction at domain walls in insulating $Pb(Zr_{0.2}Ti_{0.8})O_3$ thin films. Adv. Mater. **23**, 5377 (2011)
42. Schröder, M.: Conducting domain walls in lithium niobate single crystals. Adv. Funct. Mater. **22**, 3936 (2012)
43. Zhao, T., Scholl, A., Zavaliche, F., Lee, K., Barry, M., Doran, A., Cruz, M.P., Chu, Y.H., Ederer, C., Spaldin, N.A., Das, R.R., Kim, D.M., Baek, S.H., Eom, C.B., Ramesh, R.: Electrical control of antiferromagnetic domains in multiferroic $BiFeO_3$ films at room temperature. Nat. Mater. **5**, 823 (2006)
44. Lebeugle, D., Colson, D., Forget, A., Viret, M., Bataille, A.M., Gukasov, A.: Electric-field-induced spin flop in $BiFeO_3$ single crystals at room temperature. Phys. Rev. Lett. **100**, 227602 (2008)
45. Idlis, B.G., Usmanov, M.S.: Effect of domain structure on the energy spectrum of narrow-gap ferroelectric semiconductors. Pis'ma Zh. Eksp. Teor. Fiz. **56**(5), 268–271 (1992)
46. Xiao, Y., Shenoy, V.B., Bhattacharya, K.: Depletion layers and domain walls in semiconducting ferroelectric thin films. Phys. Rev. Lett. **9**(5), 247603 (2005)
47. Gureev, T.M.Y., Tagantsev, A.K., Setter, N.: Structure and energy of charged domain walls in ferroelectrics. 18th IEEE ISAF Proceedings, Xian (2009)
48. Aird, A., Salje, E.K.H.: Sheet superconductivity in twin walls: experimental evidence of WO3-x. J. Phys. Condens. Matter **10**, L377 (1998)
49. Scullin, M.L., et al.: Acta Mater. **58**, 457 (2010)
50. Gopalan, V., Dierolf, V., Scrymgeour, D.A.: Defect–domain wall interactions in trigonal ferroelectrics. Annu. Rev. Mater. Res. **37**, 449–489 (2007)
51. Shilo, D., Ravichandran, G., Bhattacharya, K.: Investigation of twin-wall structure at the nanometre scale using atomic force microscopy. Nat. Mater. **3**, 453–457 (2004)
52. Lee, W.T., Salje, E.K.H., Bismayer, U.: Influence of point defects on the distribution of twin wall widths. Phys. Rev. B **7**(2), 104116 (2005)
53. Salje, E.K.H., Zhang, H.: Domain boundary engineering. Phase Transit. **82**, 6 (2009)
54. Zeng, H.R., et al.: Domain wall thickness variations of ferroelectric BaMgF4 single crystals in the tip fields of an atomic force microscope. Phys. Status Solidi (RRL) **2**, 3 (2008)
55. He, L., Vanderbilt, D.: First-principles study of oxygen-vacancy pinning of domain walls in $PbTiO_3$. Phys. Rev. B **6**(8), 134103 (2003)
56. Lee, Y.-H., Wu, J.-M., Lai, C.-H.: Influence of La doping in multiferroic properties of $BiFeO_3$ thin films. Appl. Phys. Lett. **88**, 042903 (2006)
57. Yang, C.-H., Kan, D., Takeuchi, I., Nagarajan, V., Seidel, J.: Doping $BiFeO_3$: approaches and enhanced functionality. Phys. Chem. Chem. Phys. **14**, 15953–15962 (2012). doi:10.1039/C2CP43082G
58. Qi, X., Dho, J., Tomov, R., Blamire, M.G., MacManus-Driscoll, J.L.: Greatly reduced leakage current and conduction mechanism in aliovalent-ion-doped $BiFeO_3$. Appl. Phys. Lett. **86**, 062903 (2005)
59. Kim, J.K., Kim, S.S., Kim, W.-J., Bhalla, A.S., Guo, R.: Enhanced ferroelectric properties of Cr-doped $BiFeO_3$ thin films grown by chemical solution deposition. Appl. Phys. Lett. **88**, 132901 (2006)
60. Ko, K.T., Jung, M.H., Lee, J.H., Woo, C.S., Chu, K., Seidel, J., Chu, Y.H., Jeong, Y.H., Ramesh, R., Park, J.H., Yang, C.-H.: Concurrent transition of ferroelectric and magnetic ordering around room temperature. Nat. Commun. **2**, 567 (2011)
61. Ramirez, M., et al.: Spin-charge-lattice coupling through resonant multimagnon excitations in multiferroic $BiFeO_3$. Appl. Phys. Lett. **94**, 161905 (2009)

62. Ramirez, M.O., et al.: Two-phonon coupling to the antiferromagnetic phase transition in multiferroic $BiFeO_3$. Appl. Phys. Lett. **92**, 022511 (2008)
63. Zhou, J., Trassin, M., He, Q., Tamura, N., Kunz, N., Cheng, C., Zhang, J., Liang, W.-I., Seidel, J., Hsin, C., Chu, Y.-H., Wu, J.: Directed assembly of nanoscale phase variants in highly strained $BiFeO_3$ thin films. J. Appl. Phys. **112**, 064102 (2012)
64. Yang, S.-Y., Seidel, J., Byrnes, S.J., Shafer, P., Yang, C.-H., Rossell, M.D., Yu, P., Chu, Y.-H., Scott, J.F., Ager III, J.W., Martin, L.W., Ramesh, R.: Above bandgap voltages from ferroelectric photovoltaic devices. Nat. Nanotechnol. **5**, 143 (2010)
65. Seidel, J., Fu, D., Yang, S.-Y., Alarcòn-Lladò, E., Wu, J., Ramesh, R., Ager, J.W.: Efficient photovoltaic current generation at ferroelectric domain walls. Phys. Rev. Lett. **107**, 126805 (2011)
66. Seidel, J., Yang, S.-Y., Alarcòn-Lladò, E., Ager, J.W., Ramesh, R.: Nanoscale probing of high photovoltages at 109° domain walls. Ferroelectrics **433**, 123 (2012)
67. Kudo, A., Miseki, Y.: Heterogeneous photocatalyst materials for water splitting. Chem. Soc. Rev. **38**, 253 (2009)
68. Reyren, N., Thiel, S., Caviglia, A.D., Fitting Kourkoutis, L., Hammerl, G., Richter, C., Schneider, C.W., Kopp, T., Rüetschi, A.-S., Jaccard, D., Gaboy, M., Muller, D.A., Triscone, J.-M., Mannhart, J.: Superconducting Interfaces Between Insulating Oxides. Science **317**, 1196–1199 (2007)
69. Logvenov, G., Gozar, A., Bozovic, I.: High-Temperature Superconductivity in a Single Copper-Oxygen Plane. Science **326**, 699–702 (2009)
70. Ye, J.T., Inoue, S., Kobayashi, K., Kasahara, Y., Yuan, H.T., Shimotani, H., Iwasa, Y.: Liquid-gated interface superconductivity on an atomically flat film. Nat. Mater. **9**, 125–128 (2010)
71. Takahashi, K.S., Kawasaki, M., Tokura, Y.: Interface ferromagnetism in oxide superlattices of $CaMnO_3/CaRuO_3$. Appl. Phys. Lett. **79**, 1324–1326 (2001)
72. Koida, T., Lippmaa, M., Fukumura, T., Itaka, K., Matsumoto, Y., Kawasaki, M., Koinuma, H.: Effect of A-site cation ordering on the magnetoelectric properties in [$(LaMnO_3)$m/$(SrMnO_3)$ m]n artificial superlattices. Phys. Rev. B **66**, 144418 (2002)
73. Chakhalian, J., Freeland, J.W., Habermeier, H.U., Cristiani, G., Khaliullin, G., Veenendaalvan, M., Keimer, B.: Orbital reconstruction and covalent bonding at an oxide interface. Science **318**, 1114–1117 (2007)
74. Lyuksyutov, I., Pokrovsky, V.: Ferromagnet-superconductor hybrids. Adv. Phys. **54**, 67–136 (2005)
75. Tsymbal, E.Y., Kohlstedt, H.: Tunneling across a ferroelectric. science **313**, 181 (2006)
76. Zhuravlev, M.Y., Sabirianov, R., Jaswal, S.S., Tsymbal, E.Y.: Giant electroresistance in ferroelectric tunnel junctions. Phys. Rev. Lett. **94**, 246802 (2005)
77. Dagotto, E.: Nanoscale Phase Separation and Colossal Magnetoresistance. Springer, New York (2003)
78. Salafranca, J., Yu, R., Dagotto, E.: Conducting Jahn-Teller domain walls in undoped manganites. Phys. Rev. B **8**, 245122 (2010)
79. Goltsev, A.V., Pisarev, R.V., Lottermoser, T., Fiebig, M.: Structure and interaction of antiferromagnetic domain walls in hexagonal $YMnO_3$. Phys. Rev. Lett. **90**, 177204 (2003)
80. Gareeva, Z.V., Zvezdin, A.K.: Private communication (2011)
81. Daraktchiev, M., Catalan, G., Scott, J.F.: Landau theory of domain wall magnetoelectricity. Phys. Rev. B **81**, 224118 (2010)
82. Maksymovych, P., Seidel, J., Chu, Y.-H., Baddorf, A., Wu, P., Chen, L.-Q., Kalinin, S.V., Ramesh, R.: Dynamic conductivity of ferroelectric domain walls. Nano Lett. **11**, 1906 (2011)
83. He et al.: Magnetotransport at domain walls in $BiFeO_3$. Phys. Rev. Lett. **108**, 067203 (2012)
84. Hong, J., Catalan, G., Fang, D.N., Artacho, E., Scott, J.F.: Topology of the polarization field in ferroelectric nanowires from first principles. Phys. Rev. B **81**, 172101 (2010)
85. Balke et al.: Enhanced electric conductivity at ferroelectric vortex cores in $BiFeO_3$. Nat. Phys. **8**, 81 (2012)
86. Seidel, J.: Domain walls as nanoscale functional elements. J. Phys. Chem. Lett. **3**, 2905 (2012)

87. Bea, H., Paruch, P.: Multiferroics: a way forward along domain walls. Nat. Mater. **8**, 168–169 (2009)
88. Skumryev, V., Laukhin, V., Fina, I., Marti, X., Sanchez, F., Gospodinov, M., Fontcuberta, J.: Magnetization reversal by electric-field decoupling of magnetic and ferroelectric domain walls in multiferroic-based heterostructures. Phys. Rev. Lett. **106**, 057206 (2011)
89. Catalan, G., et al.: Domain wall nanoelectronics. Rev. Mod. Phys. **84**, 119 (2012)

Chapter 14
Syntheses and Properties of Some Bi-Containing Compounds with Noncentrosymmetric Structure

Wen-Dan Cheng, Chen-Sheng Lin, Lei Geng, Zhong-Zhen Luo, Wei-Long Zhang, and Hao Zhang

Abstract A genetic engineering method is employed to design the second harmonic generation (SHG) materials with infrared transparency. The compounds of Ba_2BiInA_5 (A = S, Se, Te) are constructed in views of "genome" BiA_5 pyramid and InA_4 tetrahedron. Then, the crystal structures of these compounds are predicted or reproduced to show their non-centrosymmetry based on global optimization evolutionary methodology. Thirdly, the ab initio computations of band structures and simulations of optical properties are carried, and the nonlinear optical figure of merit in views of optical transparent range and the SHG parameters are surveyed for these crystals. Finally, we provide the substance evidences by the experimental synthesis, crystal structural determinations, and optical measurements for Ba_2BiInA_5 (A = S, Se) compounds.

Abbreviations

AHO	Anharmonic oscillator
BFO	$BiFeO_3$
BiBO	BiB_3O_6
BSCCO	$Bi_2Sr_2Ca_{n-1}CunO_{2n+4+x}$
CSD	Cambridge Structural Database

W.-D. Cheng (✉) • C.-S. Lin • Z.-Z. Luo • W.-L. Zhang • H. Zhang
State Key Laboratory of Structural Chemistry, Fujian Institute of Research on the Structure of Matter, Chinese Academy of Sciences, Fuzhou, Fujian 350002, China
e-mail: cwd@fjirsm.ac.cn

L. Geng
School of Physics and Electronic Information, Huaibei Normal University, Huaibei, Anhui 235000, China

H. Li and Z.M. Wang (eds.), *Bismuth-Containing Compounds*, Springer Series in Materials Science 186, DOI 10.1007/978-1-4614-8121-8_14,

DFT	Density Functional Theory
DOS	Density of state
FBB	Fundamental building block
FOM	Figure of merit
GW_0	Green's function
LCB	Lowest conduction band
MVB	Maximum of the valance band
NCS	Noncentrosymmetric
OPO	Optical parametric oscillator
PED	Partial electron density
SHG	Second harmonic generation
USPEX	Universal Structure Prediction: Evolutionary Xtallography
VASP	Vienna Ab-initio Simulation Package

14.1 Introduction

Elemental bismuth may exist naturally and bismuth metal has been known from ancient times. Bismuth (Bi) is stable to both dry and moist air at ordinary temperatures, and it forms trivalent and pentavalent compounds and the trivalent ones are more common. Bi has commercial applications including pharmaceuticals, pigments, and cosmetics. In the early 1990s, researchers began to evaluate bismuth as a nontoxic replacement for lead in various applications. Bi-containing compounds have applications in medicine [1–3], cosmetics, pigments [4–7], etc. Bismuth strontium calcium copper oxide, or BSCCO (generalized chemical formula $Bi_2Sr_2Ca_{n-1}Cu_nO_{2n+4+x}$), was the first high-temperature superconductor which did not contain a rare earth element [8, 9]. Bismuth telluride, Bi_2Te_3, is a semiconductor and an excellent thermoelectric material, and the diodes made by Bi_2Te_3 are used in mobile refrigerators, coolers, and as detectors in infrared spectrophotometers [10, 11]. Many bismuth alloys have low melting points and are found in specialty applications, such as Bi–Cd–Pb–Sn alloy which melts at 70 °C and is used in automotive and aviation industries. For noncentrosymmetric (NCS) compounds including Bi element, such as bismuth ferrite or BFO ($BiFeO_3$) is classed as rhombohedral belonging to the space group R_{3c} in the room-temperature phase [12], and is one of the most promising multiferroic materials. The multiferroic originates from the coupling of antiferromagnetic and ferroelectric [13–15]; however, $BiMnO_3$ multiferroic results from the coupling of ferromagnetic and ferroelectric [16, 17]. BiB_3O_6 (BiBO) compound belongs to space group C_2 [18], and is a newly developed nonlinear optical crystal [19]. It possesses large effective nonlinear coefficient, high damage threshold and is non-hydroscopic, and efficiently frequency doubling the 946 nm to produce blue laser at 473 nm [20]. The stereochemically active lone pair electron of Bi^{3+} cation is also an important effect to the enhancement of SHG (second harmonic generation) response for NCS crystal material $Cd_4BiO(BO_3)_3$ [21].

In this chapter, our interest is mainly concentrated in the NCS compounds including Bi. This is because the NCS crystals can exhibit a variety of technologically important physical properties, which include piezoelectricity, pyroelectricity, ferroelectricity, and SHG response [22]. The structural symmetry of material affects its property. There are the crystallographic interrelationships between physical properties and NCS crystal classes [23, 24]. Among 32 crystal point groups [25], 20 NCS groups (21 NCS crystal classes excluding point group 432 (O group)) can exhibit the piezoelectric or SHG effect. It is noted that all pyroelectric materials have SHG behavior but the converse is not true. These 20 point groups are sometimes called piezoelectric point groups. Ten of the piezoelectric point groups possess a unique polar axis and may exhibit pyroelectric or ferroelectric effect along this unique axis. These ten polar point groups are: 1, 2, m, 2 mm, 4, 4 mm, 3, 3 m, 6, and 6 mm. All ferroelectric crystals belong to one of these ten polar point groups. In the following discussions, we only care about the SHG response of NCS crystal materials. This is because SHG materials are an important and wide application in the laser spectroscopy, optical communications, optical computing, bio-imaging, manufacture of microstructure materials, and other fields [26–28].

14.2 Computational Designs of NCS Compounds

14.2.1 Optimization of Material Gene and Constructions of Virtual Compounds

What is the definition of material gene? It connotes a fundamental building block (FBB) toward a larger purpose of dimension and function in nonbiological contexts. In fact, a gene is a locatable region of genomic sequence and basic instruction of a sequence of nucleic acid in a living organism. However, the gene in non-biological material is not a life activity. The genes in bio-system and in material are both FBB as growing a large dimension system. The different meanings between the gene or genome applied in biological and non-biological systems give the detailed descriptions in "Material Genome Initiative for Global Competitiveness" [29]. The FBB having NCS structure is considered as a gene of NCS materials. They are distortion octahedron, tetrahedron, triangular, pentagonal pyramid, triangular bipyramid, square pyramid, and other NCS structure geometry, as shown in Fig. 14.1. For the Bi-containing compounds, the coordination structure of Bi^{3+} including a lone-electron pair is generally a square-pyramidal or distorted 6-coordinated configuration, and it finds that In^{3+} is a 4-coordinated structure from the investigations and analysis of structural data in CSD (Cambridge Structural Database) and refs. [30, 31]. Accordingly, we consider that the NCS units of BiA_5 pyramid and InA_4 tetrahedron were taken as genome of SHG materials and Ba^{2+} ion as charge-balancing species to construct virtual compounds of Ba_2BiInA_5 (A = Se, Te) [32].

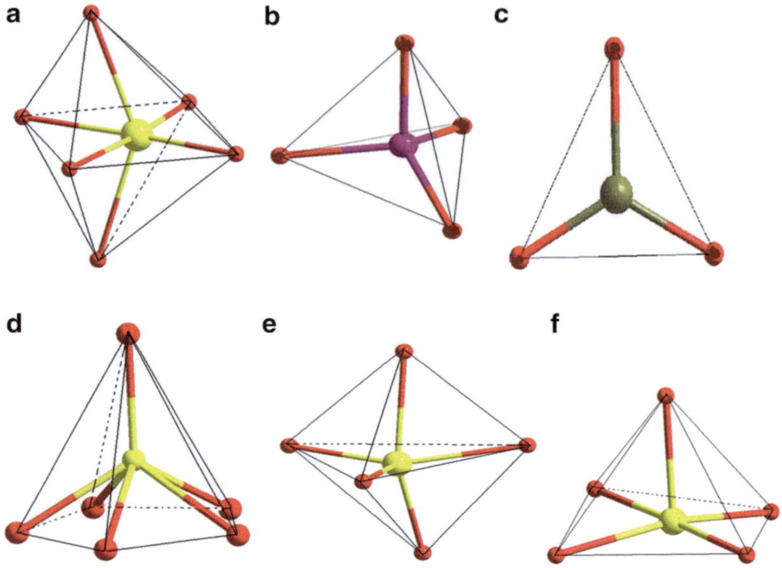

Fig. 14.1 Noncentrosymmetric gene structures. (**a**) Distortion octahedral, (**b**) tetrahedron, (**c**) triangular, (**d**) pentagonal, (**e**) triangular pyramid, and (**f**) square pyramid

14.2.2 Predictions of Crystal Structures for Compounds Ba_2BiInA_5 (A = Se, Te)

For constructing model compounds Ba_2BiInA_5 (A = Se, Te), we performed extensive structure searches based on global optimization using the ab initio calculations performed in VASP (Vienna Ab-initio Simulation Package) code [33] and evolutionary methodology coded in USPEX (Universal Structure Prediction: Evolutionary Xtallography) [34, 35] to look out the most stable structure. The first step of the algorithm is to randomly generate an ensemble of structures (first generation). These structures are relaxed by gradient-based minimization with external ab initio code such as VASP [33], and the worst ones in energy are discarded. Then, the next generation is generated according to several possible variation operations: heredity, permutation, lattice mutation, and atom mutation. Finally, the most stable structure will be found after tens of generations for a dozen atoms/cell system. In our calculation, we set heredity operation as 50 %, permutation as 10 %, atom mutation as 30 %, and lattice mutation as 10 % probability in generating the next generation. The number of formula units/simulation cell, *z*, was set to be 4. The random samplings were selected to do the unbiased initialization in the search space. The candidate structures of the first generation were set to 60 and 100 for the Ba_2BiInA_5 (A = Se, Te), respectively. Each structure produced by USPEX was locally optimized by the ab initio calculations. No spin polarization is applied in the optimization calculations due to the fact that there is no unpaired electron in these compounds. The global minimum (stable structure) on the energy landscape was found within ~2,500 and ~2,400 candidate structures for the Ba_2BiInA_5 (A = Se, Te), respectively. The relation between the energy and structure numbers is given in Fig. 14.2. For $Ba_2BiInSe_5$ compound, the structure searching calculations converged at 39th generation and the most stable structure emerged as the 2033rd predicted structure. For $Ba_2BiInTe_5$, searching calculations converged at 41st generation and the most stable structure emerged as the 1842nd predicted structure. In order to systematically compare the same homologues, we also search the most stable structure of compound Ba_2BiInS_5, and the searching process is also given in Fig. 14.2c.

The prediction of crystal structure will reveal the atomic arrangements in crystal and give crystallographic data, such as unit cell parameters and group features, for virtual or existence compound. The most stable structures of Ba_2BiInA_5 serial materials by predictions belong to the $Cmc2_1$ space group, which has NCS that is a precondition for the SHG application. The $Cmc2_1$ space group belongs to square millimeter crystal class and polar group. The latter symmetry is also a precondition for pyroelectricity or ferroelectricity applications. At the following, we have not made further discussions about the polar properties of Ba_2BiInA_5 materials. The lattice constant and the bond length of the structure predicted by DFT (Density Functional Theory) calculation are in good agreement with experiments (shown in Table 14.1). Hereafter the discussion is based on the optimized structure. The bismuth and chalcogenide form a distorted tetragonal pyramid and connect with each other by edge-sharing. These BiA_5 pyramids pointing their cones to the same direction are connecting to the InA_4 tetrahedral through the bottom chalcogenide (Fig. 14.3a) and form an infinite one-dimensional chain ${}_1\infty[BiInA_5]^{4-}$. Each ${}_1\infty[BiInA_5]^{4-}$ chain is surrounded by six neighboring ${}_1\infty[BiInA_5]^{4-}$ chains and shows the highest packing ABAB type along the *c* axis (Fig. 14.3b). The cations Ba^{2+} are surrounding a ${}_1\infty[BiInA_5]^{4-}$ chain and formed a distorted hexagonal arrangement. The shortest bond length between chalcogenide and bismuth is Bi–A3 where A3 is the cones of BiA_5 pyramids. The second shortest bond length is Bi–A2 where A2 is at the edged-sharing of BiA_5 pyramids chains. The longest Bi–A bond is the Bi–A1 where A1 is a μ_3 bonding-type which connected with the neighbored BiA_5 pyramids and InA_4 tetrahedral polyhedron. Here, we also employ USPEX code to reproduce the crystal structure of Ba_2BiInS_5 which has been reported [31] in order to make systematical comparisons of the homologue. It is shown in Table 14.1 that the bonding Bi–A is ranging from 2.62 to 2.95 Å for Ba_2BiInS_5, 2.76 to 3.07 Å for $Ba_2BiInSe_5$, and 2.98 to 3.27 Å for $Ba_2BiInTe_5$. The Bi–A4 distance is about 0.2 Å longer than Bi–A1, for example, Bi–Se4 is 3.324 Å while Bi–Se1 is 3.078 Å for $Ba_2BiInSe_5$.

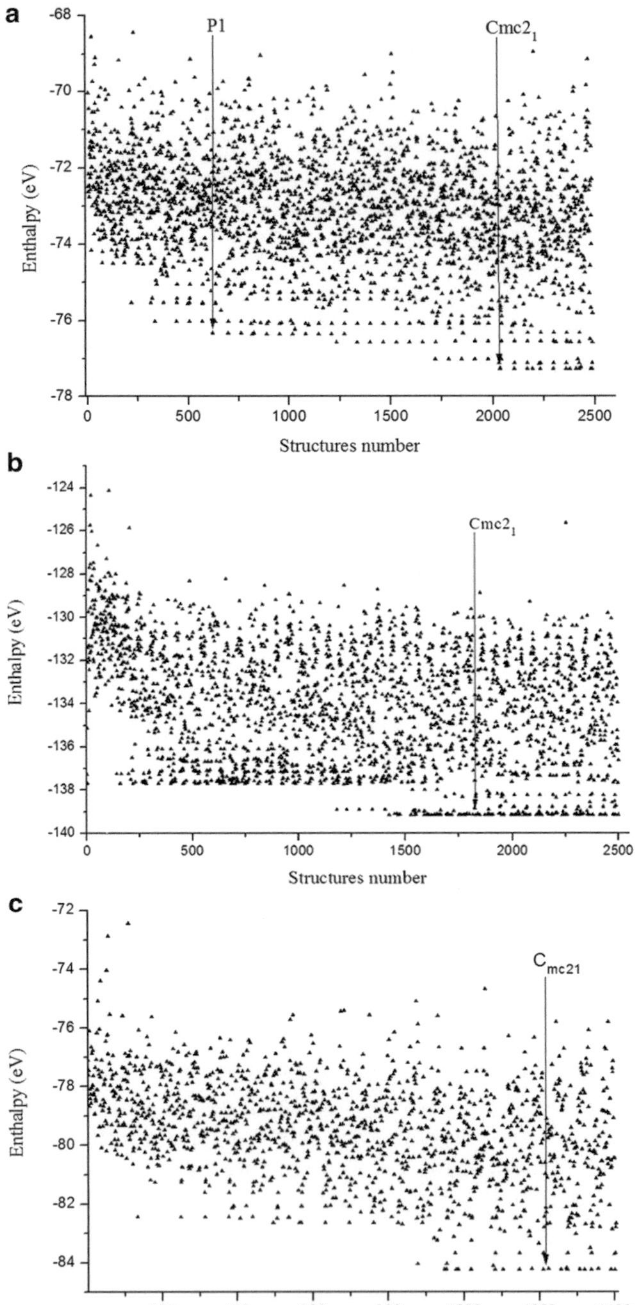

Fig. 14.2 Enthalpy of all the searched structure for Ba_2BiInA_5 (A = S, Se, Te). The most stable structures of $Ba_2BiInSe_5$ (**a**), $Ba_2BiInTe_5$ (**b**), and Ba_2BiInS_5 (**c**) emerge around at 2033rd, 1842nd and 1337th structure, respectively, and they belong to $Cmc2_1$ space group

Table 14.1 The calculated and experimental lattice constants (Å) and selected bond lengths (Å) of Ba_2BiInA_5 (A = S, Se, Te) species

Species		*a*	*b*	*c*	Bi–A1	Bi–A2	Bi–A3	Bi–A4
A = S	Cal.	4.218	18.676	12.984	2.951	2.803	2.624	3.174
	Expt.	4.254	18.252	12.685	2.959	2.794	2.581	3.212
A = Se	Cal.	4.366	19.400	13.414	3.078	2.937	2.764	3.324
	Expt.	4.328	19.117	13.181	3.051	2.890	2.706	3.326
A = Te	Cal.	4.366	19.400	13.414	3.274	3.136	2.987	3.486

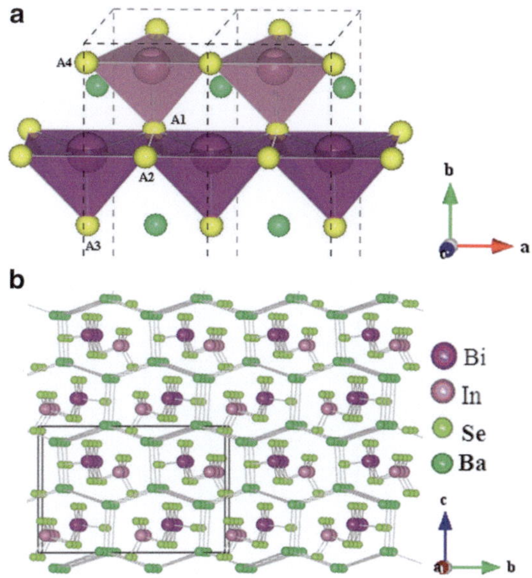

Fig. 14.3 The predicted crystal structure based on the evolutionary algorithm USPEX. (**a**) The predicted crystal structure shows the bonded chain of $BiSe_5$ and $InSe_4$ polyhedron and the *dash line* is the unit cell boundary. (**b**) Four adjacent cells of $Ba_2BiInSe_5$ along the *a*-axis and the *line* linking Ba atom is guide for eye

14.3 Theoretical Band Structures and Optical Properties for Ba_2BiInA_5 (A = S, Se, Te)

14.3.1 Calculations of Band Structures

The electronic structure calculations were performed with the plane-wave based on VASP code [33]. For the calculations on the stable structures of Ba_2BiInA_5 crystals, the plane-wave cutoff energy was set to be 300 eV, and the Monkhorst–Pack scheme *k*-point grid density of $4 \times 1 \times 2$ was used [32]. The obtained band structures reveal the transparent range of light wavelength, and they are taken as the input parameters of the optical simulation. The band structures of the Ba_2BiInA_5 homologue are very similar to each other. Figure 14.4 gives the plots of the band structure and

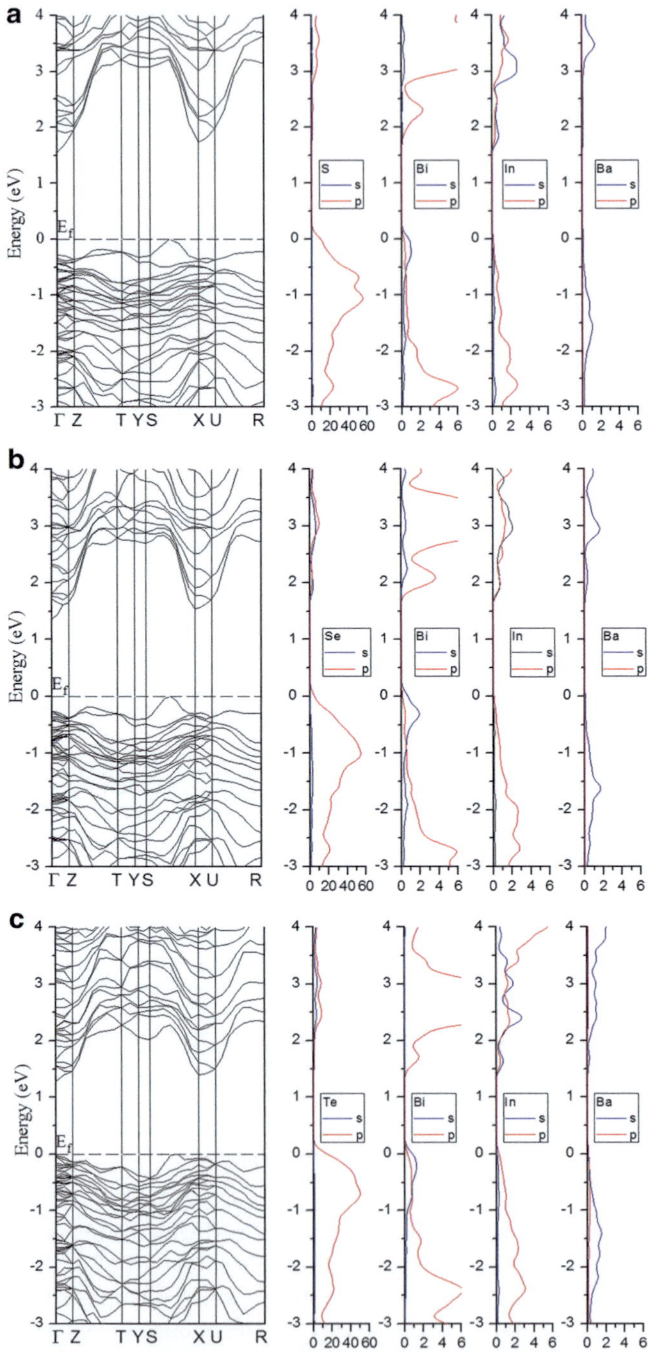

Fig. 14.4 The calculated energy band structure and projected density of state for Ba_2BiInA_5 (A = S (**a**), Se (**b**), Te (**c**))

Table 14.2 The calculated or experimental band gaps (eV) and optical parameters (*d*: pm/V) for Ba_2BiInA_5 (A = S, Se, Te) species

Species	DFT	GW_0	Expt.	ε_{xx}	ε_{yy}	ε_{zz}	d_{31}	d_{32}	d_{33}	FOM
A = S	1.267	2.372	1.550	7.22	7.23	7.87	13.7	13.7	16.6	10.4
A = Se	1.030	1.871	1.400	8.13	8.06	8.72	24.0	23.5	28.1	26.1
A = Te	0.668	1.282		10.21	10.17	11.62	74.9	74.2	98.6	191.0

projected density of state (DOS) for Ba_2BiInA_5 (A = S, Se, Te). The maximum of the valance band (MVB) is at the reciprocal *k* point (–0.25, 0.5, 0), the middle of the line from *S* point (–0.5, 0.5, 0.0) to *X* point (0.0, 0.5, 0.0). The lowest conduction band (LCB) is at the *Γ* point. The different location of MVB and LCB indicated that Ba_2BiInA_5 possesses an indirect band gap. The predicted band gap by DFT method is 1.27, 1.03, and 0.67 eV for Ba_2BiInA_5 (A = S, Se, Te), respectively. Compared to the available experimental values (1.55 eV for Ba_2BiInS_5 and 1.40 for $Ba_2BiInSe_5$), DFT values are underestimation for the band gap as expected. To predict a reasonable band gap range for the compound $Ba_2BiInTe_5$, we do the quasi-particle GW_0 (Green's function) type [36] calculation for these materials. As shown in Table 14.2, the GW_0 values are overestimation for the band gaps of Ba_2BiInS_5 and $Ba_2BiInSe_5$. We can take the DFT band gap as the lower bound and the GW_0 band gap as the upper bound, and the predicted band gap of $Ba_2BiInTe_5$ is in the range from 0.67 to 1.28 eV. Due to the lack of experimental band gap for $Ba_2BiInTe_5$ compound and the reason that in general the GW_0 result is more reliable than the DFT result, we select GW_0 band gap in the band structure and optical parameter calculations. Moreover, a more conservative estimation will be obtained using the gap of 1.28 eV. The experimental band gap or the GW_0 estimation band gap is used in the calculations of DOSs and optical properties, which is 1.55, 1.40, and 1.28 eV for Ba_2BiInA_5 (A = S, Se, Te), respectively. The decreasing of band gap from S to Te in this Ba_2BiInA_5 serial is due to the increasing metallic property of the chalcogenide element. The DOS analysis shows that the main component between −3.0 eV and Fermi energy is the chalcogenide atom *np* ($n = 3, 4, 5$) orbital (Fig. 14.4) and small sp composition from Bi, In, and Ba atoms. The top of the valence bands is mainly composed of A *np* orbital mixing with small Bi 6*s* orbitals for the Ba_2BiInA_5 (A = S, Se, Te), while for the LCB, it is mainly composed of Bi *p* orbital, In *sp* orbital, and some of A *np* orbital. The partial electron density (PED) from the Fermi level (0.0 eV) to −1.0 eV indicates the antibonding interactions between the Bi 6*s* and A *np* orbitals (A = S, Se, Te), and the bonding interactions between the Bi 6*s* and A *np* states localize at energy from −10.0 to −13.0 eV (Fig. 14.5). These findings provide the evidence of the Bi lone pair $6s^2$ electron participating interactions. It reveals that the Bi lone pair $6s^2$ electron is stereochemically active in this homologue, and the stereochemical activity reduces when the coordinate atoms around change from S to Se to Te.

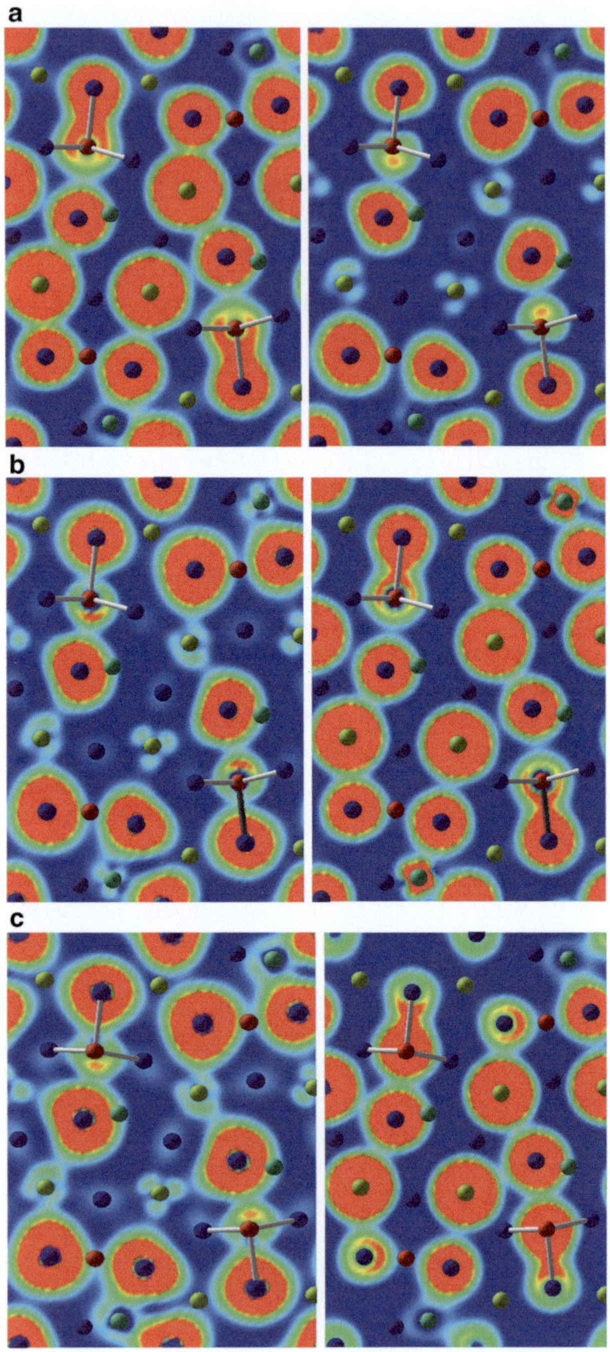

Fig. 14.5 The PED maps from −1.0 eV to the Fermi level (*left*) and from −10.0 to −13.0 eV (*right*); the electron density is represented from *blue* (0.0 e/Å^3) to *red* (0.08 e/Å^3) for compound Ba_2BiInA_5 (A=S (**a**), Se (**b**), Te (**c**))

14.3.2 Simulations of Optical Parameters

The calculations of linear optical properties in terms of the complex dielectric function $\varepsilon(\omega)=\varepsilon_1(\omega)+i\varepsilon_2(\omega)$ are made by using CASTEP package [37]. The imaginary part of the dielectric function ε_2 was given by

$$\varepsilon_2^{ij}(\omega)=\frac{8\pi^2\hbar^2e^2}{m^2V_{\text{eff}}}\sum_k\sum_{cv}(f_c-f_v)\frac{p_{cv}^i(k)p_{vc}^j(k)}{E_{vc}^2}\delta[E_{cv}(k)-\hbar w] \quad (14.1)$$

where $[E_{cv}(k)-\hbar\omega]=[E_c(k)-E_v(k)-\hbar\omega]$ denotes the energy difference between the conduction and valence bands at the k point with absorption of a quantum $\hbar\omega$. The f_c and f_v represent the Fermi distribution functions of the conduction and valence bands, respectively. The term $P_{cv}^i(k)$ denotes the momentum matrix element transition from the energy level c of the conduction band to the level v of the valence band at the k point in the Brillouin zones and V is the volume of the unit cell. The m, e, and $\hbar$ are the electron mass, charge, and Plank's constant, respectively. The first-order susceptibility and linear refractive index can be derived from the dielectric function. The first-order non-resonant susceptibility at low frequency region is given by $\chi^{(1)}(\omega)_{ii}=[\varepsilon(\omega)_{ii}-1]/4\pi$, and the linear refractive index $[n(\omega\rightarrow 0)]_{ii}^2=\varepsilon_1(\omega\rightarrow 0)_{ii}$. The experimental band gap or the GW_0 estimation band gap value is used in the ε calculation, which is 1.55, 1.40, and 1.28 eV for Ba_2BiInA_5 (A = S, Se, Te), respectively. The second-order susceptibilities can be expressed in terms of the first-order susceptibilities as follows [38]:

$$\chi_{ijk}^{(2)}(-\omega_3;\omega_1,\omega_2)=F^2\chi_{ii}^{(1)}(\omega_3)\chi_{jj}^{(1)}(\omega_1)\chi_{kk}^{(1)}(\omega_2) \quad (14.2)$$

where $F^{(2)}=ma/(N^2e^3)$ and it is equal to 0.50×10^{-6}, 0.59×10^{-6}, and 0.82×10^{-6} esu for Ba_2BiInA_5 (A = S, Se, Te), respectively. These expressions are derived from a classical anharmonic oscillator (AHO) model [38]. The m and e are, respectively, the electron mass and charge, and the parameter a characterizes the nonlinearity of the response and it's estimated as ω_0^2/d, where d is lattice constant and ω_0 is the atomic vibration frequency. The parameters ω_0 and d are roughly same for most solids. The SHG component d_{ij} is equal to half of the corresponding χ_{ijk} value for the consequence of historical convention. The first-order susceptibility expression derived from a classical AHO model [38] is employed to calculate the SHG coefficient d_{ij} of Ba_2BiInA_5 (A = S, Se, Te) materials. The $Cmc2_1$ space group of Ba_2BiInA_5 (A = S, Se, Te) belongs to crystal class square millimeter and has five nonvanishing SHG components. Under the restriction of Kleinman's symmetry, only three independent SHG components (d_{31}, d_{32}, and d_{33}) were considered. The calculated dielectric constants are listed in Table 14.2 for the Ba_2BiInA_5 (A = S, Se, Te) at input wavelength of 2.3 μm. The calculated d_{ij} components are also listed in Table 14.2 at a wavelength of 2.3 μm, respectively. The frequency-dependent d_{ij} components are plotted in Fig. 14.6. It is found from the calculated results that the nonlinear effects enhance as the A ionic radius increases or metallic characteristic strengthens for the Ba_2BiInA_5 (A = S, Se, Te) homologue, and the $Ba_2BiInTe_5$ has the largest SHG response in this homologue.

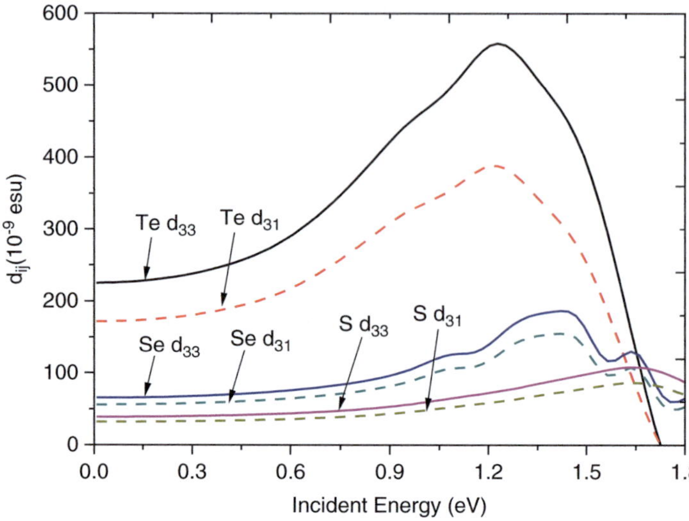

Fig. 14.6 The calculated frequency (energy)-dependent d_{ij} components for Ba_2BiInA_5 materials (A = S, Se, Te)

This is because the valance band is mainly composed of the chalcogen's *p* orbital (Fig. 14.4) of this serial compounds and the ionization potential (IP) values decrease in the order S, Se, and Te. Thus, the band gap decreases from Ba_2BiInS_5 to $Ba_2BiInTe_5$. According to (14.1) and (14.2), smaller band gap leads to larger second-order susceptibilities. The SHG coefficient of $Ba_2BiInTe_5$ is estimated to be 98 pm/V using the GW_0 band gap or even larger when using the DFT-underestimated band gap. It is often customary to describe the efficiency of SHG material in terms of a figure of merit (FOM) and it is defined as $FOM = d_{eff}^2/(n_\omega^2 n_{2\omega})$ [39]. Accordingly, the FOM is directly proportional to the effective nonlinear coefficient d_{eff} and inversely dependent on the refractive index of material. The calculated FOMs are also listed in Table 14.2 for the Ba_2BiInA_5 (A = S, Se, Te) homologue, in which the $\langle d_{ij} \rangle$ is approximate to d_{eff}. It is found that the FOM value or the NLO efficiency is increasing with anionic radius varying from S to Se to Te.

14.4 Experimental Evidences

14.4.1 *Syntheses and Measurement of Materials Ba_2BiInA_5 (A = S, Se)*

Compounds Ba_2BiInA_5 (A = S, Se) were prepared from stoichiometric amounts of Ba, Bi, In, and S or Se by a solid-state reaction method in evacuated silica tubes, respectively [31, 32]. The starting materials of Ba, Bi, In, and S or Se were mixed roughly and loaded into a graphite crucible which was subsequently placed into a predried silica tube and then sealed under 10^{-2} Pa, respectively. The sample including S was placed into a vertical furnace and heated to 973 K within 30 h (holding for 30 h), subsequently heated to 1,223 K at the rate of 10 K/h (holding for 30 h) and then slowly cooled down to room temperature at the rate of 2 K/h, and a lot of black slab-shaped Ba_2BiInS_5 was found in the tube. For the sample including Se, it was heated to 1,273 K within 50 h (holding for 30 h) by a temperature controller. The samples were then slowly cooled to 773 K over 150 h, followed by cooling to room temperature over 50 h. Bulk black crystals of $Ba_2BiInSe_5$ were obtained. Both of the compounds are stable in air and moisture conditions. X-ray energy dispersive spectroscopy (EDS, Oxford INCA) analysis on a field-emission scanning electron microscope (FESEM, JSM6700F) on each sample indicated a Ba/Bi/In/A molar ratio of 2.07:1.12:1.03:4.96 for A = S and 1.91:1.07:1.06:5.96 for A = Se, which are in agreement with their stoichiometric proportions from the results of single-crystal X-ray structure analyses.

Crystalline samples Ba_2BiInA_5 (A = S, Se) were selected for indexing and intensity data collection [31, 32]. Samples of the compounds Ba_2BiInA_5 (A = S, Se) were collected on Mercury CCD and Saturn724+ diffractometers equipped with graphite-monochromated Mo Kα radiation ($\lambda = 0.71073$ Å) at the temperature of 293 K, respectively. The Rigaku CrystalClear program was utilized for diffraction images collection and data procession. Lorentz and polarization corrections were also applied for data reduction. The crystal structures were solved by direct methods and refined by full-matrix least-squares on F^2 using SHELXL-97 [40]. Systematic absence conditions of the data suggest their possible space groups which were used for further structure solution and refinement. The final refined solution was checked by using the program PLATON, and no other symmetry elements were suggested [41]. Crystallographic data and structural refinements information are summarized in Table 14.3. The atomic coordinates and equivalent isotropic displacement parameters are listed in Table 14.4, and the important bond lengths are shown in Table 14.5. Further details of the crystal structure investigations may be obtained from the Supporting Information of refs. [31, 32].

14.4.2 Structural Features

Both of compounds Ba_2BiInA_5 (A = S, Se) crystallize in the orthorhombic space group $Cmc2_1$, and they are isostructural [31, 32]. Here, we only give the structural descriptions in detail for Ba_2BiInS_5. The crystal structure consists of an infinite one-dimension $_1\infty[BiInS_5]^{4-}$ anionic polymeric chains with charge-compensating Ba^{2+} ions arranged in six arrays around each $_1\infty[BiInS_5]^{4-}$ chain along the crystallographic *a*-axis as shown in Fig. 14.7a. The $_1\infty[BiInS_5]^{4-}$ chain is built up of an edge-shared BiS_5 tetragonal pyramid chain and a corner-shared InS_4 tetrahedra chain, and the two kinds of chains are further interconnected with each other through sharing corner to form the whole $_1\infty[BiInS_5]^{4-}$ anionic chain, shown in Fig. 14.7b. There are two crystallographically nonequivalent barium atoms, one

Table 14.3 Crystal data and structural refinement results for Ba_2BiInA_5 (A=S, Se)

Empirical formula	Ba_2BiInS_5	$Ba_2BiInSe_5$
f.w.	758.78	993.26
T (K)	293(2)	293(2)
Wavelength (Å)	0.71073	0.71073
Crystal system	Orthorhombic	Orthorhombic
Space group	$Cmc2_1$ (No. 36)	$Cmc2_1$ (No. 36)
a	4.2535(17)	4.328(4)
b	18.252(7)	19.117(16)
c	12.685(5)	13.181(10)
V (Å^3), Z	984.8(7), 4	1090.6(16), 4
ρ_{calc} (g cm^{-3})	5.118	6.049
μ (mm^{-1})	28.996	41.915
$F(000)$	1,296	1,656
θ ranges (deg)	$2.23 \leq \theta \leq 27.46$	$2.13 \leq \theta \leq 27.47$
Transmission, T_{min}/T_{max}	0.0519/0.0977	0.001/0.015
R_1[a], wR_2[b] ($I>2\sigma$)	0.0361, 0.0774	0.0297, 0.0506
R_1, wR_2 (all data)	0.0380, 0.0778	0.0411, 0.0547
GOF on F^2	1.019	0.648

[a] $$R_1 = \frac{\sum \left\| F_o \right| - \left| F_c \right\|}{\sum \left| F_o \right|}$$

[b] $$wR_2\left(F_o^2\right) = \left[\frac{\sum w\left(F_o^2 - F_c^2\right)^2}{\sum w\left(F_o^2\right)^2}\right]^{1/2}$$

Table 14.4 Atomic coordinates and equivalent isotropic displacement parameters U_{eq}[a]

Atom	x	y	z	U_{eq}
Ba_2BiInS_5				
Ba(1)	−1.0000	0.1294(1)	0.6749(1)	13(1)
Ba(2)	−1.0000	−0.1117(1)	0.5517(1)	12(1)
Bi(1)	0	0.2178(1)	0.3876(1)	17(1)
In(1)	−0.5000	0.0550(1)	0.3142(1)	16(1)
S(1)	−1.5000	0.0039(3)	0.6409(5)	13(1)
S(2)	−0.5000	0.2157(3)	0.5304(5)	14(1)
S(3)	0	0.3584(3)	0.3656(4)	12(1)
S(4)	−0.5000	0.1844(3)	0.2327(4)	16(1)
S(5)	−1.0000	0.0430(4)	0.4165(5)	34(2)
$Ba_2BiInSe_5$				
Ba(1)	−1.0000	0.6284(1)	0.5204(1)	14(1)
Ba(2)	−1.0000	0.3872(1)	0.6447(1)	13(1)
Bi(1)	0	0.2820(1)	0.3085(1)	16(1)
In(1)	−0.5000	0.4409(1)	0.3873(1)	13(1)
Se(1)	0	0.1413(1)	0.3302(1)	13(1)
Se(2)	−0.5000	0.3124(1)	0.4657(1)	15(1)
Se(3)	−0.5000	0.2837(1)	0.1632(2)	14(1)
Se(4)	−0.5000	0.5035(1)	0.5573(2)	12(1)
Se(8)	0	0.4556(1)	0.2772(1)	14(1)

[a]U_{eq} (Å$^2 \times 10^3$) is defined as one third of the trace of the orthogonalized U_{ij} tensor

Table 14.5 Selected bond lengths (Å) for compound Ba_2BiInA_5 (A = S, Se)

	Distances		Distances
Ba_2BiInS_5			
Ba(1)–S(2)	3.160(6)	Ba(2)–S(1)	3.638(7)
Ba(1)–S(5) × 2	3.203(4)	Bi(1)–S(4)	2.582(5)
Ba(1)–S(4) × 2	3.223(4)	Bi(1)–S(2) × 2	2.794(4)
Ba(1)–S(1)	3.304(7)	Bi(1)–S(3) × 2	2.959(4)
Ba(1)–S(3) × 2	3.399(4)	Bi(1)–S(1)	3.212(4)
Ba(2)–S(5) × 2	3.156(4)	In(1)–S(5)	2.447(6)
Ba(2)–S(2) × 2	3.219(4)	In(1)–S(1) × 2	2.501(4)
Ba(2)–S(4) × 2	3.229(4)	In(1)–S(3)	2.578(6)
Ba(2)–S(3)	3.476(6)		
$Ba_2BiInSe_5$			
Ba(1)–Se(1) × 2	3.321(2)	Bi(1)–Se(1)	2.706(3)
Ba(1)–Se(3) × 2	3.324(2)	Bi(1)–Se(3) × 21	2.890(2)
Ba(1)–Se(8) × 2	3.746(3)	Bi(1)–Se(2)	3.051(2)
Ba(2)–Se(3) × 2	3.276(4)	In(1)–Se(8) × 2	2.620(2)
Ba(2)–Se(1) × 2	3.310(2)	In(1)–Se(8)	2.620(2)
Ba(2)–Se(2)	3.507(2)	In(1)–Se(2)	2.665(3)

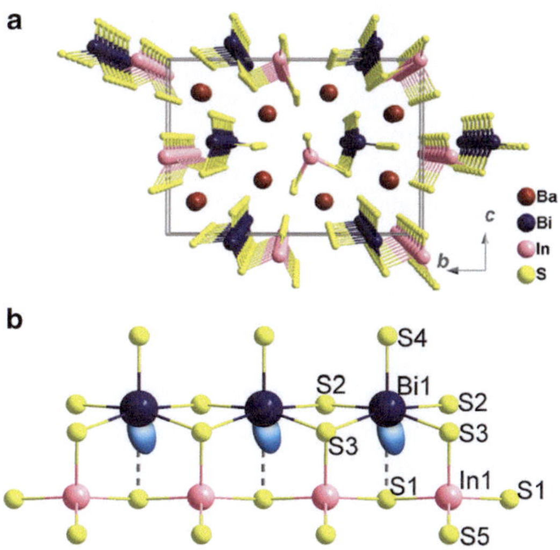

Fig. 14.7 (**a**) Crystal structure of compound Ba_2BiInS_5 viewed along the [100] direction; (**b**) 1D $_{1}\infty[BiInS_5]^{4-}$ chain in compound Ba_2BiInS_5. The *blue* lobes on Bi atoms represent the lone pairs

bismuth atom, one indium atom, and five sulfur atoms in the asymmetric unit. The two symmetrically nonequivalent Ba^{2+} cations in Ba_2BiInS_5 are both coordinated by eight S atoms with bond lengths in the range of 3.160(6)–3.399(4) Å for Ba(1) and 3.156(4)–3.638(7) Å for Ba(2). The bond lengths of Bi–S vary from 2.582(5) to 2.959(4) Å, featuring the irregular tetragonal pyramid coordination geometry. Considering the next neighboring S(1) atoms with bond length 3.212(7) Å, a seriously distorted octahedron indicates completeness. The unique In^{3+} cation in the asymmetric unit is coordinated by four S atoms in the tetrahedral arrangement with bond lengths distorted from 2.447(6) to 2.578(6) Å, while the S–In–S angles range from 92.39(19)° for S(1)–In(1)–S(4) to 116.5(3)° for S(5)–In(1)–S(5). One of the most interesting features about the compound Ba_2BiInS_5 is that the BiS_5 tetragonal pyramids and neighboring InS_4 tetrahedra are aligned with *cis* arrangement through sharing edge to form the *cis*-$_1\infty[BiInS_5]^{4-}$ chain structure (Fig. 14.7b). This arrangement leads to the Bi^{3+} lone pair electrons in the constructive alignment fashion within the *cis*-$_1\infty[BiInS_5]^{4-}$ chain.

14.4.3 Optical Properties

The infrared spectra of polycrystalline samples Ba_2BiInA_5 (A = S, Se) were recorded on a PerkinElmer Spectrum One FT-IR spectrometer in the range of 450–7,000 cm^{-1}, and they exhibit a wide infrared transmission range from near-IR up to 25 μm with no obvious optical absorption peaks, which suggests that both compounds may be useful in the mid-IR and even to far-IR ranges. The UV–vis–NIR diffuse reflectance spectra were carried out on a PerkinElmer Lambda 900 UV–vis–NIR spectrometer equipped with an integrating sphere over a 300–2,000 nm wavelength range at room temperature [31, 32]. The optical absorption spectra were converted from diffuse reflectance spectra using the Kubelka–Munk function, $\alpha/S = (1 - R)^2/2R$, where R is the diffuse reflectance, and R and S are the Kubelka–Munk absorption and scattering coefficients [42]. The UV–vis–NIR optical absorption spectra measurement results indicate that compounds Ba_2BiInA_5 (A = S, Se) are both semiconductors with optical band gaps of about 1.55 and 1.40 eV, respectively (Fig. 14.8).

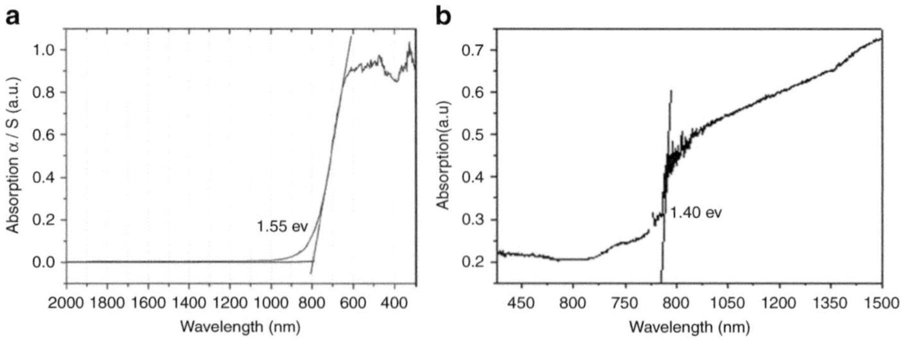

Fig. 14.8 UV–vis–NIR optical absorption spectra converted from diffuse reflectance spectra using the Kubelka–Munk equation for polycrystalline samples of (**a**) compound Ba_2BiGaS_5 and (**b**) compound $Ba_2BiInSe_5$

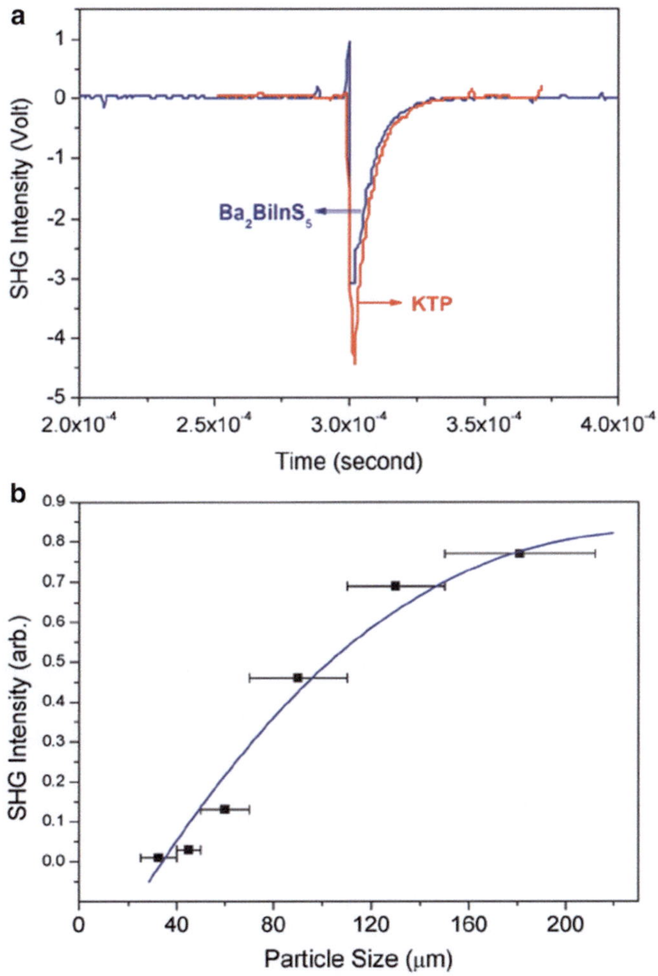

Fig. 14.9 (**a**) The SHG intensity (*blue line*) of a Ba_2BiInS_5 sample compared with that (*red line*) of KTP, and (**b**) the SHG intensity against the different particle sizes of Ba_2BiInS_5 samples (*blue line* is for eyeguide)

Powder SHG intensity was measured by means of the Kurtz–Perry method [43]. The powder samples of Ba_2BiInS_5 were irradiated with a 2.05 μm Q-switched laser, and the SHG signal at 1.025 μm was detected. It revealed that the SHG signal is similar to that of the KTP standard with a similar grain size (200 μm) at the incident wavelength of 2.05 μm (Fig. 14.9a) [31]. From the plots of SHG signal intensity vs. particle size of Ba_2BiInS_5 powders as shown in Fig. 14.9b, it is found that for the particle size less than 120 μm, the SHG intensity increases gradually with increasing particle size, and for the particle size larger than 120 μm, the SHG intensity is essentially independent of particle size. This feature suggests that compound Ba_2BiInS_5

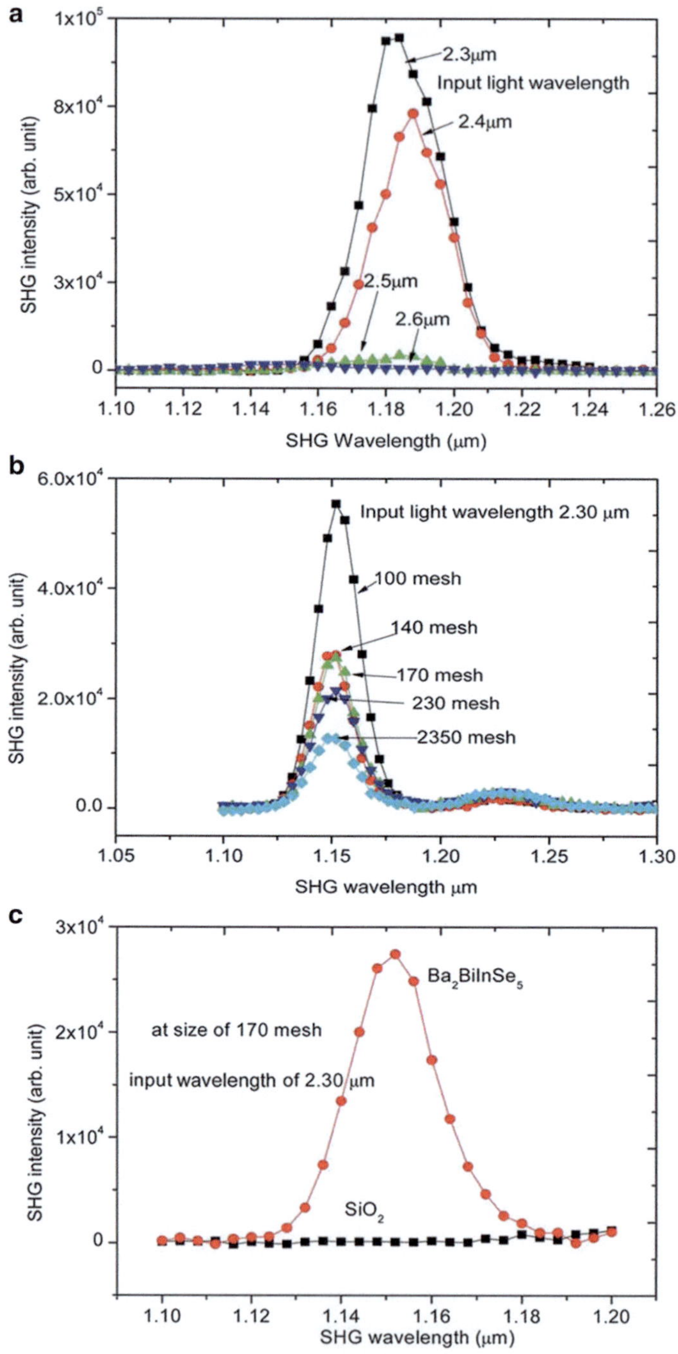

Fig. 14.10 The measurement SHG response of $Ba_2BiInSe_5$ by the Kurtz powder method. (**a**) The SHG signal intensity at the same size powder with incident wavelengths from 2.3 to 2.6 mm. (**b**) The SHG intensity at the same irradiation wavelength with different size of powder. (**c**) The relative SHG intensities of $Ba_2BiInSe_5$ and fused-SiO_2 powders at the same size

can achieve type I phase-matching which makes Ba_2BiInS_5 promising for practical applications. The SHG response of $Ba_2BiInSe_5$ was measured by the femtosecond tunable Ti:sapphire laser system (Model: Spitfire PRO F-XP; Manufacturer: Spectra Physics) [32]. The tunable wavelength of the idler beam from OPO (optical parametric oscillator) was used as the fundamental wavelength λ ranging from 2.30 to 2.60 μm, and the Kurtz powder method was also employed to determine the SHG intensity. The input (fundamental) light pulses were generated by an OPO with a pulse width of <120 fs and a repetition frequency of 1 kHz. Sieved powder samples of $Ba_2BiInSe_5$ with the same particle size were irradiated with wavelengths 2.30–2.60 μm generated by the OPO device. The SHG signal from 1.15 to 1.25 μm was detected at different incidence wavelengths, and it was found from Fig. 14.10a that the strongest SHG response of $Ba_2BiInSe_5$ is at an input wavelength of 2.30 μm. The SHG signal intensity vs. particle size of $Ba_2BiInSe_5$ powders are plotted in Fig. 14.10b at an irradiation wavelength of 2.30 μm. It was found that the SHG intensity increases gradually with increasing particle size (mesh reducing). This suggests that the $Ba_2BiInSe_5$ can achieve type I phase-matching conditions [43]. The relative SHG intensities of $Ba_2BiInSe_5$ and fused-SiO_2 powders are plotted in Fig. 14.10c at the same size mesh of 170 and irradiation wavelength of 2.30 μm. It estimates that the SHG intensity of $Ba_2BiInSe_5$ is about 230 times larger than SiO_2.

14.5 Conclusions

We conclude that a genetic engineering method was employed to design the SHG materials Ba_2BiInA_5 (A = S, Se, Te) through the road map of four steps. First, the virtual compounds were constructed in terms of gene optimization; then, the crystal structures of virtual compounds were predicted by evolutionary methodology coded in USPEX, and the Ba_2BiInA_5 (A = S, Se, Te) compounds are found to be NCS with $Cmc2_1$ space group; thirdly, the optical parameters were simulated in views of the band structures obtained from the ab initio calculations, and it is found that a FOM of $Ba_2BiInTe_5$ is the largest one among the Ba_2BiInA_5 (A = S, Se, Te) homologues; finally, the experimental evidences were provided by the syntheses and characterizations of crystal structures, and measurements of linear and nonlinear optical properties of Ba_2BiInA_5 (A = S, Se). We believe that material gene engineering can be developed to design new materials according to preset physical properties, such as the polar materials with pyroelectric or ferroelectric effect. Our work shows that experimental development guided by theoretical modeling reduces the need for time-consuming and expensive trial-experiments to characterize structures and properties of crystal materials.

Acknowledgments This investigation was based on work supported by the National Natural Science Foundation of China under project 20773131 and 21101156, the National Basic Research Program of China (No. 2007CB815307), and foundation of Fujian Provincial Key Laboratory of Theoretical and Computational Chemistry.

References

1. Hammond, C.R.: The Elements, in Handbook of Chemistry and Physics, 81st edn, pp. 4–1. CRC press, Boca Raton (2004). ISBN 0-8493-0485-7
2. USA 1540117, Gustav Giemsa: Manufacture of Bismuth Tartrates
3. Parnell, R.T.G., Surgeon-Commander, Royal Navy: Bismuth in the treatment of syphillis. J. R. Soc. Med. **17(War sect)**, 19–26 (1924). PMC2201253
4. Maile, F.J., Pfaff, G., Reynders, P.: Effect pigments—past, present and future. Prog. Org. Coat. **54**, 150–163 (2005)
5. Pfaff, G.: Special Effect Pigments: Technical Basics and Applications. Vincentz Network GmbH, Hannover (2008)
6. Tücks, A., Beck, H.P.: The photochromic effect of bismuth vanadate pigments: investigations on the photochromic mechanism. Dyes. Pigm. **72**, 163–177 (2007)
7. Müller, A.: Yellow Pigments, Coloring of Plastics: Fundamentals, Colorants, Preparations, pp. 91–93. Hanser, Munich (2003)
8. Maeda, H., Tanaka, Y., Fukutumi, M., Asano, T.: A new high-T_c oxide superconductor without a rare earth element. Jpn. J. Appl. Phys. **27**, L209–L210 (1998)
9. Subramanian, M.A., et al.: A new high-temperature superconductor: $Bi_2Sr_{3-x}Ca_xCu_2O_{8+y}$. Science **239**, 1015–1017 (1998)
10. Satterthwaite, C.B., Ure, R.: Electrical and thermal properties of Bi_2Te_3. Phys. Rev. **108**, 1164–1170 (1957)
11. Venkatasubramanian, R.: Recent Device Developments with Advanced Bulk Thermoelectric Materials. 3rd DOE Thermoelectric Application Workshop, Mar 2012
12. Catalan, G., Scott, J.F.: Physics and applications of bismuth ferrite. Adv. Mater. **21**, 2463–2485 (2009)
13. Choi, T., Lee, S., Choi, Y.J., Kiryukhin, V., Cheong, S.-W.: Switchable ferroelectric diode and photovoltaic effect in $BiFeO_3$. Science **324**, 63–66 (2009)
14. Spaldin, N.A., Cheong, S.-W., Ramesh, R.: Multiferroics: past, present, and future. Phys. Today. **63**, 38 (2010)
15. Lebeugle, D., Colson, D., Forget, A., Viret, M., Bonville, P., Marucco, J.F., Fusil, S.: Room-temperature coexistence of large electric polarization and magnetic order in $BiFeO_3$ single crystals. Phys. Rev. B. **76**, 024116–024118 (2007)
16. Son, J.Y., Kim, B.G., Kim, C.H., Cho, J.H.: Writing polarization bits on the multiferroic $BiMnO_3$ thin film using kelvin probe force microscope. Appl. Phys. Lett. **84**, 4971–4973 (2004)
17. Cheong, S.-W., Mostovoy, M.: Multiferroics: a magnetic twist for ferroelectricity. Nat. Mater. **6**, 13–20 (2007)
18. Fröhlich, R., Bohatý, L., Liebertz, J.: Die Kristallstruktur von Wismutborat, BiB_3O_6. Acta. Crystallogr. C. **40**, 343–344 (1984)
19. Hellwig, H., Liebertz, J., Bohaty, L.: Exceptional large nonlinear optical coefficients in the monoclinic bismuth borate BiB_3O_6 (BIBO). Solid. State. Commun. **109**, 249–251 (1998)
20. www.crylight.com/nlo_crystals/BIBO-Crystal_7.html
21. Zhang, W.-L., Cheng, W.-D., Zhang, H., Geng, L., Lin, C.-S., He, Z.-Z.: A strong second-harmonic generation material Cd4BiO(BO3)3 originating from 3-chromophore asymmetric structures. J. Am. Chem. Soc. **132**, 1508–1509 (2010)
22. Setter, N., et al.: Ferroelectric thin films: review of materials, properties, and applications. J. Appl. Phys. **100**, 051606–051646 (2006)
23. Halasyamani, P.S., Poeppelmeier, K.R.: Noncentrosymmetric oxides. Chem. Mater. **10**, 2753–2769 (1998)
24. Glazer, A.M., Stadnicka, K.: On the use of the term 'absolute' in crystallography. Acta. Crystallogr. **A45**, 234–238 (1989)
25. Hahn, T.: International Tables for Crystallography. D. Reidel, Dordrecht (1983)

26. Carriles, R., et al.: Imaging techniques for harmonic and multiphoton absorption fluorescence microscopy. Rev. Sci. Instrum. **80**, 081101–081123 (2009)
27. Campagnola, P.J., Clark, H.A., Mohler, W.A., Lewis, A., Loew, L.M.: Second-harmonic imaging microscopy of living cell. J. Biomed. Opt. **6**, 277–286 (2001)
28. Agullo-Lopez, F., Cabrera, J.M., Agullo-Rueda, F.: Electrooptics: Phenomena, Materials and Applications. Academic, New York (1994)
29. http://materialsinnovation.tms.org/genome.aspx
30. Fachinformationszentrum Karlsruhe, 76344 Eggenstein-Leopoldshafen, Germany (fax: (t49)7247_808_666; e-mail: crysdata@fiz-karlsruhe.de)
31. Geng, L., Cheng, W.-D., Lin, C.-S., Zhang, W.-L., Zhang, H., He, Z.-Z.: Syntheses and characterization of new mid-infrared transparency compounds: centric Ba_2BiGaS_5 and acentric Ba_2BiInS_5. Inorg. Chem. **50**, 5679–5686 (2011)
32. Lin, C.-L., Luo, Z.-Z., Cheng, W.-D., Zhang, H., Zhang, W.-L.: Designs of SHG materials with mid-infrared transparency based on genetic engineering for Ba_2BiInA_5 (A=Se, Te). J. Mater. Chem. **22**, 21713–21719 (2012)
33. Kresse, G., Furthmüller, J.: Efficient iterative schemes for ab initio total-energy calculations using a plane-wave basis set. Phys. Rev. B **54**, 11169–11186 (1996)
34. Oganov, A.R., Glass, C.W.: Crystal structure prediction using ab initio evolutionary techniques: principles and applications. J. Chem. Phys. **124**, 244704–244715 (2006)
35. Oganov, A.R., Lyakhov, A.O., Valle, M.: How evolutionary crystal structure prediction works—and why. Acc. Chem. Res. **44**, 227–237 (2011)
36. Shishkin, M., Kresse, G.: Implementation and performance of the frequency-dependent GW method within the PAW framework. Phys. Rev. B **74**, 035101–035113 (2006)
37. Segal, M.D., Lindan, P.J.D., Probert, M.J., Pickard, J., Hasnip, P.J., Clark, S.J.: First-principles simulation: ideas, illustrations and the CASTEP code. J. Phys. Condens. Matter. **14**, 2717–2744 (2002)
38. Boyd, R.W.: Nonlinear Optics, pp. 21–32. Academic, New York (1992)
39. Nie, W.: Optical nonlinearity: phenomena, applications, and materials. Adv. Mater. **5**, 520–545 (1993)
40. Sheldrick, G.M.: SHELXL-97. University of Göttingen, Göttingen (1997)
41. Spek, A.L.: Single-crystal structure validation with the program PLATON. J. Appl. Crystallogr. **36**, 7–13 (2003)
42. Wendlandt, W.W., Hecht, H.G.: Reflectance Spectroscopy. Interscience, New York (1966)
43. Kurtz, S.K., Perry, T.T.: A powder technique for the evaluation of nonlinear optical materials. J. Appl. Phys. **39**, 3798–3813 (1968)

Chapter 15
Bismuth(V)-Containing Semiconductor Compounds and Applications in Heterogeneous Photocatalysis

Xiaofeng Chang, Mohammed Ashraf Gondal, Zain Hassan Abdallah Yamani, and Guangbin Ji

Abstract Heterogeneous photocatalysis technology has become a promising approach to solve energy and environmental problems since 1972. The exploration and search for active semiconductor photocatalysts specially for solar energy applications is one of the most challenging tasks. The bismuth(V)-containing photocatalysts have gained great significance due to their unique electronic/energy band structures. The preparation techniques, physical–chemical properties, photocatalytic activities, and photostabilities of the bismuth(V)-containing photocatalysts (as represented by ilmenite-type $NaBiO_3$) are reviewed in this chapter. Some of these photocatalysts have excellent catalytic performance in water purification, disinfection of water as well as splitting of water for H_2 production. But nevertheless, their instabilities under light exposure during photocatalysis process have been gradually realized and further studied. The trend of bismuth(V)-containing photocatalyst research is also prospected in the end of this chapter.

X. Chang (✉) • G. Ji
Department of Applied Chemistry, College of Materials Science and Technology, Nanjing University of Aeronautics and Astronautics, Nanjing 211100, China
e-mail: changxf@nuaa.edu.cn; gbji@nuaa.edu.cn

M.A. Gondal (✉)
Laser Research Group, Physics Department, Center of Excellence in Nanotechnology, King Fahd University of Petroleum and Minerals, Dhahran 31261, Saudi Arabia
e-mail: magondal@kfupm.edu.sa

Z.H.A. Yamani
Center of Excellence in Nanotechnology, King Fahd University of Petroleum and Minerals, Dhahran 31261, Saudi Arabia
e-mail: zhyamani@kfupm.edu.sa

H. Li and Z.M. Wang (eds.), *Bismuth-Containing Compounds*, Springer Series in Materials Science 186, DOI 10.1007/978-1-4614-8121-8_15,

15.1 Introduction

15.1.1 Fundamental of Heterogeneous Photocatalysis

Heterogeneous photocatalysis process has become a promising technology to solve energy and environmental problems since 1972 [1]. The dream for "low-carbon and green life," as well as the pioneer achievements by scholars [2–4] in this field, is inspiring and instigating great enthusiasm to develop highly efficient photocatalysts to improve the quantum efficiency under visible light (380–780 nm) which represents around 40 % of the total light spectrum of the solar radiations [5].

The principle of semiconductor photocatalysts is illustrated in Fig. 15.1 [6, 7]. An electron–hole pair is, respectively, generated in the conduction and valence bands (CB, VB) after absorption of the photon with photo energy equal or higher than the band gap value (E_g) of semiconductor. The photogenerated electron and hole could immigrate to the reactive surface sites followed by reduction process (reduce an acceptor) or oxidation process (oxidize a donor), respectively (process c and d in Fig. 15.1).

In a typical photocatalyst, the surface-bound $^{\bullet}OH_{ads}$ radical is produced through the oxidation of OH^- (a.q.) which can effectively oxidize the adsorbed pollutants. The CB of the semiconductor has a certain potential and hence the electrons possess certain reduction capacity which depends on the band edge potential. The photogenerated electron are picked up by the oxygen thus producing peroxy radical, hydroxyl radical, and a series of species possessing strong oxidation capacity, which can oxidize the organics present in water.

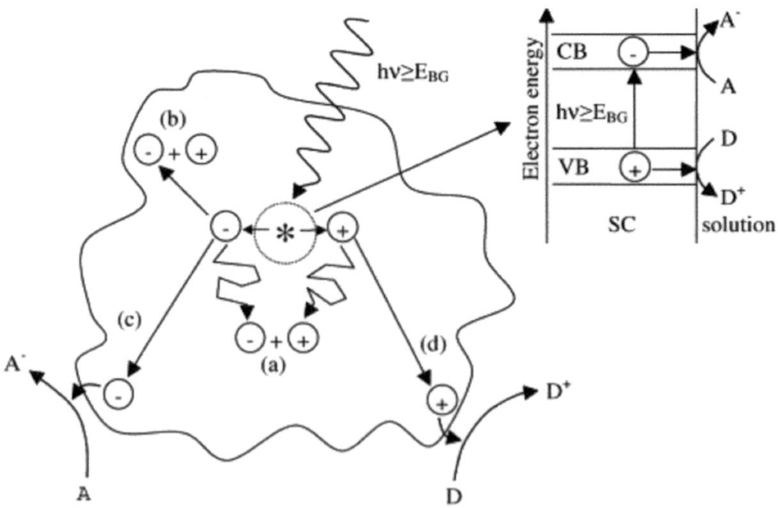

Fig. 15.1 Schematic illustration of the main processes on an irradiated photocatalyst particle (reprinted from ref. [7])

Such photocatalytic process can be adopted in splitting of water for H_2 production and CO_2 reduction as well [8, 9], but requires photogenerated electrons with higher energy than the electrochemical potential of H_2 evolution and CO_2 conversion. However, the practical quantum efficiency for CO_2 reduction during a photocatalytic process is very low (usually less than 1 %), because of the possible process such as the recombinations of the carries by the attractive force between them (process a in Fig. 15.1) and by the defects in the semiconductor, as well as the trapping in reactive surface sites which might take 10–100 ns for holes and some hundreds of picoseconds for electrons [6] (process b in Fig. 15.1). But recently, it has been realized that the low solar photocatalytic efficiency is not associated with light absorption but more likely with the charge transfer kinetics [9]. Therefore, the exploration and search for active semiconductor photocatalysts remains to be one of the most challenging tasks for solar energy utilization.

15.1.2 Motivations for Developing Bismuth(V)-Containing Semiconductor

In an oxide semiconductor, it is well known that the metal orbits which contribute to the CB, determines the band gap of the semiconductor, as the band edge potential of the VB (which is composed of O_{2p} states) has the fixed energy at around 3 eV. For example, the CV and VB of anatase TiO_2, which is thought to be one of the most typical oxide semiconductor, is composed of Ti_{3d} (at approximately −0.2 eV) and O_{2p} (at approximately 3 eV), therefore the band gap energy of anatase TiO_2 is 3.2 eV [10]. According to the Planck's equation, anatase TiO_2 could work under UV light with wavelength shorter than 387 nm. It should be noted that only around 4 % of the UV exposure can reach to the surface of the earth because of the blocking effect by the ozone layer in the upper atmosphere. However, the visible light which spans 380–780 nm represents around 40 % of the total light spectrum of the solar radiations (and the remainder is infrared which accounts for ~56 %). Thus exploration and development of visible-light response photocatalysts attract a lot of attention of research community worldwide (IR-response semiconductor is not considered to be working efficiently because of small band gap which may not facilitate the separation of photogenerated carries).

On the basis of the abovementioned facts, the substitution of metal atom in the semiconductor is essential to control the band gap energy. Exploring metal cations filled with s^2 might be a suitable strategy for the development of visible-light response photocatalysts because of the following advantages (as shown in Fig. 15.2): (1) the broad hybrid orbital composed of metal cations filled with s^2 and O2*p* can have "narrow" the band gap of the corresponding semiconductor; (2) as the *ns* orbitals of such metal cations could form the top of VB by overlapping with O2*p* orbital thus increase the VB width so that improve the photogenerated hole mobility and lifetime [11].

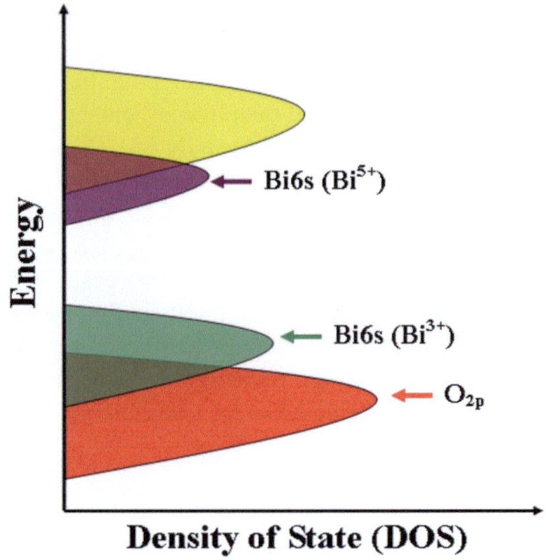

Fig. 15.2 Schematic diagram showing the electronic structure of Bi^{3+}/Bi^{5+}-modified semiconductor oxide (adapted from ref. [11])

Previously, the Bi(III) contained oxide semiconductor has been widely investigated as the broad hybrid orbital composed of Bi_{6s} and O_{2p}. The preparation and photocatalytic activity of such catalysts, like $BiVO_4$ [12], Bi_2WO_6 [13, 14], and $CaBi_2O_4$ [15] photocatalysts, have been widely studied. The vast majority of electrons in these Bi(III) are in the total movement to become a part of the VB of the respective semiconductor. However, compared to those empty *d*-orbital of the transition metal cations, the empty 6*s* orbital has a much lower energy, and Bi^{5+} ion with the 6*s* orbital has a significant impact on the CB of the corresponding semiconductor. The research progress of Bi(V)-containing semiconductors and their application in heterogeneous photochemistry is comprehensively reviewed.

15.2 Microstructure and Optical Property

15.2.1 Microstructure

15.2.1.1 Ilmenite-Type

There are two kinds of ilmenite-type oxides with formula of ABO_3 in different combinations of metal cations, i.e., $A^{2+}M^{4+}O_3$ and $A^{+}M^{5+}O_3$. The weak magnetic iron titanium oxide ($FeTiO_3$) crystal is a typical ilmenite-type oxide which belongs to the former combination of $A^{2+}M^{4+}O_3$. Two typical Bi(V)-containing semiconductors of $NaBiO_3$ and $AgBiO_3$ belong to the latter one of $A^{+}M^{5+}O_3$ which is not as common as the former one. The reports on crystal structure characterization of $AgBiO_3$ are rare.

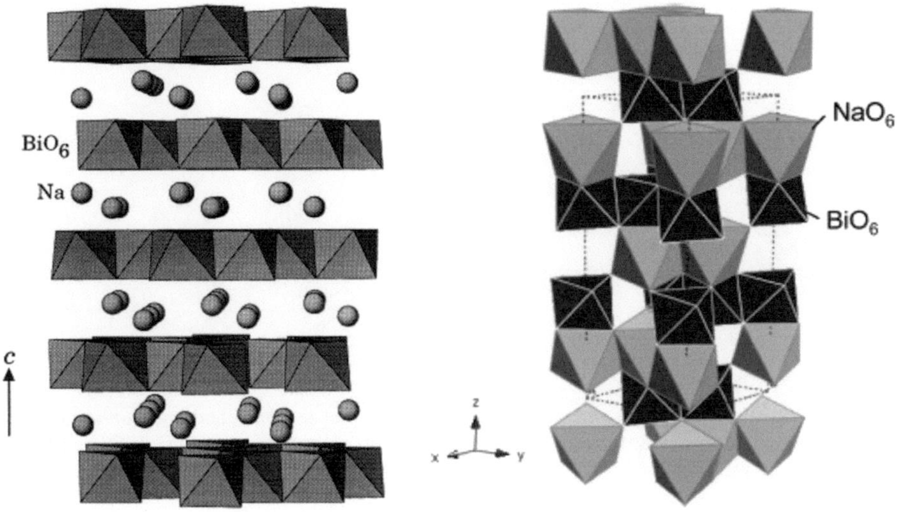

Fig. 15.3 Crystal structure of $NaBiO_3$ from different view directions (reprinted from refs. [16, 17])

By using powder neutron diffraction (PND) data, the $NaBiO_3$ and $AgBiO_3$ structures were refined as ilmenite-type under an ambient condition and the details of $NaBiO_3$ crystal structure were described by Kumada et al. [16], that is, the NaO_6 (AgO_6) octahedra and BiO_6 octahedra are stacked alternately along the c axis in a layered structure (as displayed in Fig. 15.3).

The chemical formula of commercial $NaBiO_3$ (such as Wako Pure Chemical Industries Ltd., Japan) has been confirmed as $NaBiO_3{\cdot}2H_2O$ by X-ray diffraction (XRD) method. A few investigations clearly proved that the hydrated $NaBiO_3$ changes to the ilmenite-type structure after dehydration over 140 °C [18, 19], and it can be found that the strong XRD peak at ~12 °C (Cu Kα irradiation) disappears after dehydration. Recently, Dias et al. [20] further determined the crystal structure of $NaBiO_3$ by measuring the phonon spectra via Raman and infrared spectroscopies. The conducted group theory calculations confirm the trigonal *R*3 to be the most possible space group to describe such perovskite oxides but with highly distorted structure because of the very small tolerance factor ($t=0.793$) [20]. By means of density functional theory (DFT), it was predicted by Liu et al. [21] that the phase of $NaBiO_3$ can be transformed from ilmenite phase (R-3 No. 148) to the orthorhombic phase (Pnma No. 62) under hydrostatic pressure.

15.2.1.2 Perovskite Structure ($BaBiO_3$)

In a typical perovskite oxide structure with the chemical formula of ABO_3, the A and B sites cations often consist of an alkaline-earth metal element (a rare earth element) and transition-metal element, respectively. It is well known that an ideal

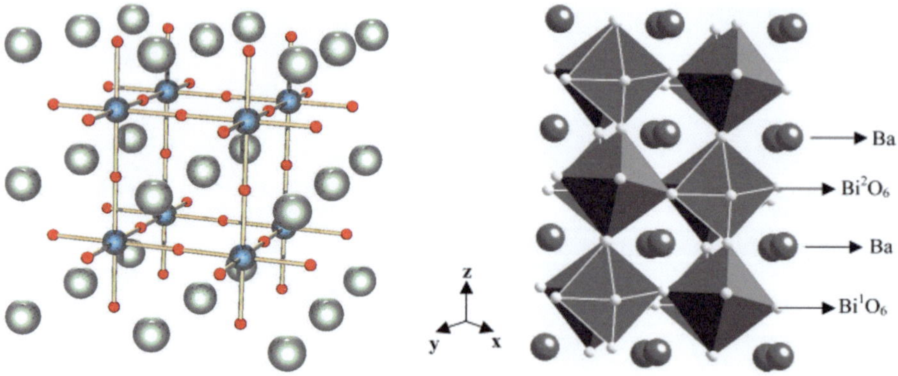

Fig. 15.4 *Left*: Ideal cubic perovskite structure for ABO_3 (the *red*, *blue*, and *green spheres* represent O-atoms, B-atoms, and A-atoms, respectively); *right*: crystal structure of $BaBiO_3$ perovskite (reprinted from ref. [22, 23])

perovskite structure should have stoichiometry as well as cubic structure consisting of corner-sharing BO_6 octahedra [22], as shown in Fig. 15.4. The structural stability of perovskite oxides can be evaluated by calculating the Goldschmidt's tolerance factor (t) which describes the deviation of the perovskite cubic structure,

$$t = \frac{r(A) + r(O)}{\sqrt{2}(r(B) + r(O))}$$

where $r(A)$, $r(B)$, and $r(O)$ are ionic radii of the respective ions but the Shannon's ionic radii have been widely employed for the tolerance factor determination. An ideal cubic structure can be expected if $0.79 < t < 1.1$ [3]. It is worth noticing that in the case of $BaBiO_3$ semiconductor, the radii of Bi ions are 1.03 and 0.76 Å for Bi^{3+} ion (with coordination number of 6) and Bi^{5+} ion (with coordination number of 6) [24]. According to the crystal structure characteristic as reported by Tang et al. [23], the Goldschmidt's tolerance factor of $BaBiO_3$ can be calculated at 0.904, indicating the relative stability of such $BaBiO_3$ oxides.

The thermal stability of such perovskite oxide has been studied systematically as well. The crystal phase of $BaBiO_3$ transforms from monoclinic at room temperature to rhombohedral at 300 °C and to cubic at ~500 °C. Further temperature increasing from 600 to 1,015 °C leads to the oxygen lost in $BaBiO_3$ and the formation of oxygen-deficient phases such as $BaBiO_{2.88}$, $BaBiO_{2.83}$, $BaBiO_{2.75}$, and $BaBiO_{2.55}$ [25].

The X-ray photoelectron spectrum of the $Bi4f_{7/2}$ levels in $BaBiO_3$ was examined by Kostikova et al. as well [26]. It was found that the XPS of $Bi4f_{7/2}$ levels contains two lines at 158.4 and 157.5 eV, respectively, indicating the ordering of trivalent and pentavalent bismuth atoms in $BaBiO_3$.

The Fourier transform-infrared spectroscopy (FT-IR) characterization is an efficient way to reveal the ordering existence of Bi^{3+}–O^{2-}–Bi^{5+} cross-linkages in $BaBiO_3$ crystal. The IR absorption by both the Bi^{3+}–O^{2-} bonds and Bi^{5+}–O^{2-} bonds are

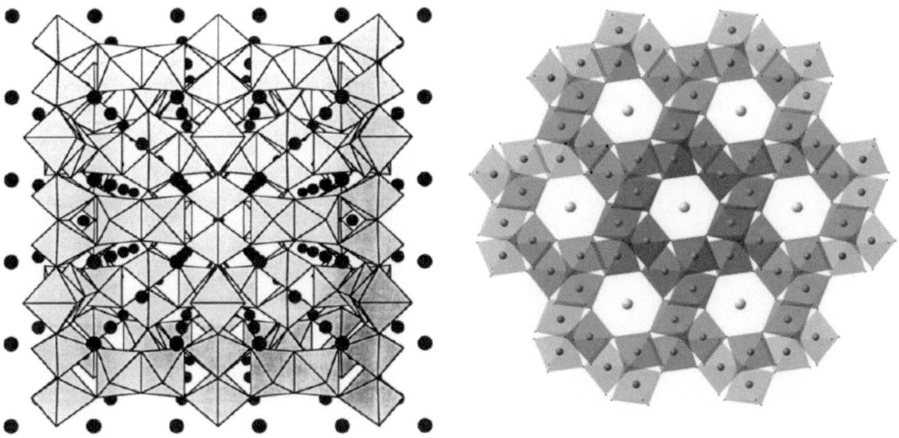

Fig. 15.5 Crystal structure of $KBiO_3$ (*shaded circles* represent potassium atoms located along the [111] direction, reprinted from [29]); *right*: oxygen atoms are at the corner of the octahedron around Bi^{5+} while K^+ cations reside within the tunnels, reprinted from ref. [30]

almost the same because of the indiscernible valence states of Bismuth cations in IR spectroscopy [27]. Yaremchenko et al. [28] detected only one strong IR absorption band centered at 467 nm in their $BaBiO_3$ sample, which is attributed to the stretching vibration of the Bi–O bonds coming from the Bi^{3+}–O^{2-}–Bi^{5+}. However, the band at 467 nm disappeared and an evolving band can be observed at 535–536 cm^{-1} after substituting Bi^{3+} with La^{3+}, which is thought to be the Bi^{5+}–O^{2-} bonds coming from Bi^{3+}–O^{2-}–La^{3+} bond. Meanwhile, after substituting Bi atom with Pr^{4+} in $BaBiO_3$, two absorption bands can be detected at 458–567 and 525–543 cm^{-1}, which is attributed to the Bi–O stretching vibrations in Bi^{3+}–O^{2-}–Bi^{5+} and Bi^{5+}–O^{2-}–Pr^{4+}, respectively. The similar trend was observed by Lakshminarasimhan et al. [11] as well. It was found that new IR absorption bands span from 540 to 520 cm^{-1} appeared after substituting Bi^{3+} with La^{3+}, which is considered as the district stretching vibration of O^{2-}–Bi^{5+} in $Bi^{5+}O_6$ octahedron.

15.2.1.3 Tunnel Structures

$KBiO_3$ and $LiBiO_3$ exhibit tunnel structures and isostructure with $KSbO_3$, which are different from ilmenite oxides of $ABiO_3$(A=Na, Ag). However, both the $KBiO_3$ and $LiBiO_3$ with tunnel structures and ilmenite-type oxides of $ABiO_3$(A=Na, Ag) are formed by edge-sharing octahedra. Such structures consist of BiO_6 octahedra pairs which are edge-shared to form Bi_2O_{10} clusters, resulting in the formation of tunnel structure (as shown in Figs. 15.5 and 15.6). For $KBiO_3$, the potassium ion partially occupies three sites in the tunnel structure. Thermal gravimetric (TG) analysis results indicated that a sudden weight loss takes place above 500 °C, suggesting the decomposition of $KBiO_3$ into K_2O and Bi_2O_3 [29]. The Goldschmidt's tolerance

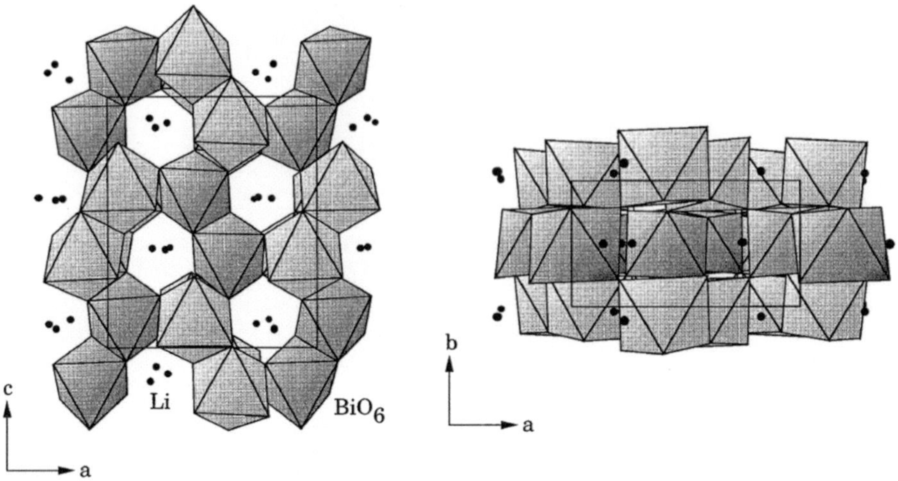

Fig. 15.6 Crystal structures of $LiBiO_3$ with different view directions (reprinted from ref. [31])

factor (t) of $KBiO_3$ and $LiBiO_3$ was calculated at 0.867 and 0.710 (as shown in Fig. 15.7), respectively, indicating that only $KBiO_3$ can be considered as a stable perovskite oxide.

Up to now, four kinds of lithium bismuth oxides containing pentavalent bismuth have been discovered, that is, Li_3BiO_4 [32], Li_7BiO_6 [33], Li_5BiO_5 [34], and $LiBiO_3$ [31]. Among them, $LiBiO_3$ has attracted more attention due to its potential applications in heterogeneous photocatalysis. By means of neutron powder diffraction data, the structure of $LiBiO_3$ oxides with high purity was characterized and reported for the first time in 1996. The as-prepared $LiBiO_3$ crystal which crystallizes in the orthorhombic system was found to have the isostructure with $LiSbO_3$, because of the fact that array of hexagonally close-packed oxygen atoms with cations occupying two-thirds of the octahedral sites can be observed in both structures. Thermal gravimetric and differential thermal analysis (TG-DTA) results show that $LiBiO_3$ is decomposed to $LiBiO_2$ after heating up to 300 °C [31].

15.2.1.4 Trirutile and Lead Antimonite Structure [35]

Figure 15.8 depicts the trirutile and lead antimonite ($PbSb_2O_6$) structure with the same chemical formula of ABi_2O_6 where A is a divalent cation. In trirutile structure (such as $MgBi_2O_6$ and $ZnBi_2O_6$), the two crystallographic sites with similar chemical environments are for the oxygen atoms which are coordinated to two Bi cations and one A cation (such as Mg^{2+} and Zn^{2+}), meanwhile retaining the planar coordination structure. The trirutile structure has the same space group with rutile ($P4_2/mnm$) but has a tripled c-axis in the unit cell.

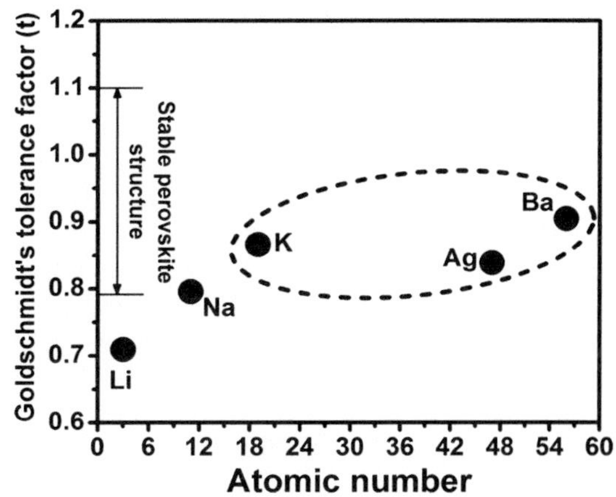

Fig. 15.7 Calculated Goldschmidt's tolerance factors of various pentavalent bismuthates

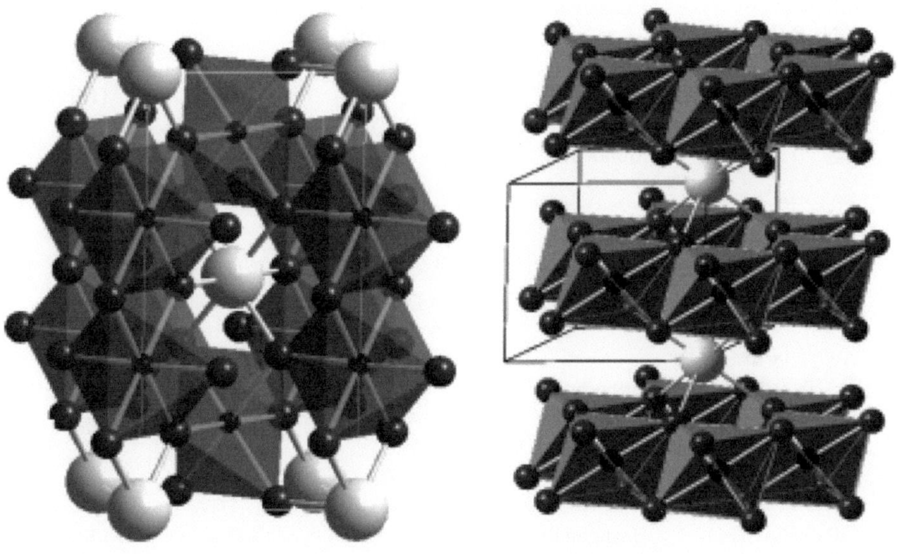

Fig. 15.8 Crystal structures of the trirutile oxides (*left*) and lead antimonite-type oxides (*right*). The A and Bismuth cations are shown as *larger light gray spheres* and *smaller black spheres* within the octahedra, respectively. The oxide ions are shown as *smaller spheres* at the vertices of the octahedra (reprinted from ref. [35])

In $PbSb_2O_6$ type structure (such as $SrBi_2O_6$, $BaBi_2O_6$, and $PbBi_2O_6$), planar layers are composed of edge-sharing BiO_6 octahedra, where two out of every three octahedral holes are filled. One out of every three octahedral holes in adjoining layers is also occupied by A cations (such as Sr^{2+}, Ba^{2+}, and Pb^{2+}) and the octahedral holes are filled by A cations by sitting above and below the vacant octahedral site in the Bi_2O_6 layers (Table 15.1).

Table 15.1 Reported crystal data for typical Bi(V)-containing crystals

Formula	Crystal system	Lattice parameters	Characterization method	References
$LiBiO_3$	Orthorhombic	$a=8.8278(3)$ Å $b=4.9135(2)$ Å $c=10.6914(3)$ Å	PND	[31]
Li_5BiO_5	Monoclinic	$a=9.9640(2)$ Å $b=4.1480(2)$ Å $c=5.9409(2)$ Å $\alpha=\beta=109.49(2)°$	TOF-ND[a], XRD	[34]
$NaBiO_3$	Hexagonal	$a=5.5672(4)$ Å $c=15.989(1)$ Å	PND	[16]
α-Na_3BiO_4	Monoclinic	$a=5.87(1)$ $b=6.69(6)$ $c=5.65(0)$ $\beta=109.8°$	XRD	[36]
$Na_{0.75}Bi_{0.25}O_4$ (β-Na_3BiO_4)	Hexagonal	$a=3.32141(9)$ Å $c=16.4852(5)$ Å	HR-XRD	[36]
$KBiO_3$	Cubic	$a=10.0194(6)$ Å	XRD	[29]
$AgBiO_3$	Hexagonal	$a=5.641(1)$ Å $c=16.118(2)$ Å	PND	[16]
$BaBiO_3$	Monoclinic	$a=6.19$ Å $b=6.15$ Å $c=8.68$ Å $\beta=90.16°$	XRD	[23]
$BaBi^{3+}{}_{0.5-x}La^{3+}{}_xBi^{5+}{}_{0.5}O_3$ ($x=0$, 0.1 and 0.3)	Monoclinic	6.18 Å $\leq a \leq$ 6.21 Å 6.12 Å $\leq b \leq$ 6.18 Å 8.64 Å $\leq c \leq$ 8.72 Å $89.8° \leq \beta \leq 90.2°$	XRD	[11]
$Na(Bi_{0.08}Ta_{0.92})O_3$	Orthorhombic	$a=5.4900$ Å $b=7.7741$ Å $c=5.5029$ Å	XRD	[37]
$Na_{0.04}$ $Sr_{0.48}BiO_3{\cdot}0.48H_2O$	Hexagonal	$a=5.536(2)$Å $c=55.371(2)$Å	XRD TG-DTA	[38]
$Na_{1-2x}Ba_xBiO_3{\cdot}nH_2O$ ($0.11 \leq x \leq 0.36$)	Hexagonal	0.564 Å $\leq a \leq$ 5.54 Å 16.03 Å $\leq c \leq$ 16.85 Å	XRD TG-DTA	[38]
$MgBi_2O_6$	Tetragonal	$a=4.826$ Å $c=9.719$ Å	XRD	[39]
$ZnBi_2O_6$	Tetragonal	$a=4.84135(6)$ Å $c=9.7506(2)$ Å	XRD	[39]

[a]Time-of-flight neutron diffraction

15.2.2 Optical Absorption Performance

Since it is difficult to measure the transmission spectroscopy of metal oxide materials, diffuse reflectance spectroscopy (DRS) has been widely adopted to evaluate the optical absorption properties. The optical absorption near the absorption edge for a semiconductor usually obeys the following equation [40]:

$$K(hv - E_g)^n = ahv$$

in which h, ν, α, and K are the Planck constant, light frequency, absorption coefficient, and constant, respectively. The exponent n depends on the transition involved and is defined for crystalline semiconductors as follows:

$n = 1$ for allowed direct transitions
$n = 3/2$ for forbidden direct transitions
$n = 2$ for allowed indirect transitions
$n = 3$ for forbidden indirect transitions

The values of n can be confirmed by plotting $\ln(\alpha hv)$ vs. $\ln(hv - E_g)$ ($\lg(\alpha hv)$ vs. $\lg(hv - E_g)$, using the approximate E_g value) because n is the slope of the straightest (tangent) line near the absorption edge. Therefore the band gap (E_g) of as-prepared semiconductor can be determined by plotting $(\alpha hv)^{1/n}$ vs. (hv) as the E_g is the intercept on the hv axis. The reported E_g data of various Bismuth(V)-containing semiconductors are summarized in Table 15.2.

15.3 Preparation Technique

15.3.1 Ion-Exchange Reaction

Ion-exchange (IE) process has been employed to prepare various pentavalent bismuthates with chemical formula of ABO_3, by using ilmenite-type $NaBiO_3$ as the ion exchanger which is a kind of cation exchanger that can exchange positively charged ions such as Ag^+, Li^+, Sr^{2+}, and Ba^{2+}.

Kumada et al. [16, 18, 31] carried out a variety of investigations on the synthesis and microstructure characterization of ilmenite-type pentavalent bismuthates (such as $AgBiO_3$ and $LiBiO_3$). It was found that $AgBiO_3$ can be prepared during one day through a very simple IE process at 70 °C and ambient pressure by using $NaBiO_3$ and $AgNO_3$ as the ion exchanger and silver cation source (with a molar ratio of Ag/Bi = 2–4), respectively. Recently, Yu et al. [45] used similar reaction process but different molar ratio of Ag/Bi (≈0.93), reaction temperature (R.T.), it was found that the black colored solid which is composed of metallic silver and $Bi_2O_2CO_3$ was produced very quickly (complete in seconds), indicating that molar ratio of Ag/Bi, where reaction temperature and reaction time may play an important role in the preparation of $AgBiO_3$.

$LiBiO_3$ oxides with ilmenite structure was once tried to be prepared by ion-exchange method in 1993 by stirring $NaBiO_3 \cdot nH_2O$ in 3 M LiCI solution at 60 °C; however, it was not very successful because the ion-exchanged products were partially decomposed to Bi_2O_3 in the reaction. However, after 3 years, the preparation of $LiBiO_3$ compound with high purity was reported by this group. $LiBiO_3$ crystals with similar structure of $LiSbO_3$ were successfully prepared through

Table 15.2 The transition type and determined band gap values of Bismuth(V)-containing semiconductors

Semiconductor	Transition type	E_g/eV	References
$LiBiO_3$	N.A.	1.63	[10]
	N.A.	1.8	[30]
	N.A.	1.7	[41]
$NaBiO_3$[a]	N.A.	2.6	[17]
	N.A.	2.53	[10]
	Indirect	2.45	[19]
	N.A.	2.7	[35, 42]
$NaBiO_3{\cdot}nH_2O$	Indirect	2.62	[43]
	Direct	2.71	
$NaBiO_3{\cdot}nH_2O$[b]	Indirect	1.92	
	Direct	2.26	
Na_xBiO_3	Indirect	1.50	
	Direct	1.77	
Na_xBiO_3[c]	Indirect	1.45	
	Direct	2.03	
$KBiO_3$	N.A.	2.04	[10]
	N.A.	2.1	[30]
$KBiO_3{\cdot}1.45H_2O$	Indirect	1.9	[43]
	Direct	2.1	
$KBiO_3{\cdot}1.45H_2O$[d]	Indirect	1.82	
	Direct	2.15	
$AgBiO_3$	N.A.	0.87	[10]
	N.A.	0.8	[35]
	Indirect	2.5	[44]
$BaBiO_3$	Indirect	2.05	[23]
$Na(Bi_{0.08}Ta_{0.92})O_3$	N.A.	2.88	[37]
$MgBi_2O_6$	N.A.	1.61	[10]
	N.A.	1.8	[35, 39, 42]
$SrBi_2O_6$	N.A.	1.93	[10]
	N.A.	2.0	[35, 42]
$BaBi_2O_6$	N.A.	1.93	[10]
	N.A.	2.6	[35, 42]
$ZnBi_2O_6$	N.A.	1.53	[10]
	N.A.	1.7	[35, 39, 42]
$PbBi_2O_6$	N.A.	1.92	[10]

[a]After heating commercialized $NaBiO_3{\cdot}2H_2O$ at 175 °C for 2 h
[b]After acidification
[c]After acidification
[d]After the acidification

hydrothermal-induced ion-exchange process, using $NaBiO_3$ and LiOH as the raw materials. A few current investigations clearly show that various pentavalent bismuthates with the chemical formula of both ABO_3-like (such as $Na_{0.04}Sr_{0.48}BiO_3{\cdot}0.48H_2O$) and AB_2O_6 (such as $SrBi_2O_6$), can be prepared by such similar IE process but under hydrothermal conditions.

15.3.2 Oxidation–Reduction Process

Pan et al. [46] prepared $NaBiO_3$ of powder with high purity through oxidation–reduction (O–R) reaction route by using $Bi(OH)_3$ and NaClO as a reducing and oxidizing agent, as shown in the reaction below:

$$2NaOH + Cl_2 = NaClO + NaCl + H_2O$$
$$Bi(OH)_3 + 3NaOH = Bi(OH)_3 + 3NaNO_3$$
$$Bi(OH)_3 + NaClO + NaOH = NaBiO_3g2H_2O + NaCl$$

The purity of $NaBiO_3$ can be determined based on the reaction between BiO_3^- and Fe^{2+} shown as below:

$$BiO_3^- + 2Fe^{2+} + 6H^+ = Bi^{3+} + 2Fe^{3+} + 3H_2O$$

Thus the purity of $NaBiO_3$ can be confirmed by potassium dichromate titration process (for Fe^{2+} determination).

The preparation of anhydrous $KBiO_3$ was first reported by Jansen [47]. In his process he used 500–600 °C and an oxygen pressure of 1,000–2,000 atm to obtain potassium bismuthate as a red powder. Here, $KBiO_3{\cdot}nH_2O$ compound was prepared in liquid phase through oxidation of bismuth nitrate by bromine in hot KOH [48, 49]. Similar approach was proved to be available for the preparation of $NaBiO_3$ by Eberl as well (but with a little intermediates of excess sodium and remaining Bi(III)-content [42]). Thus it might be considered a general method for the preparation of various pentavalent bismuthates including $NaBiO_3$ and $KBiO_3$. The general preparation process can be illustrated as follows:

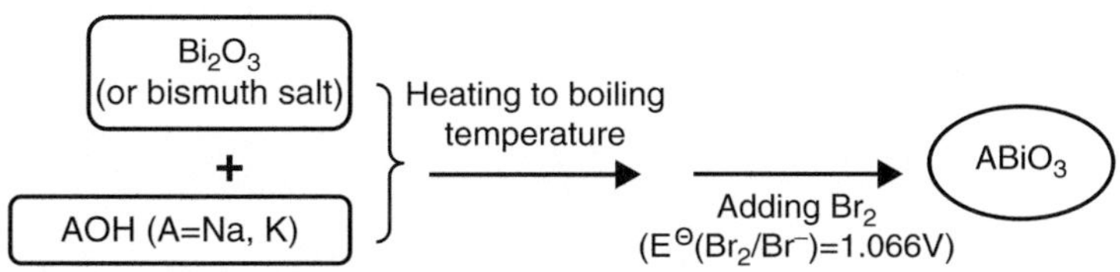

Scheme 15.1 Schematic illustration for the preparation process of ilmenite-type pentavalent bismuthates ($NaBiO_3$ and $KBiO_3$)

Chen et al. [50] prepared $NaBiO_3$ crystals through another O–R route under solventhermal process by using $Bi(OR)_3$ and NaOR (R=Pri, CMe_2Et,$CH_2CH_2OCH_3$) as the starting materials. It is well known that the oxidation of Bi(III) into Bi(V) is difficult because of the weak reduction ability of Bi(III). But in their reaction process, the reduction ability of Bi(III) was considered to be improved due to the strong basic condition caused by the starting materials. Meanwhile, the solventhermal environment could be benefits for the oxidation process which is difficult to take place at ambient temperature and pressure.

15.3.3 Solid State Reaction

The solid state (SS) reaction is one of the most widely adopted routes for the preparation of complex oxides from a mixture of starting materials (such as oxides or carbonate) as the solid can react together under high temperature. This method had been widely used for the preparation of Bi(V)-containing compounds. The solid reactants for the preparation are usually adopted as the nitrites/carbonate salts and oxides (peroxides). The bismuth source is usually adopted as Bi_2O_3 or $Bi(NO_3)_3$, indicating that a reduction–oxidation process occurs in the solid state route.

15.3.4 Complex Sol–Gel Process

Complex sol–gel (CS-G) method has also been attempted to prepare $BaBiO_3$ with perovskite structure [23], by using ethylenediaminetetraacetic acid (EDTA) and citric acid (CIT) as the chelating agent for Bi^{3+} and Ba^{2+}, respectively. As illustrated in Scheme 15.2, the complex sol–gel process involves the preparation of sol and xerogel, dissociation of ligand and complex (endothermic reaction), and finally the formation of perovskite phase at high temperature (exothermic reaction).

As a member of the polyamino carboxylic acid family of ligands, $EDTA^{4-}$ which possesses six dissociation equilibrium in total ($pK_1=0.9$, $pK_2=1.6$, $pK_3=2.0$, $pK_4=2.67$, $pK_5=5.16$, $pK_6=10.24$) has been widely selected to bind to a metal cation by its amines and carboxylates. EDTA is a candidate ligand for chelating Bi^{3+} as well because of the stability constants (lgK) of $EDTA^{4-}$–Bi^{3+} can reach to as high as 27.9 [51].

In complex sol–gel process, the excess amount of EDTA is necessary to prepare the EDTA-Bi complex sol because of the acid effect coefficient which is defined as follows (in case of EDTA):

$$\alpha_{Y(H)} = \frac{[Y']}{[Y]} = \frac{[Y^{4-}]+[HY^{3-}]+[H_2Y^{2-}]+[H_3Y^{-}]+[H_4Y]+[H_5Y^{+}]+[H_6Y^{2+}]}{[Y^{4-}]}$$

$$= 1+\frac{[H^+]}{K_6}+\frac{[H^+]^2}{K_6K_5}+\frac{[H^+]^3}{K_6K_5K_4}+\frac{[H^+]^4}{K_6K_5K_4K_3}+\frac{[H^+]^5}{K_6K_5K_4K_3K_2}+\frac{[H^+]^6}{K_6K_5K_4K_3K_2K_1}$$

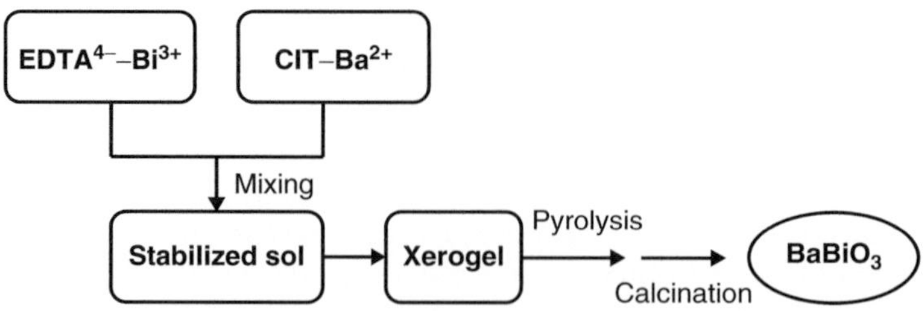

Scheme 15.2 Schematic illustration for the preparation process of perovskite-type $BaBiO_3$

in which [*Y*] is the concentration of $EDTA^{4-}$ ion and [*Y*′] is the total concentration of both $EDTA^{4-}$ ion and its six kinds of protonated forms. $a_{Y(H)}$ will be equal to 1 if all the protons are dissolved; however, the effective chelating process takes place only between Y^{4-} and metal cation. Therefore the coefficient of $a_{Y(H)}$ should be less than 1 in practical situation. Since Bi^{3+} is very easy to hydrolyze and the solubility product constant of $Bi(OH)_3$ is extremely low ($K_{sp}=4\times10^{-31}$), Bi^{3+} cannot exist in the aqueous solution unless strong acid conditions exist. But after the chelate complex formation of $EDTA^{4+}$–Bi^{3+}, Bi^{3+} can be very stable in aqueous solution near neutral pH condition. CIT is also thought to be an excellent chelating agent, even if the stability constants ($\lg K$) of CIT-based complex ($\lg K = 11$–15) is much lower than that of EDTA-based complex.

15.3.5 Electrochemical Process

The pentavalent bismuthates of β-Na_3BiO_4 [36] and $KBiO_3$ [29] were prepared via similar electrochemical process, i.e., the crystals were obtained on platinum anode through electrooxidation of alkali hydroxide melts containing Bismuth(III) oxide Bi_2O_3, meanwhile a sacrificial reductant (such as ZnO, $ZnCl_2$, and CuCl) is essential in the electrochemical (EC) process to balance the electron inventory of the whole oxidation–reduction process (Table 15.3).

15.4 Electronic/Band Structure

15.4.1 Typical Bismuth(V)-Containing Semiconductors

15.4.1.1 ABO_3 Type

As one of the most typical Bismuth(V)-containing materials, the energy band structure (EBS) and density of states (DOS) of $NaBiO_3$ were investigated by Liu et al. [21], during their study on the pressure-induced phase transition of $NaBiO_3$. By means of the DFT using the generalized gradient approximation, their DOS results (including partial density of states, PDOS) of $NaBiO_3$ (at ambient pressure) indicated that the *s*/*p* orbitals from Bismuth and the *p* orbital from Oxygen are the main constituents of the covalent bond and CB. Kako et al. [17] also found that there is a large dispersion in the hybridized *sp* orbitals in the CB of $NaBiO_3$, which might be beneficial for the high mobility of photogenerated electrons on the sp bands. Li et al. [35] reported the photocatalytic hydrogen production (from aqueous methanol) enhancement by substitution of Ta^{5+} with Bi^{5+} in $NaTaO_3$ photocatalyst. It was deduced about the formation of a lower energy of CB as well as the hybridization of Bi6*s*/Bi6*p* and Ta5*d* orbitals. This could happen that the Bi6*s*/Bi6*p* mainly contribute to

Table 15.3 Preparation techniques for Bismuth(V)-containing semiconductor compounds

Compound	Preparation technique	Starting materials	Reaction condition	References
$LiBO_3$	IE	$NaBiO_3 \cdot nH_2O$, LiOH	Hydrothermal at 120–200 °C, 7 days	[30, 31, 41]
(Li–Na)BiO_3	IE	$NaBiO_3 \cdot nH_2O$, LiCI solution	60 °C, 1 atm	[18]
$NaBiO_3$	O–R	$Bi(OH)_3$, NaClO, NaOH	20 °C–30 °C, 1 atm	[46]
		Bi_2O_3, NaOH, Br_2	1 atm, *T*: N.A.	[43]
		$Bi(OR)_3$, NaOR (R=Pri,CMe_2Et, $CH_2CH_2OCH_3$)	Solventhermal at 240 °C	[50]
α-Na_3BiO_4	SS	Na_2O_2, Bi_2O_3	*T*–600 °C (O_2 flow), 12 h	[36]
β-Na_3BiO_4	EC	Bi_2O_3, NaOH, LiOH, ZnO	330–350 °C constant current density = 1 mA/cm^{-2}, 18–42 h	[36]
$Na_{0.04}$ $Sr_{0.48}BiO_3 \cdot 0.48H_2O$	IE	$NaBiO_3 \cdot nH_2O$, $SrCl_2$	Hydrothermal at 90 °C, 2–20 days	[38]
$Na_{1-2x}Ba_xBiO_3 \cdot nH_2O$ ($0.11 \leq x \leq 0.36$)	IE	$NaBiO_3 \cdot nH_2O$, $BaCl_2$	Hydrothermal at 90 °C, 2–20 days	[38]
$KBiO_3$	O–R	Bi_2O_3, KOH, Br_2	Boiling point of 50 % KOH a.q. (~125 °C), 1 atm	[30, 43]
	EC	KOH, Bi_2O_3, $ZnCl_2$ (CuCl),	Working potential = 1.1 or 0.8 V	[29]
$AgBiO_3$	IE	$NaBiO_3$, $AgNO_3$	70 °C, 1 atm, ≥24 h	[16, 44]
		$NaBiO_3 \cdot nH_2O$, $AgNO_3$	Hydrothermal at 70 °C	[10]
$BaBiO_3$	CS-G	$Ba(NO_3)_2$, $Bi(NO_3)_3$, CIT, EDTA	dried at 393 K for 10 h and calcined at 923 K for 5 h	[23]
	SS	$Ba(NO_3)_2$, $Bi(NO_3)_3 \cdot 5H_2O$	850 °C, 24 h	[11]
		$Bi(NO_3)_3 \cdot 5H_2O$, $BaCO_3$	1,070 ± 1,270 K, 15 ± 30 h	[28]
		$Ba(NO_3)_2$, Bi_2O_3	700 °C, 9 h	[25]
		$BaO_2/Ba(NO_3)_2$, Bi_2O_3	900–1000 °C, 24 h,	[52]
$Ba_{1-x}M_xBiO_3$ (M=K, Rb)	SS	$BaCO_3$, K_2CO_3, Rb_2CO_3, Bi_2O_3	900 °C, 2 h	[53]
$BaBi_{1-x}M_xO_{3=\delta}$ (M=La, Pr)	SS	$Bi(NO_3)_3$, $La(NO_3)_3$, Pr_6O_{11}, $BaCO_3$	1,070 ± 1,270 K, 15 ± 30 h	[28]
$SrBi_2O_6$	IE	$NaBiO_3 \cdot nH_2O$, $SrCl_2$	Hydrothermal at 90–130 °C, 1 week	[10]
$BaBi_2O_6$	IE	$NaBiO_3 \cdot nH_2O$, $BrCl_2$	Hydrothermal at 90–130 °C, 1 week	[10]
$PbBi_2O_6$	IE	$NaBiO_3 \cdot nH_2O$, $Pb(NO_3)_2$	R.T., 1 atm	[10]
$MgBi_2O_6$	IE	$NaBiO_3 \cdot nH_2O$, $MgCl_2$	Hydrothermal at130 °C	[10, 54]
$ZnBi_2O_6$	IE	$NaBiO_3 \cdot nH_2O$, $Zn(NO_3)_2$	Hydrothermal at 90 °C	[10, 54]

the bottom of CB in $Na(Bi_{0.25}Ta_{0.75})O_3$, which was confirmed also by their DOS results. The lower CB and hybrid orbitals were considered as the main reason for the smaller band gap of the series of solid-solution $Na(Bi_xTa_{1-x})O_3$ ($x=0-0.10$) compared with $NaTaO_3$.

Besides $NaBiO_3$, the EBS and DOS of others typical Bismuth(V)-containing semiconductors (i.e., $AgBiO_3$, $LiBiO_3$, $KBiO_3$) have been investigated as well. According to the results from Takei et al. [10], the CB and VB of these bismuthate are mainly composed of O2*p* and Bi6*s* orbitals. But the details of their DOS results are not similar for different compounds. For the DOS results of $AgBiO_3$, the broad Ag4*d* orbital above the O2*p* is considered as the main reason for the electronic conductivity of $AgBiO_3$. For $LiBiO_3$ crystal, it is interesting to find that the hybrid orbitals (which are composed of Bi6*s* and O2*p*) do not emerge. For $KBiO_3$, the relative low CB was found as the formation of the hybrid orbital which is composed of Bi6*s* and O2*p*. The lower CB might be the main reason for the photodegradation performance over $KBiO_3$. But more importantly, it was observed from the DOS plots that two types of oxygen atoms (edge-shared and corner-shared oxygen) can be found because of the splitting of O2*p*. This might be another reason for the low band gap of red-colored $KBiO_3$ ($E_g=2.04$ eV).

15.4.1.2 AB_2O_6 Type

The DOS results of Bismuth(V)-containing semiconductors with the AB_2O_6 type (e.g., $SrBi_2O_6$, $MgBi_2O_6$, $ZnBi_2O_6$, $BaBi_2O_6$, and $PbBi_2O_6$) have been reported by Mizoguchi et al. [35] and Takei et al. [10] in 2004 and 2011.

For the semiconductors with lead antimonite structure ($SrBi_2O_6$, $BaBi_2O_6$, and $PbBi_2O_6$), the orbital of Pb6*s* could be observed at around −2 eV, which is the possible reason of the lowest band gap among these three kinds of semiconductors ($E_g(PbBi_2O_6)=1.92$ eV, $E_g(SrBi_2O_6)=E_g(BaBi_2O_6)=1.93$ eV). For the semiconductors with trirutile structure ($MgBi_2O_6$ and $ZnBi_2O_6$), a broad VB was found from the DOS plots of $ZnBi_2O_6$, possibly resulting the lower band gap (1.53 eV) than $MgBi_2O_6$ (1.61 eV). It was reported by Mizoguchi et al. [35] that the electronic structure of $MgBi_2O_6$ is qualitatively very similar to $MgSb_2O_6$ and $ZnSb_2O_6$ with trirutile crystal structures. By the linear muffin-tin orbital (LMTO) calculations with the atomic sphere approximation (ASA) including the combined correction (CC), the energy ranges of four Bi6*s* to O2*p* σ and Bi6*s* to O2*p* σ^* bands were, respectively, found to be −6.5 to −9.4 eV and 0.8–5.0 eV, indicating their much more narrower energy ranges compared with $MgSb_2O_6$ which possibly caused by the improved overlap of Bi6*s* and O2*p* with respect to that of Sb5*s* and O2*p* orbitals.

15.4.2 *Mixed-Valence Bismuth Containing Semiconductors*

The development of mixed-valence Bismuth (Bi^{3+} and Bi^{5+}) containing photocatalyst is now considered to be another efficient way to improve the mobility and

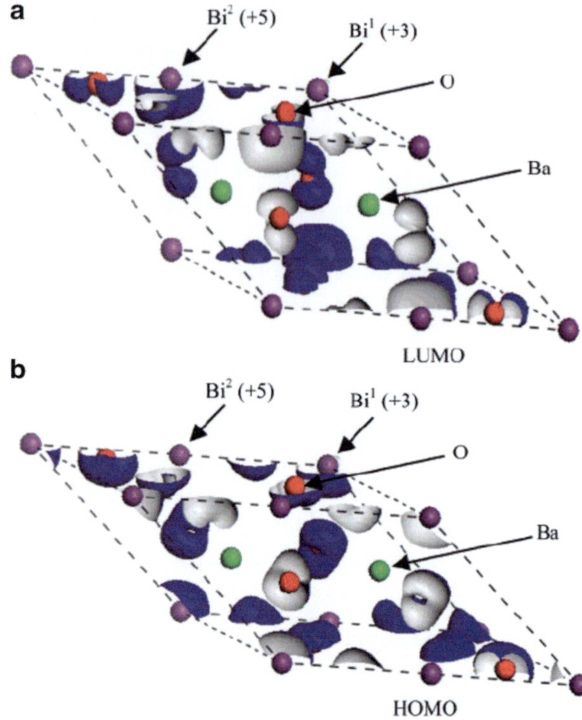

Fig. 15.9 (**a**) Density contour maps of the bottom orbital of the conduction band (LUMO) and (**b**) density contour maps of the top orbital of the VB (HOMO) for $BaBiO_3$ (reprinted from ref. [23])

lifetime of the photogenerated carries in the corresponding semiconductor, because of the effective modification of CB (to decrease band gap by forming a lower the CB position) and VB (to improve the hole mobility by the increase in the width of VB) by Bi^{3+} and Bi^{5+}, respectively. On the other hand, basically, the spatial orientation of the suborbitals (such as *s*, *p*, and *d* orbitals) has great effect on the photogenerated electron transfer process from VB to CB as well. Therefore, it can be deduced that the existence of Bi^{3+} (Bi6*s*) and Bi^{5+} (Bi6*s*) in the semiconductor might be beneficial for the migration of the photogenerated carries as the less localized characteristics. However, such report on the mixed-valence Bismuth containing photocatalysts which Bi^{3+} and Bi^{5+} coexist in the same crystal lattice is rare, even imagine that such materials had been prepared and characterized almost 36 years ago [55].

Tang et al. [23] reported the EBS and DOS of $BaBiO_3$ for the first time in early 2007. Their DOS results proved that the Bi6*s* and Bi6*s*/Bi6*p* are the predominant composition of the top of the VB and the bottom of the CB. Moreover, their density contour maps for the LUMO (corresponding to CB) and HOMO (corresponding to VB) further confirm the existence of Bi6*s* in both LUMO and HOMO (as shown in Fig. 15.9). Not only such unique characteristics could narrow the band gap, improve mobility and life time of the photogenerated carries, but facilitate the electron transition because of the much lower barrier formed, which should be the possible reason for the excellent photocatalytic performance of $BaBiO_3$.

From the contrary view point, Lakshminarasimhan et al. [11] demonstrated the significance of mixed-valence Bismuth in the photocatalytic activity enhancement. Their DOS results clearly show that the partial substitution of Bi^{3+} with La^{3+} in $BaBi^{3+}_{0.5}Bi^{5+}_{0.5}O_3$ crystal can reduce the Bi6*s* contribution to the top of VB and the CB band width, leading to a lower mobility of photogenerated carries, and therefore, a lower photocatalytic activity of La^{3+} substituted $BaBi^{3+}_{0.5}Bi^{5+}_{0.5}O_3$ crystal.

15.5 Photocatalytic Applications

15.5.1 Removal of Various Contaminates from Aqueous Solution

In early 2004, Zou et al. [56] publicized the photocatalytic application in the removal of methyl blue on pentavalent bismuthates ($NaBiO_3$, $LiBiO_3$, and $AgBiO_3$) from aqueous solution in their patent registered in China. The results reported in their patent that three kinds of pentavalent bismuthate photocatalysts demonstrated excellent decomposition activity to methyl blue under visible light irradiation. When the wavelengths of the irradiated light from Xenon lamp is greater than λ_x ($\lambda_x = 420$, 480, 500, and 580 nm), the time for the complete decomposition of methyl blue was less than 40 min. The shortest time required was just 8 min under optimal conditions at $\lambda_x = 420$ nm by $NaBiO_3$ photocatalyst, indicating the potential application of $NaBiO_3$ in heterogeneous photocatalysis.

As a matter of fact, the possible application in chemical synthesis of $NaBiO_3$ was reported for the first time in early 1950. A few fission process with good yield on $NaBiO_3$ (as oxidant) was discovered such as the fission of αβ-glycols to aldehydes or ketones [57]. Recently $NaBiO_3$ has been applied in organic compound synthesis such as glycol cleavage, conversion of acyloins to α-diketones, oxidative halogenation, and chemoselective oxidation [58–62].

The first formal report on $NaBiO_3$ photocatalyst in academic journal was published by Kako et al. [17]. Dehydrated $NaBiO_3$ was prepared by heating commercial product of $NaBiO_3 \cdot H_2O$ for 5 h. The calculated band gap value of as-prepared $NaBiO_3$ shows smaller value of 2.6 eV (indirect band gap) compared with typical trivalent bismuthates compound such as (2.8–2.9 eV), Bi_2WO_6 (2.8 eV), and $Bi_2W_2O_9$ (3.0 eV). Their photocatalytic decomposition of 2-propanol and methylene blue was investigated from air and aqueous solution, respectively. It was interesting to find that $NaBiO_3$ can convert 2-propanol into acetone effectively under full arc or visible light irradiation (irradiated from Xenon lamp), but cannot mineralize 2-propanol into CO_2 and H_2O at irradiation time less than 1 h, which might be due to the degradation pathway. On the other hand, $NaBiO_3$ exhibits superior photooxidation activity to methylene blue than N-doped TiO_2 and $BiVO_4$ photocatalysts which are considered to be the most popular photocatalysts to date.

Polycyclic aromatic hydrocarbons (PAHs) are produced mainly by the burning of fossil fuels such as crops and other natural substances. Parts of the PAHs,

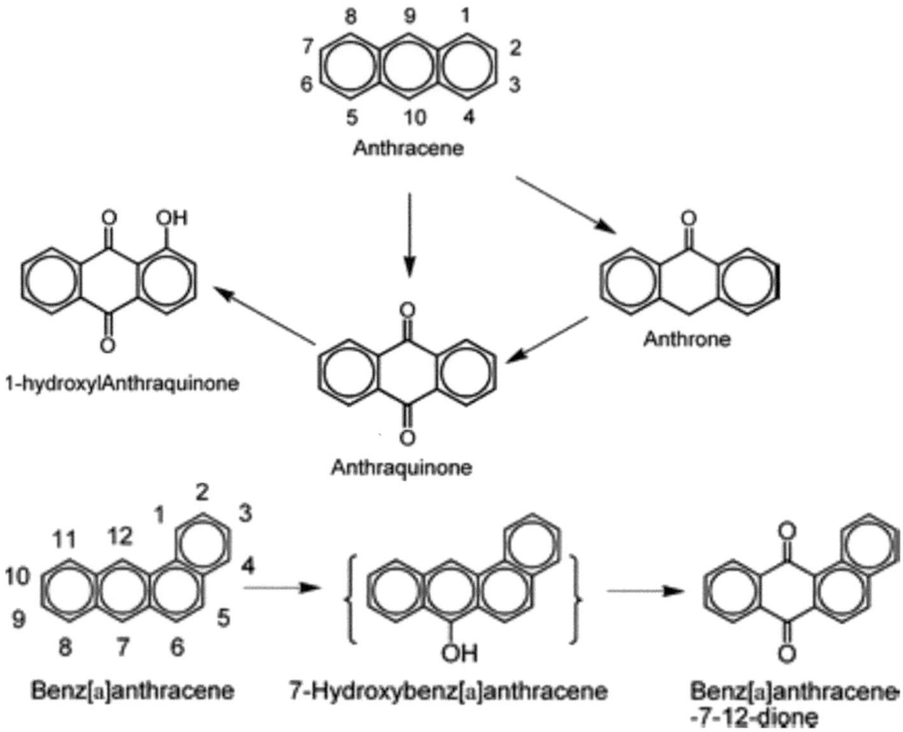

Fig. 15.10 Degradation pathway of ANT and Bz[a]A in $NaBiO_3$ suspended solution under visible light irradiation (reprinted from ref. [19])

represented by chlorinated polycyclic aromatic hydrocarbons (ClPAHs) widely presented in the environment, are considered as a great potential threat to human health. Kou et al. [19] reported the possible photocatalytic application of PAHs removal on $NaBiO_3$ from aqueous phase under visible light irradiation, by selecting anthracene (ANT) and benzo[a]anthracene (Bz[a]A) as target pollutants. It was found that under certain conditions, photocatalytic conversion rates of ANT and Bz[a]A on $NaBiO_3$ can reach to over 50 % and up to 100 % within 8 and 1 h, respectively. During the process of photocatalytic conversion of ANT and Bz[a]A, the intermediate products were detected and each contain compounds of anthraquinone (the main degradation product), anthrone,1-hydroxy-anthracene and anthracene-7,12-dione, etc., using gas chromatography–mass spectrometry (GC/MS). The results also showed that there is a similar degradation pathway in the $NaBiO_3$ photocatalytic conversion process of ANT and Bz[a]A (as shown in Fig. 15.10).

Yu et al. [63] studied the photocatalytic degradation pathway and mechanism of Rhodamine B dye on $NaBiO_3$ under visible light irradiation. They separated and confirmed all intermediate products of N-bit de-ethylation and other small-molecule compounds by means of high performance liquid chromatography (HPLC), liquid chromatography–mass–mass spectrometry (LC/MS/MS) and GC/MS. Two kinds of isomers of different contents (*N,N*-dimethylamino-rhodamine DR and *N-dimethyl-N′*-dimethyl-rhodamine EER), generated in photocatalytic reaction, might be related to the electron density of the dye molecules. Chromophore split and N-bit de-ethylation effect may be two competing mechanisms in the process of Rhodamine B photocatalytic decomposition. Yu et al.'s experiments confirmed the division process of chromophore was more dominant compared with N-bit de-ethylation effect.

Pentachlorophenol (PCP) and its sodium salt (PCP-Na) had been widely used as the insecticide to protect wood and textile worldwide; however, its toxicity, endocrine EDC, and organism gathering effect caused people's attention. Previous studies have shown that commercial grade PCP-Na contained more toxic polychlorinated dibenzo-*p*-dioxins (PCDD) and polychlorinated dibenzofurans (PCDF). We previously [64] carried out a systematic study on the kinetics of $NaBiO_3$ photocatalytic kinetics of PCP-Na on $NaBiO_3$ in visible light and conducted a preliminary analysis and discussion about its photocatalytic mechanism based on the EBS of $NaBiO_3$. It was found that photocatalytic process of $NaBiO_3$ decomposing PCP-Na obeyed the Langmuir-Hinshelwood first-order kinetics process. The amount of catalyst, initial pH value of PCP-Na aqueous solution, and the initial concentration produced a significant impact on the kinetic process. At the same time, we noticed experimentally that, there is almost no obvious impact on the photocatalytic kinetic process purged with nitrogen. It is indicated by theoretical calculation that the CB edge potential of $NaBiO_3$ is closer to the electrochemical potential of O_2/O^{2-}, thus the overpotential of oxygen anions produced by photogenerated electron reduction of O_2 is smaller, resulting in the photocatalytic decomposition process; therefore, oxygen anions' oxidation organics is not the leading step in the whole process.

Apart from $NaBiO_3$, other pentavalent bismuthates such as $LiBiO_3$, $KBiO_3$, and $AgBiO_3$ have also been attempted in the heterogeneous photocatalytic application. Kikugawa et al. [41] investigated the photocatalytic degradation of methylene blue over as prepared $LiBiO_3$ under white fluorescent light exposure. Their results show that methylene blue can be completely decolorized after 4-h reaction, and the mineralization efficiency can be reached to 70 %. The highest apparent photoefficiency (PE) of 0.15 % could be obtained under 420 nm monochromatic light irradiation. The photocatalytic activity of $LiBiO_3$ was further confirmed by Ramachandran et al. [30]. Two cationic dyes of Rhodamine B and Methylene blue have been used as model compounds. Under UV and solar radiations, it was found that $LiBiO_3$ exhibits better photocatalytic activity of Methylene blue than Rhodamine B in terms of both degradation efficiency and reaction kinetics constant.

The photocatalytic degradation of various anionic dyes, i.e., Orange G (OG), Amido black 10B (AB10B), Alizarin cyanine green (ACG), Indigo carmine (IC), and Coomassie brilliant blue R 250 (CBBr) over $KBiO_3$ from aqueous phase were investigated by Ramachandran et al. [30] as well. It was found that the overall photocatalytic degradation of CBBr and ACG are higher than that of OG and AB10B, possibly because of the more reactive of CBBr and ACG where sulfonyl group is attached to the benzene ring, but less reactive naphthalene leads to the less reactive of OG and AB10B.

On the contrary, it was found by Eberl that no photocatalytic degradation of 4-Chlorophenol [43]. $KBiO_3$ did not exhibit the photocatalytic mineralization performance, and powder sample after exposure changed into yellow.

In recent years, Bi(III)/Bi(V)-containing mixed-valence semiconductor compounds with a single phase structure have also drawn people's attention. Compared with the semiconductor compound simply containing Bi(V), the energy band/electron structure of this kind of compound could improve the carrier's mobility and life, thus increase the heterogeneous photocatalytic activity. Tang et al. [23] prepared $BaBiO_3$ photocatalyst containing mixed-valence Bi(III)/Bi(V) by complex sol–gel method. The result shows that the photocatalytic activity of $BaBiO_3$ is superior to other Bi-containing visible light photocatalysts such as $ZnBi_{12}O_{20}$ and $CaBi_2O_4$. By selecting acetaldehyde and methyl blue as model compound, after 1-h reaction under the lighting conditions of $\lambda \geq 440$ nm and $\lambda \geq 420$ nm, the photocatalytic decomposition efficiency can reach to 80 % and 100 %, respectively. For its reason, mainly because crystal Bi(III) and Bi(V) effectively controlled the VB and CB of semiconductor, respectively. Theoretical calculation based on the electron structure in semiconductor, it was found that photogenerated electrons and holes all have better mobility because of better dispersion of CB and VB in the crystal which can then improve the lifetime of the carrier. Transitions of photogenerated electrons from VB to CB are influenced by the sub-shell electron orbital orientation to a great extent. During the photocatalytic process of $BaBiO_3$, the photoelectron jumps from 6*s* orbit of Bi^{3+} to 6*s* orbit of Bi^{5+}. Therefore the energy barrier of *s*–*s* transition is the lowest as the ball symmetrical characteristic of *s* orbit. Thus it can promote the photocatalysis reaction.

Eberl [43] applied Na_xBiO_3 (the content of Bi(V) accounted for more than 90 %) to the photocatalytic decomposition experiment of 4-Chlorophenol and it showed that 4-Chlorophenol was completely mineralized within 1 h under the lighting condition of $\lambda > 455$ nm. It is probably because of the fact that the common existence of Bi(III) and Bi(V) in photocatalytic materials made photogenerated electron–hole get more effective separation, and then improve the carrier's life.

Lakshminarasimhan et al. [11] prepared $BaBi^{3+}{}_{0.5-x}La_xBi^{5+}{}_{0.5}O_3$ photocatalyst which Bi^{3+} was partly displaced by La^{3+} ($x=0$, 0.1 and 0.3, the ionic radius of La^{3+} and Bi^{3+} were 1.061 A and 1.020 A). All of these catalysts were dark brown. It was found that absorption band centered at 680 nm (about 1.82 eV) in the diffuse reflection spectrum of $BaBi^{3+}{}_{0.5}Bi^{5+}{}_{0.5}O_3$ disappeared due to the displacement of less content of La^{3+} (namely, $x=0.1$). According to Mizoguchi's point of view [65], this absorption band might be attributed to the carrier transition from Bi^{3+} ($6s^2$) to Bi^{5+} ($6s^0$). Thus it is not difficult to find, the displacement of La^{3+} which can hinder the differentiation of $Bi^{3+}–O^{2-}–Bi^{5+}$ valence bond. This conclusion is also confirmed by Fourier transform-infrared absorption spectroscopy (FT-TR). By using 4-Chlorophenol as a target pollutant, the photocatalytic investigation showed that the photocatalytic activity of $BaBi^{3+}{}_{0.5-x}La_xBi^{5+}{}_{0.5}O_3$ weakened with the increase of the amount of displacement of La^{3+}, and $BaBi^{3+}{}_{0.5-x}La_xBi^{5+}{}_{0.5}O_3$ with no substitution showed the highest photocatalytic activity because the photogenerated hole had stronger oxidation capacity. The displacement between Bi^{3+} and La^{3+} destroyed the valence bond differentiation, which led to the decrease of photocatalytic activity.

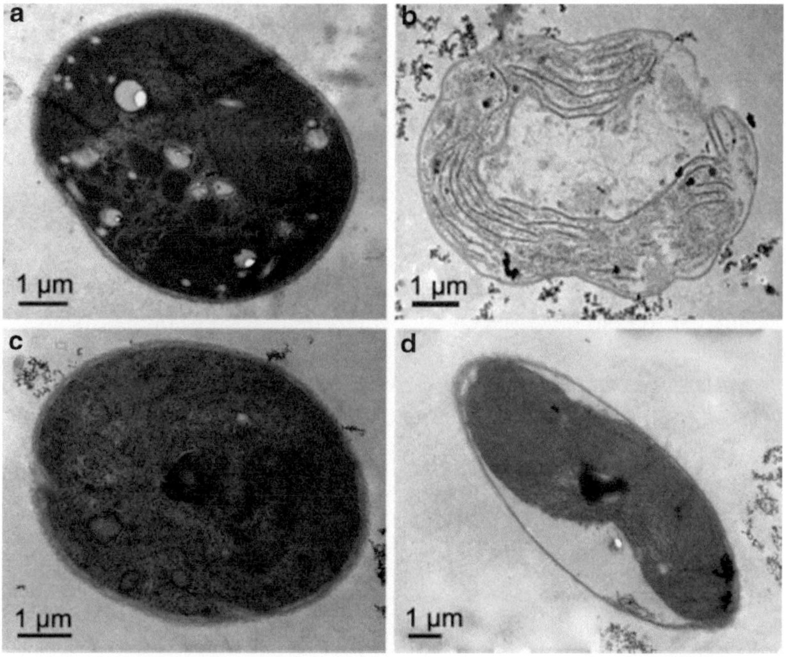

Fig. 15.11 TEM of *M. aeruginosa* with or without $AgBiO_3$ ((**a**) control sample; (**b–d**) $AgBiO_3$ treatment sample, reprinted from ref. [44])

15.5.2 Disinfection from Water

Up to now, to the best of our knowledge, there is no report on the photocatalytic removal of contaminants on $AgBiO_3$ compound (a visible light-responsive catalyst that absorbs strongly in the visible light spectrum) from air or aqueous phase. However a new attempt, that is, the photocatalytic inhibition of the growth of cyanobacteria on $AgBiO_3$ under natural light irradiation was made by Yu et al. [44].

Microcystis (blue-green algae) is one of the most common toxic cyanobacteria in fresh water which includes the harmful algal bloom *Microcystis aeruginosa*. Microcystis bloom which is one of the most objectionable characteristics of eutrophication in tropical and subtropical waters is thought to harm the ecological environment, local economies, and even human health besides its unfavorable effects on the landscape. Yu et al. [44] found that the as-prepared $AgBiO_3$ compound which is imagined to be able to produce more hydroxyl radicals than $NaBiO_3$, can effectively inhibit the growth of *M. aeruginosa* meanwhile damaged its cell walls and membranes, resulting to the electrolyte and cytoplasmic leakage and finally the death of the bacteria (as shown in Fig. 15.11).

15.5.3 Photocatalytic Hydrogen Production

The photocatalytic splitting of water requires a higher (more negative) CB band edge potential of semiconductor compared with H^+ reduction potential. However, the report on the relevant application of Bismuth(V)-containing semiconductor is rare, due to the lower CB band edge potential of such semiconductor. Ultraviolet photoelectron spectroscopy (UPS) [66] and flatband potential determination-based photoelectrochemical techniques [67–69] were two important methods to test the electrochemical potentials of CB and VB. Recently, a theoretical method based on the absolute (Mullikan) electronegativity was widely used to predict band position of semiconductors and such simple method is in good agreement with the experimental results for semiconductor compounds [23, 70–72]. The CB and VB band edge (E_{CB}/E_{VB} vs. standard hydrogen electrode, SHE) of the semiconductor at the point of zero charge can be calculated by the equation as follows (the error of this method is around 0.2 eV [73]):

$$E_{CB} = X - E^{C} - \frac{1}{2} E_g E_{VB} - E_{CB} = E_g$$

where X is the absolute electronegativity of the semiconductor which can be calculated as the geometric mean of the Mulliken electronegativity of the constituent atoms. E^C is the energy of free electrons on hydrogen scale (≈4.5 eV) and E_g is the band gap of the semiconductor which can be calculated by DRS measurements. Even if the calculated E_{CB} (4.32 eV vs. vacuum, −0.18 V vs. SHE) is close to the H_2 evolution potential, however, no H_2 evolution was observed by Tang et al. [23], possibly due to the practical CB is lower than H_2 evolution potential or the overpotential of the material is not enough for H_2 production.

Li et al. [37] investigated the H_2 production from aqueous methanol solutions over Bi(V) substituted $NaTaO_3$ solid-solution ($Na(Bi_xTa_{1-x})O_3$) under visible light irradiation ($\lambda > 400$ nm). No H_2 evolution was found for $NaTaO_3$ under visible light irradiation because of the relative wide band gap value (3.91 eV). The substitution of Bi(V) in $NaTaO_3$ could decrease the band gap as well as the CB band edge; therefore, over amount substitution might not be beneficial for the H_2 production because a much lower CB band edge will be formed. The $Na(Bi_{0.08}Ta_{0.92})O_3$ photocatalyst which was considered to be the optimal composition in the solid-solution exhibited the highest activity of H_2 evolution (59.48 mmol h^{-1} g^{-1}) under visible light irradiation (Table 15.4).

15.6 Photostability

The photostability evaluation of any catalyst is of great importance in fundamental study as well as due to other engineering aspects. Up to now, however, such investigations and discussions on the possible corrosion mechanism are very rare.

Table 15.4 Photocatalytic degradation/conversion of various contaminates on Bismuth(V)-containing photocatalysts

			Reaction condition			Activity evaluation			
Photocatalyst	Target pollutant	Light source	C_{cat}[a]	C_0[b]	t	Degradation/ conversion %	k/min^{-1}	TOC removal (TOC%) or PE %	References
$LiBiO_3$	Methylene blue	White fluorescent light 0.36 mW/cm^2	0.3 g/100 mL	13.2 mg/L	4 h	~100 %	N.A.	TOC% ≈ 70 % PE% = 0.15 % (420 nm)	[31]
	Methylene blue	Xenon lamp 300 W	3 g/L	16 mg/L	40 min	~100 %	N.A.	N.A.	[10]
	Phenol	λ>420 nm		N.A.	2 h	N.A.	N.A.	TOC% ≈ 72 %	
	Rhodamine B	High pressure mercury vapor lamp 125 W	0.1 g/100 mL	50 mg/L	2 h	57 %	0.00513 (first 20 min)	N.A.	[30]
	Methylene blue			20 mg/L		~83 %	0.00102 (first 20 min)	N.A.	
	Rhodamine B	Solar radiation ~975 W/m^2		50 mg/L		~34 %	0.0024	N.A.	
	Methylene blue			20 mg/L		~97 %	0.01473	N.A.	
$NaBiO_3$	2-Propanol (g)	Xenon lamp ~3 mW	0.4 g	600 ppm	1 h	N.A. (~47 ppm acetone evolved)	N.A.	N.A.	[17]
	Methylene blue	Xenon lamp 300 W	0.3 g/100 mL	16 mg/L	10 min	~100 %	0.42	N.A.	
	Methylene blue	Xenon lamp 300 W	3 g/L	16 mg/L	40 min	~100 %	N.A.	N.A.	[10]
	Phenol	λ>420 nm		N.A.	2 h	N.A.	N.A.	TOC% ≈ 82 %	
	Sodium pentachlorophenate	Xenon lamp 500 W λ>400 nm	0.1 g/150 mL	50 mg/L	1 h	90.50 %	0.03638	N.A.	[64]
	4-*t*-Octylphenol	Xenon lamp 500 W λ>400 nm	0.3125 g/L	29 mg/L	1 h	~90 %	~0.03741	N.A.	[74]
	4-Chlorophenol	Xenon lamp 150 W λ>455 nm 950 ± 100 W/m^2	0.9 g/L	2.5×10^{-4} mol/L	2 h	N.A.	N.A.	TOC% ≈ 65 %	[43]
	Anthracene	Xenon lamp 300 W	0.3 g/60 mL	3 mg/(20 mL H_2O+40 mL Actone)	10 h	~85 %	N.A.	N.A.	[19]
	Benz[a]anthracene	λ>420 nm	0.05/60 mL		80 min	~100 %	N.A.	N.A.	
	Rhodamine B	Xenon lamp 750 W λ>400 nm	1 g/L	20 mg/L	30 min	~100 %	N.A.	N.A.	[63]

(continued)

Table 15.4 (continued)

Photocatalyst	Target pollutant	Light source	Reaction condition C_{cat}[a]	C_0[b]	t	Activity evaluation Degradation/conversion %	k/min^{-1}	TOC removal (TOC%) or PE %	References
$KBiO_3$	Methylene blue	Xenon lamp 300 W	3 g/L	16 mg/L	40 min	~100 %	N.A.	N.A.	[10]
	Phenol	λ>420 nm		N.A.	2 h	N.A.	N.A.	TOC% ≈ 55 %	
	Orange G	High pressure mercury vapor lamp 125 W	0.1 g/100 mL	50 mg/L	2 h	~26 %	0.00144	N.A.	[30]
	Amido black 10B					~22 %	0.0017	N.A.	
								N.A.	
	Alizarin cyanine green					~49 %	0.00102	N.A.	
								N.A.	
	Indigo carmine					~65 %	0.00326	N.A.	
	Coomassie brilliant blue R 250			15 mg/L		~80 %	0.01043	N.A.	
	Orange G	Solar radiation ~975 W/m^2		50 mg/L		~0 %	0	N.A.	
	Amido black 10B					~28 %	0.00345	N.A.	
								N.A.	
	Alizarin cyanine green					~55 %	0.00958	N.A.	
								N.A.	
	Indigo carmine					~94 %	0.02319	N.A.	
	Coomassie brilliant blue R 250			15 mg/L		~60 %	0.01585	N.A.	
	4-Chlorophenol	Xenon lamp (150 W) λ>455 nm 950 ± 100 W/m^2	N.A.	2.5×10^{-4} mol/L	N.A.	~0 %	0 %	0 %	[43]
$BaBiO_3$	Methylene blue	Xenon lamp 300 W λ>420 nm	0.3 g/100 mL	15.3 mg/L	2 h	~100 %	N.A.	N.A.	[23]
	Acetaldehyde	Xenon lamp 300 W λ>440 nm	0.8 g	837 ppm	40 min	N.A.	N.A.	TOC% > 80 % PE = 0.2 % (437.6 nm)	
					3.3 h	~100 %	N.A.	N.A.	
	4-Chlorophenol	Xenon lamp 300 W λ>420 nm	0.5 g/L	100 µM	4 h	~27 %	N.A.	N.A.	[11]

[a]Photocatalyst dosage
[b]Initial concentration of target contaminant

By using methylene blue as model compound, Kako et al. [17] found that the photooxidation activity of $NaBiO_3$ declined only a little bit after the reaction was cycled for 6 times, which was demonstrated by XRD measurements as there was almost no significant changes observed in the catalyst sample after the reaction.

On the basis of photooxidation to methylene blue, it was found by Kako et al. [17] that the photodegradation of methylene blue on $NaBiO_3$ decreased slightly after six subsequent cycling reaction and the corresponding XRD signals of $NaBiO_3$ were still detectable after recycling, indicating the good photostability of the $NaBiO_3$ catalyst. Another investigation reported by Yu et al. is that little decrease in photoactivity can be found after five subsequent cycling degradation of Rhodamine B over $NaBiO_3$ and the corresponding XRD peaks of $NaBiO_3$ were detectable after recycling. Similar conclusions were made by Yu et al. as well [63].

But unfortunately, the instability of $NaBiO_3 \cdot nH_2O$ under long time exposure was found by Eberl [43]. It was found that the color of $NaBiO_3$ gradually changes from yellow to brown as the time of the photocatalytic reaction was extended. XRD results showed that after a long time (about 20 h) of photocatalytic reaction, $NaBiO_3 \cdot nH_2O$ was transformed into $(BiO)_2CO_3$ crystal finally. Moreover, Sepulveda-Guzman et al.'s study also showed the poor stability of $NaBiO_3$ under electron beam exposure. Selected Area Electron Diffraction (SAED) results showed that $NaBiO_3$ can be partially reduced to metallic Bismuth particles under electron beam irradiation [75].

Eberl's investigation [43] on the photostability of Na_xBiO_3 under light exposure showed that the problem of poor stability of this photocatalyst still exists. It was found that the mineralization rate of 4-Chlorophenol was close to 0 after photoreaction cycled more than 26 h, and α-Bi_2O_3 was thought as a possible corrosion product by XRD analysis.

Our previous study [74] also clearly suggested that $NaBiO_3$ crystal was extremely unstable under acid conditions in photocatalysis. Different degrees of $NaBiO_3$ corrosion in 4-*t*-Octylphenol solution (target contaminant) containing hydrogen chloride were found under visible light irradiation. XRD results showed that after 1 h photocatalytic reaction, $NaBiO_3$ crystal was corroded into compounds including Bi^{3+} such as Bi_2O_3 when the initial pH is 4.29 and finally was corroded into BiOCl with tetragonal system structure when the initial pH is around 2.13. Additionally, it was found by our group that $NaBiO_3$ can react with hydrogen halide aqueous solution (HCl, HBr, and HI a.q.) very easily at ambient conditions, meanwhile producing the oxidation product of hydrogen halide were proved to be the corresponding halogen gas (Cl_2 and Br_2) or solid (I_2). Therefore it can be preliminary concluded that $NaBiO_3 \cdot nH_2O$ should be neglected as a suitable photocatalyst driven by visible light.

The stability of the $BaBiO_3$ in photocatalytic decomposition of acetaldehyde gas was tested by Tang et al. [23] under exposure to air for 2 months and cycling the catalytic reaction for 10 times (each time exposure duration of 2 days). It was found that there is no apparent change in its photocatalytic activity and the bulk crystal structure and optical absorption, indicating the excellent photostability of as-prepared $BaBiO_3$ in photocatalysis from air. However, they found that such complex oxide material exhibits far better stability in organic solvent compared with that in aqueous solution.

15.7 Future Directions and Prospectus

Although so far, there are a lot of research reports about Bi(V)-containing photocatalysts, some basic scientific problems, especially the basic structure-activity role of Bi(V) in the photocatalyst, still remains to be studied deeply by means of in-depth experiments and theories. According to the previous literature reports, most Bi(V)-containing photocatalysts have better photocatalytic activity in visible light, but the instability of this kind of catalysts in the photocatalytic process greatly hinders the long-term practical applications. Therefore restraining the chemical instability of the catalyst in water environment and photocatalytic reaction, while making full use of the advantage of Bi(V) for electron/EBS in semiconductors, will be a new problem for researchers in material physics, material chemistry, catalytic chemistry, and other areas in the coming years.

Acknowledgements The support by King Fahd University of Petroleum and Minerals is gratefully acknowledged. Xiaofeng Chang likes to extend his thanks to undergraduate students Mr. Qi Su and Ms. Yaling Chen (from College of Materials Science and Technology, Nanjing University of Aeronautics and Astronautics) for their help in manuscript preparation.

References

1. Fujishima, A., Honda, K.: Electrochemical photolysis of water at a semiconductor electrode. Nature **238**, 37–38 (1972)
2. Asahi, R., Morikawa, T., Ohwaki, T., Aoki, K., Taga, Y.: Visible light photocatalysis in nitrogen doped titanium oxides. Science **293**, 269–271 (2001)
3. Zou, Z.G., Ye, J.H., Sayama, K., Arakawa, H.: Direct splitting of water under visible light irradiation with an oxide semiconductor photocatalyst. Nature **414**, 625–627 (2001)
4. Chen, X.B., Liu, L., Yu, P.Y., Mao, S.S.: Increasing solar absorption for photocatalysis with black hydrogenated titanium dioxide nanocrystals. Science **331**, 746–750 (2011)
5. Fu, Q.: Solar radiation. In: Holton, M, Pyle, J. (eds.) Encyclopedia of Atmospheric Sciences, Acadamic Press, 1859–1863
6. Hoffmann, R.M., Martin, T.S., Choi, W., Bahnemannt, W.D.: Environmental applications of semiconductor photocatalysis. Chem. Rev. **95**, 69–96 (1995)
7. Mills, A., Lee, S.K.: A web-based overview of semiconductor photochemistry-based current commercial applications. J. Photochem. Photobiol. A **152**, 233–247 (2002)
8. Kudo, A., Miseki, Y.: Heterogeneous photocatalyst materials for water splitting. Chem. Soc. Rev. **38**, 253–278 (2009)
9. Roy, C.S., Varghese, K.O., Paulose, M., Grimes, A.C.: Toward solar fuels: photocatalytic conversion of carbon dioxide to hydrocarbons. ACS Nano **4**, 1259–1278 (2010)
10. Takei, T., Haramoto, R., Dong, Q., Kumada, N., Yonesaki, Y., Kinomura, N., Mano, T., Nishimoto, S., Kameshima, Y., Miyake, M.: Photocatalytic activities of various pentavalent bismuthates under visible light irradiation. J. Solid State Chem. **184**, 2017–2022 (2011)
11. Lakshminarasimhan, N., Park, Y., Choi, W.: Role of valency ordering on the visible light photocatalytic activity of BaBi3+0.5Bi5+0.5O3. Chem. Phys. Lett. **452**, 264–268 (2008)
12. Kudo, A., Omori, K., Kato, H.: A novel aqueous process for preparation of crystal form-controlled and highly crystalline $BiVO_4$ powder from layered vanadates at room temperature and its photocatalytic and photophysical properties. J. Am. Chem. Soc. **121**, 11459–11467 (1999)

13. Tang, J., Zou, Z., Ye, J.: Photocatalytic decomposition of organic contaminants by Bi_2WO_6 under visible light irradiation. Catal. Lett. **92**, 53–56 (2004)
14. Zhang, C., Zhu, Y.F.: Synthesis of square Bi_2WO_6 nanoplates as high-activity visible-light-driven photocatalysts. Chem. Mater. **17**, 3537–3545 (2005)
15. Tang, J.W., Zou, Z.G., Ye, J.H.: Efficient photocatalytic decomposition of organic contaminants over $CaBi_2O_4$ under visible-light irradiation. Angew. Chem. Int. Ed. **43**, 4463–4466 (2004)
16. Kumada, N., Kinomura, N., Sleight, A.W.: Neutron powder diffraction refinement of ilmenite-type bismuth oxides: $ABiO_3$ (A=Na, Ag). Mater. Res. Bull **35**, 2397–2402 (2000)
17. Kako, T., Zou, Z.G., Katagiri, M., Ye, J.H.: Decomposition of organic compounds over $NaBiO_3$ under visible light irradiation. Chem. Mater. **19**, 198–202 (2007)
18. Kumada, N., Kinomura, N.: Preparation of ilmenite type oxides via ion-exchange reaction. Mater. Res. Bull. **28**, 849–854 (1993)
19. Kou, J.H., Zhang, H.T., Li, Z.S., Ouyang, S.X., Ye, J.H., Zou, Z.G.: Photooxidation of polycyclic aromatic hydrocarbons over $NaBiO_3$ under visible light irradiation. Catal. Lett. **122**, 131–137 (2008)
20. Dias, A., Moreira, R.L.: Crystal structure and phonon modes of ilmenite-type $NaBiO_3$ investigated by Raman and infrared spectroscopies. J. Raman Spectrosc. **41**, 698–701 (2010)
21. Liu, X.J., Wu, Z.J., Hao, X.F., Xiang, H.P., Meng, J.: First-principles comparative study of the pressure-induced phase transition of $NaSbO_3$ and $NaBiO_3$. Chem. Phys. Lett. **416**, 7–13 (2005)
22. http://en.wikipedia.org/wiki/Perovskite_(structure)
23. Tang, J.W., Zou, Z.G., Ye, J.H.: Efficient photocatalysis on $BaBiO_3$ driven by visible light. J. Phys. Chem. C **111**, 12779–12785 (2007)
24. David, R.L.: Handbook of Chemistry and Physics. CRC Press, Boca Raton (2006–2007)
25. Sunarso, J., Liu, S., Lin, Y.S., da Costa, J.C.D.: Oxygen permeation performance of $BaBiO_{3-}\delta$ ceramic membranes. J. Memb. Sci. **344**, 281–287 (2009)
26. Kostikova, G.P., Korol'kov, D.V., Kostikov, Y.P.: Valence states of lead and bismuth atoms in the high-temperature superconductor $BaPb_{1-x}Bi_xO_3$. Russ. J. Gen. Chem. **71**, 1010–1012 (2001)
27. de Hair, T.W., Blasse, G.: Determination of the valency state of bismuth in $BaBiO_3$ by infrared spectroscopy. Solid State Commun. **12**, 727–729 (1973)
28. Yaremchenko, A.A., Kharton, V.V., Kovalevsky, A.V., Lapchuk, N.M., Naumovich, E.N.: EPR spectra and electrical conductivity of perovskite-like $BaBi_{1-x}Ln_xO_{3-}\delta$ (Ln=La, Pr). Mater. Chem. Phys. **63**, 240–250 (2000)
29. Nguyen, T.N., Giaquinta, D.M., Davis, W.M., zur Loye, H.C.: Electrosynthesis of $KBiO_3$: a potassium ion conductor with the $KSbO_3$ tunnel structure. Chem. Mater. **5**, 1273–1276 (1993)
30. Ramachandran, R., Sathiya, M., Ramesha, K., Prakash, A.S., Madras, G., Shukla, A.K.: Photocatalytic properties of $KBiO_3$ and $LiBiO_3$ with tunnel structures. J. Chem. Sci. **123**, 517–524 (2011)
31. Kumada, N., Takahashi, N., Kinomura, N.: Preparation and crystal structure of a new lithium bismuth oxide: $LiBiO_3$. J. Solid State Chem. **126**, 121–126 (1996)
32. Blasse, G.: On the structure of some compounds $Li_3Me^{5+}O_4$, and some other mixed metal oxides containing lithium. Z. Anorg. Allgem. Chem. **331**, 44–50 (1964)
33. Nomura, E., Greenblatt, M.: Ionic conductivity of Li_7BiO_6. J. Solid State Chem. **52**, 91–93 (1984)
34. Greaves, C., Katib, S.M.A.: The structures of Li_5BiO_5 and Li_5SbO_5 from powder neutron diffraction. Mater. Res. Bull **24**, 973–980 (1989)
35. Mizoguchi, H., Woodward, P.M.: Electronic structure studies of main group oxides possessing edge-sharing octahedra: implications for the design of transparent conducting oxides. Chem. Mater. **16**, 5233–5248 (2004)
36. Vensky, S., Kienle, L., Dinnebier, R.E., Masadeh, A.S., Billinge, S.J.L., Jansen, M.: The real structure of Na_3BiO_4 by electron microscopy, HR-XRD and PDF analysis. Z. Kristallogr. **220**, 231–244 (2005)

37. Li, Z.H., Wang, Y.X., Liu, J.W., Chen, G., Li, Y.X., Zhou, C.: Photocatalytic hydrogen production from aqueous methanol solutions under visible light over $Na(Bi_xTa_{1-x})O_3$ solid-solution. Int. J. Hydrogen Energy **34**, 147–152 (2009)
38. Kumada, N., Kinomura, N., Sleight, A.W.: Ion-exchange reaction of Na^+ in $NaBiO_3{\bullet}nH_2O$ with Sr^{2+} and Ba^{2+}. Solid State Ionics **122**, 183–189 (1999)
39. Mizoguchi, H.N., Bhuvanesh, S.P., Woodward, P.M.: Optical and electrical properties of the wide gap, n-type semiconductors:$ZnBi_2O_6$ and $MgBi_2O_6$. Chem. Commun. **39**, 1084–1085 (2003)
40. Butler, M.A.: Photoelectrolysis and physical properties of the semiconducting electrode WO_3. J. Appl. Phys. **48**, 1914–1920 (1977)
41. Kikugawa, N., Yang, L.Q., Matsumoto, T., Ye, J.H.: Photoinduced degradation of organic dye over $LiBiO_3$ under illumination of white fluorescent light. J. Mater. Res. **25**, 177–181 (2010)
42. Sheets, W.C., Stampler, E.S., Kabbour, H., Bertoni, M.I., Cario, L., Mason, T.O., Marks, T.J., Poeppelmeier, K.R.: Facile synthesis of BiCuOS by hydrothermal methods. Inorg. Chem. **46**, 10741–10748 (2007)
43. Eberl, J.: Visible light photo-oxidations in the presence of bismuth oxides. Dissertation, Friedrich-Alexander-Universität Erlangen-Nürnberg (2008)
44. Yu, X.J., Zhou, J.Y., Wang, Z.P., Cai, W.M.: Preparation of visible light-responsive $AgBiO_3$ bactericide and its control effect on the *Microcystis aeruginosa*. J. Photochem. Photobiol. B **101**, 265–270 (2010)
45. Yu, K., Yang, S.G., Boyd, S.A., Chen, H.Z., Sun, C.: Efficient degradation of organic dyes by $BiAg_xO_y$. J. Hazard. Mater. **197**, 88–96 (2011)
46. Pan, J.Q., Sun, Y.Z., Wan, P.Y., Wang, Z.H., Liu, X.G.: Preparation of $NaBiO_3$ and the electrochemical characteristic of manganese dioxide doped with $NaBiO_3$. Electrochim. Acta **51**, 3118–3124 (2006)
47. Jansen, M.: Preparation of anhydrous $KBiO_3$. Z. Naturforsch. B **32**, 1340–1341 (1977)
48. Scholder, R., Stobbe, H.: Über bismutate. Z. Anorg. Allgem. Chem. **247**, 392–414 (1941)
49. Trehoux, J., Abraham, F., Thomas, D.: Synthese et caracterisation de nouvelles phases du diagramme (K ou Na)-Bi-O. Mater. Res. Bull **17**, 1235–1243 (1982)
50. Chen, D.R., Meng, X.J., Li, J.X., Sun, S.X., Xu, R.R.: Synthesis of anhydrous $NaBiO_3$ polycrystal powders by solvothermal method. Chin. J. Inorg. Chem. **13**, 109–111 (1997)
51. Sun, Y.Q.: Analytical Chemistry. Science Press, Beijing (2003)
52. Klinkova, L.A., Nikolaichik, V.I., Barkovskii, N.V., Fedotov, V.K.: Thermal stability of the perovskite $BaBiO_3$. J. Solid State Chem. **146**, 439–447 (1999)
53. Mattheiss, L.F., Gyorgy, E.M., Johnson Jr., D.W.: Superconductivity above 20 K in the Ba-K-Bi-O system. Phys. Rev. B **37**, 3745–3746 (1998)
54. Kumada, N., Takahashi, N., Kinomura, N., Sleight, A.W.: Preparation of ABi_2O_6 (A=Mg, Zn) with the trirutile-type structure. Mater. Res. Bull **32**, 1003–1008 (1997)
55. Cox, D.E., Sleight, A.W.: Crystal structure of $Ba_2Bi^{3+}Bi^{5+}O_6$. Solid State Commun. **19**, 969–973 (1976)
56. Zou, Z.G., Chen, Y.F., Ye, J.H.: China Patent, Application No.: ZL03158264.8 (2005)
57. Rigby, W.: Oxidations with sodium bismuthate: glycol fission and related reactions. J. Chem. Soc. **1950**, 1907–1913 (1950)
58. Postel, M., Dunach, E.: Bismuth derivatives for the oxidation of organic compounds. Coord. Chem. Rev. **155**, 127–144 (1996)
59. Floresca, R., Kurihara, M., Watt, D.S., Demir, A.: Cleavage of unsaturated .alpha.-ketols to .omega.-oxo-.alpha.,.beta.-unsaturated acids. J. Org. Chem. **58**, 2196–2200 (1993)
60. Lena, J.I.C., Fernandez, E.M.S., Ramani, A., Birlirakis, N., Barrero, A.F., Arseniyadis, S.: Mild Protocols for generating molecular complexity: a comparative study of hetero-domino reactions based on the oxidant and the substitution pattern. Eur. J. Org. Chem. **2005**, 683–700 (2005)
61. Muathen, H.A.: Oxidative halogenation of aromatic compounds with metal halides and sodium bismuthate. Helv. Chim. Acta **86**, 164–168 (2003)

62. Mukhopadhyay, C., Tapaswi, P.K., Sarkar, S., Drew, M.G.B.: *N*-Methylthiomethylation of benzimidazoles with DMSO and their chemoselective oxidation to sulfoxides with $NaBiO_3$. ARKIVOC **9**, 393–406 (2011)
63. Yu, K., Yang, S.G., He, H., Sun, C., Gu, C.G., Ju, Y.M.: Visible light-driven photocatalytic degradation of rhodamine B over $NaBiO_3$: pathways and mechanism. J. Phys. Chem. A **113**, 10024–10032 (2009)
64. Chang, X.F., Ji, G.B., Sui, Q., Huang, J., Yu, G.: Rapid photocatalytic degradation of PCP-Na over $NaBiO_3$ driven by visible light irradiation. J. Hazard. Mater. **166**, 728–733 (2009)
65. Mizoguchi, H., Kawazoe, H., Hosono, H., Fujitsu, S.: Charge transfer band observed in bismuth mixed-valence oxides. $Bi_{1-x}Y_xO_{1.5+\delta}$ (x = 0.3). Solid State Commun. **104**, 705–708(1997)
66. Xu, B., Zhang, W.F., Liu, X.Y., Ye, J.H., Zhang, W.H., Shi, L., Wan, X.G., Yin, J., Liu, Z.G.: Photophysical properties and electronic structures of the perovskite photocatalysts Ba3NiM2O9 (M=Nb, Ta). Phys. Rev. B **76**, 125109 (2007)
67. Matsumoto, Y., Omae, M., Watanabe, I., Sato, E.: Photoelectrochemical properties of the Zn-Ti-Fe spinel oxides. J. Electrochem. Soc. **133**, 711–716 (1986)
68. Matsumoto, Y.: Energy positions of oxide semiconductors and photocatalysis with iron complex oxides. J. Solid State Chem. **126**, 227–234 (1996)
69. Brahimi, R., Bessekhouad, Y., Bouguelia, A., Trari, M.: $CuAlO_2/TiO_2$ heterojunction applied to visible light H_2 production. J. Photochem. Photobiol. A **186**, 242–247 (2007)
70. Yao, W., Ye, J.: A new efficient visible-light-driven photocatalyst $Na_{0.5}Bi_{1.5}VMoO_8$ for oxygen evolution. Chem. Phys. Lett. **450**, 370–374 (2008)
71. Long, M., Cai, W., Cai, J., Zhou, B., Chai, X., Wu, Y.: Efficient photocatalytic degradation of phenol over $Co_3O_4/BiVO_4$ composite under visible light irradiation. J. Phys. Chem. B **110**, 20211–20216 (2006)
72. Lin, X., Xing, J., Wang, W., Shan, Z., Xu, F., Huang, F.: Photocatalytic activities of heterojunction semiconductors $Bi_2O_3/BaTiO_3$: a strategy for the design of efficient combined photocatalysts. J. Phys. Chem. C **111**, 18288–18293 (2007)
73. Butler, M.A., Ginley, D.S.: Prediction of flatband potentials at semiconductor-electrolyte interfaces from atomic electronegativities. J. Electrochem. Soc. **125**, 228–232 (1978)
74. Chang, X.F., Huang, J., Cheng, C., Sha, W., Li, X., Ji, G.B., Deng, S.B., Yu, G.: Photocatalytic decomposition of 4-*t*-octylphenol over $NaBiO_3$ driven by visible light: Catalytic kinetics and corrosion products characterization. J. Hazard. Mater. **173**, 765–772 (2010)
75. Sepulveda-Guzman, S., Elizondo-Villarreal, N., Torres-Castro, D.F.A., Gao, X., Zhou, J.P., Jose-Yacaman, M.: In situ formation of bismuth nanoparticles through electron-beam irradiation in a transmission electron microscope. Nanotechnology **18**, 335604 (2007)

Index

H. Li and Z.M. Wang (eds.), *Bismuth-Containing Compounds*, Springer Series in Materials Science 186, DOI 10.1007/978-1-4614-8121-8,

MIX
Papier aus verantwortungsvollen Quellen
Paper from responsible sources
FSC® C105338

If you have any concerns about our products,
you can contact us on
ProductSafety@springernature.com

In case Publisher is established outside the EU,
the EU authorized representative is:
Springer Nature Customer Service Center GmbH
Europaplatz 3, 69115 Heidelberg, Germany

Printed by Libri Plureos GmbH
in Hamburg, Germany

Springer
Tokyo
Berlin
Heidelberg
New York
Barcelona
Budapest
Hong Kong
London
Milan
Paris
Santa Clara
Singapore